Australian Shepherds of the Mountain West

Kris Toft

Rowe Publishing

ISBN 13: 978-1-939054-18-0
ISBN 10: 1-939054-18-4

Cover Photos: (left) One of the earliest known photos of Jay Sisler, Shorty, and Stub. Photo courtesy of Les Burlile. (right) Silverledge Slide Me To Da Bar, CD, STDs. Photo courtesy of Jean Taylor.

Note from publisher: This book is published posthumously. Statements and opinions expressed by the author are not necessarily those of the publisher and its affiliates. Neither shall be liable for erroneous information.

1 3 5 7 9 8 6 4 2

Printed in the United States of America

Published by

Rowe Publishing

www.rowepub.com

Stockton, Kansas

Dedication

To one of the finest and undoubtedly the single most important person the Australian Shepherd has ever known, Jay Sisler

Foreword

Since this book is being published posthumously it seems important for the reader to know something about the author as well as the purpose and original intent for writing it.

Kris had a passion for animals since she was two-years old and it became her life when she purchased her first horse and subsequently her first dog, an Australian Shepherd named Hannah. As was typical of Kris, she jumped right into that world with both feet. She showed dogs, learned what was important about breeding and whelping litters, and ultimately became an AKC judge for instinct. She traveled around the country for a time. She remained involved in the breed in many ways till the end of her life.

Her 'day job' as a medical research assistant in the Department of Surgery at the University of Missouri Medical School was a natural for her. Her patients were animals and it was then that research became so important and interesting to her. With her rapidly increasing interest in the breed, she began to keep records about dogs both hers and others. As a result of that passion she self-published three pedigree books, as well as numerous articles for Australian Shepherd magazines through the years. She had an uncanny understanding and communication with dogs and mentored many dog owners, including me. Her ability to recognize and understand behavior in these animals was legendary. And her deep interest in genetics enabled her to seek out good breeding dog partners. It commanded her interest for the rest of her life.

But the real reason for the research and writing of this book was her concern, along with the concern of other people, that the history along with myriads of pictures of the early western working Aussies and their owners and breeders was soon to be lost to time and human lifespans. To Kris, right was right and wrong was wrong, and her mission was to set the record straight. I once heard her speaking to someone who asked why she wrote about the western mountain dogs and her reply was "they were rough and tough and excellent working dogs with great intelligence." In addition, her respect and awe of the Wood's line of dogs has carried her far in her breeding and genetic research.

With this in mind Kris set out to contact, personally, as many survivors of those early days as possible. Many of them such as Jay Sisler and Fletcher Woods are now deceased but she found surviving family members who still had stories to tell and pictures to lend so the history could be told. She contacted many fellow Aussie dog owners who had kept their own records, pictures, stories and were even keepers of records of deceased dog owners of note. She searched newspaper archives, museums and even the Cowboy Hall of Fame for any bits of information to help her weave it all together. To know Kris is to know that she would not give up until she found the answers and the people who could provide them.

It was a definite labor of love when you realize she was fighting cancer the whole time. But she persevered, declared it finished, and sent it off to the publisher. Her disease made it difficult for her to edit but she was able to see it through with the help of family and friends. Kris died on October 27, 2013, and left us this legacy. She wrote it in her own words and that becomes the treasure of this book. She could back up everything she wrote and that makes it timeless.

Kris didn't get to see it published but she knew it was done and that it would eventually be made public and besides, she was already moving on to the next book. Only time will tell if that book will reach publication. We'll see…

Her loving and proud Mother,
Kay Toft

Acknowledgments

As you will notice throughout this book, the hand of the Creator was busy intervening at many points in the history of the mountain-type Aussie. Good, sound dogs were placed into the hands of the right people at the right moments over and over again. You will also find His fingerprints throughout this book, continuously giving me the right words to tell the story of this stellar family of Australian Shepherds. He has also blessed me with contacts to a wonderful family of foundation breeders who have graciously consented to tell me their stories for preservation in this volume. You will never know how grateful I am for their generosity.

Those without whom this book would have been impossible:

Casey Masterson Newman—My very good friend who provided motivation when I needed it most, contacts with fascinating people who likely would never have talked to me alone, and most of all gave me my start with these unique dogs.

Jean Taylor—The other person who greeted my proposal for this project with unbridled enthusiasm and who provided wonderful historical details and gorgeous photos. I am forever in her debt for "loaning" the dog of a lifetime—Tigger—to me.

Les Burlile—Jay Sisler's nephew and a gold mine of first hand knowledge about Jay's life and his foundation dogs. Thank you for your generosity to a total stranger.

Jerry and Sharon Rowe—The people everyone told me HAD to be included. "Everyone" was more right than they could have guessed.

Dick and Leslie Sorensen—Leslie's Virgo tendencies (like mine!) for recording history in wonderful photographic detail and "packratting" every piece of paper about Aussies she could get her hands on provided a wealth of content for this book. Dick's memory and amusing storytelling added so much personal detail and color to this book.

Melissa Couch—Old friends are the best friends. Melissa rescued much of Cathy Jones' priceless memorabilia after her death and, even though it was painful, shared her memories for this book.

Mary Wood—Fletcher Wood's spicy widow who welcomed Casey and me into her home and shared her vivid memories for preservation in this book.

Leila Green—A fine pioneer lady of sharp mind and steel resolve.

Raymie Jones-Brown—Cathy Jones' daughter who graciously shared her mother's records, photos, and memories for this book.

Ruth McGuire—Another hardy breed pioneer of great character and big heart.

Linda Rorem—For encouragement and for sharing her wealth of knowledge on herding dogs. The puzzle pieces really came together with your help.

Kyle Trumbull-Clark—My "partner-in-crime" for editing and always great advice.

All praise goes to God for planting the desire to write this book in my head and heart, inspiring me often with enthusiasm from my friends, and guiding me to the right people for information and photos. It was a fun project. I hope you enjoy the reading as much as I enjoyed the writing.

Contents

Introduction

Since 1978, I've loved, owned, and trained Australian Shepherds of several bloodlines. My fascination with the foundation dogs was fed by the pages of the *Aussie Times*. My eyes were drawn over and over to dogs of a particular type within the breed. The type is now known generically as the "Woods" type after the man who concentrated it into an early strain or bloodline in the 1950s and early 1960s, Fletcher Wood. Mention of dogs called "Australian shepherds (or sheep dogs)" are found in public records of mountain and coastal areas of the west beginning around the mid-1800s. Descendants of these dogs founded the breed we now call Australian Shepherd.

The family now called "Woods" has survived as a remnant of these original dogs. The term "Woods" has expanded over the years to refer to dogs of a general type which may not have been bred by Mr. Wood nor even encompass dogs of his breeding. For the sake of this book, this designation will be used to represent a type of Aussie rather than just the Wood bloodline. It might even be more correctly called the Sisler-type because the foundation of the Wood dogs, themselves, was none other than a son of the famed Sisler's Shorty—Wood's Jay. The Sisler dog type was common in the mountainous livestock producing areas of this country.

Many breed books are touted as "complete" works on the breed, but nearly always gloss over this particular type of dog. So, I make no apologies for the bias of this book. The purpose is to describe as best I can, with the help of others personally familiar with this family of dogs, the background, lineage, type, and character of this special group of dogs within the Aussie breed.

Every effort has been made to accurately document the pedigrees and legacy of the dogs discussed in this book. In some cases, misspellings may creep in due to various spellings found in old records. In many cases, the original owners of the dogs in question are no longer available to provide the correct information. In addition, incorrect pedigree information abounds and, where possible, the correct information has been verified by the actual owner of the dog and is included in this volume.

In some cases an intelligent and sensible theory has to be formed regarding the background of some dogs. These theories are explored and the evidence presented in order to allow the reader to make their own informed decision about the origin of a dog. Accurate records were hard to obtain or do not exist about many of the earliest dogs and we must observe the context of the breed and the early breeders in order to reach a logical conclusion regarding the heritage of these dogs. Even "official" records from various registries are rife with verifiable errors. Finding the truth has been a difficult and elusive process.

Kris Toft

chapter 1

Emergence of an American Breed

The background of the Australian Shepherd is uniquely and solely American. Like American people, the Aussie is a melting pot of a variety of breeds and types of dogs, many of which will never be known for certain. The breed was recognized first as a kind of dog known for blue merle color and a bob tail. Its usefulness and versatility on farms and ranches in the American West are the traits which determined whether it survived or was discarded.

The Australian Shepherd's association with Basque sheepherding people has become a fanciful legend within the breed. Although it is likely that the early spread of Aussie-type dogs was tied closely to the movement of sheep flocks throughout the western United States, it is unlikely that dogs of the Pyrenees mountain area bordering France and Spain—homeland of the Basque people—had more than a minor influence on the breed. From reports of those with intimate knowledge of the Pyrenean Shepherd, the body type and character is vastly different from the mountain-type of Aussie. In addition, the Pyrenees dogs were mostly sheep dogs while the mountain-type Aussie was predominantly developed as a cattle working dog. So, while it is possible the odd Pyrenean Shepherd type found its way into the gene pool of the Australian Shepherd, it is not likely the influence has been very great. Although "little blue dog" is a popular historical description of the Aussie, it is mainly used in reference

Pre-1918 photo of an Australian Shepherd with two hunters in Washington State. From Author's collection

to alleged "Basque" background dogs. If any Basque breed influence exists in the Aussie it is likely confined to lines which produce smaller types of Aussies for sheep work. Mountain dogs were inclined to be a larger size, which was a preferred type for working cattle in the areas which formed this breed.

It is more likely the background of the Australian Shepherd breed itself, including the Wood-type dogs, has a more British base. In the early to mid 1800s the British Isles were home to many generic "collie"-type dogs. Purebred dog breeding had not really become established for working types of dogs. People used dogs and bred dogs for strictly utilitarian purposes. The Border Collie, as we know it today, was not established yet and the base stock for that breed and many other "collie" breeds was a variable stew of genetics. Some of these generic-type dogs found their way to the East Coast of the United States and became known as "English shepherds"—though no such "breed" exists in England. Names at that time were more descriptive than representative of pure breeding. Likewise, dogs were exported from Britain to Australia, where there is accurate documentation of many of these exports. Eventually sheep—and dogs—from Australia found their way to the West Coast of the United States. Those dogs were called "Australian shepherds" in the same descriptive manner of the "English" shepherd. It is very likely these two types of dogs (the "English" and "Australian" shepherd) were closely related in their original homeland.

The breed owes a debt of gratitude as big as the West itself to Jay Sisler. A humble and modest man, without Mr. Sisler the breed would not have attained the stature it now enjoys. Sisler was not a prolific breeder but the attention he brought to the Australian Shepherd throughout the country—not only through rodeo appearances but also via the budding Walt Disney films—cemented a type and direction for the Australian Shepherd which has served it well. What little breeding practiced by Sisler, and his brother Gene, scattered the mountain-type of Aussie throughout the United States. Many dogs with "unknown" at the far right side of their pedigree likely descend from puppies the Sislers sold, traded or gave away through the years.

Venerable horseman and National Western Stock Show ringmaster Fletcher Wood was directly influenced by his close friend, Jay Sisler. Wood became a major vehicle through which the priceless qualities of Jay's original dogs were preserved and spread around the country. Wood, an intuitively great breeder of Australian Shepherds, set a type so quickly and thoroughly it still comes through in our dogs today, even after generations of dilution. Wood practiced intensive inbreeding, culling, and selection which stamped his bloodline in just a few generations. From Sisler's Shorty to Wood's Jay to the fountainhead of his bloodline—the intensely inbred Wood's Stubby (or Stubbie as it is sometimes written)—this type has lived on today.

Merle Australian Shepherd, early 1900s, Oregon. From Author's collection

Sadly, though, this type has become increasingly uncommon in the breed. Times change and with it, our breed. The uncanny intelligence, independent thinking, personal judgment and nobility of the Wood-type of dog have been replaced with soft, pliant, less challenging dogs. People do not have the once-common daily contact with animals found in the past. The respect and common sense developed through these interactions is now lacking and our breeds of dogs and livestock have been temperamentally dampened for the safety of this naive populace.

A small pool of these dogs still remains among those who need strong working dogs. A few reside with those who prefer the old-fashioned type of dog merely because we couldn't have it any other way. Soundness and uncommon intelligence still exist in the breed if one wishes to find it.

THE GROUNDWORK TO A BREED

Development of the Western livestock industry paralleled the spread and popularity of the Australian shepherd—spelled with a lower case "s" prior to formation of a formal registry (written accounts from that era uniformly make this distinction). The farm collie or shepherd-type of

Black and white Australian Shepherd, early 1900s, California. From Author's collection

Western rancher with his Australian Shepherd, circa 1910. From Author's collection

dog was a fairly common member of the ranching and rural community from the gold rush days of the 1850s onward. Shepherd being spelled with a lower case "s" appears to refer to a generic type of dog that could be prefaced with a descriptive adjective such as "Australian."

Many are the stories told of acquiring a dog "from a neighbor" or "at a livestock or horse sale." Pedigree was irrelevant, even if it was known. More than one Australian shepherd was obtained from a dog pound—even a very well-known sire! Whatever the source, the look, character, and type of the dog spoke for itself. People recognized a particular kind of dog which was defined as "Australian" shepherd.

In the biography of Ray Eyerly—a western painter—he describes his experiences on a Montana ranch in the early 1900s. He refers to Australian shepherds on the ranches and of their great value in this quote from the book, *A Picture or Two*. "Best herdin' dog for sheep or cattle is the Australian shepherd. During lambing, the herder separates the sheep into three different flocks and keeps them apart. It's the dog's job to see to it that the three groups don't move in together. The sheep will lie down to rest, and then pretty soon get up to graze. The lambs are still sleepin' maybe, and they get left by the flock. A lamb will follow anything. So that old dog just ranges out when he hears the lamb bleatin' and gets its attention. Then he slowly trots back to the flock with the lamb following."

This type of dog was most common in the Western United States. More specifically the mountainous areas of the West—Wyoming, Montana, Idaho, Washington, Oregon, and Northern California.

This type could be found in Utah, Nevada, Colorado, and southwestern Canada as well. The true origin will likely never be known, nor is it likely these dogs were "purebred" in any sense. Working dogs begat working dogs.

Despite this random beginning the dogs remained surprisingly uniform in look, character, and type. Easily recognizable "Aussies" were photographed time and time again. They were owned by the wealthy and common man alike. The only criteria for the perpetuation of their genes was their ability to work or be useful to their owners. No single source of dogs went into the creation of the breed which eventually reached capital "S" status. Dogs which pleased their masters produced more dogs which had to measure up to this personal standard.

So the foundation was laid for the dogs which are known today as "Woods"-type Australian Shepherds. Originally, this book was to be a recounting of the "Woods"-type of dog—a specific bloodline created by the masterful Fletcher Wood. As the research unfolded, it became clear Mr. Wood was being credited with many more dogs than actually traced solely to his program. His type was so firmly set that many dogs with very little actual "Woods" blood, but closely resembling the Wood dogs, were credited to this origin. This identity has become a general reference term in the breed and includes many dogs of similar type—some with similar bloodlines, some with different bloodlines, but all with a common look and character.

In speaking to a number of early breeders, it became clear a more general background dog—which certainly includes the "Woods"-bred dogs—was

1901 photo of the Smith family of Mesa, Colorado. The Smiths were ancestors of Casey Masterson Newman and ranched extensively in western Colorado with holdings in Mesa, Fruita, and Grand Junction. They settled in the area in the 1880s and raised cattle there until the 1920s and 1930s. The Aussie type farm dog was a common sight on western ranches at this time. Photo courtesy Casey Masterson Newman

Merle Australian Shepherd with Oregon cowboy, early 1900s. From Author's collection

Solid black Australian Shepherd ranch dog. Mid-1900s. From Author's collection

Northern California ranch Aussie, 1920s. From Author's collection

a more accurate description of the type of dogs referenced. Many founding dogs were of unknown ancestry but were strikingly uniform in so many traits they crossed well and blended together as one. Those who remember the early days first-hand agree this general-type of dog was found primarily in the mountain West. So, for the purpose of this book, "Woods" and "mountain-type" will be used synonymously. I hope Mr. Wood would have found this acceptable and perhaps even given it his blessing.

Fletcher Wood began his breeding program with a dog from the renowned Jay Sisler. Wood's Jay was a son of Sisler's Shorty—Jay's favorite dog—and was named after Sisler. Wood's Jay was a very dark blue merle with minimal white markings—just a spot on the chest and on his paws. Like many of Sisler's dogs, he had no copper markings. Fletcher Wood's personal preference was for a very dark colored blue merle dog with brown eyes and he bred and culled ruthlessly for this in addition to preserving the character of Wood's Jay—who forever remained his ideal.

Very dark Aussies with no white or copper are known as "self"-colored dogs. It was this coloration which became Fletcher Wood's trademark. It was also an easily visible trademark of other dogs and other early bloodlines, and, in fact, a fairly common trait among many of the early dogs of the mountain west. White markings were not unknown and were even preferred by Jay Sisler. He favored well-marked dogs for his rodeo act since their white markings showed up well in dark arenas and were attractive to the general public. However, the minimally marked puppies showed up frequently in his litters—hence Wood's Jay and the red merle self-colored dogs chosen by Gene Sisler for his own rodeo act. Lack of variety was never a problem in the early dogs.

Merle Australian Shepherd, early 1900s, Oregon. From Author's collection

About the same time Fletcher Wood was creating his bloodline, other early breeders were collecting their foundations and starting breeding programs. Jerry and Sharon Rowe and Dick Sorensen began the J Bar D breeding program with similar-type dogs—some even tracing to dogs with Wood's breeding. Dick Sorensen's dog, Sorensen's Gunsmoke, contributed to the background of these kennels and credit should be given to the broad type being bred and preserved at that time. Tom and Jean Taylor happened to favor the Wood-type dog. Terry Martin, with the Slash V dogs, used some dogs with Wood's background to found her line. The look and character of the dogs produced by these early breeders still tended to be similar.

Cathy Jones made famous the red "Woods" dog with Jones' Reddy Teddy. Ted's grandsire, Piz's Joseph, sometimes described as a "red cropout," was merely a revival of dogs which were not uncommon in earlier times. Cathy apparently shared Gene Sisler's attraction to the red dogs a quarter century earlier. Tom and Jean Taylor were largely responsible for bringing this original-type of Aussie to the east coast of the United States. Not only the Australian Shepherd, but also the Wood-type dog, was almost completely unknown in the east until Tom brought them home from his travels around the country as a horse trainer. Jean stayed home in Pennsylvania tending the stock and raising puppies from the dogs Tom picked up on his travels.

Aussies became a popular horse show dog through the appearance of Sisler's dogs at rodeos and Fletcher Wood's dogs at horse shows. Now, a few breeders still preserve the old mountain-type of dog today. Others are crossing these dogs with working dogs of other bloodlines to produce top-notch working dogs. It is these outcrosses which serve to prove the timelessness and prepotency of the old mountain-type of dogs. Over and over again the dominant traits of the mountain-type dog come through the breedings, even though they are diluted by the outcross. Perpetuation of this type speaks to a uniformity in the foundation dogs which can still be saved if breeders choose to preserve it.

Black and white Australian Shepherd, pre-1918, British Columbia, Canada. From Author's collection

chapter 2

Defining a Type

What makes a "Woods" or "Mountain-Type" Australian Shepherd distinct from other types of Aussies? Color is certainly the first and most obvious trait people will notice. The dark, self-colored dog nearly always contains old mountain-type breeding in it's pedigree. The prepotence of this blood is sometimes truly amazing! From a few ancestors four or more generations back, the look and character of this type dog can still be very obvious. Because the type was easily recognized over 100 years ago, it is likely people were practicing some sort of rudimentary selection even then. Culling was common as people in the west did not keep large kennels of dogs on ranches or in their backyards just for the purpose of breeding more and more dogs. The genes for a very specific look and temperament were selected in the most primitive manner possible—survival of the fittest.

Dog shows in the late 1800s and early 1900s were a thing for wealthy folks. The breeds appearing in dog beauty contests were usually well established in other countries and imported to the United States to be perpetuated here. Common working dogs without "papers" weren't highly thought of and their ancestry was not a concern. As Sharon Rowe puts it, "The Aussie was considered a 'redneck' dog."

Many people who owned Australian shepherds lived in remote areas. They required a dog who could do anything they wanted it to do around

Australian Shepherd crossing Burnt River, Oregon, bridge, circa 1907. From Author's collection

their ranch. Even urban areas were of another era—when dogs were expected to be dogs. The early Aussie was a working dog on ranches but his work included more than just herding stock. Much of the Aussie's job on a ranch involved guard duty. Herd dogs in sheep camps not only helped move stock or keep it together, but also alerted sheepherders to intruders of all varieties. The Australian shepherd guarded against coyotes, rustlers, wolves, bear, and anything else with ill intent. If the herdsman wasn't around to take charge of the situation himself, the dog was expected to use its own judgment in handling the situation. And these dogs did. From warnings to a full fledged attack if necessary, the dogs were expected to read any situation and take whatever action was called for.

The Australian shepherd's later popularity as a horse show dog stems more from this employment than any need for a stock dog. Taken to an extreme, this use of Aussies was nearly their undoing. Dogs were left in trucks, trailers, and stalls to guard horses and tack at shows. Some of the dogs took their job a little too seriously or were encouraged beyond common sense to be aggressive no matter what. The Wood's dogs still suffer today from the image of an overly aggressive dog which will bite anything that moves.

The Woods-type of dog has been branded with a poor image due to the excesses of a handful of owners. Make no mistake, this type of dog can and will take action when under threat. They are no different from any dog in this respect. What distinguishes them from other dogs, however, is their ability and willingness to take charge when nobody else will. They do not wait submissively for direction from their owner when a situation calls for action or control. An owner who expects a dog to be unaware of its surroundings and meekly accepting of any and all affronts to its dignity or sense of justice will create an out of control menace. But given the chance, these dogs also show an uncanny degree of judgment born of experience and proper education. A properly handled Wood's dog, given guidance and shown the ropes of the modern world will discern, on its own, when to act and when to give control to the boss—if one exists.

The old mountain-type of dog knew how to take care of him- or herself, property, owners, and territory. They not only knew, but were expected to do this without direction. They were often deliberately left unsupervised for this very purpose. They were the only ranch and home security system available. They retain those valued characteristics today. Unfortunately, the world has passed them by. Those enchanted by their look or by the romance of a Hollywood west that never was, acquire a dog which requires from them a personal responsibility for which they are ill-prepared. This uncanny alertness and judgment has not been diminished by the generations. These dogs are still constantly on the lookout for threats in their environment and still prepared to do what is necessary to handle them. Sadly, their actions are not always tempered by equally alert and responsible owners. These priceless qualities have become legal liabilities. And, so, their usefulness is being quickly lost—replaced by a caricature of what the dog once was. The same can be said of many breeds and the world is a poorer place for it.

On a happier note, this type does still survive. Ranches still exist where a good dog is needed and his brainpower is an advantage. This type of dog is still found and raised as he was in the past—with little emphasis on lineage and much value placed on performance. And his type survives much as it has in the past. The mountain Aussie tended to be a fairly large and substantial dog. Not heavy and coated like today's show animals but strong and hardy. Legends about "little blue dogs" were probably referring

Washington State ranch dog, an Australian Shepherd, circa 1910. From Author's collection

to a different type of dog used under different conditions. Early breeders preferred a dog which could stand up to strenuous conditions and even today many branches of this type of dog are fairly large. Les Burlile describes his uncle Jay Sisler's first dogs as weighing from 75-80 pounds and having large, broad heads. It would have been impossible for Jay's dogs to perform the feats they did if this was 75-80 pounds of fat or coat.

This is not to say all mountain-type dogs were this large. The type had a lot of variety which extended to their size as well. But the size of the dog was not what really defined it. It is the size and scope of its heart and brain which are more definitive. These are the dogs of legend who would literally give their life for something or someone else. They possess a nobility and primitive type of behavior which is closer to that of wild canids than the milquetoast temperament predominating in the breed today. Assertive confidence defines the founding Australian shepherds' character, which is distinctly different from aggression. A no nonsense disposition is bound with a sometimes sophisticated sense of humor. Even puppies, while still engaging in occasional silliness, display a mature sensibility and self possession not commonly seen in the breed today.

Tud Kent, well known Idaho dog musher, training his team for the 1926 American Dog Derby. This team has an Australian shepherd lead dog and wheel dogs. From Author's collection

chapter 3

Jay Sisler THE MAN

The influence of Jay Sisler on the Australian Shepherd breed is incalculable. His genius as an animal trainer still thrills and amazes audiences through the recordings we can view today. Well ahead of his time in the methods he used, his positive way with animals—horses and other creatures as well as dogs—created rodeo animal acts still spoken of with awe. Appearing at small and large rodeos, fairs, and western shows across the nation, he entertained audiences with acts that were not only amusing but also marvels of creativity. His talent as a trainer and the visual appeal of his two gifted dogs, Shorty and Stub, introduced the Aussie to the "outside world" in a way that was never equaled by anyone.

Jay Howard Sisler was born in the heart of the region which forged the mountain-type Australian Shepherd. Gem County, Idaho, is centered in the mountain northwest of the United States. The crown jewel of Gem County, Emmett, lies in the emerald green Payette River valley northwest of Boise near the Oregon border. This is ranching and agricultural country well irrigated by the Payette. During Jay's early life, the town was well known for its abundant fruit orchards. This typical ranching community was a self sustaining area well populated with crops, sheep, cattle, horses, and dogs. Jay's ancestors settled and helped develop Emmett two generations before he was born. The ancestors of the Aussie were largely Pacific Northwest and Pacific coast dogs found as far inland as the Rocky

Jay Sisler with his second Greyhound, Sparky—the dog who portrayed "Greta" in the *The Wonderful World of Disney's Greta, the Misfit Greyhound*. This photo was taken in Jay's front yard at his second Emmett, Idaho, ranch. Photo by Jay's sister, June Sisler. Used with permission of Mr. Don Coates from the Patricia Highsmith Collection of the Swiss National Library.

Mountains and northern Great Plains—coincidentally prime cattle and sheep country. It is this cradle which united the man and the dogs which founded a breed.

On Jay's birthday, February 20, 1926, the Aussie was not yet a breed but was a common, recognized type of dog in the American West. Jay grew up working hard on the ranch and showed an early aptitude as an animal trainer. By the time he was five years old he had aspirations to be an Alaskan musher, training his father's bird dogs to pull him around in a cart in preparation for that career. The cow dogs on the place were also drafted as substitute huskies and taught not only to work cattle but also to pull sleds, wagons, or any other load Jay could create.

In 1939, when Jay was about 13 years old, he got the first of what he called his "blue dogs." Like many of his later dogs and typical of most Aussies of that day, this dog had no known ancestry. Jay's first blue dog would stamp his breed preference for life. The Aussie type was well known and established, so plenty of them were available for Jay to practice his trade. Hungry for knowledge about dog training, Jay joined 4-H and while attending the 1939 fair in Caldwell, Idaho, he met a man who had a dog act. Wishing to learn his secrets he asked how the Shelly, Idaho, man trained his dogs. "Just like this" was the cagey reply as the Shelly man snapped his fingers. Jay's education as a dog trainer, like his knowledge of many other subjects, would have to be self taught.

Idaho has a long rodeo history in which Jay and his brother Gene were immersed from an early age. Between ranch chores, both boys turned to horse training as a way to earn money, showing great talent at early ages. Jay's nephew, Les Burlile, remembers Jay as a better than average bronc rider. He says Jay knew that being thrown would just make a horse worse the next time he got on, so he not only avoided the pain of a sudden encounter with the ground but also managed to teach the horse something good from the experience. Racing horses was a common side attraction at most rodeos. Jay and Gene shared a love of horse racing with good friend and employer David Little. David was the son of an Emmett immigrant and "Idaho Sheep King," Andy Little, and Jay was working for him when he and Gene decided to enter a relay race with Little.

One wonders what the Aussie breed would look like today if not for the fortunate accident which happened to Jay on that fateful day in 1948. He was 22 years old when he and Gene entered the three lap "Pony Express" relay. Speculation is all we have left regarding the site of this race, but we do know it was held in Emmett, probably somewhere on the western edge of town. A team consisted of three horses, each racing one lap with the rider jumping off to mount another horse at the end of each lap. The Sisler brothers had only two horses they wanted to run that day so they made a deal with David Little. Little had two teams and extra horses, so he loaned one spare horse to the Sisler team, ridden by Gene, while Jay would ride one of Little's teams. Little had two horses he wanted to race head to head to see which was the fastest. He and Jay agreed whoever was ahead at the last lap would hold up so they could race the two fastest Little horses together.

Little was well ahead of Jay at the final lap and waited quite awhile until his uncle, George Cruikshank, raced up on his other horse. As they took off together, Little shouted to his uncle asking what had happened to Jay. It was while changing horses for the last lap that Jay's fortuitous accident occurred. As he dismounted into the hustle and confusion, Jay fell and an excited horse stepped on him. His ankle was broken badly enough he wasn't able to run to his last mount. As the race continued on without him, Jay was taken to the hospital.

One of the earliest known photos of Jay Sisler and Shorty at Buzzard Gulch—the cow camp where he recovered from his broken ankle and began his dog training career. Courtesy of Les Burlile

After he was patched up and sent home on crutches, Jay accompanied Gene to a cow camp about ten miles northwest of Emmett to stay while his ankle healed. A 1948 cow camp was a dull and quiet place for a hardworking cowboy who couldn't work. Since he couldn't ride or train horses, Jay alleviated his boredom by reviving his childhood interest—training dogs. He and Gene had a litter of Australian shepherd puppies at the camp. Jay hobbled around on crutches teaching them whatever came to his fertile mind. Yet another providential stroke was at work in the form of two prodigal male puppies (described as "collies" by the Littles) who began to distinguish themselves from the rest of the litter—Shorty and Stub. Some of the pairs' earliest chores were fetching items for Jay—including his horse—and acting as early "service dogs" by doing things for Jay which he couldn't do with the broken ankle. The pups were even trained to walk between his legs while he was still on crutches! His earliest rodeo act was a take-off on his accident—he entered the arena pretending to have a broken leg and the dogs performed a routine of "helping" him do chores.

Jay also dabbled in some sheep herding training while at the ranch using an old tire placed in his corral. The dogs would take direction and move sheep from side to side around the tire and gather sheep into the corral for Jay. By the time his ankle had healed, Jay's budding stars were well on their way to fame and glory.

Very early photo of Jay, Shorty (left), and Stub (right) at one of his first performances. Jay wore the black hat for only a short time at the beginning of his career and his dress was casual until his sister began sewing his fancy costumes. Photo by Gene Sisler, courtesy of Les Burlile.

A visiting neighbor of Jay's was treated to the world's first informal backyard show featuring Shorty and Stub. At the end of the "show" he offered Jay his first rodeo job, performing his act about 20 miles away in prophetically named Star, Idaho. Jay's family and friends supported this new venture and Jay had hopes the money might pay his entry fees so he could compete in rodeo events. On Easter Sunday in 1949, the trio debuted to a thoroughly delighted crowd. The young dogs already had a wide repertoire. This first show included jumping rope in a variety of ways, playing leap frog, weaving in and out of Jay's legs, jumping over each other, working a teeter-totter, and walking on their hind legs. When Jay would lie on the ground, the dogs would pull a blanket over him, then each would roll himself up in a blanket of his own. In later years Jay would remark this was the only performance in which his dogs weren't at their best. The site was indoors on a freshly waxed floor and his one year old dogs were not prepared, nor was Jay, for this variable!

The Star rodeo paid only $10, but in 1949 that was a lot of money to young Jay. "As we left the arena," he said, "we were offered contracts for two more rodeos. I had to go 400 miles to that second show, but the feller offered me $50 a performance and there were two performances. Right then I thought I had the world by the tail with a downhill pull, 'cause I'd never made any more than $6 a day until then." In later years, Jay's fee increased substantially to $150 per show. He says, "The most we ever got

Behind the scenes at an early rodeo with Jay, Shorty and Stub. Photo by Gene Sisler, courtesy of Les Burlile.

was $1,600 for four performances we did with the Gunsmoke Show." Riding rodeo rough stock for a living was fast becoming a faded memory.

Jay's act with Shorty and Stub grew steadily in complexity and popularity after this rodeo. Performing next at The Dalles, Oregon, Jay got his first taste of life on the road. Uncertain when he left home where his next job might come from, he seemed to always find another employer at the end of each day. He muses, "Each place I would go, someone would see me and hire me for still another place." He adds, "I had no booking agent."

With so many shows, Jay needed to keep the act fresh. He immediately began adding to the act and polishing it up. Many of his routines involved props which he always made by hand himself—a teeter-totter which had a base that doubled as a balancing bar plus the simple props like blankets, a piggin' string which doubled as a leash, jump rope, and many others. He began to dress up his own appearance, too. His sister, June, began creating fancy hand-sewn western shirts with matching pants in the style of many western performing stars. Jay's costumes were always custom made by June and were almost always blue with white accents to compliment his dogs. For his first few shows he wore a black cowboy hat with his ranch clothes but it wasn't long before he switched to the white hat of a "good guy" to go with his fancy new duds. June's talents didn't end with her sewing machine. Jay needed promotional photographs for the media and many of his most famous posed photos were taken by June against the backdrop of the mountains around Emmett. As his fame spread he added one other accessory to his outfit. An Arizona artisan named Charlie McAlexander created a belt buckle for Jay which is seen in all the photos ever taken of him later in his career. The buckle shows Jay, Shorty, and Stub performing their most well known stunt—the two dogs balancing on a narrow bar lifted waist high by Jay. McAlexander remained a lifelong friend of Sisler and created many whimsical sculptures for Jay of all his performing dogs. One such piece was inside a gallon jug. Jay is seen inside the jug holding on for dear life to the tail of a Greyhound trying to jump out of the top of the jug!

In keeping with his new professionalism, Jay joined the Rodeo Cowboys Association in 1951. His schedule that year totaled 20 shows given in Idaho, Oregon, Utah. and Nevada. He spent his first full year performing with the dogs and his horse, Cindy, on the big time rodeo circuit in 1952. The season kicked off a long way from home at the February rodeo in El Paso, Texas. In a promotional interview in the *El Paso Herald Post*, Jay's humility shines, "The dogs are Australian shepherds who could learn more than the number of tricks they have already mastered, if it weren't for my lack of imagination." To promote the El Paso rodeo, he, Shorty, and Stub toured the areas schools—hitting 8 schools in 3 days and most likely giving the students a preview of his act at each stop. Jay's modest demeanor and soft spoken style would be mentioned time and time again when describing him. Both are traits needed for greatness in an animal trainer. After months on the road, Jay's 1952 schedule reached it's pinnacle with his first appearance at Madison Square Garden, followed by his TV debut on Ed Sullivan's *Toast of the Town*.

Jay, Shorty, and Stub during a performance at the John Day, Oregon, rodeo on September 23, 1949. Photo by DeVere Helfrich, courtesy of Donald C. and Elizabeth M. Dickinson Research Center, National Cowboy and Western Heritage Museum.

After only one year as a full time rodeo performer, Jay's reputation had grown nationwide. In 1953, All-Iowa Fair Manager, Andy Hanson, had heard the reports of this topnotch new animal act. He pulled out all the stops to bring Jay east, saying, "Several rodeo people told me, 'If you want something really different, get Sisler.' I've never seen the act but if it's half as good as everyone says it is, it's still the best act we ever had at our rodeo." Charles Story, rodeo chairman of the Ogden (Utah) Pioneer Days, relates, "Sisler has been besieged with stage and television offers since his performances last year. But the shy young man has chosen to remain with rodeo, and Ogden feels fortunate in having him again this year." In addition, until at least 1953, Jay also included a

Jay practicing behind the bleachers with Shorty and Stub at a very early performance. Photo by Gene Sisler, courtesy of Les Burlile.

little known border collie in his act. This dog was used to pen sheep and calves as a demonstration dog in his act. This full act kept audiences happy and entertained with it's richness and variety.

Jay's success enabled him to buy his own first home and begin ranching for himself. His first house on the northeast edge of Emmett still stands today, although it has become part of the town and the property is no longer used for ranching. Jay, always the self-made man, did much of the building himself with the help of Gene, nephew Les Burlile, and their uncle. They first built the basement of the house and lived in that basement for many years until Jay could afford to finish building the house on top. He fenced the place for livestock and began his cattle herd with two Hereford cows. Since Jay spent nearly 9 months of every year on the road, Gene and Les lived and worked on the ranch, caring for the livestock and raising dogs. In 1957, the Idaho newspaper, *Messenger Index*, wrote a feature article titled, "The House the Dogs Built" about this ranch and the dogs who earned the money for it.

Between 1953 and 1955, Jay encountered a wonderful problem. He was often overbooked or double booked. He could not meet the demand for appearances at every rodeo that wanted him. He had started training a second set of dogs as back-ups to Shorty and Stub just in case anything ever happened to either of them. The show, now his livelihood, had to go on. Jay's brother, Gene, stepped in with these understudies, named Tom and Jerry, to take up the slack in Jay's burgeoning schedule. Jay, as the original star, would appear at the bigger rodeos or livestock shows. Gene would work the smaller rodeos with Tom and Jerry. When Tom was just beginning his career as a rodeo performer, he contracted distemper. Although he survived the disease, which took many canine lives in that era, he developed a severe tremor in his back legs, permanently disabling him. After this loss, Gene added the two red merle males, Rock and Rye, to his act with Jerry. Later he would establish his own style by choosing only red and red merle Aussies to populate his act.

Gene wasn't completely relegated to the small time circuits. Texas rodeo producer Dan Coates arranged a tour to several Caribbean nations in 1957. Gene handled this trip as it was during the era Jay was beginning to make films with Larry Lansburgh. Stops on this tour included

Cuba, the Dominican Republic, and possibly other island nations.

The first meeting between Jay Sisler and Larry Lansburgh, like many of Jay's contacts, followed one of Jay's performances at a rodeo. Lansburgh was so impressed by the act he asked Jay to join his fledgling film crew to train dogs for animal movies. This historic friendship left the Aussie breed with priceless film records of the founding dogs which might otherwise never have been made. All because the budding television industry was looking for talent and Lansburgh was trying to fill that need.

Shorty, Jay, and Stub in 1950 at the Tygh Valley, Oregon, rodeo. Jay now sports a white hat but hasn't yet begun wearing the fancy costumes of later years. Photo by DeVere Helfrich, courtesy of Donald C. and Elizabeth M. Dickinson Research Center, National Cowboy and Western Heritage Museum.

In 1956, when Shorty and Stub were at the peak of their careers, Lansburgh made his first short film called *Cowdog* which eventually became the opening half of *Stub, the Best Cowdog in the West*. *Cowdog* was first released to theaters November 6, 1956, as a live-action film and nominated for an Oscar under the category of "Short Subjects, Two Reel" but did not win. A few years later, Lansburgh shot more film of Shorty, Stub, Queen, and Jay's first Greyhound, Blondie, in the informal setting of a schoolground in Jay's Emmett neighborhood. Lansburgh apparently made regular trips to visit Jay in his hometown and made "home movies" of his dogs over time. In the 1990s, a video was eventually made combining the film of Jay's original dogs with that of his later dogs, Silver, John, Joker, Doc, Sparky (Greta in the Lansburgh movie), plus dogs trained by Jay for Wilbur Plaugher. This video was put on the internet and became an instant sensation with dog trainers around the world. Jay narrates the video, telling amusing stories about the dogs and describing their individual talents and personalities.

Larry Lansburgh was a prolific film maker for Disney and saw in Jay just the partner he needed to help put his ideas on celluloid. Jay remembers Walt Disney as a perfectionist, saying "Larry would sweat blood when we showed Disney the rough cuts. If you saw Disney's finger tapping, you knew there'd be changes. If his name was going to be on it, he wanted it perfect. But he was always pleasant to me." This description of Disney is a nearly perfect portrayal of Jay himself. By design, his rodeo acts and, indeed, nearly every aspect of his own life were as perfect as humanly possible. The Sisler name equaled quality.

The custom made belt buckle created by Arizona artist Charlie McAlexander. It shows Shorty, Jay, and Stub in their signature pose—the dogs balancing on their hind legs on a bar held by Jay. This buckle was one of Jay's most treasured possessions and he wore it with his show costumes and throughout his life. Courtesy of *The Idaho Statesman*

Not many dog breeds are fortunate enough to have a professional filmmaker capture the most influential dogs in breed history "in living color." Although Stub receives top billing in the Lansburgh movie, it is Shorty who was to become the prime breeding dog in Australian Shepherd history. All Aussie lovers should treasure this excellent glimpse of the breed's past.

Lansburgh developed the idea for the film *Cowdog* from an incident which actually happened to him 20 years earlier. In 1936, he took part in one of many roundups in California's Santa Ynez Valley to retrieve a wild Brahma cow which had invaded an area herd. The main difference between real life and the film was the switch of the villain's sex from cow to bull. Lansburgh jokes, "Actually, we should have called this picture 'Bull Dog', but I am afraid this would have been somewhat misleading." He intended to make this picture for years but it wasn't until he met Jay Sisler, Shorty, and Stub the idea became reality.

Jay comments on this film, "Cow dogs don't get much credit in the movies, but in brush country, where a pony can't work, a cowman couldn't get along without dogs." Lansburgh called Sisler's dogs, "the finest I've ever seen." His script for stars Shorty and Stub (and "co-star" Queen) kept to reality versus the kind of films Disney sometimes made anthropomorphizing the stars. Lansburgh emphasizes, "In our animal stories we try to keep the animals acting like animals. We think animals are wonderful just as animals, without making them into half-humans. In our stories animals use their own instincts to solve their own problems."

Cowdog is also a kind of Hollywood enhanced version of Jay's life. Jay meets a rodeo promoter at a picnic who asks him to entertain the group with his dogs, Shorty, Stub, and Queen. In 1956, Shorty and Stub were actually veteran performers while Queen had only recently joined the act. Queen's most impressive trick in the movie is walking on her hind legs carrying a plate in her mouth to the grill to get another piece of steak for lunch. She then carries the plate with "seconds" on it back to the table to eat! The final amusement in this scene shows all three dogs being allowed to sit at the table to eat their lunch. And they each wait for Jay's permission to eat!

Shorty and Stub, of course, are the real stars of this segment of the film. When the rodeo promoter asks to see Jay's dogs perform, he runs them through a portion of his real rodeo act. They begin by leaping sideways over each other's backs—a common entrance for the dogs in the actual rodeo act. Jay's doffing of his cowboy hat is the signal to spin in circles on their hind legs—enjoying the applause of the audience.

Three youngsters at the start of a career—L to R: Shorty, Jay, and Stub. Around 1952, Jay's new professionalism started to take shape. He is probably wearing one of the first of many costumes hand sewn for him and brother Gene by their sister June. He now has a white hat and is posing at home. Photo by June Sisler, courtesy of Les Burlile.

Many of Jay's tricks required amazing strength from his dogs and he was always careful when starting a new dog to bring them along slowly to allow time for building muscle. In *Cowdog*, Queen climbs on Shorty's back, then sits up and begs while balancing on his back. Shorty stands absolutely still the entire time Jay is instructing Queen to sit up. After she is released, Shorty spins on his hind legs then jumps over the backs of Stub and Queen on his hind legs alone!

Another amazing feat of strength and balance is one of Jay's most famous tricks. The dogs were first taught to carry any one of their legs as if lame. Next they learned to carry two legs on one side of their body. Jay didn't consider either of these feats to be very difficult for a dog. What he considered most difficult was having the dogs tuck their hind legs up under their bodies and walk on their front legs alone. Shorty and Stub both were masters at this trick and show it off in this film.

The story's barbecue demonstration ends with Jay and the three dogs jumping rope together until Slim Pickens (playing Windy Bill) appears and pretends to shoot at Jay. Pickens then "escapes" by climbing a nearby tree while Shorty and Stub chase him down. The two dogs leap into the air until they manage to grab his pants and rip them off. The two

Shorty and Stub performing one of the most difficult tricks Jay ever created—walking on their front legs. The dogs are about 2 or 3 years old here and have matured and developed the strength needed to master this feat. 1951 Redmond, Oregon, rodeo. Photo by DeVere Helfrich, courtesy of Donald C. and Elizabeth M. Dickinson Research Center, National Cowboy and Western Heritage Museum.

dogs play tug of war with the pants to the delight of the small audience.

The premise of this movie is the roundup of a rogue Brahma bull which has found his way into the purebred Hereford herd of one of Jay's neighbors. The neighbor has invited everyone around to help with the roundup in the Santa Ynez valley of California. After the barbecue, the hands gather in an arena to help the rancher's daughter decide which of her horses to ride. She displays her reining horse—losing her hat in the process—and Shorty performs an old "service dog" trick by retrieving her hat and delivering it to her while she remains mounted.

On the roundup the group splits up with the girl taking Shorty and Stub. Of course they are the ones to find the bull first. While attempting to flush the bull from the brush, the girl is knocked from her horse by a tree limb. Stub "revives" her by licking her face, leads her horse back to her by the reins, and retrieves her lost hat. The roundup is an overnight campout and features another of the dog's cutest tricks—going to bed. Shorty and Stub first cover Jay by picking up the corners of his blanket and pulling it up to his chin. They put themselves to bed by lying down on their own blankets, grabbing a corner of the blanket in their mouth and rolling over to wrap themselves in it.

There are several instances of the dogs working cattle in the movie. Fortunately this gives us some idea how these founding dogs worked. Both appear to want to circle stock and fetch to the handler but they are mainly used as heelers. As was common in that era, dogs were mainly left to figure out the job at hand on their own and they are shown being worked in the movie with very few commands. It is likely this is the reason Aussies became popular with stockmen. Their intelligence and their strong connection with people drove them to work on figuring out what their person wanted and how to get the job done.

The last part of this segment of *Cowdog* shows Jay and the three dogs performing at a rodeo. Shorty and Stub begin the performance by "leapfrogging" into the arena. One dog bows, the other places it's front paws on the back end of the bowing dog and jumps over it. Then the jumping dog bows and the formerly bowing dog leaps over him and on and on. Queen enters by walking across the arena on her hind legs then sits in a begging position when she reaches Jay.

Now surrounded by the town of Emmett, Gene Sisler poses with Shorty and Stub in front of the pasture that was behind Jay's first home, "The House the Dogs Built." This photo was probably taken in the mid-1950s—notice how Stub's blue merle color is starting to darken with age. Gene is also wearing a June Sisler handmade costume. In the pasture behind Gene—seen just over his right shoulder—are Jay's first two horned Hereford cows, the beginning of his ranch operation. Photo by June Sisler, courtesy of Les Burlile.

The short film *Cowdog* was the first half of the show which would later be expanded for television and renamed *Stub, the Best Cowdog in the West*. In 1972, after almost 20 years of working with Jay Sisler, Lansburgh decided to develop his first successful story into a feature length film for Disney's long running Sunday night show—*The Wonderful World of Disney*. After a visit to Sisler, Lansburgh rounded up Rex Allen and Slim Pickens to return for the sequel/ending. The two old pals, Allen and Pickens, agreed instantly to return and the team was united again. Allen wrote new songs for the film to tie the two sections of the show together logically.

As narrator, Allen begins the second half by saying, "Usually a man buys a dog house. In this case, dogs bought a man's house." The setting is Jay's second ranch in Emmett, Idaho, purchased with the earnings from his rodeo dog act. The entire second half of the movie was filmed on this Sisler ranch—which is still in the Sisler family today. Allen says it is 17 years later and, in reality, it was!

Jay was retired from the rodeo by this time and had not been living at this ranch for very long when the movie was filmed. He purchased the ranch from the first Emmett native to become a senator in the state legislature and always referred to it as the "Old Bane Place." At the time of purchase, the ranch was nothing but meadows, swamps, and sagebrush. Situated four and a half miles west of Emmett on the Payette River, Jay saw the potential of this property. The lower valley is on the Payette flood plain and once cleared would become fertile, productive acreage. Above the valley on what is known as a "bench" in this area was a perfect site to build the home Jay would live in for the rest of his life. Safe from

Probably the only photo of Jay's first two "understudy" dogs, Tom and Jerry, working together. Jerry performed later with Gene Sisler but Tom's bout with distemper left him permanently disabled. L to R: Jerry, Jay, and Tom in 1953 at the San Angelo, Texas, rodeo. Photo by DeVere Helfrich, courtesy of Donald C. and Elizabeth M. Dickinson Research Center, National Cowboy and Western Heritage Museum.

floods with a panoramic view of Emmett, the Payette valley and the mountains beyond, Jay leveled the area and built his last house. An elderly couple occupied the original 1800s era homestead in the valley below and Jay allowed them to remain in their home for the rest of their lives.

In 1972, the house was so new and the property so recently renovated it appeared somewhat stark to director Lansburgh. In order to give a more settled feeling to the scenes at Jay's home, he brought in a number of potted trees to set around the house. Lansburgh also contributed the "star" of the second half. The dog who played "Stub the Third" was not one of Jay's dogs but was, in reality, a dog named "Hymp" (pronounced "Hemp") which belonged to Lansburgh. Les Burlile remembers the dog well and states he looked very much like a Border Collie. Lansburgh was a hobby breeder of Border Collies, so it is very possible this dog was a Border Collie or at least a part-Border Collie. Hymp's style of working sheep in the movie is very reminiscent of a Border Collie.

"Stub the Third" is the hero of segment two of the movie. Supposedly the heir to the Shorty and Stub rodeo act, this black and white dog is cast as a performing washout. The story opens with Jay in a training session with his last group of performing dogs. Though he was retired, Jay continued to train dogs for the rest of his life. Wilbur Plaugher's rodeo clown career continued after Jay's retirement and grew to include dog acts using dogs trained by Sisler. This scene shows Sisler's Doc weaving through Jay's legs forward, then Jay walks backward and Doc weaves back end first through his legs. The dogs and Jay play jump rope with Jay and Sisler's John holding the rope—Jay turns the rope while John holds his end—and Doc jumps. Silver, the dog star of *Run, Appaloosa, Run* is also seen in this movie but he does not play a major role. One other attractive dark blue dog is seen in this movie but never identified.

"Stub the Third" is portrayed as a dog uninterested in being a rodeo star, much preferring to be a working ranch dog. A goat appears to disrupt

1957 Rodeo Tour organized by rodeo announcer and promoter Dan Coates. Gene Sisler was invited on this tour due to Jay's obligations with Larry Lansburgh and Disney at the time. From L to R: Jimmie Adams—Roman Rider and Trick Roper who performed with the 101 Ranch Wild West Show in 1965, Clark Shultz—rodeo clown for 29 years and performer with the 101 Wild West Rodeo in 1960, Pat Howell, J. R. Anil, Janette Plunkett—101 Ranch Wild West show trick rider in 1962, Mercado (Micardo)—local host, Joan Bradley, Dan Coates, Gene Sisler, J.W. Stoker—1999 National Cowboy Hall of fame inductee, 2011 Pro Rodeo Hall of fame inductee, stunt rider for Roy Rogers "King of the Cowboy Trick Ropers," PRCA Entertainer of the Year 1985 and 1986. Used with permission of Mr. Don Coates, from the Patricia Highsmith Collection of the Swiss National Library.

the training session and Stub springs into action, chasing it away. Slim Pickens' role in this segment of the movie is as one of Jay's neighbors. (Jay's nephew Les Burlile, a former rodeo rough rider, has a cameo role in this movie as a stunt double for Slim Pickens.) Slim and Jay arrange to help each other move cattle with the help of Jay's "nephew" Tim. Stub and Tim are portrayed as best buddies although in doing chores around the ranch it is Doc who does the work with Tim. Jay hands Doc the strings from a hay bale and has him carry them to a barrel and drop them in. Tim

Jay with his two best dogs, Shorty (closest to camera) and Stub, plus their newest member, Queen (behind Jay). Jay was well established as a performer when this photo was taken at the 1955 Pendleton, Oregon, rodeo. Photo by DeVere Helfrich, courtesy of Donald C. and Elizabeth M. Dickinson Research Center, National Cowboy and Western Heritage Museum.

drives a tractor with a bale of hay to pasture. Doc rides on the bale, then jumps off and opens the gate by putting his front feet on it and walking on his hind legs while pushing it open.

Jay, Tim, Stub, and Doc ride out to the irrigated fields to change the water. It's Doc who crawls under the fence and picks up the drain covers and moves them to allow water to flow into new fields. Stub only cools off in a canal. However, Stub soon redeems himself as Jay, Slim, and Tim ride off to move cattle. They encounter a Basque sheepherder who can't get his sheep out of some brush. Jay sends Stub to bring them out. Stub does a marvelous job of circling the sheep and bringing them down the mountain in a much more controlled manner than the methods Shorty and Stub used on cattle.

The dogs don't appear to be used much on the cattle as they are moved but the last trick for Stub is a daring water rescue of nephew Tim.

At a water crossing, Tim is swept from his horse. Slim and Jay attempt rescues but are unable to get their ropes out to Tim to pull him to shore. Tim manages to catch a large rock and climb up on it but the swift current is too much for the men. Jay and Slim tie their ropes together and Slim takes Stub upstream. He hands the rope to Stub, who carries it in his mouth while swimming out to Tim on the rock. Tim is rescued by the two men while Stub swims to shore on his own. Rex Allen sings a ballad about great cow dogs as the three gather up their herd and the dogs move them out. The full version of *Stub, the Best Cowdog in the West* was first aired in 1974 on NBC as the 10th episode of the 21st season of *The Wonderful World of Disney*.

Sandwiched between the two halves of *Stub, the Best Cowdog in the West* was another big film credit for Jay and his dogs, *Run, Appaloosa, Run*. This film obviously revolves around an Appaloosa horse but one of Jay's later dogs, Silver, plays a significant role in the movie. Released in the summer of 1966, the story was filmed in northeastern Washington state and proposed to tell the story of the Nez Perce Indians and their Appaloosa horses. Silver saves the life of an orphaned colt after his mother dies. He is shown "fighting" a cougar which is trying to catch the newborn. The foal jumps into a river running from the cougar and Silver has to swim out and herd him back to shore. Much is made of the "bond" between dog and horse after this and Silver participates in halter breaking by leading the young colt around himself.

Silver is a background player for the rest of the movie, riding in the Jeep and

Jay with his first Greyhound, Blondie, in 1956 at the Pendleton, Oregon, rodeo. It never occurred to Jay a Greyhound couldn't do all the same tricks his Aussies did, even the most difficult like walking on their front legs. He considered Blondie the best of his three Greyhounds. Photo by DeVere Helfrich, courtesy of Donald C. and Elizabeth M. Dickinson Research Center, National Cowboy and Western Heritage Museum.

The August 1955 cover of *The Western Horseman* magazine featured Jay with Shorty and Queen. Jay's mare, Cindy, was once a part of the act until her tragic death in a trailer accident. Jay is wearing another June Sisler creation—a departure from his usual blue and white garb. Cindy's saddle and hackamore were both handmade by Jay. Photo by Clarence Coil, courtesy of *Western Horseman*.

Jay with Silver, Doc, and Jet at his Emmett ranch. Photo by Clarence Coil, courtesy *Western Horseman*.

accompanying the star, Adele Palacios, to her rodeo trick riding performances. Palacios, as Mary Blackfeather, forms a strong bond with the orphaned foal, training him so he brings a huge price for the tribe at the annual sale. She eventually is able to buy him back to ride in the big Omak Suicide Race. Silver has one more important role during the race. Two horses and riders compete as a relay team and must pass a baton during the race. During the race, Mary drops the baton and Silver retrieves the baton and Mary uses her tricking riding skills to lean down on the run to take it from Silver. After winning the race, Silver jumps into the saddle and rides with Mary on her victory lap in front of the cheering crowd.

Run, Appaloosa, Run was another Lansburgh produced and directed film and Jay made more contacts with the top trainers of his day. Although the legendary show jumping trainer Jimmy Williams is credited in the film with training the horse star, Holy Smoke, he likely only trained the horse for the jumping scenes. The horse was owned by Larry Lansburgh's wife and spent most of his early life on Jay's ranch. Les Burlile remembers using him on a daily basis to bring in cattle, irrigate, and do many chores around the ranch. He was ridden by both Jay and Les so he was well trained and fit for the movie before shooting began. The colt, which was cast as a young Holy Smoke, was raised by another Emmett native and his real name was Christians Ohadi. When the film premiered in Emmett, this horse was a part of the festivities, housed in a pen outside the theater for the public to admire. Rodeo clown Wilbur Plaugher has a co-starring role playing himself in the film.

Interestingly, Silver was not the only Sisler dog appearing in this film although it takes some close observation to spy the other dogs. Sisler's John appears momentarily during the training of Holy Smoke. The horse is supposedly reluctant to enter water after his trauma as a foal. Mary tells Silver to get him into the river. The real Silver begins circling, barking and feinting at the horse's heels as if to bite. In a few frames of a tight close-up, a dog is captured actually making contact with the horse's heels. The markings confirm it is Sisler's John!

One other Australian Shepherd appears during a scene where the horse is being attacked by a bucking bull. It doesn't appear to be John but it is a very different looking dog from Silver (although it supposedly IS Silver). It is most likely another of Jay's dogs.

Sisler's John while at Weldon Heard's farm. Courtesy of Suzanne Anaya

Jay had a number of other film credits but his Aussies had no more starring roles. The Aussies did have small parts in another Disney film. Silver appears briefly in *Ida the Offbeat Eagle*—barking at the story's star and frightening her away from his post on a river dam.

In 1956, about the same time *Cowdog* was being filmed, Jay trained the dog who starred in the Oscar winning Disney film *Wetback Hound*. Yet another Lansburgh production with Rex Allen narrating, *Wetback Hound* was released in 1957 as a 20-minute short subject. As with the short *Cowdog*, this story was trimmed for a 1962 rebroadcast to promote the film *Big Red*, then later expanded to 48 minutes to run as a feature on *The Wonderful World of Disney*, airing in the 5th season. For the 15th season (1968), Jay trained yet another horse to star in the enjoyable *Brimstone, the Amish Horse*.

Season 9 of *The Wonderful World of Disney* featured a Sisler dog sporting a tail. Jay's second greyhound, Sparky, starred in *Greta, the Misfit Greyhound*. Filmed in 1962 and aired February 3, 1963, once again the team of Lansburgh, Allen, and Sisler produced a winner. (Jay's 10-year-old real life nephew, Les Burlile, appears in this film as the boy throwing food out a car window for Greta.) Much of this film was shot in Jay's hometown of Emmett, Idaho. The scene of Greta's abandonment is a road leading into Emmett near the racetrack.

Jay says of adding a Greyhound to his string, "They told me I was wasting my time when I tried to retain this greyhound too. They said a greyhound is too narrow between the eyes to learn anything." The storyline of Greta is a racing greyhound who isn't all that interested in racing. When she finally jumps into the infield during a race to take a shortcut to the lure, she is disqualified from the track. Her desperate owner abandons her beside a desert road and Greta is on her own.

There are several exciting sequences of Greta coursing hare and coyotes during the film but the hand of Jay Sisler is more than evident in the wide array of tricks Greta performs. Jay's greyhounds were known for their jumping ability. His first greyhound, Blondie, could clear Jay's six foot tall, cowboy hatted head while he stood nearly at his full height. Sparky demonstrates this ability in Greta by jumping over a bull's back

The Greyhound who started Jay's love affair with the breed, Blondie. Pictured doing a wonderfully creative trick only she did—the wind-up toy. Photo from the 1956 Pendleton, Oregon, rodeo. Photo by DeVere Helfrich, courtesy of Donald C. and Elizabeth M. Dickinson Research Center, National Cowboy and Western Heritage Museum.

to escape a threat. Sparky also speaks on command, stops and poses during Rex Allen's narration, walks and trots in different directions on command, carries a leg as if lame, sits up and "waves," walks on her hind legs, and lies down and covers her head with her paws when there is trouble. In the story, Greta is adopted by a Basque sheepherder after proving her worth saving the flock from coyotes. She is put to work carrying packs and pulling a travois with the herder's supplies—no easy feat for a breed of dog bred to do the opposite of carrying or pulling heavy loads!

Although films like *Greta, the Misfit Greyhound* often had simplistic storylines, Lansburgh insisted on remarkable accuracy in details. The man who played the sheepherder who adopted Greta was a real Idaho Basque named John "Tacolo" Chacartegui. Lansburgh went to great lengths to portray a realistic shepherd's camp and lifestyle and showed the sheepman using border collies to work his flock. Just as in *Stub, the Best Cowdog in the West*, the Basques were not portrayed using Aussies or Pyrenean-type dogs.

Lansburgh relates an incident that occurred during filming, "Greta, who belongs to a friend of mine, cut her foot on a piece of glass several weeks after we started production. We had to wait three months until it was cured because she was in all the early close-up shots. However, we made use of her limping in a fight with a bear. She wasn't acting then; she was really limping."

Jay had quite a long and very close relationship with Lansburgh and Disney. From the time Lansburgh made his first short film, *Cowdog*, he says, "Since then, Jay has helped me on all the pictures I have made, with and without dogs." Lansburgh considered Jay, "...the finest dog trainer in the world." Jay not only trained Sparky for *Greta, the Misfit Greyhound*, he also helped with the other animals appearing in the story—bear, mountain lions, raccoons, and other animals. Les recalls a special knack Jay possessed for observing the wild animals and working their natural behavior into a scene. Many of these animals were owned and trained by a Washington trainer named Beebe, but it was Jay who worked their talents into the show.

Sisler also assisted Lansburgh on the 1969 movie, *Hang Your Hat on the Wind* and has a credit in the film, *A Man Called Horse*. His talents didn't end with animal training. Lansburgh also praises Jay for his creativeness with mechanical props, his imagination for story ideas, and his carpentry skills! Jay's wife, Joy, relates he was never idle in retirement at home, either. Whether he was working on his own house, making a saddle, or even trying to make his own boots, Jay kept himself busy. Joy is quick to say he wasn't a "Border Collie" type of personality, just industrious. She says, "He was a quiet, solid man and the Aussies suited him."

This industrious side of Jay was not something which suddenly appeared during his movie career, it was an integral part of his character. He had an interest in literally everything but he never did anything halfway. Les Burlile has not forgotten a miserably hot summer day working near the Payette River building fence on Jay's second ranch. He and his companion were supposed to be digging post holes to Jay's exact specifications—36 inches deep, not an inch more or less. They were to be set level and in a perfectly straight line. But the heat and youthful casualness led Les to slack off and stop digging short of a yard. The level was soon abandoned, too. After setting a line of these "close enough" posts, Jay appeared to inspect the work. His disappointment in the job led to an object lesson for Les and his helper. Jay made them pull up every post and redo the job until it was exactly right—the only thing worse than putting the posts in the ground in the first place! The payoff came many years later in another memorable lesson. When the Payette River had a historic flood and swept away the possessions of a great many Emmett residents, they were able to find most of them hung up downstream in Jay's still standing fence—the fence which had been built right. Les credits Jay with his own attitude toward his superb quality saddlemaking and leatherwork. Burlile's work is in high demand due to his attention to detail—just as Jay's work with trained dogs has remained the standard against which all others are measured.

Jay on the front porch of his ranch overlooking the Payette River valley in Emmett, Idaho. His attention to detail and insistence on doing things right show in the neat appearance of his ranch as well as the training of his dogs. Photo by Clarence Coil, courtesy of *Western Horseman*.

1960 National Western Stock Show, Denver, Colorado. This is probably Jay's second Greyhound, Sparky. The Aussie is unidentified but probably one of many dogs Jay was trying out between the retirement of Shorty and Stub and the formation of his second troupe of performers. Photo by DeVere Helfrich, courtesy of Donald C. and Elizabeth M. Dickinson Research Center, National Cowboy and Western Heritage Museum.

By 1969 Jay had shown his dogs at the National Western Stock Show 6 times, the Calgary Exhibition and Stampede, 5 times, Madison Square Garden, Boston Garden Rodeo, 18 times, the Cow Palace in San Francisco, the 1962 World's Fair in Seattle, the American Royal Livestock Show, the International Livestock Show, and state fairs in Florida, Ohio, Tennessee, Iowa, Arkansas, Minnesota, and Wisconsin, in addition to shows all over Canada. In 1953, Jay performed 43 times at Madison Square Garden alone. The last day, he did a matinee, dashed across town for the *Ed Sullivan Show* and made it back to the Garden in time for his evening encore show. 1959, 1961, and 1962 found him touring with the Roy Rogers western show—a live traveling show which was a favorite at the nation's larger fairs. A 1959 Roy Rogers advertising mailer featured "Jay Sisler and His Miraculous Trained Animals." In 1960, Jay's act had star billing on *The Roy Rogers Show*. Roy and Dale hosted *The Dinah Shore Show* in 1961 and brought along Jay and his Aussies for that appearance, too. Rogers was not just Jay's boss while traveling with the Sons of the Pioneers show, but also a friend. When Jay's mother died during one of Rogers western shows, it was Rogers who paid Jay's airfare home.

He and his Aussies were also in a 1954 segment of the television show *You Asked for It*. For over 20 years, Jay entertained audiences of all sizes across the country—from California to New York and Texas into Canada. His genial nature made him friends wherever he went.

Jay's long career in the film industry parallels his rodeo show career. His first film credit for *The Wetback Hound* coming in the 5th season of *The Wonderful World of Disney* (1959), *Greta, the Misfit Greyhound* in season 9 (1963), *One Day at Teton Marsh* and *Ida the Offbeat Eagle* in season 11 (1965), *Run, Appaloosa, Run* in the 14th season (1967), *Brimstone, the Amish Horse* in the 15th season (1968), *Hacksaw* in the 18th season (1971), then *Stub, the Best Cowdog in the West* in

Gene Sisler with his first three performing Australian Shepherds. Rye in front of Gene closest to the camera, Rock next to him, Jerry jumping behind Gene. Gene is wearing a costume handmade by sister June and jumping rope in front of the two custom made trucks he and Jay used to travel to performances. The campers were both made by Jay and Gene themselves and contained built in kennels for the dogs. Courtesy of Les Burlile

L to R—Jerry (sitting on the ground), Rye, Gene, and Rock displaying Jay's signature pose sitting on a bar lifted to his waist. 1959 Yuma, Arizona, rodeo. Photo by DeVere Helfrich, courtesy of Donald C. and Elizabeth M. Dickinson Research Center, National Cowboy and Western Heritage Museum.

the 21st season (1974). Jay even recalls an acquaintance with Hollywood's most famous animal star, Lassie. In one episode of *Lassie*, one of Jay's dogs stood in for the star and performed a scene he couldn't! From the release of *Cowdog* in 1956 coming full circle to the release of *Stub, the Best Cowdog in the West* in 1974, Jay's film career was a testament to his sound training methods, his insistence on doing things well and his wonderful skill. He says, "This thing, that started by accident, has paid for my ranch and my cattle, let me see a lot of country I might not have seen otherwise, and it has made a lot of good friends for me."

Sisler's career as a rodeo specialty act did not languish while he worked on films. Numerous appearances every year kept him grounded in the arena which gave him his start. When scheduling conflicts arose, Jay's brother, Gene, filled the gaps with his own troupe of trained Aussies to help out. The two brothers bred and trained dogs together and would trade dogs back and forth so their acts remained nearly identical. Gene would even take Shorty and Stub on the road if Jay was detained elsewhere. The rodeo always knew what it was getting when either performer was hired. If Jay couldn't make a rodeo due to other commitments, brother Gene could step in and the appreciative audience didn't seem to care what color the dogs were!

In 1956, Gene had three Australian Shepherds in his act named Rock (red merle), Rye (red merle), and Jerry (blue merle). The *Tri-City Herald*, August 26, 1956, describes them as "...display(ing) unbelievable feats of balance and finesse between events." Announcing the showing of *Cowdog* at the Liberty Theater, the *Walla Walla Union-Bulletin* of December 4, 1956, states the Sislers have been touring the rodeo circuit for six years. Gene followed closely in Jay's footsteps with his own act and is described as a top trainer in his own right. Nephew, Les Burlile, says Jay started Gene's dogs, but after Gene took them he would have had to continue the training and kept the dogs "show-ready" since Jay was on the road for the majority of every year. Gene seems to have toured the smaller rodeos in nearby Washington and Oregon with Jay taking the bigger shows such as Calgary, the National Western, and the Cow Palace. It is doubtful any of Gene's audiences felt cheated since his dogs were every bit as talented and well trained as Jay's.

Jay Sisler and ASCA President, Gwen Stevenson, had met and he was aware of the budding interest in forming a club for the Australian Shepherd. Mrs. Stevenson includes a letter from the first President of the Australian Shepherd Club of America, Eloise Hart, in her book, *The Australian Shepherd's Origin and History*. Quoting Mrs. Hart, "Our little club continued with great enthusiasm (sic). Jay Sisler visited one of our matches, and his dog 'Queenie' stole the show. At the 'Sit-Stay' command, all the dogs were sitting for three minutes, with one minute to go. Queenie got up on her hind legs and walked to Jay! The other dogs were dumbfounded."

Sisler's fame was also noted in a 2001 Jane Burnett Smith murder mystery titled, *Always Standing By*. Smith is from a Montana ranching and rodeoing background and fashioned a character in this mystery after Jay. Her description of Queenie, Stubby, and Shorty as "diminutive blue and white dogs" partially corresponds to reality and she describes their

Queen (standing in center) waiting her turn to work while other dogs are performing. 1956 Pendleton, Oregon, rodeo. Photo by DeVere Helfrich, courtesy of Donald C. and Elizabeth M. Dickinson Research Center, National Cowboy and Western Heritage Museum.

act as sheep herding demonstrations rather than the specialized acts they were. The character named "Tooley" claims his dogs to be descendants of Sisler Aussies but uses them as sheep herding dogs rather than as a comedy act like Jay's. Although Jay Sisler was known for his entertainment acts, his background in sheepherding, through his experience working for the Little family, may have entered the act from time to time.

Jay's rodeo career eased to an end in 1971. After 20 years and three quarters of a million miles of driving, his life came full circle as he retired to his ranch only a mile and half from where he was born. Jay's life of hard work was catching up to him and he was finding traveling and getting in and out of a pickup to be too much trouble. He reminisces, "Traveling with the dogs was also getting harder and harder. When I first started traveling, I would stop just about anywhere and let the dogs run. But with today's interstate highways, you can only stop at rest areas and then the dogs must be leashed. My dogs were seldom leashed." Although he never stopped training dogs—many for other rodeo performers such as bullfighting clown, Wilbur Plaugher—he was now doing it at home on his own 300 acres.

Jay purchased the ranch he had once worked on in 1959. Overlooking the valley of his hometown, he could now enjoy his favorite view, Freezeout Hill, on a daily basis. Spotting this major landmark was when Jay knew he was truly home. He reminisces, "Whenever I saw old Freezeout Hill, I wanted to shout."

This ideal stock raising country borders the Payette River to the south and includes bottomland and the bench above the river. The original homestead house built in the late 1800s still stands on the ranch. Starting with 2 horned Hereford cows in the 1950s, Jay eventually grew his herd to 100 head of crossbred cattle. Some sources cite Jay's start in cattle with 20 Shorthorns but this is a mis-remembrance of an occurrence at his second ranch. His neighbor raised Shorthorns and one day he found they had crossed the river onto his property. After returning them, Jay considered what he had seen and bought 10 to add to his own Herefords to help increase their milk production. After a few years, this experiment in outcrossing was ended when Jay found the Shorthorns ate more than his own Herefords, got too fat, and would not produce as well. He and Wilbur Plaugher were experimenting with some Brahma crosses in their herds shortly before he died.

Jay never tried to keep more cattle than he could care for himself. In keeping with his lifelong high standards, his emphasis in raising cattle was on the quality, not the quantity he kept or raised. Jay's self-made life didn't change after retirement. He always ran the ranch himself, tending to the chores and management firsthand. Technology was added when it proved useful and Jay participated in many university feeding trials and experiments. The quality of his own philosophy was proven time and time again when his weanling calves would weigh in higher than most ranchers yearlings. He was a longtime member and past President of the Payette River Cattlemen's Association and remained active in this group until his death.

One of the most famous photos of Jay with Shorty, Queen, and Stub. 1955 Pendleton, Oregon, rodeo. Photo by DeVere Helfrich, courtesy of Donald C. and Elizabeth M. Dickinson Research Center, National Cowboy and Western Heritage Museum.

This attention to detail and high quality was a reflection of the honest and frugal life Jay led. He was generous to anyone and everyone who needed his help but sparing when it came to his own lifestyle. About 10 years prior to his death the fair skinned, red haired Jay was diagnosed with a very rare skin cancer on his face. He never believed in insurance but his personal thrift allowed him to pay for all of his treatment—surgery and chemotherapy—out of his own pocket and he was cured. Truly a rare man!

Jay never gave up dog training and always kept at least one Aussie around to practice with. His dogs continued to be given to him by friends but by the time he retired he no longer had any dogs of his original bloodline. Nevertheless, his enjoyment of training never waned, even though after the birth of his daughter he became less of a perfectionist and his interest grew in teaching young people how to train. He realized the dogs would not be as obedient as he might have liked but accepted his new role of training new trainers with his usual good cheer.

In 1978, at the age of 52, he married his wife Joy, a native New Yorker. In 1979, their only child, daughter Maggie, was born. Life on the ranch was a quiet change from Jay's 20 years on the road but in his mind the respite was his real life's dream. Ranching was always a true love for Jay and retirement was just the beginning of another great adventure.

Jay was able to concentrate on developing his ranch and his cattle herd as he had always envisioned. For the most part his place was self sustaining and he raised his own feed for his cattle and horses, irrigating alfalfa fields near the river and grass hay on the bench above the river. He didn't rely on much hired help, preferring to continue working as he had always done either from horseback or on a tractor.

It was while baling hay on July 8, 1995, Jay Sisler was lost to this world. Daughter Maggie was putting her horses in the barn when she looked up to see her Dad in a nearby field. When she looked again, the tractor was still moving but her father was not in sight. When she found him, his Aussie, Darby, was by his side but it was too late to help Jay. He was only 69 years old. He now lies for all time in the Emmett cemetery not far from his ranch.

The rodeo world did not forget Jay Sisler. At the time of his death he had been scheduled to appear in late July 1995 at the Snake River Stampede in Nampa, Idaho, as part of a historical tribute to rodeo. The 45-year-old stadium was going to be replaced with a new arena in 1996 and Jay would be returning to a place he had performed at four times—in 1953, 1958, 1966, and 1969.

In a long overdue honor, Jay was posthumously inducted into the Pro Rodeo Hall of Fame in Colorado Springs, Colorado, on August 10, 2002, for his contributions to the sport as a Specialty Act entertainer. From the Pro Rodeo Hall of Fame biography, "Sisler is remembered as a phenomenal animal trainer with unique acts not seen before or since." This has to rank as one of the all time great understatements in history. Internet video of Jay's acts continues to amaze viewers in spite of all the "new and improved" training techniques at their disposal.

Jay's widow and daughter continue to run his Emmett ranch with the help of Maggie's husband. The ranch is in transition to organic certification and now produces hay, cattle, horses, and a variety of fowl including ducks, geese, chickens, and chukkars.

The final words of Jay's obituary sum up the well-lived life of this admirable and noble man: "Jay's God-given talent to train animals was shared generously and enjoyed by many. He loved his family, friends, and neighbors, and we all were blessed by his life." No person has ever done more good for the Australian Shepherd, nor likely ever will.

chapter 4

Jay Sisler
THE TRAINER AND SHOWMAN

Jay Sisler was a natural animal trainer of the highest degree. The two most critical qualities necessary for success in training—dedication and a true love for one's work—were found in abundance in Jay. The 1930s and 40s were well ahead of the era of widespread obedience clubs and training seminars. Jay was a man who learned by doing. He had a gentleness and empathy for his students that created a close-knit team. His dogs were companions first and he was always proud that his show dogs were good buddies, too.

Training horses was Sisler's stock in trade at first. Although not much is known about his early days as a horse trainer, he and brother Gene both began with horses and both are said to have been fair rough riders. It's likely Sisler tried his dog training techniques on horses when he discovered how well they worked. Nephew, Les Burlile, confirms this was what he saw in Sisler's mare Cindy—part of his original act with Shorty and Stub. Cindy, like the dogs, showed an uncommon loyalty to Jay and would do anything he asked. One of her most notable tricks was getting down on her knees, putting her head between Jay's legs and lifting her head to slide him down her neck and onto her back.

In keeping with Jay's humble nature, he credits his first two dogs, especially Shorty, with much of his success. Les Burlile remembers Shorty as a dog who was "all business." He goes on to describe Shorty as an extremely loyal dog who lived to work. Stub, on the other hand, was a "more playful" personality—a trait which likely eliminated him from the future gene pool of the breed. It appears Shorty was the dog most used for breeding, which suggests the serious dog with a strong work ethic was most desirable to the earliest breed fanciers—especially Jay. He always believed it was important to have a dog which meshed with his own personality. He saw in dogs all the character traits he found in people, saying, "Some will even cheat you." An honest dog was one which always impressed him, probably due to his own impeccable honesty. In an interview published in *The Western Horseman*, July 1969, Jay confesses, "Probably if I hadn't had real good dogs when I started, I'd never have learned to train dogs. I'm sure that some of the dogs I've trained since would have discouraged me before I ever got started."

Shorty "buries" Jay while Stub stands guard. This routine was always a favorite part of the act. 1949 John Day, Oregon, rodeo. Photo by DeVere Helfrich, courtesy of Donald C. and Elizabeth M. Dickinson Research Center, National Cowboy and Western Heritage Museum.

Sisler never stopped learning about animal training. His career as a performer spanned over 20 years and the act was expanded and perfected during that time. This description of his act is found in a promotional article in *The Montana Standard*, July 1969, "In a fast moving few minutes, the dogs romp through a variety of tricks with a personality quite their own, showing amazing canine intelligence cleverly combined with comedy." Even in retirement, Jay still trained and practiced with dogs around his place. He kept dogs in training for other performers such as Wilbur Plaugher. Although Wilbur remained a rodeo clown, he added animal acts to his repertoire and many of his animals—not just dogs—were trained by Jay. A bronze statue at the Pro Rodeo Hall of Fame in Colorado Springs depicts Plaugher herding ducks into a burlap sack with a Sisler-trained Border Collie named Fox.

Jay's training philosophy was based on starting a dog at a young age and gaining it's confidence. Consistent teaching, plenty of patience, and many hours of practice was the only way to create a strong bond and a flawless performance. His first lessons were similar to any obedience trainer of today—sit, stay, and the meaning of "no." Practical lessons were taught which Jay used in his everyday life and when on the road. One newspaper reporter describes Jay discussing the meaning of "good-bye" to his dogs. Jay spoke the word a little too loudly and Shorty and Stub leapt to their feet and headed for their trailer. Jay immediately called the two back but Stub came a bit reluctantly. He explained that Stub would get his feelings hurt when he gave the "goodbye" command because, "He thinks I don't want him around."

Sisler expected undivided attention from his dog. Rodeos are noisy, exciting events with many temptations for a working ranch dog. The audience adds to the general ruckus. A dog which missed too many of Jay's cues would ruin the act. Jay emphasizes, "I teach them to listen to me and to pay attention; those are the fundamentals, and I just go on from there." Sisler naturally directed his dogs using a soft tone of voice, even while carrying on a conversation. Jay needed to rely on hand signals and strong body language to guide the dogs through the act. The noise—or potential for noise—at a public exhibition left him no other choice. The dogs' focus was simply increased out of necessity. The combination of these factors created the seamless perfection which marked Jay's acts.

Bronze statue depicting Wilbur Plaugher with a Jay Sisler trained dog named Fox performing his own rodeo act of herding a flock of ducks into a burlap bag. This statue can be seen at the Pro Rodeo Hall of Fame in Colorado Springs, Colorado. Photo by Author

It wasn't all work for Jay's dogs. As he said many times, his dogs were also his buddies and treated as pets and friends when not working. He often stayed with Fletcher and Mary Wood when he was in Colorado and he always had a truck load of dogs, those in the act and those he was training for the future. For some reason, he never brought enough feed dishes to feed all the dogs at once. Mary says he usually only had four dishes and often had eight to ten dogs with him. But this was never a problem and Mary marveled at how well trained these dogs were. Jay would prepare food for all of them and divide it into the four dishes he had. He would put the dishes on the ground and call the names of four dogs who would each go to a different pan and eat. When Jay thought they had had enough he would call off those four dogs and call the names of four more to take their turn! This continued until every dog had eaten and all the food was

Jay works Blondie while Shorty, Queen, and Stub wait patiently for their turn, 1957, Phoenix, Arizona, rodeo. Photo by DeVere Helfrich, courtesy Donald C. and Elizabeth M. Dickinson Research Center, National Cowboy and Western Heritage Museum.

gone with never a cross word between the dogs and always with perfect obedience to Jay's commands!

Mary remembers another Sisler visit when he was performing at the National Western Stock Show. His first Greyhound, Blondie, became sick during the show. They had just finished their first performance and Blondie just hadn't worked up to her normal standard. Jay didn't have time to take her to the vet so Mary volunteered to go. The vet determined Blondie had some swelling on her brain and had probably been accidentally kicked at some point. Mary took her home to let her rest and recover and thought nothing of bringing her into her house. Blondie recognized a good thing immediately and hopped up on the couch, resting her head on a nearby pillow. Unbeknownst to Mary this was Blondie's first time in a house! When Jay returned to the Wood home that evening he caught his greyhound making herself at home and apologized profusely to Mary for her "bad manners." Mary loved dogs and didn't mind a bit and Blondie always remembered Mary after this episode with special affection. Her loving care brought Blondie back to her normal self and she was well enough to do her part in that evening's performance.

Jay likens training to building a house one block at a time, increasing the dog's knowledge and abilities slowly and logically. "Once the dog has many small lessons down pat… (you) put them together into something that becomes an act," says Jay. The importance of many small steps is repeated by him many times. He says, "…you have to cut the steps down small enough that the dog can gain a little every day, and that way you can keep encouraging and rewarding him. That way he'll keep his interest in the act and he'll keep trying." Dog trainers, sound familiar? In an interview in the July 31, 1965, issue of the Colorado Springs *Gazette Telegraph*, he states, "I've never used choke chains or punishment in training my dogs." He goes into more detail in this interview, saying, "Usually, like teaching a dog to scale a high fence, I start out with a very small barrier and reward the dog when he jumps over it. I just don't believe in punishment but then I'm a self-taught trainer." He could still teach many of us a thing or two. He concludes, "I think it is important that you get a dog that wants to do things for you—that likes to do tricks. The other kind will only obey as long as you hold a whip over him."

The trick that led Jay to an understanding of how to teach difficult things one small step at a time. Photo by June Sisler, courtesy of Les Burlile.

Sisler started his dogs as freshly weaned puppies. His rewards were varied—from petting to food to verbal praise. Long before Karen Pryor and "clicker" training, Jay grasped the concept of randomization and reduced his food rewards to strengthen the dog's response to him. This is what makes animal acts appear to be "magic" to most audiences. The dog responds immediately and enthusiastically to every cue, never needing enticement and enjoying the performance as much as the humans. Another goal of Jay's methods was to display as much of each dog's own personality to the audience as possible. If the dog was naturally shy he would have him act shy in the routine. Or he would have an aggressive dog appear aggressive. Each dog was also taught as many of the stunts as possible. The group of dogs could exchange roles easily, perform as a unified group, or display a special skill as Sisler needed for each show.

One statement in Jay's *The Western Horseman* interview neatly sums up his methods, "I find that if I can get him to do the action himself, he'll realize that he is doing it. If I do it for him, he'll wait for me to use manual means to position him." At the time of this interview, obedience classes still taught mostly force methods. Even in 1969, Jay was way ahead of everybody else. In another interview, Jay describes his training as "unorthodox," saying he didn't teach his dogs to "heel" but instead used "stay back." If needed, he would block the dog with a piece of brush if they tried to run past him. He clarifies that with advanced tricks such as cartwheels or other tricks he might help the dog to learn by gently rolling the dog over with his hands until he understands the goal.

Jay's wife, Joy, remembers, "He really had the ability to think things through and solve problems." She describes his training regimen as clear and logical, focusing on only one thing at a time. He kept training sessions

Many hours of patient practice go into creating an act like this with several dogs. Jay with Shorty (closest to camera), Stub (on the far side of Shorty), and Queen (behind Jay). 1955 Pendleton, Oregon, rodeo. Photo by DeVere Helfrich, courtesy Donald C. and Elizabeth M. Dickinson Research Center, National Cowboy and Western Heritage Museum.

short and frequent and allowed his dogs to bond with him as companions for the rest of the day. Rewards were praise and leftover pancakes he carried around with him all day. The dogs rarely knew the feel of a leash.

New dogs were normally introduced into Jay's act when they were about a year of age. By this time they were able to perform simple tricks and ready to start learning to perform in public. A dog's first performance was always an eye opening occasion for Sisler. Although they might be perfectly trained at home, he was never quite sure how they would react to displaying their talents for an audience. "More than once I spent an

Gene Sisler performing an impressive variation of the bar balancing trick with the blue merle Jerry! His red merle, Rock, watches and waits for his turn to work. 1959 Yuma, Arizona, rodeo. Photo by DeVere Helfrich, courtesy Donald C. and Elizabeth M. Dickinson Research Center, National Cowboy and Western Heritage Museum.

Shorty, Jay, and Stub at one of their first rodeos performing the difficult bar balancing trick. It looks like even Jay was struggling with this one at first! 1949 John Day, Oregon, rodeo. Photo by DeVere Helfrich, courtesy Donald C. and Elizabeth M. Dickinson Research Center, National Cowboy and Western Heritage Museum.

entire year training a dog to bring into the act, only to find out it had stage fright and wouldn't perform in front of the public." He continues, "...some dogs are more dependable than others, some more honest. But each one had his own personality." So many of Jay's best tricks required great strength and balance he did not want to rush the dogs until they had matured and fully developed their muscles. Once the novice dogs proved to be show material and learned to listen, obey and do the simple tricks in front of an audience, they were ready to begin learning more advanced parts of the act.

Jay's trademark pose—two dogs balanced on their hind legs on opposite ends of a metal bar he held in the middle and lifted—required a great deal of time and practice. Jay considered this skill and walking on their front legs the two most difficult ones he taught. The bar required a keen sense of balance and co-ordination with two partners. Walking on the front legs required an immense amount of sheer strength. As with all his training, Sisler taught these tricks slowly, logically, and in small steps. He explains, "I teach them to sit on the ground, then on a two by four, then on the bar." He continues, "After that, I train them to stand (on their hind legs) on the ground. From there they learn to stand on the two-by-four because they have to get used to something which keeps them from moving their feet around. Then, before I put them on the bar, I usually start them standing on my arm. That way I can help to balance them." This slow, methodical progression created a dog who understood what was required, trusted his partners, and always worked with enthusiasm.

A preparatory skill to walking on the front legs was a trick Jay considered fairly easy for the dogs to perform—walking on both left or both right feet. To train this, Jay merely began by tapping one leg with a stick and saying, "Pick it up." The dog would quickly learn to pick up the foot on voice command only. Once this was accomplished, Jay would ask for the other foot on the same side and the dog soon mastered carrying both legs on one side.

Jay discovered if a dog was having trouble understanding or mastering a skill, a layoff of one to two months would give the dog time to absorb the lesson. Often, his student would pick up the trick right after the vacation. Jay never became discouraged, displaying a perpetual optimism that things would improve. His experience with his first two dogs probably taught him persistence and ingenuity. This served him well with later dogs which may have had less innate talent.

Gene Sisler with the red merle, Rye, performing the incredible feat of walking on his front legs. 1959 Yuma, Arizona, rodeo. Photo by DeVere Helfrich, courtesy Donald C. and Elizabeth M. Dickinson Research Center, National Cowboy and Western Heritage Museum.

The February 18, 1955, *San Antonio Light* describes a Sisler routine that must have been absolutely hysterical. By now Shorty and Stub were seasoned performers and this act combined a number of old tricks with some new genius from Jay. The dogs run through some regular tricks, then move into re-enactment of a mini-tragicomedy. The old cover-up-Jay-for-the-night routine is performed, with the dogs rolling themselves into their own blankets. "Daybreak" occurs next and one dog awakens early to untie Jay's horse and lead him over. Sisler rises to mount his horse and is "shot" by a rustler nearby. The dogs attack the "rustler" in the old "Stub" movie routine of running him up a tree and tearing off his breakaway pants. Once the rustler has been dispatched, one dog remains on guard against his return while the other checks on Jay. The dog supposedly finds Jay dead and proceeds to bury him "in the best tradition of the range." But the story has a happy ending. The guarding dog returns to "camp" and finds signs of life in poor Jay. The story describes it this way, "With amazing realism, he applies artificial respiration, the cowboy revives, mounts his horse and all ends happily." With, most assuredly, roars of laughter and applause from the audience!

The world renowned "Horse Whisperer," Monty Roberts, writes in the Foreword to British dog trainer Jan Fennell's book, *Tales from the Dog Listener*, about Jay Sisler. Roberts saw Jay in rodeos in the early 1950s—the era of Shorty and Stub. He was impressed by the love Jay had for his dogs and even more impressed by the love the dogs had for him. He remembers seeing Jay work with half a dozen dogs at once, each obediently waiting with "eyes sparkling and feet prancing in place" for their turn to be Jay's star. Roberts, like all of Jay's audiences, was drawn to

Stub revives Jay with "artificial resuscitation" while Shorty stands lookout in the background. 1949 John Day, Oregon, rodeo. Photo by DeVere Helfrich, courtesy Donald C. and Elizabeth M. Dickinson Research Center, National Cowboy and Western Heritage Museum.

the ingenuity and originality of Jay's acts. He recalls a night in 1955 when Jay and his dogs arrived at a pub-style restaurant near Roberts' university. He says the owner knew Jay well enough to let him bring the dogs into the restaurant's bar. (Les Burlile takes great issue with this portion of Roberts story, emphatically insisting Jay never drank in his life and would never have entered a bar for any purpose. Perhaps Roberts embellished this telling of his encounter with Jay for the benefit of the readers of Fennell's book—maybe catering to the "wild west" image many Europeans hold of America. Other portions of Roberts' version of this story, such as the names of dogs, are somewhat questionable also. The awe with which Roberts viewed Jay's training is consistent with that of many others, even if the setting and names are not entirely true. Burlile's depiction of Jay as abhorring liquor are much more consistent with all other descriptions of his straight-as-an-arrow character than Roberts version of a drinking and womanizing Sisler. (The inclusion of Monty Roberts' tale in this book is only to cite another example of Jay Sisler's immense influence in the world of animal training.)

Roberts, sitting nearby, took in an impromptu show. As part of Jay's normal conversation he would put the dogs through a plethora of tricks. Roberts was amazed at the way the dogs interacted with each other and the bar's customers. "Blue, go to sleep" was Jay's order and "Blue" would "flatten himself in the middle of the saloon" says Roberts. (It isn't known if the names are real in Roberts' descriptions as this would have been at the time Shorty, Stub, and possibly Queen who were the main part of Jay's act—however these could have been other dogs he was in the process of training—and, of course, the bar setting is also a questionable detail.)

Roberts continues telling about the stunts Jay's dogs performed in the bar that night. "Freddy, take my jacket and cover Blue up. It's going to get cold," Jay would murmur and Freddy would cover Blue with his jacket. Roberts tells of Jay's instruction to "Freddy" to find a pretty girl and bring her over to have a drink with him. This dog would sort through the girls in the room, find the most attractive and lead her gently to Jay with her hand in his mouth. (The possibility exists the girl was someone Jay and the dogs knew—something which would be much more believable than a Jay Sisler who roamed bars using his clever dogs to pick up girls. Jay's natural sense of humor would readily have created such a situation to amuse onlookers!) Roberts lists many of the familiar tricks Jay used in the act—walking tightropes, sitting on each other's backs, "dancing," "singing," riding horses, leading a donkey around, and forming obstacles for his greyhounds to jump.

Jay with Shorty and Stub at the 1950 Tygh Valley, Oregon, rodeo. Photo by DeVere Helfrich, courtesy Donald C. and Elizabeth M. Dickinson Research Center, National Cowboy and Western Heritage Museum.

The fact remains that Roberts was mesmerized by Jay's acts and chose to single him out of all the animal trainers in his wide experience to describe for the forward of Fennell's book. He was equally charmed by the feats they performed and intrigued by how Jay got them to do it. He knew Jay never raised his voice nor struck the dogs, even though he was unaware of exactly how Jay did his basic training. But the man who eventually earned two doctorates in behavioral science did know the lesson of Jay Sisler and his phenomenal dogs was this—that an animal will do more for a human willingly than the human could ever achieve by force.

Probably the toughest schedule the Sisler dogs were asked to work was the National Western Stock Show, held every January in Denver, Colorado. The 1960 program indicates a total of 18 performances given over 9 days—2 performances per day every day. At one point in his career he gave 43 performances in 28 days in cities all across the country. He was even working a rodeo the day his mother died. Today we consider our dogs "versatile" at a National Specialty if they perform in a half dozen events over a week's time. Seldom is our performance in front of audiences of thousands and never does our paycheck depend on our performance. Jay and his dogs had to get it right every time and please the audience or he would not be asked to return. Failure was not an option, nor was a half-hearted try. The dogs and the man never failed to please the audience.

In 1960, Jay had been absent from the National Western for three years. His show that year consisted of the next generation of dogs following Shorty and Stub. This era proved the act was not just the fluke of two exceptional dogs but the result of good dogs plus a man who knew

Jay Sisler and one of his later dogs, Sisler's (Wood's) Panda, March 1963. Courtesy of Rowe

how to get the most from them. Not only did Jay maintain the quality of his act, he grew and refined the act every year. Audiences enjoyed Jay's creativity time and time again, never growing bored with an old act replayed over and over. The same was true of the dogs. Their fertile minds continued to be tested and entertained as well.

Jay's very first act with Shorty and Stub included his mare Cindy. Although he was as fine a horse trainer as he was a dog trainer, Jay eventually found training dogs to be quicker and easier than training horses. He said, "Any time you have a minute, you can just call a dog and put him through a few tricks; but, with a horse, you have the problem of going out to the barn, catching him, and putting equipment on him. All this takes quite a bit more time for each lesson." After losing Cindy in a Colorado trailer accident, Jay's act became a mostly "dogs only" performance although there were times a horse was used as part of his act. In his artificial resuscitation skit, a horse was one of the props for him and the dogs. The *Chilicothe Constitution-Tribune* reported on October 2, 1965, that Jay had created a special act for that year's American Royal in Kansas City which included a horse. Billed as a special "surprise" for the showgoers, this performance shows Sisler did not limit himself entirely to his dogs. During his 20 plus years as a rodeo performer he had a total of 17 dogs in his acts (and many more who were trained but never measured up to the standard needed to go on the road). Australian Shepherds predominated but he later included a series of retired racing Greyhounds, too.

Jay's versatility as an animal trainer knew no bounds. The Lansburgh partnership brought him together with a wide variety of wild and domestic animals. The most critical part of being a Hollywood animal trainer is getting any animal to behave in the desired way on cue. Working with cantankerous beasts such as otters and beavers could make even "acting natural" a challenge! In *Run, Appaloosa, Run*, Sisler's Silver "fights" a mountain lion. In *Greta, the Misfit Greyhound*, Jay's Greyhound "Greta" has an encounter with bears. Raccoon, eagle, donkey, no animal was too difficult for Jay. He always deciphered the mind of his student and drew out their best performance willingly.

Jay was never one to limit the potential of his dogs, saying, "Maybe I shouldn't say this, but I think a dog could be taught to read. I mean he could be trained to recognize a written word and act on it, and to tell one word from another. That's reading isn't it?" With dogs like Shorty and Stub that was a distinct possibility.

Dog training was not merely a living for Jay Sisler but a passion which he pursued to the very end of his life. The last Australian Shepherd he owned, a dark blue merle dog named Darby, learned many of the tricks of his famous predecessors. "Come shine my boots" was his cue to weave through Sisler's legs. "March, Darby" indicated a fancy strut, both forward and backward. Darby skipped rope and jumped through Sisler's arms as well as helped move cattle, lead horses, open gates, chase geese from the alfalfa fields, and dozens of other helpful chores first taught to Jay's Aussies almost 50 years before.

Darby was not the only dog to be trained by Jay in his retirement. His many rodeo and movie friends called often for advice or help with training their own dogs—for acts or their ranch. Sisler gladly met all requests to share his love of training, even that of his own daughter and her friends.

Jay with one of his horses, Silver (left) and John (right). Photo by June Sisler

Jay in 1995 at his ranch with his last dog, Darby. He continued to train dogs after his retirement just for the fun of it. Courtesy of *The Idaho Statesman*

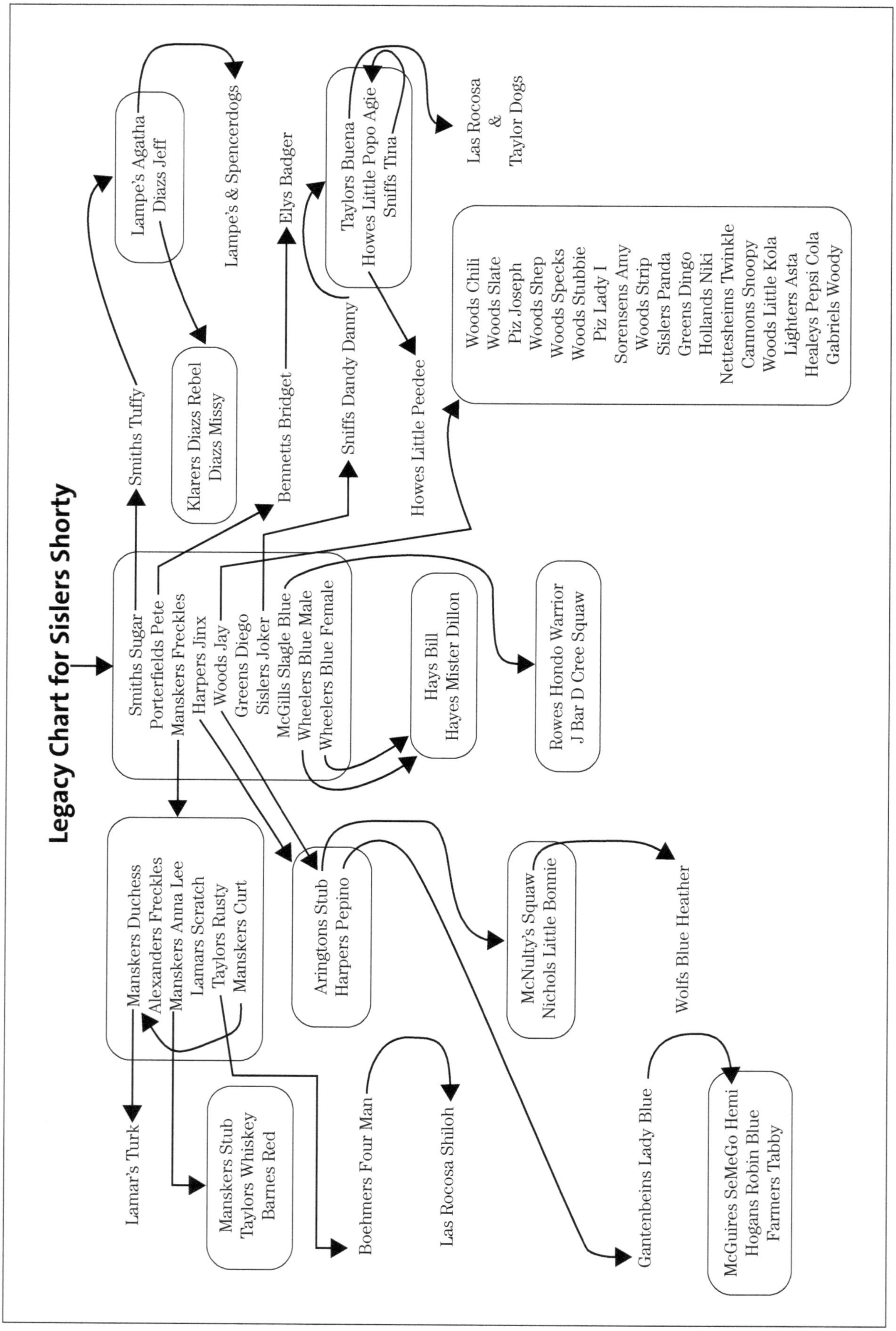
Legacy Chart for Sislers Shorty
Smiths Sugar
Porterfields Pete
Manskers Freckles
Harpers Jinx
Woods Jay
Greens Diego
Sislers Joker
McGills Slagle Blue
Wheelers Blue Male
Wheelers Blue Female
Manskers Duchess
Alexanders Freckles
Manskers Anna Lee
Lamars Scratch
Taylors Rusty
Manskers Curt
Lamar's Turk
Manskers Stub
Taylors Whiskey
Barnes Red
Boehmers Four Man
Las Rocosa Shiloh
Aringtons Stub
Harpers Pepino
Gantenbeins Lady Blue
McGuires SeMeGo Hemi
Hogans Robin Blue
Farmers Tabby
Hays Bill
Hayes Mister Dillon
McNulty's Squaw
Nichols Little Bonnie
Wolfs Blue Heather
Rowes Hondo Warrior
J Bar D Cree Squaw
Smiths Tuffy
Lampe's Agatha
Diazs Jeff
Lampe's & Spencerdogs
Klarers Diazs Rebel
Diazs Missy
Bennetts Bridget
Elys Badger
Sniffs Dandy Danny
Taylors Buena
Howes Little Popo Agie
Sniffs Tina
Las Rocosa
&
Taylor Dogs
Howes Little Peedee
Woods Chili
Woods Slate
Piz Joseph
Woods Shep
Woods Specks
Woods Stubbie
Piz Lady I
Sorensens Amy
Woods Strip
Sislers Panda
Greens Dingo
Hollands Niki
Nettesheims Twinkle
Cannons Snoopy
Woods Little Kola
Lighters Asta
Healeys Pepsi Cola
Gabriels Woody

chapter 5

The Sisler Dogs

Jay Sisler was one man. But that was all it took to make all the difference to the Australian Shepherd breed. Although "breeder" wasn't the main hat he wore, the visibility he gave the breed through his acts and eventually through the scattering of puppies throughout the West had a profound and permanent effect. Everybody who saw Shorty and Stub wanted a dog just like them. And many were lucky enough to have their desire fulfilled. There was no registry. Pedigrees were word of mouth. Pure breeding was practiced casually, if at all. But the influence of the Sisler dogs is still felt today, even though it will likely never be fully recognized.

Sisler was proud of the fact none of his dogs were registered. However, in later years, some breeders who appreciated what Jay had in his dogs managed to register a few. Jay's dogs were unregistered in large part because there was no registry for most of the years he owned, trained, and exhibited his dogs. His first blue dog was born in 1939—almost 20 years before a club to recognize the Aussie as a breed was even formed.

Jay's love for the Aussie was summed up neatly in an interview in the Colorado Springs *Gazette Telegraph*, "An unusually modest showman, Sisler says only that he prefers the Australian shepherd for training, without downgrading other breeds. 'I've gotten better results with these dogs,' Sisler said, adding, 'But then I'm not a professional trainer.'" Surely anyone who ever enjoyed his acts would argue with that statement!

Shorty, Jay, and Stub at one of their first performances, 1949, John Day, Oregon, rodeo. Photo by DeVere Helfrich, courtesy Donald C. and Elizabeth M. Dickinson Research Center, National Cowboy and Western Heritage Museum.

Keno was the first "good dog" Jay says he owned. Acquired in 1944 or '45, Keno was Jay and brother Gene's main breeding dog. His most famous sons, of course, were Sisler's Shorty and Sisler's Stub. Jay states Keno was given to him by a man named Myron Whitely. Whitely told Jay his sire was a blue dog but little seems to be known of Keno's dam. Jay describes Keno's color as "...too dark to be sable and also (he) had a redder coat. He had no merling." Nephew, Les Burlile, remembers Keno as "colored like a lion," neither red nor merled but more a solid colored yellow or tan. Burlile confirms Keno was a producer of all colors of Aussies which suggests he may have been an "e locus recessive" dog. These dogs can be genetically black or merle but appear to be a solid yellow/tan with black pigmentation on noses and eye rims. Burlile further describes Keno as a fairly large dog, weighing 85 to 90 pounds and having a large, square head.

Not much is known about Keno's personality. He was an older dog when Les Burlile was a child and had become a typical quiet elder by then. Like all the Sisler dogs, Keno was used to work cattle when he was young. Burlile credits Keno with Shorty and Stub's conformation as they took after their sire in both size and head type. Because the memories of both Jay Sisler and his nephew, Les Burlile, focus on Keno it is probably safe to assume he was the prepotent parent of this cross and likely responsible for the intelligence and temperament of Shorty and Stub. Jay called him his "first good dog" and since he held Shorty and Stub in such high esteem, they were probably their father's sons. Les describes Keno as very long lived although his exact age at death is unknown. He tells that one day Keno just disappeared. This is an interesting trait that appears to be somewhat common to the mountain dogs. When they are sick or dying they will take it upon themselves to retreat from the world. Burlile owns a dog today who will hide when she becomes ill. This seems to be a very primitive behavior in keeping with the basic primal type of the founding dogs. Burlile remembers finding Keno's remains not far from home six months to a year later. He had crawled up into some brush in a draw not far from where Jay would eventually own "The House the

Dogs Built" and died. They identified him from the color of the hair that was left on his bones.

Burlile recalls another Keno son owned by his grandfather. This solid black dog's name was Jack and he describes him as looking like a bear. His grandfather and Jack worked for a cattle association which grazed cattle in the Idaho high country timber. Grandpa and Jack worked 2000-2500 cow-calf pairs alone all summer. Gathering in the fall took a few extra hands but one good dog and his human pal could care for a lot of cattle by themselves.

Amusing yet amazing, Sisler's Shorty and Sisler's Stub perform another phenomenal trick. Photo by June Sisler

Jack was a kids' playmate as well as a working dog. Uncle Jay's childhood aspirations to be a musher emerged to build a dogsled and harness for the kids. Les remembers all the kids harnessing Jack to sled and heading to the high country to find snow. Jack would pull them around and be their playmate all day long. He was akin to Stub in temperament, showing a more playful attitude about life than Shorty. When the weather was good for sledding, Jack became part of a team of six to seven dogs pulling the sled and assorted children—who steered them like a team horses, using "gee" and "haw." As the kids grew older, Jay's later dogs were also pressed into service for sled dog duty. Silver, Doc, Joker, and many others took their turn pretending to win the Iditarod—even in the summer when wheels replaced sled runners!

Gene Sisler was the actual owner of the dam of the two dogs who put Jay on the map. She was called by many names—"Blue," "Star," "Blue Starr," and "Old Blue"! It's obvious what color she was but little else is known about her. Burlile remembers her as just a blue bitch but "nothing fancy." He does confirm her pups got their color from Blue Starr. Les describes her as smaller than Keno—about the size of a Border Collie—though with conformation typical of an Aussie. Her relative anonymity is often the fate of mothers of the famous. Gene purchased Blue Starr at a local livestock auction in 1947. Her background is as much a mystery as

Gene Sisler, like his brother Jay, kept dogs for the remainder of his life. Pictured at home in his old age with two of his last companions. Courtesy Les Burlile

Keno's. She was just another of the many farm collie-type dogs common throughout the Western United States.

What is exceedingly well known about Keno and Blue Starr is the niche they filled in the history of the Australian Shepherd breed. Their two famous sons, Shorty and Stub, launched Jay, and the breed, on a path leading to one of the most popular American-made breeds in canine history. Born in March of 1948, Shorty and Stub were dark blue merle dogs with showy white trim and no copper markings. Les Burlile speaks to Keno's prepotence in describing Shorty and Stub as large dogs of about 70 pounds each with Keno's broad head.

Another bitch owned by Gene Sisler was responsible for producing Gene's red merle Aussie show dogs sired by Sisler's Shorty as well as Fletcher Wood's Jay. Trixie was a blue merle bitch acquired from Parr Norton, who was living in Oregon when Trixie was given to Gene. Jay described her as "...blue and black, with little, if any, white on." Les remembers a "smokey gray with lots of black" appearance. Both of these descriptions are very suggestive of the look of dogs Fletcher Wood eventually bred. Most of Gene's red merles seem to have favored the self coloration of Trixie. Les Burlile's memories of Trixie and her offspring are of a more painful variety. He says Trixie had a somewhat "cranky" disposition and would bite strangers if given half a chance. Burlile describes Gene's dogs as extremely loyal to him and great performers in his rodeo acts. However, they also seem to have inherited their dam's penchant for biting and Les has the scars to prove it! He tells of a time Gene asked him to let his dogs out of a trailer. Les had had experience with these dogs and carried a stick to protect himself. This puny weapon proved inadequate to three dogs bolting from a trailer. With one dog on his arm, one on the seat of his pants, and the last on his leg, it took Jay to end the joke but not before Les had earned 17 stitches in his leg. Gene never had trouble with the dogs at rodeos but he was always careful to keep the dogs away from the public except while they were performing. On pedigrees, Sisler's Trixie is listed as being sired by a dog named Blue Khan and out of Emmett's Blue.

Red dogs were hardly rare in the early days of the breed. Gene Sisler's backup act may not have been as widely seen as Jay's, but his dogs were every bit as talented, learning the same act, the same tricks, as Jay's. Often these dogs came from the same litters as Jay's show dogs. The color blue being associated with the Australian shepherd seems to be a later development—possibly due to the popularity of Jay's taste in dogs and their wider exposure to the public.

Gene Sisler with Rye on the left and Rock on the right. His two red merles were well known to rodeo fans even though they have no recorded descendants in Aussie history. 1959 Yuma, Arizona, rodeo. Photo by DeVere Helfrich, courtesy Donald C. and Elizabeth M. Dickinson Research Center, National Cowboy and Western Heritage Museum.

Sisler's Shorty was forever to be Jay's favorite dog and eventually the most influential of the Sisler dogs to the background of the Wood's dogs. Even nearly 40 years after Shorty's death, Jay would grow misty-eyed when talking of him. He stated Shorty was the most honest dog he ever owned and no matter how much pressure he was under at a show he never let him down. In Jay's own words, "He (Shorty) had all the characteristics you'd approve of in a person. He had a lot of courage, but he wasn't a bully. He never let you down and always did what you asked… It wasn't just that he was of value to me. He was my right hand. All he ever wanted was to please me." According to Les Burlile, his talented but more playful brother, Stub, was not used for breeding but he still thought highly of him. He describes the pair of brothers this way, "They would try anything I asked. It seemed as if their only desire was to please me as best they could." Shorty, like all the Sisler dogs, was used to work cattle on the ranch. He was a natural heeler but would work the head when needed. His judgment of just how much power to use in any situation would later also appear in his son, Wood's Jay. He could move any cow or sheep but was gentle enough to move calves, lambs, and ducks.

There was nothing Shorty could not and would not do. Les remembers he would hunt and retrieve birds (especially pheasants). In fact, he would retrieve anything and once successfully tracked an elk Jay thought he had wounded. He was evidently a prepotent sire as these gifts were

Sisler's Shorty at the 1949 John Day, Oregon, rodeo. Photo by DeVere Helfrich, courtesy Donald C. and Elizabeth M. Dickinson Research Center, National Cowboy and Western Heritage Museum.

passed on to his descendants. Shorty was immortalized in a painting which hung in Jay's home office and on a belt buckle with Stub which Jay wore until the day he died.

Jay did not keep bitches for breeding—perhaps because of his prolonged absences from home. Gene Sisler used Shorty on his bitch Trixie to raise dogs for his own act. Gene's original dogs—Rock and Rye—were red merle Shorty sons. Jay would also breed Shorty to bitches owned by other people. Ruth McGuire remembers when people knew Jay would be in town they would hope their female would be in heat so she could be bred to Shorty. Many people planned their breedings around Jay's visits. Replacement dogs for his act often came from these litters or their descendants. Trading and bartering for dogs, or just picking up any dog that came along was the normal practice in those days. "Breeding programs" were for livestock or purebred show dogs.

Early in Jay's career he realized he would be "out of business" if anything were to happen to his two partners. A backup team of dogs was needed as understudies to the stars. Two Shorty sons, Tom and Jerry, were the first of many replacements Jay started. According to a newspaper interview, Jay claimed these two dogs were not "purebred" but were one fourth "collie." In that day and age this most likely referred to either Border Collie or just plain farm collie-type dogs of the area. Not unusual when one considers the Australian Shepherd itself is more or less the old fashioned farm collie!

Tom and Jerry were blue merles like their famous father. At about the time the two dogs were seasoned enough to easily step into the starring roles, Tom contracted distemper. He was about 3-years-old at the time and never fully recovered from the disease, developing tremors which handicapped his ability to perform in the act. Tom was eventually given away and Jerry was shown in partnership with Queen by Jay and later teamed with Rock and Rye by Gene.

Although it is likely the show dogs Tom, Jerry, Rock, and Rye were, in fact, sired by Shorty, Les also remembers another dog owned by his aunt named Skeezix. He was a red dog, about the size of Shorty and Stub. Burlile didn't recall his background or breeding but has stated he could possibly have sired the four dogs who comprised the Sisler brothers' "second generation" of show dogs. The possibility also remains that Skeezix, too, was a son of Shorty, maybe even a brother of these four dogs.

Oh, all the adventures the Sisler, Shorty, and Stub threesome must have had! In addition to traveling the country and making movies, these two dogs were Jay's best company on the road. Extremely well socialized and impeccably trained, these ambassadors for the Aussie-made fans wherever they appeared. There were never enough puppies from them to fill all the requests.

Portrait of Jay's most beloved dog, Sisler's Shorty. This painting hangs in his home. Jay's reflection in the glass speaks to the fond memories he always had of Shorty. Courtesy of *The Idaho Statesman*

Many stories of their exploits have already been told. This one has to be a favorite. Shorty and Stub once kept Jay from being attacked by a Brahma bull that broke into a dark arena during his act. In 1956, at the Fat Stock Show rodeo in Houston, Jay was performing under bright spotlights. Jay recalls, "I was working my act under the spotlight with the rest of the arena dark and was about half through when two Brahma bulls broke out of the chutes and came charging down through the arena." The spotlights had Jay completely blinded so he had no way of knowing which way to run into the darkness to escape. He says, "The announcer told me over the loudspeaker that the bulls were coming, but until they got into the light I couldn't tell where they were." The dogs forgot the act and instinctively turned the bulls and sent them back to the exit chutes. Their chores completed, they returned to find Jay still frozen in place. He concludes, "Many people thought it was part of the act, but every time I heard a chute rattle during that show, it always made me jumpy." He adds, "From that moment on I was always glad I had trained cow dogs rather than poodles."

Shorty and Stub led lives of action and adventure about which we really know very little. The untold stories of their lives on the road with Jay would likely make great reading—or perhaps a movie! Eventually the rigors of a hard working life took their toll and the two brothers were retired to Jay's Idaho ranch. In 1959, at the age of about 11 years, Shorty was lost to this world, with his brother, Stub, following in 1960. Both are buried on Jay's ranch overlooking the Payette River. It is not likely there will ever be any who match their place in Australian Shepherd history.

Sisler's Queenie holds the distinction of being one of only two female Australian Shepherds to ever appear in Jay's act. His three Greyhounds were all females but male Aussies were preferred in his and Gene's acts. Queenie was born in 1954 and her background is somewhat in doubt. One source states she was a full sister to Ely's Blue—a bitch obtained

Rye, Gene, and Rock perform the bar balancing trick made famous by Shorty and Stub. Photo by June Sisler, courtesy of Les Burlile.

from Gene Sisler by Juanita Ely. Ely's Blue was obtained by Mrs. Ely around 1952 so Queen was apparently from a later litter. It is possible Gene was the breeder or perhaps he obtained Blue from someone else, then decided to sell her and Jay obtained Queenie later from her breeder.

The only other dog to work with Shorty and Stub, Queen was forced into retirement shortly after Shorty and Stub retired due to her failing eyesight. The January 21, 1960, *Rocky Mountain News* reports her last performance at that year's National Western Stock Show in Denver. The article states nothing could be done for her condition, which might indicate this was a genetic malady. Sisler was grooming a replacement for her at the time, Dandy, who was not nearly as reliable as Queenie. In time, Dandy was taken out of the act due to his disobedience and replaced with the only other female to appear in Jay's act, Panda. Queen is pictured in the article with Jay and his first Greyhound, Blondie, who was trained to carry Queen on her back.

Queen was immortalized with Shorty and Stub in the Disney movie *Stub, the Best Cowdog in the West* before her retirement to Jay's ranch. Les Burlile describes her as a very pretty dog who was more petite than her bigger co-stars. He can't remember Jay ever having another dog as small as she was, although it is possible this was more due to her sex than anything. Burlile remembers her as less than 50 pounds with a bright silver blue coat and copper markings over her eyes and on her front legs. He describes her attitude as differing markedly from Shorty and Stub. More "businesslike" and prone to crankiness, she would sometimes nip at people. Like many females there was a "what's in it for me" tinge to her behavior and she never had the strong work ethic Shorty was known for. She doesn't appear to have been used for breeding or, if she was, no puppies were ever recorded. Les Burlile thinks she may have been spayed at an early age to prevent any problems with the act. A bitch in heat with Shorty and Stub could have caused serious problems for the act. Queen and Jerry were Jay's main show dogs when brother Gene would take Stub and Shorty on the road.

Jay had a series of dogs next, none of whom quite lived up to the standards set by Shorty and Stub. In spite of this, he maintained a high quality act while continuing to audition new talent. A dog named Sisler's Dan (written on some pedigrees as "Dandy," although Jay never called him by this name) was Queen's understudy for a time. Recorded as by "Teton" out of "Sal" he was likely a local Idaho dog Jay found near home. He doesn't seem to have lasted very long in Jay's act but he did figure into the background of the Australian Shepherd by siring Iacovetta's Shasta from a great-great-granddaughter of Sisler's Joker! Dan worked with Joker, Snip, and Greyhound Blondie.

Jerry teaming with Gene Sisler for a game of tug of war. Rye is jumping for joy while Rock looks on. 1959, Yuma, Arizona, rodeo. Photo by DeVere Helfrich, courtesy Donald C. and Elizabeth M. Dickinson Research Center, National Cowboy and Western Heritage Museum.

Sisler's Panda was not bred by Jay or Gene but her background was all theirs. Panda was a stud fee puppy obtained by Fletcher Wood and then given to Jay. The only puppy in her litter, she was sired by Wood's Jay and out of the bitch Gene sold to Mrs. Ely, Ely's Blue. Panda fit Jay's criteria for a flashy appearance perfectly. A light blue merle with abundant white markings, she appeared to be a natural for the act. For some reason she was never shown much and eventually was replaced. Jay's main criteria of trainability is likely the reason dogs were moved out of his act. After her short career with Jay she became the companion of Judy Wood, one of Fletcher's daughters.

Sisler's Joker appears to have entered the picture next. A stud fee puppy by Shorty and out of Fletcher Wood's dark blue Sisler bred female Wood's Little Blue, Joker may be one of only two Shorty sons to be a part of his act (Jerry being the other). Les Burlile remembers Joker as a blue merle with a gray tint—a more mottled gray and black instead of a clear blue. Like many of these dogs he had little white and was of the larger type, weighing around 70 pounds. Joker began his career performing with the Greyhound, Blondie, and two relatively unknown Sisler dogs, Snip and Dan. These three little known Aussies and the Greyhound Blondie were together in the late 50s and early 60s. Snip and Dan were soon replaced by Silver and John. Promotional photos show Joker appearing

Sisler's Stub—a handsome dog that was apparently not used for breeding. Jay valued mental qualities highly and Stub was no match for his brother, Shorty, in that respect. Stub is pictured here at the 1949, John Day, Oregon, rodeo. Photo by DeVere Helfrich, courtesy of Donald C. and Elizabeth M. Dickinson Research Center, National Cowboy and Western Heritage Museum.

with Jay's second Greyhound, Sparky, and a young John and Silver.

Sisler's Silver came into the act to replace Snip around 1962. Silver was another favorite of Jay's. Les Burlile remembers Silver as a big, gangly clownish dog who would do anything for anyone. He had the "playful" type of personality which led Les and his buddies to pull pranks on him. Les describes him as being like a kid who is very gullible. He was highly trained and almost perfectly obedient which added up to much amusement for Les and his friends. They would tell Silver to relax and he would collapse in a heap. They would excite Silver until he was leaping up and down, then tell him to relax in mid-air. Taking them literally,

Sisler's Queenie during a performance at the 1956 Pendleton, Oregon, rodeo. Photo by DeVere Helfrich, courtesy of Donald C. and Elizabeth M. Dickinson Research Center, National Cowboy and Western Heritage Museum.

Silver would go limp in flight and crash to the ground. Apparently loving the joke as much as the kids, Silver would always play along!

Sisler's Panda. Courtesy of Mary Wood

He was immensely loyal to Jay, as well as extremely trainable and talented. Silver garnered more movie appearances than any of Jay's other dogs. His biggest role was in *Run, Appaloosa, Run* as the companion of the female star of the show. He had a cameo appearance in *Stub, the Best Cowdog in the West* and a small part in *Ida, the Offbeat Eagle*. Burlile remembers Silver as a big dog along the lines of Keno, Shorty and Stub. He was a dark grayish colored merle without strong contrast in his coloration. He had, however, plenty of "chrome" to meet Jay's criteria for flashiness. A wide blaze, big white collar, and white legs with the white creeping up into his stifles set off Silver's dark body. Silver was a gift to Jay from the Vaughn Johnson family of Glenrock, Wyoming. Jay had given the Johnson's Silver's grandfather many years before and the Johnsons offered back this intelligent, flashy puppy to Jay for his act. This was not an unusual way for Jay to acquire potential performers. Silver's look is quite similar to Jay's original dogs and his size and type suggest a strong similarity to Shorty and Stub.

Sisler's John made his debut in Jay's act in 1963 shortly after Silver came on the scene, replacing Dan. Sisler's John was an extensively white trimmed blue merle stray who wandered up to the ranch of David Robinson, Jay's neighbor. He was given to Jay and became one of Jay's favorite dogs. Jay believed John was likely part Border Collie due to some of his behaviors. Called "honest" by Sisler, he had a work ethic similar to his predecessor, Shorty, and remained in the act for many years. He makes a cameo appearance in *Run, Appaloosa, Run* in a tight close-up heeling a horse into the river during training. He holds one end of a jump rope in *Stub, the Best Cowdog in the West* while Jay turns the rope and Doc jumps. He also appeared on the cover of *Dodge News Magazine*, November 1967, and is pictured sitting in front of a sheep in the article inside.

John was used extensively for breeding by a number of people. ASCA Stockdog Judge, Wally Butler, was a contemporary of Jay's nephew, Les Burlile, and got his start in Aussies from Jay Sisler. Les and

Sisler's (Wood's) Panda, March 1963. Blue merle and white female who spent only a short time with Jay before being returned to Fletcher Wood as a beloved pet of the Wood girls. Courtesy of Rowe

Wally attended high school and college together, both becoming Aussie breeders following college graduation. Many of Wally's original dogs were Sisler's John's get, and Les, too, owned a number of dogs who traced to John. Veterinarian Weldon Heard used John to sire a litter from an inbred Harper's Old Smokey bitch. Jay received a stud fee puppy from this cross which he called "Doc" after Heard.

John was eventually given to Wally Butler after he finished college and decided to begin breeding Aussies seriously. Butler practiced inbreeding to John and Les remembers the puppies he obtained from these crosses had a very wide range of quality. Although some were very good, many were not the quality desired. This result likely speaks to a genetically mixed background typical of many founding Aussies. During development of the Aussie, many top breeders practiced extremely close breeding to both set a type and to uncover the genetic strengths and weaknesses of their dogs. Although John played an important role in Aussie history, his status as a sire did not equal that of his predecessor Sisler's Shorty.

Sisler's Doc was one of Jay's last show dogs who joined John, Silver, and Jay's last Greyhound, Jet, in the act. He seems to have replaced Joker (or possibly the little known Cal). Les Burlile describes Doc as colored like and built very similar to his sire, John. He has Jay's preferred white trim but to a lesser extent than Silver and John. Slightly smaller than Silver, Burlile estimates him at about 65 pounds or so. Doc was another good performer for Jay, working his act up until the end of Jay's career, then he and Jet went on the road with rodeo clown Wilbur Plaugher. Doc was also a star in the second half of *Stub, the Best Cowdog in the West,* performing a number of wonderfully clever stunts and tricks disguised as chores on the ranch.

Les Burlile remembers two other dogs trained for Jay's act—one named Cal and one named Snip. Cal is described as a smaller dark blue dog with copper over his eyes and on his front feet. He wasn't shown much and, if bred, left no recorded descendants. He doesn't remember much about Snip except that he was around about the time Joker was in the act but didn't work out and was replaced in the act by Silver. Burlile recalls that by the end of Jay's career he had no dogs left of his original line from Keno and Shorty. Although this bloodline survived through Fletcher Wood and others, apparently Jay preferred to train dogs which came his way through fate.

Jay's brother, Gene, had a shorter rodeo career than Jay and about half a dozen dogs overall which performed publicly. The red merles Rock and Rye and the blue merle Jerry, all Shorty sons, formed his original troupe. Rock and Rye were like Keno and Shorty, weighing in at approximately 70-75 pounds. Burlile remembers Gene's dogs as much less social than Jay's, although as performers they were just as good. He speculates this could be due to Gene's more solitary lifestyle which could have meant less socialization for the dogs. This theory is bolstered by Mary Wood, who recalls Shorty and Stub as dogs who were well trained and socialized but still retained a healthy reserve with people. Gene also had three other dogs named Brandy, Whiskey, and Bourbon who performed in his act. Bourbon, like Rock and Rye, was a red merle and Whiskey and Brandy were red and white dogs. Les recalls Brandy was even bigger than Rock and Rye with a generous amount of white on his head. Like Rock, Les considered him untrustworthy around anyone but people he knew and respected. Jay and Gene always shared dogs and many of their dogs came from the same litters. With Jay's constant travel and public

Sisler's Panda, March 1963. Panda was trained for Jay's act but did not perform with the consistency he required. Courtesy of Rowe

Sisler's Doc balancing on the teeter fulcrum while his sire, Sisler's John waits for his cue. Courtesy of *Western Horseman*.

(Left to Right) Sisler's John, Sisler's Joker, Jay Sisler, Sparky, Sisler's Silver. Promotional photo taken in Jay's front yard by sister June, courtesy of *Western Horseman*.

exposure, he had more opportunity to teach his dogs to handle the rigors of public life and foolish humans.

A story Les tells about the last dog Jay had when he died may give us our best perspective on this overall aspect of these dogs. Jay's last dog was named Darby. Darby was really no different than any other dog he had ever owned. He accepted people and life and trusted Jay implicitly. This close relationship was something Jay had with all his dogs. The dogs' lives were safe and constant due to Jay's steady influence. Within two to three days of Jay's death, Darby became a different dog. His temperament became mistrustful and "cranky" as Les describes it. His world had come unglued and his personality reflected this loss. Darby remained at the Sisler ranch into his old age but he remained unhappy until he died. The close relationship which released such trust and creativity also crushed the spirit when broken.

Altogether Jay had 17 dogs which were shown over his long career. This number includes the three retired racing Greyhounds he obtained—Blondie, Sparky, and Jet (in that order)—along with the Aussies. Only three of these dogs, Sisler's Shorty, Sisler's John, and John's son Sisler's Joker, made significant KNOWN genetic contributions to the foundation of what was to become the purebred Australian Shepherd. It is very likely there were many other Sisler dogs which made significant UNKNOWN contributions to the breed through the multitude puppies which Jay and Gene sold or gave away to people all across the country. What we know of Jay Sisler's contribution to the Australian Shepherd is a great legacy we must treasure and appreciate.

Sisler's John, from the cover of the November 1967, *Dodge News Magazine*. Photo by well-known Boise photographer Ansgar Johnson. Courtesy of Chrysler Group, LLC

Sisler's John pictured at Jay's ranch. From *Dodge News Magazine*. Photo by well-known Boise photographer Ansgar Johnson. Courtesy of Chrysler Group, LLC

Jay in 1969 with his last group of seasoned performers. From left to right: Jay Sisler, Sisler's John, Sisler's Silver, Sisler's Doc, and his third Greyhound, the black male, Jet. Notice the light-colored collar used on Jet to make him more visible to audiences. Photo from *The Seattle Times* archive.

The front of Gene Sisler's business card. This photo was likely taken by his sister June and pictures Gene with his first three dogs, from L to R: Rye, Jerry, and Rock. Courtesy of Les Burlile

The back of Gene Sisler's business card. Courtesy of Les Burlile

Jay with his troupe of show dogs (John, Joker, Sparky "Greta," Silver). Photo taken at Jay's ranch by his sister June.

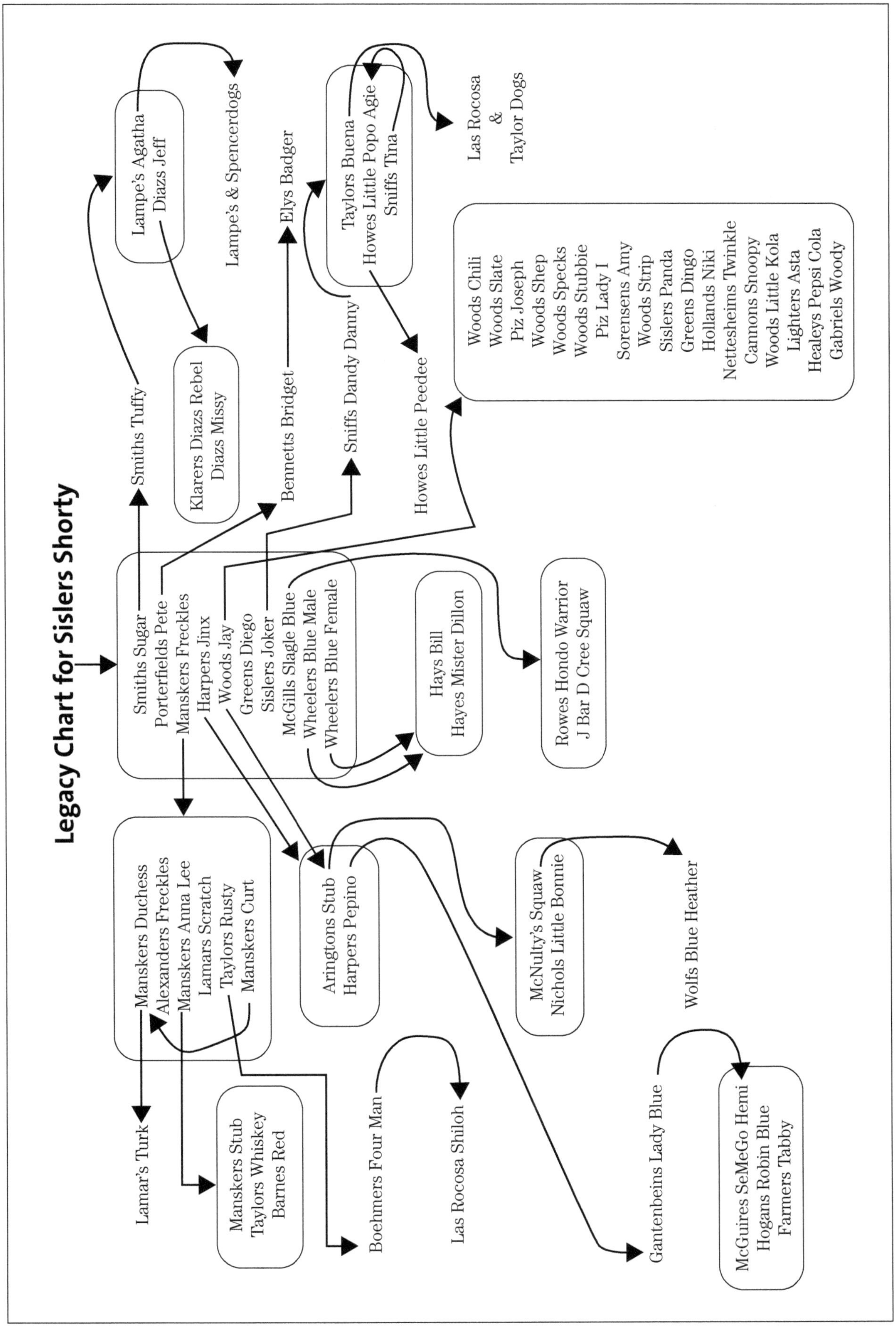
Legacy Chart for Sislers Shorty
Smiths Sugar
Porterfields Pete
Manskers Freckles
Harpers Jinx
Woods Jay
Greens Diego
Sislers Joker
McGills Slagle Blue
Wheelers Blue Male
Wheelers Blue Female
Smiths Tuffy
Lampe's Agatha
Diazs Jeff
Lampe's & Spencerdogs
Klarers Diazs Rebel
Diazs Missy
Bennetts Bridget
Elys Badger
Sniffs Dandy Danny
Taylors Buena
Howes Little Popo Agie
Sniffs Tina
Las Rocosa
&
Taylor Dogs
Howes Little Peedee
Woods Chili
Woods Slate
Piz Joseph
Woods Shep
Woods Specks
Woods Stubbie
Piz Lady I
Sorensens Amy
Woods Strip
Sislers Panda
Greens Dingo
Hollands Niki
Nettesheims Twinkle
Cannons Snoopy
Woods Little Kola
Lighters Asta
Healeys Pepsi Cola
Gabriels Woody
Hays Bill
Hayes Mister Dillon
Rowes Hondo Warrior
J Bar D Cree Squaw
Manskers Duchess
Alexanders Freckles
Manskers Anna Lee
Lamars Scratch
Taylors Rusty
Manskers Curt
Aringtons Stub
Harpers Pepino
McNulty's Squaw
Nichols Little Bonnie
Wolfs Blue Heather
Lamar's Turk
Manskers Stub
Taylors Whiskey
Barnes Red
Boehmers Four Man
Las Rocosa Shiloh
Gantenbeins Lady Blue
McGuires SeMeGo Hemi
Hogans Robin Blue
Farmers Tabby

chapter 6

The Sisler Influence **THROUGHOUT THE BREED**

Fletcher Wood ranks as the most important breeder who preserved the original Sisler lines and stamped this original type forever with his own name. Through the Shorty son, Wood's Jay, a genuine family of dogs was created which has survived to the present day. This influence was so great and widespread that Mr. Wood and the bloodline he founded will be examined in detail in their own chapter.

Red merle grandson of Lamar's Turk, owned by Walt Lamar. Courtesy of Rowe

J Bar D and Colorado are two kennel names nearly on a par with the Wood name in terms of preserving and continuing the mountain-type Aussie into the present. Jerry and Sharon Rowe with Dick and Leslie Sorensen furthered not only the registration of these dogs as purebred but also the recognition of the Australian Shepherd as a distinct breed in the dog world. Their contributions were monumental and will receive the special detailing they deserve in later chapters.

Although many of the Sisler dogs will remain forever anonymous in the background of the Australian Shepherd, a few can be credited with influence in many of today's different bloodlines. Much credit for detailing these connections belongs to Jan Haddle Davis and the research she did for the ASCA Yearbook 1957-1977 and thanks goes to her for permission to quote her work from this source. Sisler's Joker, Jay's stud fee puppy from Shorty out of Fletcher Wood's dark blue bitch Wood's Little Blue, sired the blue male named Dandy Danny. Danny sired the most influential bitch in both the Joe Taylor and Hartnagle bloodlines, Taylor's Buena. Buena, in turn, produced Wilcox's Cindy for Larry Wilcox of Alamosa, Colorado. Cindy was a huge influence in Hank Weiscamp's dogs, granddam of Martin's Adobe Susie, Martin's Adobe Gypsy and Iacovetta's Buck. Cindy was bred back to another Sisler dog, Dan, to produce Buck's dam, Iacovetta's Shasta. Martin's Adobe Gypsy took this line back to the Northwest for Harold May and Susie contributed to Steve Mansker's breeding in Oklahoma.

Lamar's Turk, very dark blue male of Sisler background who founded his own dynasty of top working dogs. Photo by Rowe

14 years of age. Photo by Rowe

Steve Mansker, like so many people, saw his first "blue dogs" in a Jay Sisler rodeo performance. Quoting Mansker, "My interest in the Aussies probably stemmed from getting acquainted with Jay Sisler back in the late forties when I was rodeoing. I told him then that I would like to buy one but didn't get around to it until about 1956 that I recall. He then sold me Freckles, who had been trained for his act but had got bred by accident while he was gone to a rodeo. She only had one pup and it turned out deaf, so I started looking for a good male that was a cow dog and found one from a lady by the name of Green at Boulder (later registered as Green's Kim). She said he was too tough and that was just what I was after. I bred Freckles to him and raised one litter which is where my Anna Lee female came from (1957)." Anna Lee later whelped Taylor's Whiskey, who was bred to Taylor's Buena a number of times—bringing the Sisler influence back together from both sire and dam.

Whiskey's sire, Mansker's (Lamar's) Turk, also brought the Sisler influence back into this line. Turk's dam, Duchess, was sired by Alexander's Curt—a son of Green's Kim and Freckles, Mansker's original female obtained from Jay Sisler. Freckles contributed another Taylor dog, Rusty, when bred back to her great grandson Turk. Mansker, Taylor, and another breeder by the name of Bud Hildreth produced many ranch dogs which were spread throughout the mountain homeland of the breed. The Las Rocosa bloodline traces its most important foundations through Whiskey and Buena. Dogs of Sisler background likely represent many of the "unknowns" at the end of today's pedigrees.

Lamar's Turk at age 14. Photo by Rowe

Lamar's Turk at age 14. Photo by Rowe

Lamar's Turk at age 14. Photo by Rowe

Lamar's Turk at age 14. Photo by Rowe

chapter 7

Fletcher Wood

If Jay Sisler was the Australian Shepherd's greatest PR man, then Fletcher Wood was its greatest banker. His masterful horseman's eye was equally astute when it came to developing a bloodline of Aussies which would not only influence nearly every other bloodline in the breed but also define the mountain-type Aussie as his own.

Born in 1925, Fletcher Wood was raised in the era when the horse was still the King of the Road. A passionate horseman, he devoted his life to everything equine. Jefferson County, Colorado, was his lifelong home and the place where his greatest influence was felt. Sometimes called "Mr. Jefferson County," Wood immersed himself in everything to do with the county, from Board member of the Soil Conservation District to Captain of the Mounted Sheriff's Posse, he was an icon of the region. In addition to raising three children with his wife, Mary, he served in the US Army, Colorado National Guard, Masons, and El Jebel Shrine.

Wood's greatest passions were kids, rodeo, and horses. For over 27 years he served as Director of Activities at the Jefferson County Fairgrounds and worked with the Westernaires riding group, headquartered at the grounds, from their founding. Serving on the County 4-H Advisory Board did not remove him from the hands-on work of helping kids. He taught horse packing and leather work and was honored with the "Friend of 4-H Award" in 1987. On November 13, 1991, following Wood's sudden death, the Jefferson County Commissioners paid a lasting tribute to him by dedicating the Jefferson County Fairground Arena to him. This was a man who left a lasting impression on the life of nearly everyone he knew.

Fletcher Wood was also a rodeo man, riding bareback broncs and bulls in his younger days. He eventually became a rodeo judge, presiding over the National Finals Little Britches Rodeo for 18 years, in addition to service at many high school and amateur rodeos. He was a member of the National Western Stock Show Association and eventually served 34 years as their Ringmaster and 3 years as the Horse Show Manager. In 2001, Steve Henry, longtime bullfighting partner of Wood's son, Kent, founded the Fletcher Wood Memorial Rodeo to honor this aspect of his life.

It was his passion for horses and his experience with a wide range of horse breeds which carried over into his extremely successful stint as a breeder of Australian Shepherds. Wood owned, trained, and showed nearly every kind of horse from Hunters and Jumpers, to Saddlebreds, pleasure and trail horses, Reining, Cutting, Team Penning, Calf Roping and Team Roping horses. He designed and managed the Golden Mesa Ranch (now known as Table Mountain Ranch), Arbloade De Oro in Conifer, Colorado, and the American Equestrian Center in Los Angeles. He held memberships in the Round Up Riders of the Rockies and the Colorado Horseman's Association. In 1982, he was named "Colorado Horseman of the Year."

Wood was also a cattleman. This, combined with a long and close friendship to Jay Sisler, led him to found a line of Australian Shepherds whose descendants are still identified with his name today. Using his well honed horseman's eye he collected the finest blue chip breeding stock. In a time before any written standards, he held a time tested and unchanging vision of the highest physical and mental breed standard in his mind. Then he practiced to the letter every principle of breeding and genetics to develop a prepotent and established bloodline in a very short period of time. His genetic investment is still paying dividends today.

Mr. Wood's diverse background in ranching and cattle and his skill as a horseman prepared him well for his role as dog breeder. The Wood family had Airedales before they had Aussies but when they met Jay Sisler it was obvious this was the breed for them. It didn't hurt that good fortune again smiled on the Aussie breed in bringing Mr. Wood a superior individual for his first Aussie and foundation sire—Wood's Jay.

Wood's Jay quickly became Fletcher's ideal and remained so throughout his time as a breeder. To preserve Jay's excellence, Wood immediately embarked on a very intensive breeding system which not only cemented Jay's type and genes, but also would reveal any genetic weakness he might have. Providence again smiled by revealing this dog to be as nearly genetically flawless as possible. Wood proved this over time by continuously breeding Jay back to his daughters and their daughters and granddaughters. His cattle breeding experience guided his dog breeding and he held to the cattle producers' theory that one only inbred by crossing father to daughter, never mother to son.

His ruthless culling of litters and breeding only the best was the ultimate road to his success. Although male puppies were sold without strings, Fletcher Wood kept close watch on his female pups. He believed too many dogs that were not worthy of breeding were being used to produce inferior puppies and he did not want to add to this mediocrity. The good females he sold to ranches and guided the rancher in breeding choices. He usually bred a female three times and would keep pick puppies from the litters. If Wood believed a bitch was not being used appropriately he would have her spayed.

Wood was familiar with the legend of a Basque foundation for the Australian Shepherd. Since not much was known about Pyrenees sheepdogs he speculated they were all blue merles and that he could create a line of Aussies which would breed "true" for blue merle color with no white or copper trim. Although this assumption is now known to be false, this was the dogma of his era and it didn't hurt that Wood personally favored a dark merle dog himself! In addition, the inheritance of white in Aussies was poorly understood at the time. Wood experienced the pitfalls of merle to merle breedings and had a strong prejudice against any white markings because of the problems now known to be associated with the double merle combination. He would tolerate a thin white strip on a dog's face but insisted all must have a black nose leather (in the blue merles—he would later accept the reds as well with the liver nose leather). Wood believed double merle deafness was similar to the deafness found in Dalmations and that along with defects associated with white, the blue eyed dogs were more likely to go blind. He based this second assumption on Jay Sisler's experience with his blue eyed bitch, Queenie, who did, indeed, go blind at about the age of 6. Sisler and Wood believed it was caused by bright arena lights, but it is possible she had one of the inherited eye diseases we now know exist in the Aussie. Later, blue-eyed dogs of Sisler's did not go blind, although he tended to select brown eyed dogs for his shows anyway, so the belief seems to have become a self fulfilling prophecy. It is also possible the general prejudice against white markings and blue eyes in horses was something Wood carried subconsciously into his dog breeding.

Regardless of the overall lack of knowledge about the breed at that time, Wood had exceptional natural instincts for producing top quality animals, whether bovine, equine, or canine. Though color was important to him (as it was to many breeders during the founding of the breed), he still put working ability at the top of his list. In an interview he conducted for the *Australian Shepherd Quarterly* in 1989, he says, "Most important, of course, was a primary working ability. Out of the first 300 pups I sold, I never sold one as a pet. They all went to ranches as working dogs." His belief that all Pyrenees sheepdogs were merles might have prompted his focus on merle to merle breedings but he also recognized the black and red dogs were normal, good, sound dogs. He continues, "I was trying to weed out the undesirables. The first was the white gene and the totally white eye. A merle eye was not so bad. Next, I bred out the black, but the black was a good, sound, usable dog." His experience with red and red merle Aussies was typical and his knowledge of cattle genetics dampened any prejudice he might have held against them. He says, "The red merle Australian Shepherd is no different than a red Angus which is the product of two Angus cattle. The same thing would apply in breeding the merle dogs. I did get some red dogs. I had some good, dark red merle dogs. In fact, the last pair of pups I kept were red merles. Jay Sisler's brother worked a pair of red merle dogs in a dog act. The red merles still have all the characteristics in their genes. I actually got some solid-colored dark red ones, too, the same as with the black. It was just that the black gene carried a red gene into it." This aspect of Australian Shepherd inheritance was clear to Fletcher Wood and he was astute enough to recognize the value of these non-blue-merle dogs.

As a cattleman, Wood was primarily selecting for cattle working dogs. Shorty and Stub and a large number of mountain-type Aussies were used for this job. The problem of finding good working bitches from which to breed was solved by his contacts in the stock world. Colorado State University sheep specialist, Murray Trump, had Australian Shepherds. Wood bred Jay to his bitch, Trump's Stubbie, which was the foundation bitch in the line which eventually produced Wood's Stubbie. (Perhaps he was named for his great granddam?) From this cross, Wood obtained Wood's Specks. Specks was, of course, a dark blue merle with a thin ring of white around her neck, a narrow blaze and white feet. She was one of Fletcher's earliest breeding bitches and seems to have been a good example of what he was searching for.

He began his concentrated breeding by putting Specks back to her sire, Jay. One result of this cross was a self blue merle bitch called Wood's Slate. Wood wasted no time cementing his type and establishing his line. Slate was finally bred back to her sire and grandsire, Jay, to produce the ultimate sire of his line, Wood's Stubbie. Although Jay sired countless dogs which also figure prominantly in breed history, it was Stubbie who propelled the "Wood's" type into the future.

February 1960 gathering of three dogs which are behind nearly every Australian Shepherd alive toda. L to R: Sisler's Shorty, Wood's Specks, Wood's Jay. Photo by Jay Sisler, courtesy Mary Wood.

Stubbie was a large, dark blue, self merle who represented the ultimate dog Fletcher Wood was attempting to produce. His unattainable goal of recreating Jay was nearly achieved with Stubbie. It was at least as close as any human could get in that age before cloning. Jean Taylor remembers a conversation she had once with Fletcher Wood when he told her, "I found that you can't reproduce an individual, but you can stabilize a type."

Jean met Fletcher at a horse show in Golden, Colorado, after she got her foundation sire Taylor's Luke the Drifter. She was walking this Stubbie son when Fletcher spotted him and stopped Jean to find out where she got him. He recognized Luke's type immediately and told Jean he was the image of Stubbie. At the time Fletcher was running cattle on about 900 acres of brushy and rocky pasture and couldn't do it without the dogs. Stubbie was the perfect dog for this work as he could handle the toughest bull but would just shove baby calves with his nose to move them.

One of those funny quirks that seem to run in families was a habit both Luke and Stubbie had of lying with one front foot folded under him. Fletcher asked Jean about this specifically during their meeting, stating he had a theory about it. These dogs were very deep chested and Wood believed it was like parking a deep keeled boat on solid ground. The chest fell so far below the elbows it took a folded leg under the elbow in order to balance and keep their bodies level.

In some respects it is possible Stubbie was a better dog than Jay. The long line of extremely close inbreeding coupled with a brutal selection process concentrated the positive traits of Jay while weeding out any negatives. Also proven in this process was the lack of any severe genetic defects. Time and further inbreeding has proven the wisdom of this process and the value of this blood.

chapter 8

Wood's Jay

Colorado was the cradle of development of the Wood bloodline and also for much of the Australian Shepherd breed as an organized and registered collection of dogs. Fletcher Wood was a well known and influential horseman and stockman who followed his passion for Australian Shepherds as thoroughly and methodically as he did his horse and cattle breeding. Other horsemen and stockmen recognized this and sought Wood for the best Aussies, too. The full scope of Wood's Jay's contribution to the breed, like his famous father's, is probably greater than what is recorded. But what is known is his legacy of soundness and ability which is unmatched in the breed.

Jay was whelped in either 1949 or 1950 according to information in the ASCA Yearbook (1954 according to his registration papers) and he came to Fletcher as a young puppy. Jay had been raised by Jay Sisler's sister and was Sisler's pick to send to Fletcher Wood. He was hauled to Colorado in the tack room of Jay's trailer and arrived unsocialized to men and children. When he was let out of the trailer and into the house he dashed behind the couch and would not come out until Fletcher left the room. Fletcher's wife, Mary, immediately took over his care and socialization out of necessity. Jay would only appear when she was around. Fletcher finally decided to crawl behind the couch and laid quietly beside Jay in order to convince him men were his friends. It took Jay about two

Three of the most famous and important dogs in the Australian Shepherd bred. L to R: Sisler's Shorty, Wood's Specks, and Wood's Jay. Specks was the foundation matron of the Wood's line by Jay and out of Trump's Stubbie. She was the great granddam of Wood's Stubbie. Photo by Jay Sisler, courtesy of Mary Wood.

The renowned Wood's Jay. Courtesy of Mary Wood

weeks at the Wood residence to decide men and children were part of his world now. Jay Sisler always regretted his canine namesake was not as highly marked with white as his sire, Shorty. He would have been a star in the act if not for his dark color. It was only his lack of white trim, the very trait which would later become the hallmark of the Wood bloodline, which brought him to Colorado to change the Australian Shepherd breed. If not for his color, the Wood bloodline might never have been.

Like all pioneers, Fletcher Wood learned about working stock dogs through trial and error. Again, his background gave him a leg up since his experience as a horse trainer delivered insights to his dog training. Wood's Jay surely taught him a few things along the way about working with a dog's natural instincts! The pair learned together how best to work cattle and when a new idea came along, Fletcher was open minded enough to give it a fair try. Like Jay Sisler, Wood developed and refined his methods over time until he knew both how to breed a good working dog and how to use it well. Others saw this success and the quality of his dogs and lined up for a puppy from the remarkable Wood's Jay.

Jay, like all good Aussies, was an all around dog. His talents ran the gamut from top working dog to top family dog. Wood describes how Jay fit into his own family, "As a family dog, the Aussie is one of the best dogs we've ever been around. In fact, old Jay is the one that taught my son how to walk. He'd crawl alongside the dog, the dog would barely move, and Kent would move with him. Pretty soon the dog was walking through the house, and Kent was hanging onto the dog and learning how to walk." Jay was 11 years old when Kent was born and was, by then, a watchful babysitter. Mary describes a common scene of the pair sitting by the highway watching traffic with Kent lying on top of old Jay. If Kent should wander

Parents | **Grand Parents** | **Great Grand Parents**

Sire: Sisler's Shorty
Reg. No. A019-5868
Color B & T ☐ B-Wht. & T ☐ B & W ☐ S ☐ S & W ☐
Coat: Curly ☐ Wavy ☐ Straight ☐
Natural Working Characteristics: Rough ☒ Ave. ☐ Mild ☐
Color-Blue, Grey
Eyes-Blue

Gr. Sire: Keno
Reg. No. A042-6390
Color B & T ☐ B-Wht. & T ☐ B & W ☐ S ☐ S & W ☐
Coat: Curly ☐ Wavy ☒ Straight ☐
Natural Working Characteristics: Rough ☐ Ave. ☐ Mild ☒
Color-Red

* * * *

Reg. No. A043-6390
Color B & T ☐ B-Wht. & T ☐ B & W ☐ S ☐ S & W ☐
Coat: Curly ☒ Wavy ☐ Straight ☐
Natural Working Characteristics: Rough ☐ Ave. ☒ Mild ☐
Color-Blue
Gr. Dam: Blue Star

Gr. Gr. Sire

Reg. No. ....
Color B & T ☐ B-Wht. & T ☐ B & W ☐ S ☐ S & W ☐
Coat: Curly ☐ Wavy ☐ Straight ☐
Natural Working Characteristics: Rough ☐ Ave. ☐ Mild ☐

* * * *

Reg. No. ....
Color B & T ☐ B-Wht. & T ☐ B & W ☐ S ☐ S & W ☐
Coat: Curly ☐ Wavy ☐ Straight ☐
Natural Working Characteristics: Rough ☐ Ave. ☐ Mild ☐

DAM

SIRE

Reg. No. ....
Color B & T ☐ B-Wht. & T ☐ B & W ☐ S ☐ S & W ☐
Coat: Curly ☐ Wavy ☐ Straight ☐
Natural Working Characteristics: Rough ☐ Ave. ☐ Mild ☐

* * * *

Reg. No. ....
Color B & T ☐ B-Wht. & T ☐ B & W ☐ S ☐ S & W ☐
Coat: Curly ☐ Wavy ☐ Straight ☐
Natural Working Characteristics: Rough ☐ Ave. ☐ Mild ☐

DAM

Wood's Jay

Male
Color - Blue
Date of Birth 1/25/54

SIRE

Reg. No. ....
Color B & T ☐ B-Wht. & T ☐ B & W ☐ S ☐ S & W ☐
Coat: Curly ☐ Wavy ☐ Straight ☐
Natural Working Characteristics: Rough ☐ Ave. ☐ Mild ☐

* * * *

Reg. No. ....
Color B & T ☐ B-Wht. & T ☐ B & W ☐ S ☐ S & W ☐
Coat: Curly ☐ Wavy ☐ Straight ☐
Natural Working Characteristics: Rough ☐ Ave. ☐ Mild ☐

DAM

SIRE

Reg. No. ....
Color B & T ☐ B-Wht. & T ☐ B & W ☐ S ☐ S & W ☐
Coat: Curly ☐ Wavy ☐ Straight ☐
Natural Working Characteristics: Rough ☐ Ave. ☐ Mild ☐

* * * *

Reg. No. ....
Color B & T ☐ B-Wht. & T ☐ B & W ☐ S ☐ S & W ☐
Coat: Curly ☐ Wavy ☐ Straight ☐
Natural Working Characteristics: Rough ☐ Ave. ☐ Mild ☐

DAM

Reg. No. A018-57178
Color B & T ☒ B-Wht. & T ☐ B & W ☐ S ☐ S & W ☐
Coat: Curly ☐ Wavy ☒ Straight ☐
Natural Working Characteristics: Rough ☒ Ave. ☐ Mild ☐
N.B.T.
Dam: Sisler's Australian Shepherd (Trixie)

Gr. Sire: Blue Khan
Reg. No. A044-6390
Color B & T ☐ B-Wht. & T ☐ B & W ☐ S ☐ S & W ☐
Coat: Curly ☐ Wavy ☒ Straight ☐
Natural Working Characteristics: Rough ☐ Ave. ☐ Mild ☐
Color - Blue

* * * *

Reg. No. A040-6390
Color B & T ☐ B-Wht. & T ☐ B & W ☐ S ☐ S & W ☐
Coat: Curly ☐ Wavy ☒ Straight ☐
Natural Working Characteristics: Rough ☐ Ave. ☒ Mild ☐
Color - Blue
Gr. Dam: Emmett's Blue

No. assigned by Registrar A416-63272

National Stock Dog Registry pedigree of Wood's Jay. Courtesy of Mary Wood via Sorensen

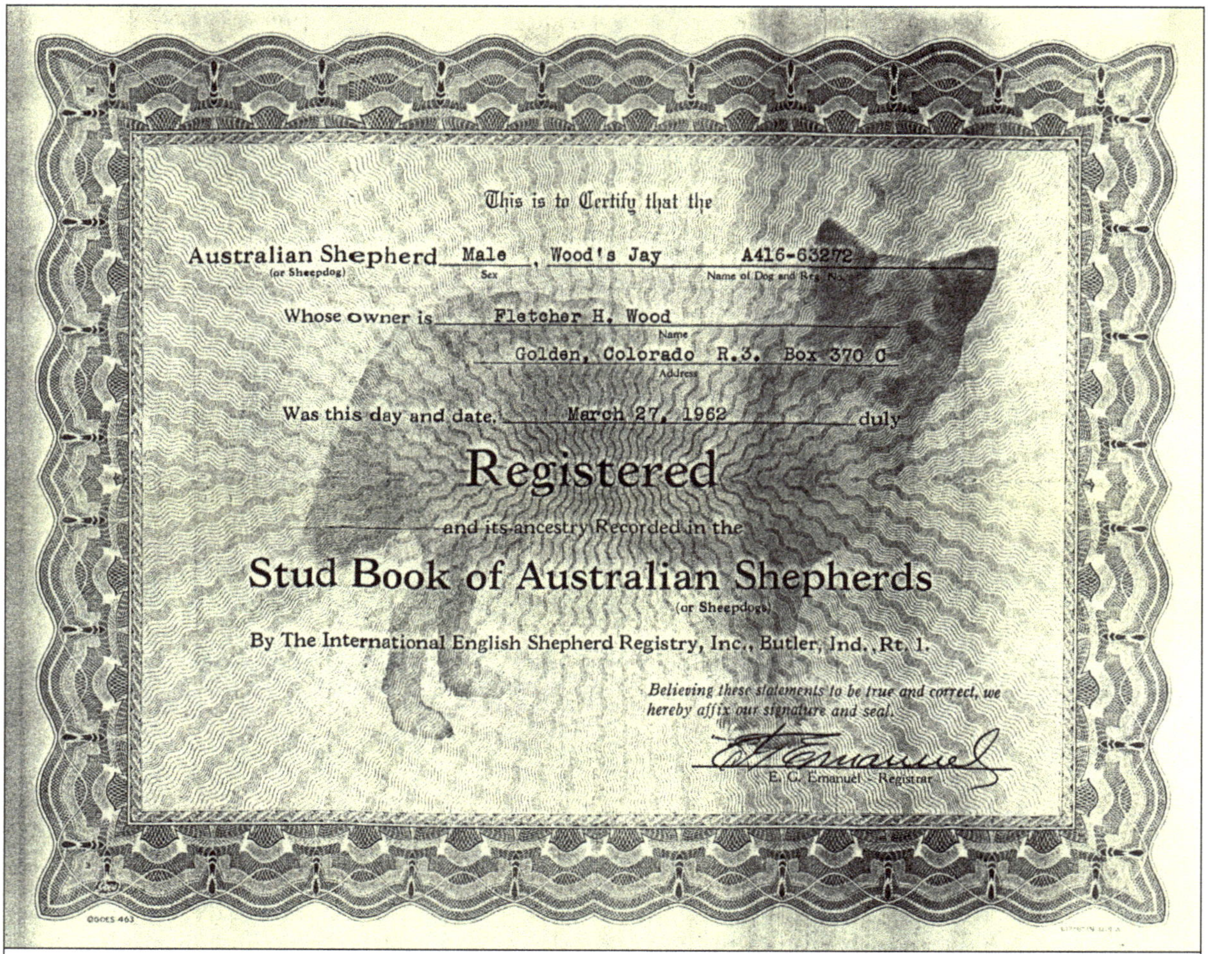

This is to Certify that the

Australian Shepherd (or Sheepdog) Male (Sex), Wood's Jay A416-63272 (Name of Dog and Reg. No.)

Whose owner is Fletcher H. Wood (Name)

Golden, Colorado R.3. Box 370 C (Address)

Was this day and date, March 27, 1962 duly

Registered

and its ancestry Recorded in the

Stud Book of Australian Shepherds (or Sheepdogs)

By The International English Shepherd Registry, Inc., Butler, Ind., Rt. 1.

Believing these statements to be true and correct, we hereby affix our signature and seal.

E. G. Emanuel - Registrar

National Stock Dog registration papers of Wood's Jay. Courtesy of Mary Wood via Sorensen

toward the road, Jay would grab him by the diaper and look for Mary to come rescue him. The guardian trait of the Aussie was present in abundance in Jay and his line. Fletcher describes their vigilance over both home and ranch, "When you bring a baby home, usually the dog will go right underneath the crib. They're going to protect the baby. If the baby cried and we didn't hear it, our dog would wake us up. When the mares started to foal, the dogs would start raising a fuss. Very seldom did I ever miss watching the mare have the foal, because the dogs had alerted me that something was wrong."

Fletcher Wood and Jay did not have the public exposure of Sisler and his dogs, but their bond was equally close. Wood describes how he acquired Jay, "...Jay (Sisler) finally shipped one down to me. When I first got the Jay dog, he had never been handled much. He's been raised out in the barn. I socialized him. He was a couple of months old. I found that if I looked at him he wouldn't come to me; he would hide. If I turned around and walked away, he would follow me wherever I went. Little by little, we finally broke down the barriers." This type of mature temperament, even in a very young puppy, is another typical trait of this line of Aussies. In the 1940s, '50s, and '60s, the Aussie was still very much a rural dog and a lack of early socialization was not the issue it can be today.

Wood's Jay in a light moment with the Wood daughters. Courtesy of Mary Wood

How Wood eventually bonded with Jay, through subtle body language and patience which allowed the dog to bond at his own pace, again points to a native understanding of animals and how to best handle them. Like the gentle taming of a horse, Wood allowed the dog to learn he was a partner without pressuring him. Emulating Sisler's way with animals, the bond he formed was greater than any which could have been made by compulsion. Wood describes his own experiences with Jay, "About the time we had Jay and I was working him so much, there was some work being done on controlling chimpanzees by thought. The person would actually think about what he wanted that animal to do, and the animal would do it. I thought, Jay and I are very close. I could actually start staring at him when he was asleep and think 'Get up and come here.' He would get up and come to me. I never could go any farther than that, but a lot of times when I was working cattle, it seemed like he would read my mind concerning what I wanted him to do. Whether it was association from doing it that way too much or whether it was due to something else, I don't know. When I saw an animal go over on one side, that's where the dogs would go." At the risk of sounding overly sentimental about the Wood or mountain-type dogs, it is astonishing how many owners will tell similar stories of what seem to be psychic abilities!

Wood had good sense about animals and the natural curiosity of a trainer. While recognizing this heightened mental acuteness, he devoted some thought to possible reasons for it. He says, "It has to be body movements. If I had an old cow that felt like fighting, I'd get a little mad at her, and those dogs would eat her up. But if I was placid, especially with

Wood's Jay with the Wood children and Mary Wood's sister. Courtesy of Sorensen

Mary Wood with Wood's Jay (right) and holding Sisler's (later Wood's) Panda. Panda was from one of several crosses of Wood's Jay and the Sisler bred Ely's Blue. Courtesy of Mary Wood

calves, they were the same way. If I had a couple of bulls fighting the dogs would get savage with them if I was uptight, too. It's the same thing with loading horses in trailers. At one time, I had a bunch of colts who would jump in the trailer when I made a hissing noise. The dogs wouldn't even have to be there."

Jan Haddle Davis quotes Fletcher, "Most of my dogs were centered around old Jay as he was what I really wanted. I guess I have produced over 600 pups since I started. Only a small number of these dogs ever got registered because they went to ranches and cowboys that crossed my path here at the stock shows." Mr. Wood's star individual swept his name and this breed into the future.

Jay matured into a large, athletic dog weighing 75-80 pounds, much like his sire and grandsire. He had the same broad, wedge shaped head and uncanny intelligence of his ancestors, too. Mary Wood lends some insight to this preference for a larger size dog. This type of dog was used in mountainous and northern areas. Winters would be long and hard and the large dogs were better at negotiating deep snow than a smaller dog. Their size was not a hindrance in other types of weather nor did it hamper their ability to work cattle effectively.

Jay's temperament and personality matched those of his sire and grandsire, showing the honesty and strong work ethic Jay Sisler and Fletcher Wood both admired and needed in their dogs. His strong loyalty had a practical side when working livestock in remote mountain pastures. Once Fletcher was gathering cattle and sent Jay to find a group the riders couldn't reach. Jay disappeared and did not return. Eventually the group couldn't wait any longer for him and Fletcher left a pair of his own leather gloves where they had last seen him and left. The next morning when they returned to look for him, he was waiting patiently with the gloves! Mary Wood describes Jay as "regal" and this trait is a common characteristic in many of these dogs.

The stories about Jay's judgment and ability to think independently are legion. Mary tells of a friend Fletcher met one day at a local horse sale. He invited him to drop by his house after the sale for a cup of coffee. Fletcher would be along to meet him once he finished up his work at the sale. Nobody was home when the friend arrived so he decided he would just wait in the kitchen until Fletcher arrived. Jay was loose in the yard and didn't know this man. He let the man out of his truck and he even got all the way to the back door but when he put his hand on the door knob, Jay had reached his limit of tolerance for this trespasser. He gently but firmly placed his mouth on the man's hand and held on until the man released the door knob. The instant he took his hand from the knob, Jay released his grip on the man's hand. When the man tried again to open the door, Jay again took his hand to stop him. This silent communication continued for several more tries by Fletcher's friend until he finally decided he could wait in his truck until the master returned.

This oil painting made of Wood's Jay still hangs in the Wood home. Courtesy of Mary Wood

A Wood family friend was visiting their barn one day and was admiring Jay. She was an artist and asked if she could come by sometime to paint a portrait of Jay for Fletcher. Mary Wood was home the day the artist returned with her canvas and they went to the barn with Jay. Mary laid a saddle blanket on top of a tack trunk and told Jay to get up on it. She had him lie down and told him to stay. The artist spent the rest of the day painting Jay from life and the obedient dog never moved from his perch! Mary eventually returned to release him, leaving the artist in awe of her patient model.

Jay had a number of Casanova qualities and did his share of roaming the neighborhood creating high quality mixed breed puppies. One cross

Sisler's (Wood's) Panda. Courtesy of Sorensen

was with a roaming husky female which produced a number of nice dark brown eyed puppies. He would not bother females when they were not ready to breed but he knew instantly when they were ready. Females brought to the Wood ranch for breeding were allowed in the house because this was where Jay preferred to breed. His favorite spot was under the kitchen table, much to Mary's consternation!

One of the first bitches Mr. Wood bred Jay to was named Ely's Blue. This is the same Ely's Blue which Mrs. Ely purchased from Gene Sisler so the breedings were all Sisler background. He states, "In the mid-50s, I was producing dogs that were mostly merle in color with little tan or white. I liked the larger dogs as they seemed to have that extra stamina for running range cattle." His goal of producing a "true" breeding blue merle line had begun. One of the first puppies from this cross was Fletcher's pick for the stud fee. She became a foundation dam for a line of Wood bred Aussies in the hands of Dr Kenneth Green. Her name was Green's Dingo.

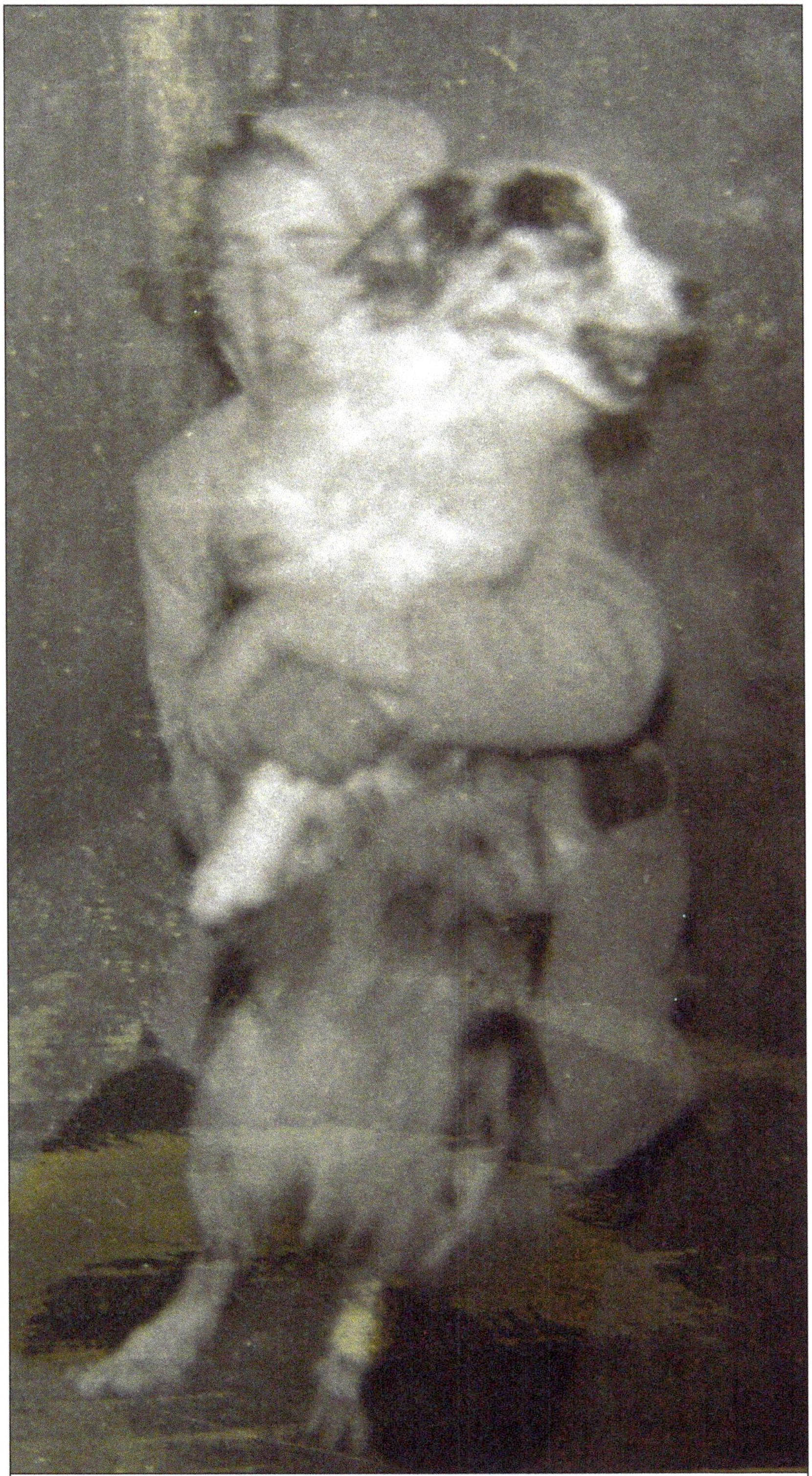

Sisler's (Wood's) Panda with her best friend, Judy Wood. Judy raised Panda before she became part of Jay Sisler's act. When she didn't work out, she came home to Judy and remained a beloved companion for the rest of her life. Courtesy of Mary Wood

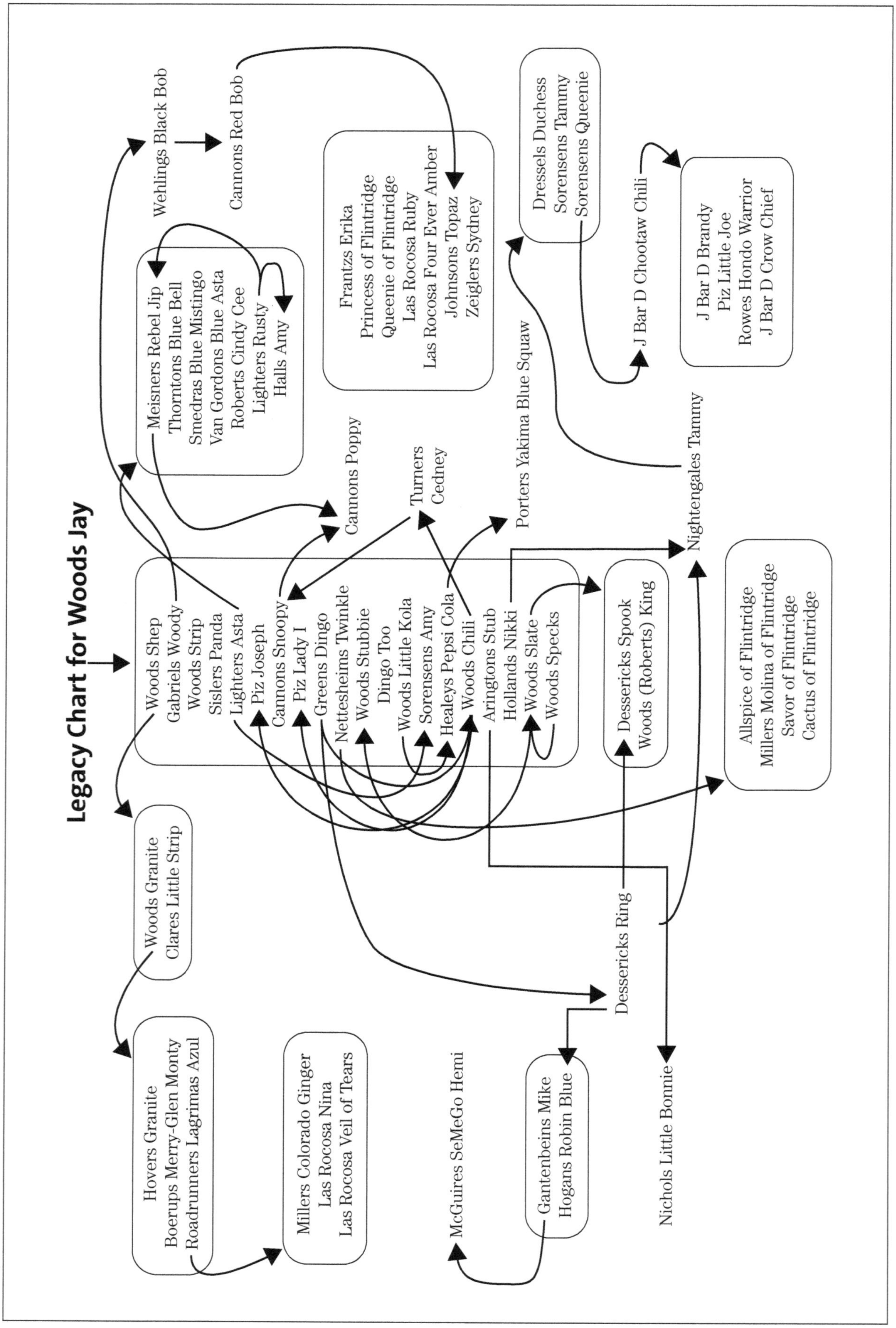
Legacy Chart for Woods Jay
Woods Shep
Gabriels Woody
Woods Strip
Sislers Panda
Lighters Asta
Piz Joseph
Cannons Snoopy
Piz Lady I
Greens Dingo
Nettesheims Twinkle
Woods Stubbie
Dingo Too
Woods Little Kola
Sorensens Amy
Healeys Pepsi Cola
Woods Chili
Aringtons Stub
Hollands Nikki
Woods Slate
Woods Specks
Woods Granite
Clares Little Strip
Hovers Granite
Boerups Merry-Glen Monty
Roadrunners Lagrimas Azul
Millers Colorado Ginger
Las Rocosa Nina
Las Rocosa Veil of Tears
McGuires SeMeGo Hemi
Gantenbeins Mike
Hogans Robin Blue
Nichols Little Bonnie
Dessericks Ring
Dessericks Ring
Dessericks Spook
Woods (Roberts) King
Allspice of Flintridge
Millers Molina of Flintridge
Savor of Flintridge
Cactus of Flintridge
Nightengales Tammy
Porters Yakima Blue Squaw
Turners Cedney
Cannons Poppy
Meisners Rebel Jip
Thorntons Blue Bell
Smedras Blue Mistingo
Van Gordons Blue Asta
Roberts Cindy Cee
Lighters Rusty
Halls Amy
Wehlings Black Bob
Cannons Red Bob
Frantzs Erika
Princess of Flintridge
Queenie of Flintridge
Las Rocosa Ruby
Las Rocosa Four Ever Amber
Johnsons Topaz
Zeiglers Sydney
Dressels Duchess
Sorensens Tammy
Sorensens Queenie
J Bar D Chootaw Chili
J Bar D Brandy
Piz Little Joe
Rowes Hondo Warrior
J Bar D Crow Chief

chapter 9

Dr. Kenneth and Lela (White) Green

Green's Dingo was owned by veterinarian Dr. Kenneth Green and his wife, Lela (White) Green. The Green and White families were among the earliest pioneer settlers of Colorado, southwest of Denver, tracing to at least the 1880s in that area. The families were neighbors and friends and Ken Green and Lela White grew up together. After their marriage in 1940, Ken and Lela ranched for over 60 years on 7,700 mountain acres, running about 200-250 cow/calf Hereford pairs on their own land and a number of forest service permit areas. Both the White and Green families used dogs to work cattle. Most of the dogs were obtained "from a neighbor" and were, like Jay Sisler's dogs, of unknown ancestry. Lela Green's earliest memory of working stock dogs is of a bob-tailed puppy named Ring which the White family obtained from the Green family around 1917. In that time, these dogs were known in the area as English Shepherds, exhibiting a typical black tri-coloration with copper trim and white face markings, white collar, and white feet. Ring was named for the full white collar around his neck and became a well known sire of good cattle working dogs for many of the local ranches. Ring lived a long and useful life on the White Ranch, working on a daily basis the rugged country ideally suited for raising good cattle. The White Ranch Open Space Park in Jefferson County, Colorado, is how this property is known today and any student of the Australian Shepherd would be well served to visit this

Dr. Kenneth Green as a young man on his horse, Ace, packing salt to his second ranch in Gilpin, Colorado. Dr. Green's father, George Septimus Green, came to Las Animas, Colorado, from Virginia in the 1890s and ranched there until 1917. This photo was taken in the late 1920s. The black tri Aussie was named Cross and was from a litter of puppies bred by Lela White's family. Cross' sire was a black tri dog named Ring—for his white collar, commonly called rings in those days. Cross' litter had 2 long tailed puppies, 2 puppies with half tails, and 2 natural bob tails. Cross was one of the half tailed puppies. Courtesy of Lela Green

Allen Green with a litter of Dingo's puppies. Before genetics were well understood, most breeders had double merle white puppies in their litters due to the mostly merle to merle breedings they practiced. Courtesy of Lela Green

area to see firsthand the actual conditions under which our breed was born.

When Ken and Lela Green were married in 1940, they and their families were well established in the livestock producing community of the region. Dr. Green first met Fletcher Wood when he became a part of the Jefferson County Fair Board. Fletcher was breeding Aussies and the cross of his Jay on Ely's (Sisler's) Blue would be repeated a number of times. One of the first puppies from this cross was the female named Green's Dingo obtained by Fletcher from Mrs. Ely as a stud fee puppy. She was given to the Green's by Fletcher Wood around 1952 or '53. Fletcher Wood proposed a partnership with Ken Green to breed Aussies but the Greens were not interested in breeding Aussies except for their own use. Dingo was the only breeding bitch the Greens owned but she and a dog they acquired later made a significant mark on the breed and the mountain-type Aussie.

The Greens esteemed Dingo as their ideal cattle dog. Mary Wood remembers her as a "working fool." Mrs. Green describes her temperament as "sweet" when relating to people. Dingo was a constant presence with the Greens around the ranch and everywhere they went. She learned to work cattle the way the Greens preferred—with an easy, subtle manner, no barking, and a slow pace. She learned her job by using her own intelligence and her own perception of her owner's wishes. A team player, Dingo would not set out on her own to bother cattle the Greens didn't want worked. If the people did nothing, Dingo followed suit.

When gathering cattle off the mountainous Green Ranch, Dingo would range out to bring strays back to the group. She was a low heeler when she needed to push a stubborn bovine but once the cattle were grouped and moving along she learned to drop to the back and conserve her energy just trailing the cows. Her primary job was helping in these cyclic moves from home pastures at Robinson Hill Ranch at the bottom of Golden Gate Canyon in the winter where the cows could be fed, to high country grazing on Forest Service permit land in May, to regular cattle drives down the mountain to market. In this area and this era, ranching was still done on horseback or on foot.

Winter feeding on the Robinson Hill Ranch in Golden Gate Canyon, Colorado. Pictured is Allen Green with Dingo. Courtesy of Lela Green

Although Fletcher Wood came to appreciate a dog which would bark on cattle, the Greens preferred a silent worker. They found a noisy dog provoked the type of cattle they had and only made their job more difficult. When moving their cattle, calves had a tendency to fall to the back of the herd. A barking dog would awaken the protective instincts of the cows, causing them to turn back for their calves and create a milling, out of control herd. Their cattle were accustomed to being worked by Dingo and she had no need to become pushy or excite the cattle.

Corral work and loading were off limits to dogs at this time. Ranchers did not have time to spend training dogs to a fine edge for close work, especially when the work only came once or twice a year. The Greens found Dingo to be most useful for gathering or at weaning time and she waited outside the corrals during branding, doctoring, or other chores. Dogs in close quarters tended to get people kicked and no rancher could afford to be out of commission nursing broken bones.

It wasn't all work for Dingo. The many hours spent riding in the mountains offered ample opportunity for adventure. Because of the vast area to be covered, most roundups were several days ride from home. Wildlife was abundant on the Green Ranch (which is now a Colorado State Park). Mrs. Green describes Dingo as a "good bird dog." She would amuse herself on these outings by tracking grouse in the mountains. Once she stumbled into a covey and as they flushed, one flew right into her mouth! Whether from surprise or a naturally soft mouth, she didn't harm the bird but just held it gently. She offered it to Ken who took it from her. So gentle was her handling of the bird that Ken released it and it flew off unharmed! Mary Wood tells similar stories of Fletcher Wood

Allen Green with Green's Dingo and a Dingo puppy. Courtesy of Lela Green

using his dogs for retrieving downed birds when he would hunt with them.

Lela Green remembers only seeing two cougars and two bears in her lifetime in the mountains. Instead, coyotes were the most prevalent problem to cattle ranchers. So brazen were these predators, husband Ken could call them and they would run right by him. The females were especially costly to the ranchers, killing calves to feed their young. Their legendary wiliness brought them together as a pack to lure a cow from her calf while other members would sneak in to take the calf.

Dingo raised four or five litters in her lifetime. She was bred to two or three different males and puppies were kept for the use of the Greens on their ranch or given to ranchers they knew. As with most dogs of this era, most remained unregistered and their names are lost to history. The sire of at least one litter was an Ely dog who produced a female the Greens called Ginger. As her name suggests, Ginger was a red dog who Mrs. Green describes as "very small" with a sweet temperament. The Greens retained this puppy but she was not bred. Ginger became a Green working ranch dog who wanted to heel horses as well as cattle. A habit discouraged by all the ranch riders!

Dingo normally produced puppies with full tails, which the Greens docked at birth, along with removing their dew claws. The Greens performed this alteration not because of a breed standard or a theory of function. They found through experience that dew claws on a mountain working dog were a useless and vulnerable appendage. Crusted snow and rocks left the legs of their working dogs bloody and sore when dew claws were torn off or hung up on these common hazards. Fletcher Wood found the same problem in his own dogs if he did not remove dewclaws and he considered them a hazard on a working dog.

Green's Dingo lived out her life on the Green ranch as a favored working dog. She was approximately 12 or 13 years old at her death. The Greens valued and needed good working dogs on their ranches. For a number of years, they attempted to improve their dogs and breed for their needs using Dingo as their foundation. They bred her to outside males such as Wood's Jay and also attempted to purchase males for their own use. One of these dogs was found through a newspaper ad, purchased from a neighbor and later made his mark on the Aussie breed when used by other breeders. This dog was Green's Kim.

A large blue merle with two striking blue eyes, Kim was of the mountain-type. Mrs. Green describes him as a "very strong" working dog but this appears to have been an understatement! She remembers an incident with Kim when they were out gathering stray cattle on the ranch. Kim latched onto a stray cow by the nose and refused to let go, finally taking her to the ground. After repeated attempts to get him to release the cow, Mrs. Green was eventually forced to dismount from her horse and physically pull him off the cow. After a year of working cattle on their mountain ranches, the Greens decided this dog was too rough and didn't work cattle the way they wanted their cattle worked. He was bred to Dingo at least once while the Greens owned him but he was eventually sold to Oklahoma cowboy Steve Mansker. Mansker was looking for a tough cow dog to use in Oklahoma ranching conditions and Kim was well suited to his needs. Used for breeding and work by Mansker, Kim founded another family of top working dogs when bred to the Sisler bitch, Freckles, which Mansker had obtained from Jay in about 1956. Mansker kept the female, Mansker's Anna Lee from the litter for his own use. Anna Lee's brother, Alexander's Curt, sired a female named Duchess who eventually became the dam of the legendary Mansker's Turk. Turk bred back to his great aunt Anna Lee produced another legend of the breed, Taylor's Whiskey.

Ken and Lela Green's son, Allen, with (left to right) Green's Dingo, Green's Kim, and another blue merle Aussie. Courtesy of Lela Green

One of Kim and Dingo's pups was sold or given to a Green acquaintance from the Jefferson County Fair Board, the Desserick family. Desserick's Ring was a black and white male who, through his son Gantenbein's Mike, was grandsire of the amazing Ch McGuire's SeMeGo Hemi UDT OTD. A common description for a white collar at that time was a "ring around the neck" and, so, Desserick's Ring, like many others, was named for this feature.

The Greens owned one other male Aussie after Kim who was also not up to their standards and they decided they were not going to be able to breed the kind of Aussie they needed for their ranch. The Greens, like so many early breeders, needed and used their dogs everyday on their ranches. These dogs never had a day off and worked as hard as their human partners. Dingo was proven as both an individual and a

Allen Green with (left to right) Wood's Jay and Green's Dingo. The Greens often kept Jay for Fletcher Wood when he traveled. Courtesy of Lela Green

brood bitch and was not found lacking. Although the Greens didn't find what they needed in her produce, they nevertheless left the breed with a number of important dogs who proved their worth into the future. In 1961, the Greens and Fletcher Wood imported some of the first Australian Cattle Dogs to this country. The Greens stopped breeding Australian Shepherds entirely once they obtained these imports. Dingo remained with them until her death.

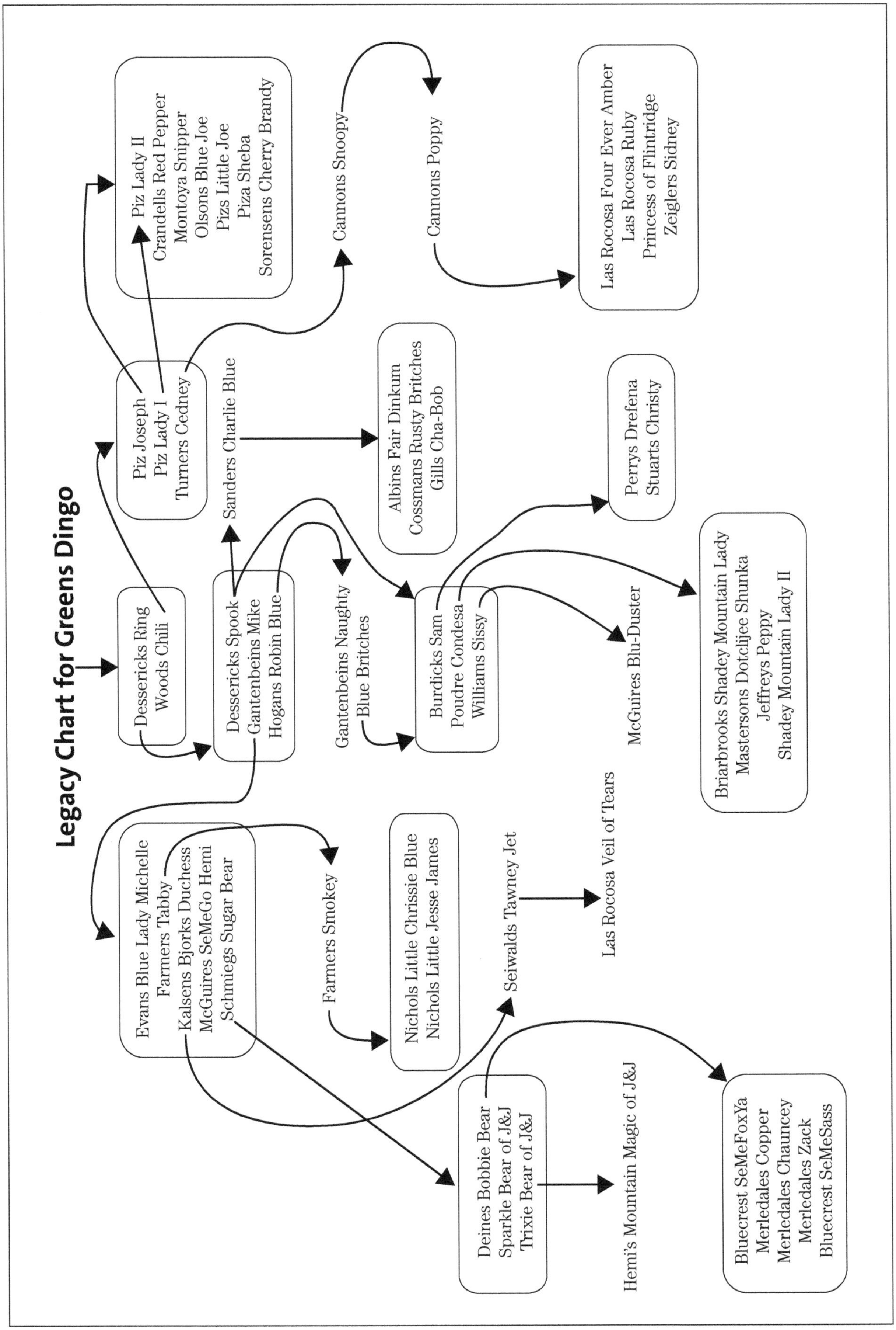
Legacy Chart for Greens Dingo
Dessericks Ring
Woods Chili
Dessericks Spook
Gantenbeins Mike
Hogans Robin Blue
Evans Blue Lady Michelle
Farmers Tabby
Kalsens Bjorks Duchess
McGuires SeMeGo Hemi
Schmiegs Sugar Bear
Piz Joseph
Piz Lady I
Turners Cedney
Piz Lady II
Crandells Red Pepper
Montoya Snipper
Olsons Blue Joe
Pizs Little Joe
Piza Sheba
Sorensens Cherry Brandy
Cannons Snoopy
Cannons Poppy
Las Rocosa Four Ever Amber
Las Rocosa Ruby
Princess of Flintridge
Zeiglers Sidney
Sanders Charlie Blue
Albins Fair Dinkum
Cossmans Rusty Britches
Gills Cha-Bob
Gantenbeins Naughty
Blue Britches
Burdicks Sam
Poudre Condesa
Williams Sissy
Perrys Drefena
Stuarts Christy
McGuires Blu-Duster
Briarbrooks Shadey Mountain Lady
Mastersons Dotclijee Shunka
Jeffreys Peppy
Shadey Mountain Lady II
Farmers Smokey
Nichols Little Chrissie Blue
Nichols Little Jesse James
Seiwalds Tawney Jet
Las Rocosa Veil of Tears
Deines Bobbie Bear
Sparkle Bear of J&J
Trixie Bear of J&J
Hemi's Mountain Magic of J&J
Bluecrest SeMeFoxYa
Merledales Copper
Merledales Chauncey
Merledales Zack
Bluecrest SeMeSass

chapter 10

Joe Piz

We do know a little about one Green's Dingo daughter who left a legacy of red Wood bred dogs, Wood's Chili. Chili was owned by Fletcher Wood and was a ranch dog for him. An inveterate cat hater, she had to live in the Wood barn because she would not stay in a pen. Her excited jumping often resulted in escapes and Wood finally gave up confining her in a pen and left her in his barn. She managed to terrorize the barn cats on a full time basis, even jumping up into the hayloft to catch them!

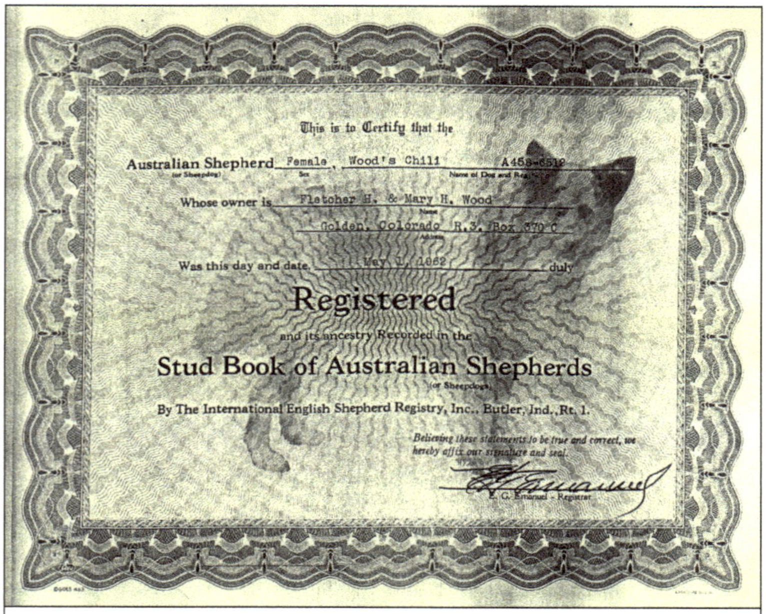
This is to Certify that the

Australian Shepherd (or Sheepdog) Female (Sex), Wood's Chili A458-6512 (Name of Dog and Reg. No.)

Whose owner is Fletcher H. & Mary H. Wood (Name)

Golden, Colorado R.3. Box 370 C (Address)

Was this day and date May 1, 1962 duly

Registered

and its ancestry Recorded in the

Stud Book of Australian Shepherds (or Sheepdogs)

By The International English Shepherd Registry, Inc., Butler, Ind., Rt. 1.

Believing these statements to be true and correct, we hereby affix our signature and seal.

E. G. Emanuel - Registrar

National Stock Dog registration for Wood's Chili. Courtesy of Mary Wood via Sorensen

A red merle female from the cross of Dingo back to her sire Wood's Jay, Chili would be bred back again to Wood's Jay to produce Piz's Joseph—patriarch of the Piz family of top cattle working dogs. Joseph's prepotency, like Wood's Stubbie's, came from

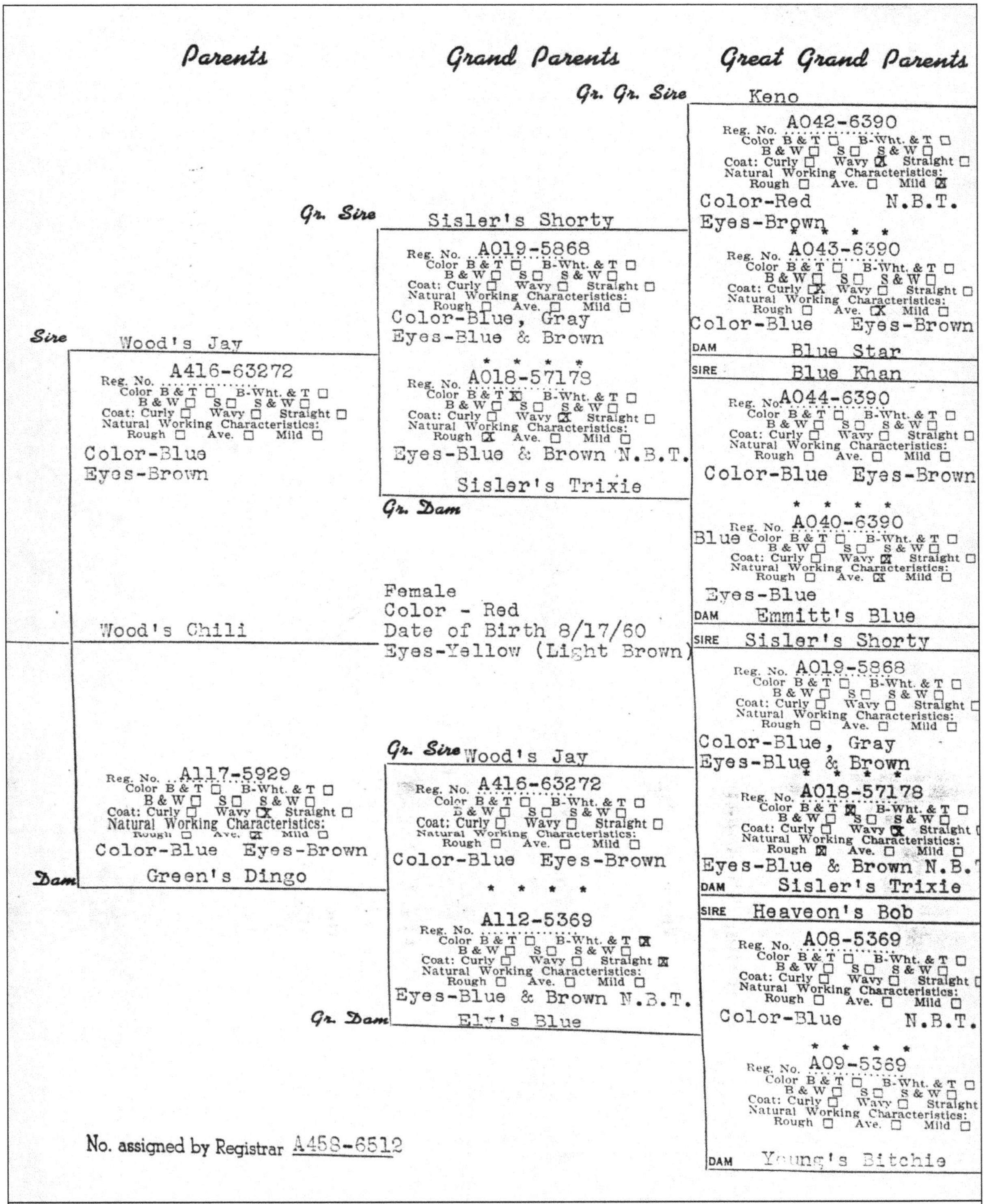

Parents | Grand Parents | Great Grand Parents

Sire: Wood's Jay
Reg. No. A416-63272
Color B & T ☐ B-Wht. & T ☐ B & W ☐ S ☐ S & W ☐
Coat: Curly ☐ Wavy ☐ Straight ☐
Natural Working Characteristics: Rough ☐ Ave. ☐ Mild ☐
Color-Blue
Eyes-Brown

Gr. Sire: Sisler's Shorty
Reg. No. A019-5868
Color B & T ☐ B-Wht. & T ☐ B & W ☐ S ☐ S & W ☐
Coat: Curly ☐ Wavy ☐ Straight ☐
Natural Working Characteristics: Rough ☐ Ave. ☐ Mild ☐
Color-Blue, Gray
Eyes-Blue & Brown

Gr. Gr. Sire: Keno
Reg. No. A042-6390
Color B & T ☐ B-Wht. & T ☐ B & W ☐ S ☐ S & W ☐
Coat: Curly ☐ Wavy ☒ Straight ☐
Natural Working Characteristics: Rough ☐ Ave. ☐ Mild ☒
Color-Red N.B.T.
Eyes-Brown

Reg. No. A043-6390
Color B & T ☐ B-Wht. & T ☐ B & W ☐ S ☐ S & W ☐
Coat: Curly ☒ Wavy ☐ Straight ☐
Natural Working Characteristics: Rough ☐ Ave. ☒ Mild ☐
Color-Blue Eyes-Brown
DAM Blue Star

Gr. Dam: Sisler's Trixie
Reg. No. A018-57178
Color B & T ☒ B-Wht. & T ☐ B & W ☐ S ☐ S & W ☐
Coat: Curly ☐ Wavy ☒ Straight ☐
Natural Working Characteristics: Rough ☒ Ave. ☐ Mild ☐
Eyes-Blue & Brown N.B.T.

SIRE Blue Khan
Reg. No. A044-6390
Color B & T ☐ B-Wht. & T ☐ B & W ☐ S ☐ S & W ☐
Coat: Curly ☐ Wavy ☐ Straight ☐
Natural Working Characteristics: Rough ☐ Ave. ☐ Mild ☐
Color-Blue Eyes-Brown

Reg. No. A040-6390
Blue
Color B & T ☐ B-Wht. & T ☐ B & W ☐ S ☐ S & W ☐
Coat: Curly ☐ Wavy ☒ Straight ☐
Natural Working Characteristics: Rough ☐ Ave. ☒ Mild ☐
Eyes-Blue
DAM Emmitt's Blue

Wood's Chili
Female
Color - Red
Date of Birth 8/17/60
Eyes-Yellow (Light Brown)

Dam: Green's Dingo
Reg. No. A117-5929
Color B & T ☐ B-Wht. & T ☐ B & W ☐ S ☐ S & W ☐
Coat: Curly ☐ Wavy ☒ Straight ☐
Natural Working Characteristics: Rough ☐ Ave. ☒ Mild ☐
Color-Blue Eyes-Brown

Gr. Sire: Wood's Jay
Reg. No. A416-63272
Color B & T ☐ B-Wht. & T ☐ B & W ☐ S ☐ S & W ☐
Coat: Curly ☐ Wavy ☐ Straight ☐
Natural Working Characteristics: Rough ☐ Ave. ☐ Mild ☐
Color-Blue Eyes-Brown

SIRE Sisler's Shorty
Reg. No. A019-5868
Color B & T ☐ B-Wht. & T ☐ B & W ☐ S ☐ S & W ☐
Coat: Curly ☐ Wavy ☐ Straight ☐
Natural Working Characteristics: Rough ☐ Ave. ☐ Mild ☐
Color-Blue, Gray
Eyes-Blue & Brown

Reg. No. A018-57178
Color B & T ☒ B-Wht. & T ☐ B & W ☐ S ☐ S & W ☐
Coat: Curly ☐ Wavy ☒ Straight ☐
Natural Working Characteristics: Rough ☒ Ave. ☐ Mild ☐
Eyes-Blue & Brown N.B.T.
DAM Sisler's Trixie

Gr. Dam: Ely's Blue
Reg. No. A112-5369
Color B & T ☐ B-Wht. & T ☒ B & W ☐ S ☐ S & W ☐
Coat: Curly ☐ Wavy ☐ Straight ☒
Natural Working Characteristics: Rough ☐ Ave. ☐ Mild ☐
Eyes-Blue & Brown N.B.T.

SIRE Heaveon's Bob
Reg. No. A08-5369
Color B & T ☐ B-Wht. & T ☐ B & W ☐ S ☐ S & W ☐
Coat: Curly ☐ Wavy ☐ Straight ☐
Natural Working Characteristics: Rough ☐ Ave. ☐ Mild ☐
Color-Blue N.B.T.

Reg. No. A09-5369
Color B & T ☐ B-Wht. & T ☐ B & W ☐ S ☐ S & W ☐
Coat: Curly ☐ Wavy ☐ Straight
Natural Working Characteristics: Rough ☐ Ave. ☐ Mild ☐
DAM Young's Bitchie

No. assigned by Registrar A458-6512

National Stock Dog Registry pedigree of Wood's Chili. Courtesy of Mary Wood via Sorensen

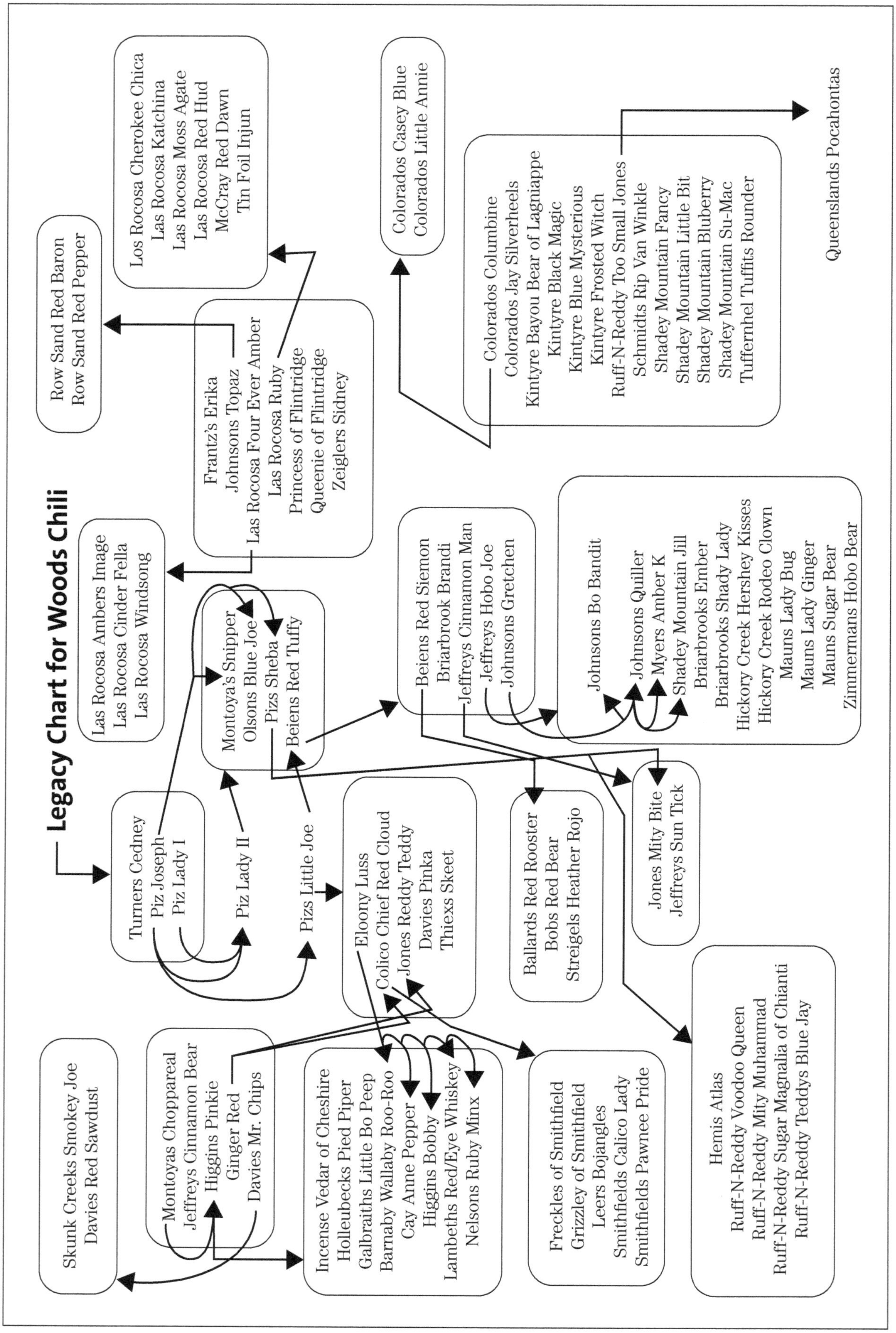
Legacy Chart for Woods Chili
Turners Cedney
Piz Joseph
Piz Lady I
Piz Lady II
Pizs Little Joe
Las Rocosa Ambers Image
Las Rocosa Cinder Fella
Las Rocosa Windsong
Montoya's Snipper
Olsons Blue Joe
Pizs Sheba
Beiens Red Tuffy
Frantz's Erika
Johnsons Topaz
Las Rocosa Four Ever Amber
Las Rocosa Ruby
Princess of Flintridge
Queenie of Flintridge
Zeiglers Sidney
Row Sand Red Baron
Row Sand Red Pepper
Los Rocosa Cherokee Chica
Las Rocosa Katchina
Las Rocosa Moss Agate
Las Rocosa Red Hud
McCray Red Dawn
Tin Foil Injun
Colorados Casey Blue
Colorados Little Annie
Colorados Columbine
Colorados Jay Silverheels
Kintyre Bayou Bear of Lagniappe
Kintyre Black Magic
Kintyre Blue Mysterious
Kintyre Frosted Witch
Ruff-N-Reddy Too Small Jones
Schmidts Rip Van Winkle
Shadey Mountain Fancy
Shadey Mountain Little Bit
Shadey Mountain Bluberry
Shadey Mountain Su-Mac
Tuffernhel Tuffits Rounder
Queenslands Pocahontas
Beiens Red Siemon
Briarbrook Brandi
Jeffreys Cinnamon Man
Jeffreys Hobo Joe
Johnsons Gretchen
Johnsons Bo Bandit
Johnsons Quiller
Myers Amber K
Shadey Mountain Jill
Briarbrooks Ember
Briarbrooks Shady Lady
Hickory Creek Hershey Kisses
Hickory Creek Rodeo Clown
Mauns Lady Bug
Mauns Lady Ginger
Mauns Sugar Bear
Zimmermans Hobo Bear
Eloony Luss
Colico Chief Red Cloud
Jones Reddy Teddy
Davies Pinka
Thiexs Skeet
Ballards Red Rooster
Bobs Red Bear
Streigels Heather Rojo
Jones Mity Bite
Jeffreys Sun Tick
Skunk Creeks Smokey Joe
Davies Red Sawdust
Montoyas Choppareal
Jeffreys Cinnamon Bear
Higgins Pinkie
Ginger Red
Davies Mr. Chips
Incense Vedar of Cheshire
Holleubecks Pied Piper
Galbraiths Little Bo Peep
Barnaby Wallaby Roo-Roo
Cay Anne Pepper
Higgins Bobby
Lambeths Red/Eye Whiskey
Nelsons Ruby Minx
Freckles of Smithfield
Grizzley of Smithfield
Leers Bojangles
Smithfields Calico Lady
Smithfields Pawnee Pride
Hemis Atlas
Ruff-N-Reddy Voodoo Queen
Ruff-N-Reddy Mity Muhammad
Ruff-N-Reddy Sugar Magnalia of Chianti
Ruff-N-Reddy Teddys Blue Jay

a similar three time cross of Wood's Jay back to his daughter/granddaughter/great-granddaughter—daughter Green's Dingo bred back to Jay to produce Wood's Chili, who was bred back to Jay to produce Piz Joseph. Proving yet again the genetic power of Wood's Jay, the Piz line was linebred on Jay through Joseph being bred to his daughters and granddaughters to produce a legendary family of red dogs most famously represented by Ch. Jones Reddy Teddy CDX CWD. Ted's legacy will be discussed in greater detail in a later chapter of this book. Joseph's full sister, Piz's Lady I also contributed greatly to the Piz line and helped concentrate the Wood's Jay blood even more with a cross back to Piz's Joseph, producing Piz's Lady II. Lady II bred to Joseph produced the grand matron Piz's Sheba, the great sire Olson's Blue Joe and another influential Piz-bred sire Montoya's Snipper. Lady II bred to the Joseph son Piz's Little Joe produced another great sire, Beien's Red Tuffy who produced a number of good dogs for the Jeffreys' Shady Mountain kennel.

Italian by heritage, Joe Piz was raised by an uncle on a Wyoming sheep ranch. His familiarity with the breed began in the 1920s and '30s during the height of the sheep ranching era of the northern Great Plains. The growth of the Australian Shepherd was taking off during this period and Joe knew many early Aussies on the ranch. Red merles, blue merles, and blacks were all common near his home and the ranchers often crossed these dogs with the local border collies.

Clare Olson's Blue Joe, dark blue merle and white male (Piz Joseph x Piz Lady II). Courtesy of Sorensen

Olson's Blue Joe as a mature dog. Courtesy of Raymie Jones

"Wilbur," red merle brother of Piz's Joseph. Courtesy of Rowe

Rare photo of the red merle male, Piz's Little Joe, affectionately called "Joey." Courtesy of Mary Wood

"Wilbur," red merle full brother of Piz's Joseph. Courtesy of Rowe

Dam of many of the best Wood type dogs, Piz's Sheba, blue merle and white. Courtesy of Raymie Jones

PEDIGREE OF Piz Joseph

Born: ______ Sex: ______
Owned by: ______

Parents	Grandparents	Great-grandparents	Great-great-grandparents
Sire: Woods Jay	Sire: Sislers Shorty	Sire: Keno	Sire
			Dam
		Dam: Blue Star	Sire
			Dam
	Dam: Sislers Trixie	Sire: Blue Khan	Sire
			Dam
		Dam: Emmetts Blue	Sire
			Dam
Dam: Woods Chili	Sire: Woods Jay	Sire: Sislers Shorty	Sire: Keno
			Dam: Blue Star
		Dam: Sislers Trixie	Sire: Blue Khan
			Dam: Emmetts Blue
	Dam: Greens Dingo	Sire: Woods Jay	Sire: Sislers Shorty
			Dam: Sislers Trixie
		Dam: (Sislers) Elys Blue	Sire: Heavrons Bob
			Dam: Youngs Bitchie

Bred by: ______ Address: ______ Phone: ______

The details given above are correct to the best of my knowledge. Signed: ______(Breeder)

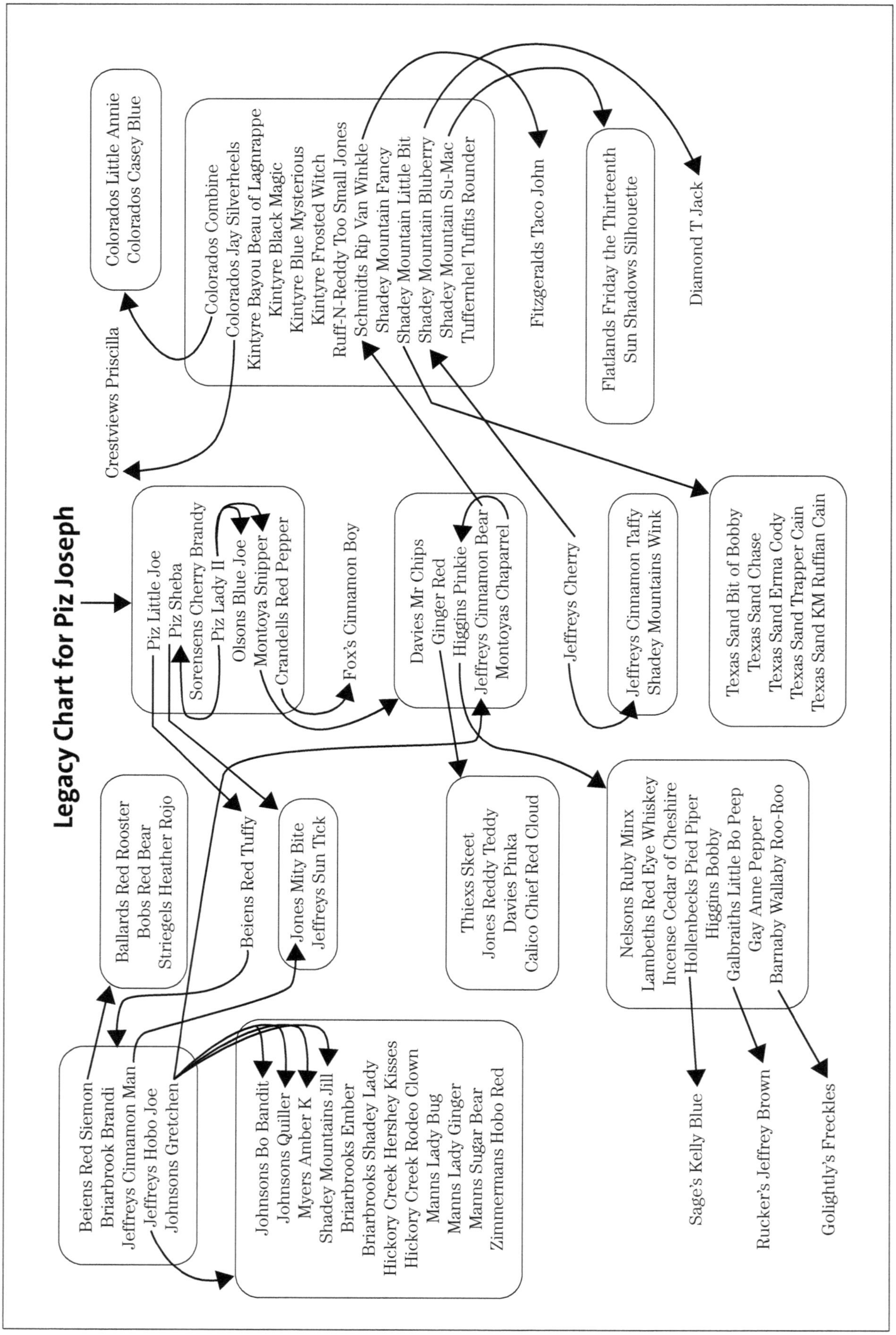
Legacy Chart for Piz Joseph
Piz Little Joe
Piz Sheba
Sorensens Cherry Brandy
Piz Lady II
Olsons Blue Joe
Montoya Snipper
Crandells Red Pepper
Crestviews Priscilla
Colorados Little Annie
Colorados Casey Blue
Colorados Combine
Colorados Jay Silverheels
Kintyre Bayou Beau of Lagnrappe
Kintyre Black Magic
Kintyre Blue Mysterious
Kintyre Frosted Witch
Ruff-N-Reddy Too Small Jones
Schmidts Rip Van Winkle
Shadey Mountain Fancy
Shadey Mountain Little Bit
Shadey Mountain Bluberry
Shadey Mountain Su-Mac
Tuffernhel Tuffits Rounder
Fitzgeralds Taco John
Flatlands Friday the Thirteenth
Sun Shadows Silhouette
Diamond T Jack
Fox's Cinnamon Boy
Davies Mr Chips
Ginger Red
Higgins Pinkie
Jeffreys Cinnamon Bear
Montoyas Chaparrel
Jeffreys Cherry
Jeffreys Cinnamon Taffy
Shadey Mountains Wink
Texas Sand Bit of Bobby
Texas Sand Chase
Texas Sand Erma Cody
Texas Sand Trapper Cain
Texas Sand KM Ruffian Cain
Ballards Red Rooster
Bobs Red Bear
Striegels Heather Rojo
Beiens Red Tuffy
Jones Mity Bite
Jeffreys Sun Tick
Thiexs Skeet
Jones Reddy Teddy
Davies Pinka
Calico Chief Red Cloud
Nelsons Ruby Minx
Lambeths Red Eye Whiskey
Incense Cedar of Cheshire
Hollenbecks Pied Piper
Higgins Bobby
Galbraiths Little Bo Peep
Gay Anne Pepper
Barnaby Wallaby Roo-Roo
Beiens Red Siemon
Briarbrook Brandi
Jeffreys Cinnamon Man
Jeffreys Hobo Joe
Johnsons Gretchen
Johnsons Bo Bandit
Johnsons Quiller
Myers Amber K
Shadey Mountains Jill
Briarbrooks Ember
Briarbrooks Shadey Lady
Hickory Creek Hershey Kisses
Hickory Creek Rodeo Clown
Manns Lady Bug
Manns Lady Ginger
Manns Sugar Bear
Zimmermans Hobo Red
Sage's Kelly Blue
Rucker's Jeffrey Brown
Golightly's Freckles

Piz's entrepreneurial spirit emerged at a young age, beginning with his first money making enterprise catching wild horses to break for the Army. In the 1920s, the US Army was paying $1.00 per head for broke horses. Joe's talent with horses was born riding bareback—learning to stick tight since falling off meant a possible encounter with rattlesnakes! Paralleling Fletcher Wood's multi-facted life, Joe served in the Merchant Marines during World War II, owned Mother's Home Bakery in Denver with his brother Bob, served in the Sheriff's Department, rodeoed, trained cutting horses and successfully raced Quarter Horses and fast cars.

Piz's Sheba, grand matron of the Piz breeding program. Pictured here in old age after a number of litters. Courtesy of Melissa Couch

It was while racing horses that Joe Piz and Fletcher Wood made the fateful encounter which would revive the red Wood's dogs. Joe had a knack with racehorses which he says Fletcher did not share. Wood had his success with horses in performance events and Piz is quick to credit him with owning some of the finest equines for sport. But Joe had the upper hand when it came to racehorses and winning a bet on a horse race with Fletcher Wood brought Joe his foundation Aussie, Piz's Joseph. Joseph brought the tried and true Wood's Jay blood to Piz's operation in a striking dark red merle package. Sired by Jay and out of Wood's Chili, Joseph embodied the same intense inbreeding Fletcher Wood practiced to produce his own great Wood's Stubbie.

Piz's Little Joe, red merle male. Courtesy of Sorensen

Piz remembers his Wood's bred dogs as bigger than the sheepdogs on the ranch in Wyoming. This description and memory is consistent with the memories of Les Burlile and others about the size and type of the mountain strain of Aussies. By adding crosses with the original Sorensen and J Bar D dogs, Piz produced a line of tough cattle working dogs which served to popularize red colored Aussies in the breed. Ch. Jones Reddy Teddy CDX CWD epitomized the type of dog Piz strove to produce and became a major producer of top cattle working dogs. The strong line-breeding behind Ted and Piz's other dogs ensured their prepotency as producers.

PEDIGREE OF Piz Little Joe

Born: ______ Sex: ______
Owned by: ______

Parents	Grandparents	Great-grandparents	Great-great-grandparents
Sire: Piz Joseph	Sire: Woods Jay	Sire: Sislers Shorty	Sire: Keno
			Dam: Blue Star
		Dam: Sislers Trixie	Sire: Blue Khan
			Dam: Emmetts Blue
	Dam: Woods Chili	Sire: Woods Jay	Sire: Sislers Shorty
			Dam: Sislers Trixie
		Dam: Greens Dingo	Sire: Woods Jay
			Dam: (Sislers) Ely's Blue
Dam: J Bar D Choctaw Chili	Sire: Rowes Warrior Bow	Sire: Ronsleys Charcoal Smoke	Sire: Ritters Streak
			Dam: Ritters Lass
		Dam: Murphys Cochina	Sire: Ronsleys Charcoal Smoke
			Dam: Ronsleys Mitzi Buck
	Dam: Sorensens Queenie	Sire: Sorensens Gunsmoke	Sire: Williams Laddie
			Dam: Jeffreys Sheila
		Dam: Nightingales Tammy	Sire: Elys Chico
			Dam: Hollands Niki

Bred by: ______ Address: ______ Phone: ______

The details given above are correct to the best of my knowledge. Signed: ______(Breeder)

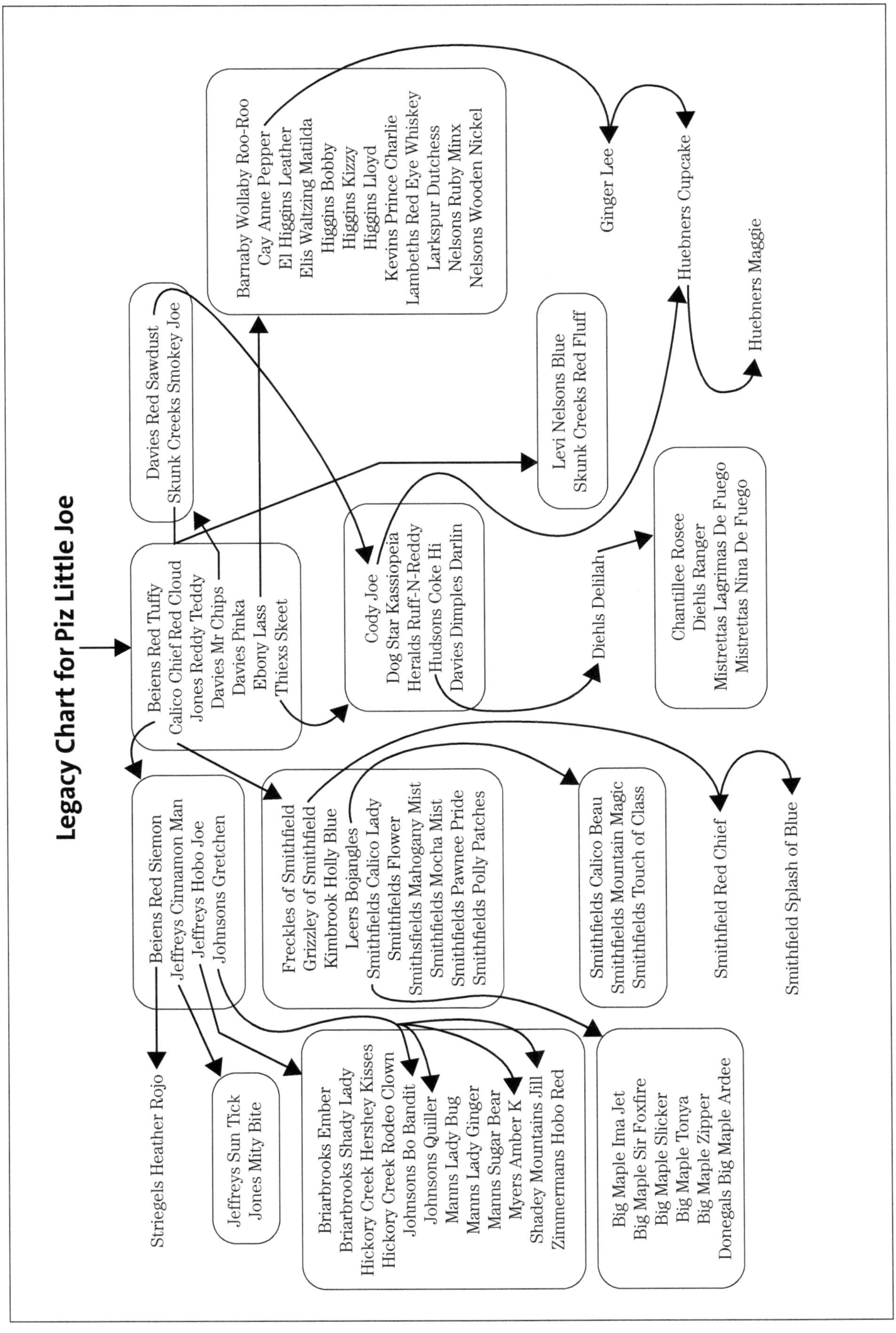
Legacy Chart for Piz Little Joe
Beiens Red Tuffy
Calico Chief Red Cloud
Jones Reddy Teddy
Davies Mr Chips
Davies Pinka
Ebony Lass
Thiexs Skeet
Davies Red Sawdust
Skunk Creeks Smokey Joe
Barnaby Wollaby Roo-Roo
Cay Anne Pepper
El Higgins Leather
Elis Waltzing Matilda
Higgins Bobby
Higgins Kizzy
Higgins Lloyd
Kevins Prince Charlie
Lambeths Red Eye Whiskey
Larkspur Dutchess
Nelsons Ruby Minx
Nelsons Wooden Nickel
Ginger Lee
Huebners Cupcake
Huebners Maggie
Levi Nelsons Blue
Skunk Creeks Red Fluff
Cody Joe
Dog Star Kassiopeia
Heralds Ruff-N-Reddy
Hudsons Coke Hi
Davies Dimples Darlin
Diehls Delilah
Chantillee Rosee
Diehls Ranger
Mistrettas Lagrimas De Fuego
Mistrettas Nina De Fuego
Beiens Red Siemon
Jeffreys Cinnamon Man
Jeffreys Hobo Joe
Johnsons Gretchen
Freckles of Smithfield
Grizzley of Smithfield
Kimbrook Holly Blue
Leers Bojangles
Smithfields Calico Lady
Smithfields Flower
Smithsfields Mahogany Mist
Smithfields Mocha Mist
Smithfields Pawnee Pride
Smithfields Polly Patches
Smithfields Calico Beau
Smithfields Mountain Magic
Smithfields Touch of Class
Smithfield Red Chief
Smithfield Splash of Blue
Striegels Heather Rojo
Jeffreys Sun Tick
Jones Mity Bite
Briarbrooks Ember
Briarbrooks Shady Lady
Hickory Creek Hershey Kisses
Hickory Creek Rodeo Clown
Johnsons Bo Bandit
Johnsons Quiller
Manns Lady Bug
Manns Lady Ginger
Manns Sugar Bear
Myers Amber K
Shadey Mountains Jill
Zimmermans Hobo Red
Big Maple Ima Jet
Big Maple Sir Foxfire
Big Maple Slicker
Big Maple Tonya
Big Maple Zipper
Donegals Big Maple Ardee

Piz's Little Joe (top). Courtesy of Sorensen

Davies' Pinka, red merle full sister to Ch. Jones Reddy Teddy CDX CWD (Piz's Little Joe x Ginger Red). Courtesy of Melissa Couch

chapter 11

Wood's Stubbie

Fletcher Wood produced legions of top Aussies by Wood's Jay. Many, if not most, were heavily linebred and inbred on Jay. Since many were not registered, it will never be known exactly how widespread their actual influence on the breed was. Wood recognized Jay's quality as an individual and a producer and his main goal eventually became to preserve his type and quality. Through his extremely concentrated inbreeding of Jay he proved not only his quality but also his genetic prepotence and "cleanness," producing generation after generation of healthy, long-lived, superior individuals.

The living pedigree of Wood's Stubbie. L to R: Sisler's Shorty, Wood's Jay, Wood's Specks, Wood's Flick and Stubbie's dam, Wood's Slate. Photo by Jay Sisler, courtesy of Mary Wood.

Perhaps the closest he ever came to reproducing Jay was in the dog named Wood's Stubbie. Stubbie was the result of Jay bred back to his daughter, then back to a daughter of that cross so he was not only Stubbie's sire, but also his grandsire and great-grandsire. The founding bitch bred to Jay to begin the line which would culminate in producing Stubbie was named Trump's Stubbie. She was a dark blue merle owned by a professor at Colorado

This is to Certify that the

Australian Shepherd (or Sheepdog) Female (Sex), Wood's Slate A457-3512 (Name of Dog and Reg. No.)

Whose owner is Fletcher H & Mary H. Wood (Name)

Golden, Colorado R.3. Box 370 C (Address)

Was this day and date, May 1, 1962 duly

Registered

and its ancestry Recorded in the

Stud Book of Australian Shepherds (or Sheepdogs)

By The International English Shepherd Registry, Inc., Butler, Ind., Rt. 1.

Believing these statements to be true and correct, we hereby affix our signature and seal.

E. G. Emanuel - Registrar

National Stock Dog Registry pedigree of Wood's Slate. Courtesy of Mary Wood via Sorensen

State University named Murray Trump. Dr. Trump was a sheep specialist at the university and used his female to work the university flock. Like so many of the breed's foundation dogs, he had obtained Trump's Stubbie from Idaho. An amusing story is told by Mary Wood about the day Trump's Stubbie was delivered to the Wood ranch for breeding to Jay. Fletcher was out for the day and Stubbie was not yet ready to breed so Dr. Trump dropped her off and would return for her after the breeding was done. Fletcher came home that evening and while admiring what was to be the founding female of his most famous line, she bit him!

Through Fletcher's selection for less and less white trim, Wood's Stubbie epitomized the look Wood was striving for. It is the same look—extremely dark color with minimal white or copper trim—which has now become referred to as the "Wood's" type throughout the breed.

Wood's Stubbie, like most mountain-type Aussies, was a large dog with a strong temperament and working ability. Dick Sorensen remembers a cattle drive he was part of with Tom Jeffries and Fletcher Wood.

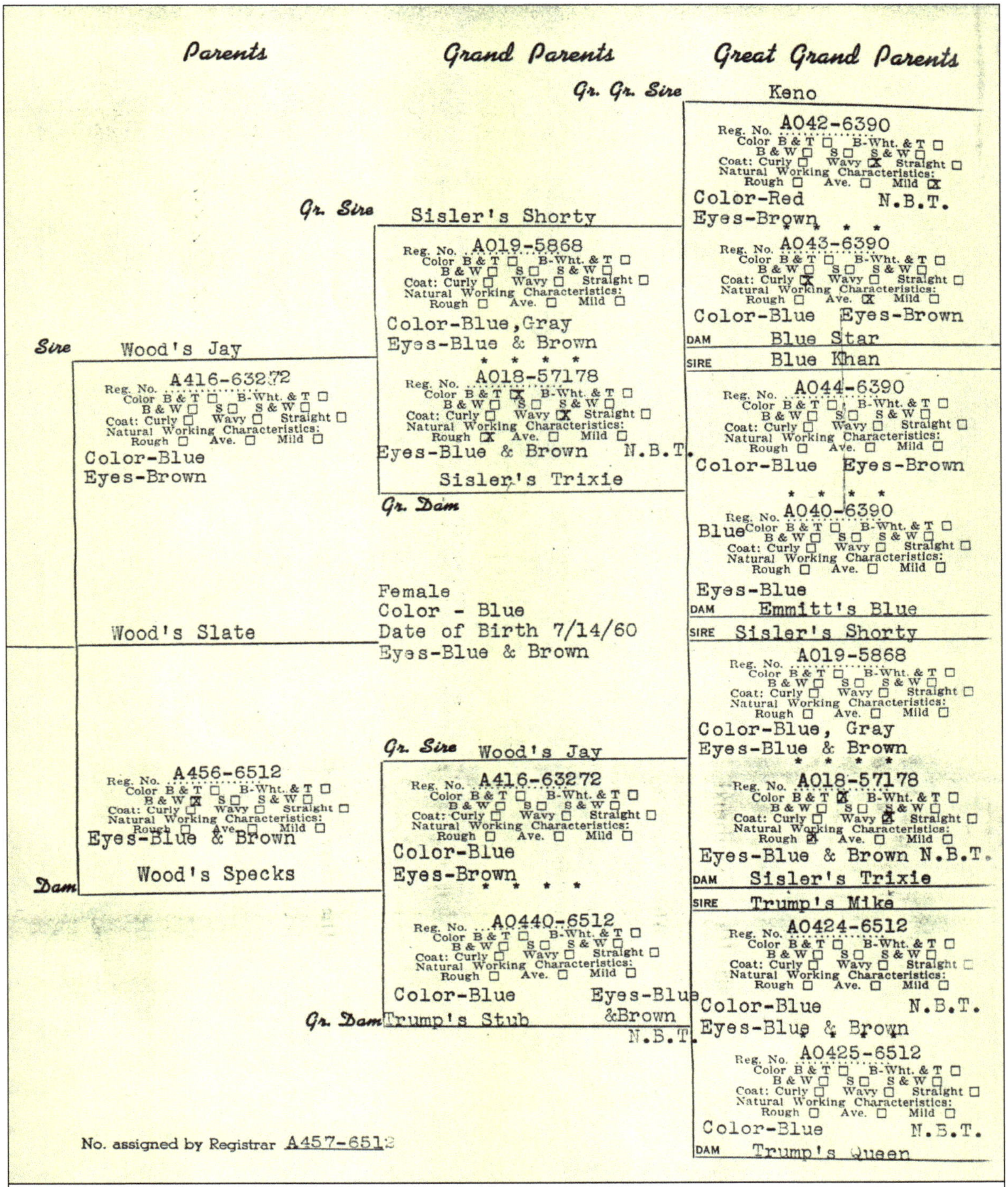

Parents | **Grand Parents** | **Great Grand Parents**

Sire: Wood's Jay
Reg. No. A416-63272
Color B & T ☐ B-Wht. & T ☐ B & W ☐ S ☐ S & W ☐
Coat: Curly ☐ Wavy ☐ Straight ☐
Natural Working Characteristics: Rough ☐ Ave. ☐ Mild ☐
Color-Blue
Eyes-Brown

Gr. Sire: Sisler's Shorty
Reg. No. A019-5868
Color B & T ☐ B-Wht. & T ☐ B & W ☐ S ☐ S & W ☐
Coat: Curly ☐ Wavy ☐ Straight ☐
Natural Working Characteristics: Rough ☐ Ave. ☐ Mild ☐
Color-Blue,Gray
Eyes-Blue & Brown

Gr. Gr. Sire: Keno
Reg. No. A042-6390
Color B & T ☐ B-Wht. & T ☐ B & W ☐ S ☐ S & W ☐
Coat: Curly ☐ Wavy ☒ Straight ☐
Natural Working Characteristics: Rough ☐ Ave. ☐ Mild ☒
Color-Red N.B.T.
Eyes-Brown

Reg. No. A043-6390
Color B & T ☐ B-Wht. & T ☐ B & W ☐ S ☐ S & W ☐
Coat: Curly ☒ Wavy ☐ Straight ☐
Natural Working Characteristics: Rough ☐ Ave. ☒ Mild ☐
Color-Blue Eyes-Brown
DAM Blue Star

Gr. Dam: Sisler's Trixie
Reg. No. A018-57178
Color B & T ☒ B-Wht. & T ☐ B & W ☐ S ☐ S & W ☐
Coat: Curly ☐ Wavy ☒ Straight ☐
Natural Working Characteristics: Rough ☒ Ave. ☐ Mild ☐
Eyes-Blue & Brown N.B.T.

SIRE Blue Khan
Reg. No. A044-6390
Color B & T ☐ B-Wht. & T ☐ B & W ☐ S ☐ S & W ☐
Coat: Curly ☐ Wavy ☐ Straight ☐
Natural Working Characteristics: Rough ☐ Ave. ☐ Mild ☐
Color-Blue Eyes-Brown

Reg. No. A040-6390
Blue
Color B & T ☐ B-Wht. & T ☐ B & W ☐ S ☐ S & W ☐
Coat: Curly ☐ Wavy ☐ Straight ☐
Natural Working Characteristics: Rough ☐ Ave. ☐ Mild ☐
Eyes-Blue
DAM Emmitt's Blue

Wood's Slate
Female
Color - Blue
Date of Birth 7/14/60
Eyes-Blue & Brown

Dam: Wood's Specks
Reg. No. A456-6512
Color B & T ☐ B-Wht. & T ☐ B & W ☒ S ☐ S & W ☐
Coat: Curly ☐ Wavy ☐ Straight ☐
Natural Working Characteristics: Rough ☐ Ave. ☐ Mild ☐
Eyes-Blue & Brown

Gr. Sire: Wood's Jay
Reg. No. A416-63272
Color B & T ☐ B-Wht. & T ☐ B & W ☐ S ☐ S & W ☐
Coat: Curly ☐ Wavy ☐ Straight ☐
Natural Working Characteristics: Rough ☐ Ave. ☐ Mild ☐
Color-Blue
Eyes-Brown

SIRE Sisler's Shorty
Reg. No. A019-5868
Color B & T ☐ B-Wht. & T ☐ B & W ☐ S ☐ S & W ☐
Coat: Curly ☐ Wavy ☐ Straight ☐
Natural Working Characteristics: Rough ☐ Ave. ☐ Mild ☐
Color-Blue, Gray
Eyes-Blue & Brown

Reg. No. A018-57178
Color B & T ☒ B-Wht. & T ☐ B & W ☐ S ☐ S & W ☐
Coat: Curly ☐ Wavy ☒ Straight ☐
Natural Working Characteristics: Rough ☒ Ave. ☐ Mild ☐
Eyes-Blue & Brown N.B.T.
DAM Sisler's Trixie

Gr. Dam: Trump's Stub
Reg. No. A0440-6512
Color B & T ☐ B-Wht. & T ☐ B & W ☐ S ☐ S & W ☐
Coat: Curly ☐ Wavy ☐ Straight ☐
Natural Working Characteristics: Rough ☐ Ave. ☐ Mild ☐
Color-Blue Eyes-Blue &Brown N.B.T.

SIRE Trump's Mike
Reg. No. A0424-6512
Color B & T ☐ B-Wht. & T ☐ B & W ☐ S ☐ S & W ☐
Coat: Curly ☐ Wavy ☐ Straight ☐
Natural Working Characteristics: Rough ☐ Ave. ☐ Mild ☐
Color-Blue N.B.T.
Eyes-Blue & Brown

Reg. No. A0425-6512
Color B & T ☐ B-Wht. & T ☐ B & W ☐ S ☐ S & W ☐
Coat: Curly ☐ Wavy ☐ Straight ☐
Natural Working Characteristics: Rough ☐ Ave. ☐ Mild ☐
Color-Blue N.B.T.
DAM Trump's Queen

No. assigned by Registrar A457-6512

National Stock Dog Registry pedigree of Wood's Slate. Courtesy of Mary Wood via Sorensen

Fletcher Wood's pick female of the cross between Wood's Jay and Trump's Stubbie, Wood's Speck, poses with her sire, Jay, and grandsire, Sisler's Shorty. Photo by Jay Sisler, courtesy of Mary Wood.

L to R: Wood's Specks, Wood's Flick, Wood's Slate. Courtesy of Mary Wood

Jeffries worked on a ranch in the Colorado mountains and at gathering time he enlisted the help of Wood and Sorensen. Wood brought Stubbie along to help and Dick remembers him as an aggressive heeling dog. In fact, when Stubbie hit the heels of a cow, the cow didn't stop until he reached the front of the herd! Stubbie drew the line when it came to working sheep. He hated them. This emotion may have emanated from Stubbie when he was in the presence of sheep. Jerry Rowe and Dick Sorensen recall the early days when everyone was trying to learn to work livestock with Aussies. Fletcher Wood was invited to an event their fledgling group was holding and he brought Stubbie. The dog was quick to demonstrate his dissatisfaction with these wooly excuses for livestock, managing to kill one sheep without even touching it.

Stubbie's guardian qualities were also in abundant supply and Dick Sorensen remembers an amusing situation one could find at the Wood home when visiting. Stubbie was tied to a log chain in the yard when not with Fletcher. The chain was long enough for him to guard nearly half the yard, including two sides of the house. By spending most of his time out of sight of the front entrance he could stealthily monitor the comings and goings. Most visitors could see the worn path in the yard which marked the limits of Stubbie's reach while chained. They learned to respect that boundary and wait at the edge for a Wood family member to arrive before proceeding to the house.

This knowledge deprived Stubbie of his sense of purpose regarding his guard duties around home. But his clever mind did not despair. Somehow he eventually discovered how to run to the end of the chain and flip his rear end just beyond the reach of the chain and his neck, using his considerable size and strength to knock the legs from under any intruder. Although perhaps not as satisfying to Stubbie as using his teeth on trespassers, he was still able to command the respect of interlopers! Wood's son, Kent, developed a good deal of competence at first aid for visitors. Even after Stubbie revised his techniques for protecting the homestead,

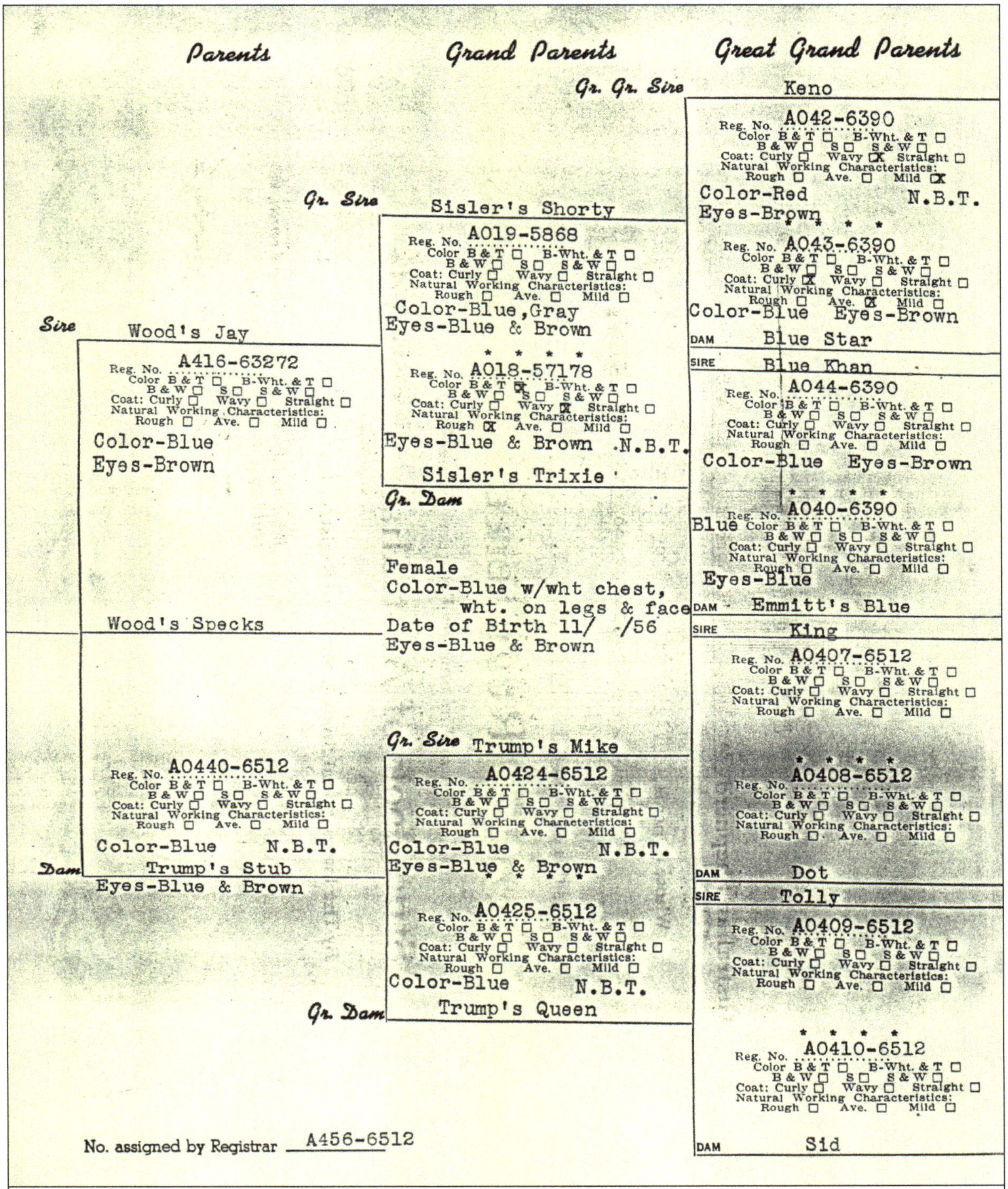

Parents | Grand Parents | Great Grand Parents

Sire: Wood's Jay
Reg. No. A416-63272
Color B & T ☐ B-Wht. & T ☐ B & W ☐ S ☐ S & W ☐
Coat: Curly ☐ Wavy ☐ Straight ☐
Natural Working Characteristics: Rough ☐ Ave. ☐ Mild ☐
Color-Blue
Eyes-Brown

Gr. Sire: Sisler's Shorty
Reg. No. A019-5868
Color B & T ☐ B-Wht. & T ☐ B & W ☐ S ☐ S & W ☐
Coat: Curly ☐ Wavy ☐ Straight ☐
Natural Working Characteristics: Rough ☐ Ave. ☐ Mild ☐
Color-Blue, Gray
Eyes-Blue & Brown

Gr. Gr. Sire: Keno
Reg. No. A042-6390
Color B & T ☐ B-Wht. & T ☐ B & W ☐ S ☐ S & W ☐
Coat: Curly ☐ Wavy ☒ Straight ☐
Natural Working Characteristics: Rough ☐ Ave. ☐ Mild ☒
Color-Red N.B.T.
Eyes-Brown

* * *

Reg. No. A043-6390
Color B & T ☐ B-Wht. & T ☐ B & W ☐ S ☐ S & W ☐
Coat: Curly ☒ Wavy ☐ Straight ☐
Natural Working Characteristics: Rough ☐ Ave. ☒ Mild ☐
Color-Blue Eyes-Brown
DAM Blue Star

* * * *

Gr. Dam: Sisler's Trixie
Reg. No. A018-57178
Color B & T ☒ B-Wht. & T ☐ B & W ☐ S ☐ S & W ☐
Coat: Curly ☐ Wavy ☒ Straight ☐
Natural Working Characteristics: Rough ☒ Ave. ☐ Mild ☐
Eyes-Blue & Brown .N.B.T.

SIRE Blue Khan
Reg. No. A044-6390
Color B & T ☐ B-Wht. & T ☐ B & W ☐ S ☐ S & W ☐
Coat: Curly ☐ Wavy ☐ Straight ☐
Natural Working Characteristics: Rough ☐ Ave. ☐ Mild ☐
Color-Blue Eyes-Brown

* * * *

Reg. No. A040-6390
Blue
Color B & T ☐ B-Wht. & T ☐ B & W ☐ S ☐ S & W ☐
Coat: Curly ☐ Wavy ☐ Straight ☐
Natural Working Characteristics: Rough ☐ Ave. ☐ Mild ☐
Eyes-Blue
DAM Emmitt's Blue

Wood's Specks
Female
Color-Blue w/wht chest, wht. on legs & face
Date of Birth 11/ /56
Eyes-Blue & Brown

Dam: Trump's Stub
Reg. No. A0440-6512
Color B & T ☐ B-Wht. & T ☐ B & W ☐ S ☐ S & W ☐
Coat: Curly ☐ Wavy ☐ Straight ☐
Natural Working Characteristics: Rough ☐ Ave. ☐ Mild ☐
Color-Blue N.B.T.
Eyes-Blue & Brown

Gr. Sire: Trump's Mike
Reg. No. A0424-6512
Color B & T ☐ B-Wht. & T ☐ B & W ☐ S ☐ S & W ☐
Coat: Curly ☐ Wavy ☐ Straight ☐
Natural Working Characteristics: Rough ☐ Ave. ☐ Mild ☐
Color-Blue N.B.T.
Eyes-Blue & Brown

SIRE King
Reg. No. A0407-6512
Color B & T ☐ B-Wht. & T ☐ B & W ☐ S ☐ S & W ☐
Coat: Curly ☐ Wavy ☐ Straight ☐
Natural Working Characteristics: Rough ☐ Ave. ☐ Mild ☐

* * * *

Reg. No. A0408-6512
Color B & T ☐ B-Wht. & T ☐ B & W ☐ S ☐ S & W ☐
Coat: Curly ☐ Wavy ☐ Straight ☐
Natural Working Characteristics: Rough ☐ Ave. ☐ Mild ☐
DAM Dot

* * * *

Gr. Dam: Trump's Queen
Reg. No. A0425-6512
Color B & T ☐ B-Wht. & T ☐ B & W ☐ S ☐ S & W ☐
Coat: Curly ☐ Wavy ☐ Straight ☐
Natural Working Characteristics: Rough ☐ Ave. ☐ Mild ☐
Color-Blue N.B.T.

SIRE Tolly
Reg. No. A0409-6512
Color B & T ☐ B-Wht. & T ☐ B & W ☐ S ☐ S & W ☐
Coat: Curly ☐ Wavy ☐ Straight ☐
Natural Working Characteristics: Rough ☐ Ave. ☐ Mild ☐

* * * *

Reg. No. A0410-6512
Color B & T ☐ B-Wht. & T ☐ B & W ☐ S ☐ S & W ☐
Coat: Curly ☐ Wavy ☐ Straight ☐
Natural Working Characteristics: Rough ☐ Ave. ☐ Mild ☐
DAM Sid

No. assigned by Registrar A456-6512

National Stock Dog Registry pedigree of Wood's Specks, Fletcher Wood's foundation female and granddam of Wood's Stubbie. Courtesy of Mary Wood via Sorensen

This is to Certify that the

Australian Shepherd (or Sheepdog) Female (Sex), Wood's Specks A456-6512 (Name of Dog and Reg. No.)

Whose owner is Fletcher H. & Mary H. Wood (Name)

Golden, Colorado R.3, Box 370 C (Address)

Was this day and date, May 1, 1962 duly

Registered

and its ancestry Recorded in the

Stud Book of Australian Shepherds

(or Sheepdogs)

By The International English Shepherd Registry, Inc., Butler, Ind., Rt. 1.

Believing these statements to be true and correct, we hereby affix our signature and seal.

E. G. Emanuel – Registrar

©GOES 463 LITHO IN U.S.A.

National Stock Dog registration for Wood's Specks. Courtesy of Mary Wood via Sorensen

young Kent would come racing from the house with Bactine® and Band-Aids® whenever he heard Stubbie sound the alarm!

Mary tells a similar story to that told about his famous sire when strangers come to his home when nobody was there. He, too, would not allow anyone to enter the home unchaperoned by taking hold of a person's hand if they touched the door knob. His home territory guarding skills notwithstanding, Dick says Fletcher would take Stubbie to work with him at the Jefferson County Fairgrounds where he was always perfectly behaved. This sort of discretion is a common hallmark of the mountain-type Aussie temperament.

Mary Wood remembers Stubbie almost as fondly as she remembers Wood's Jay. He was very similar in personality, as one would expect. He had quirks like any dog and one of those was a hatred of pajamas. When he was a toddler, the Wood's son Kent liked wearing his pajamas all the

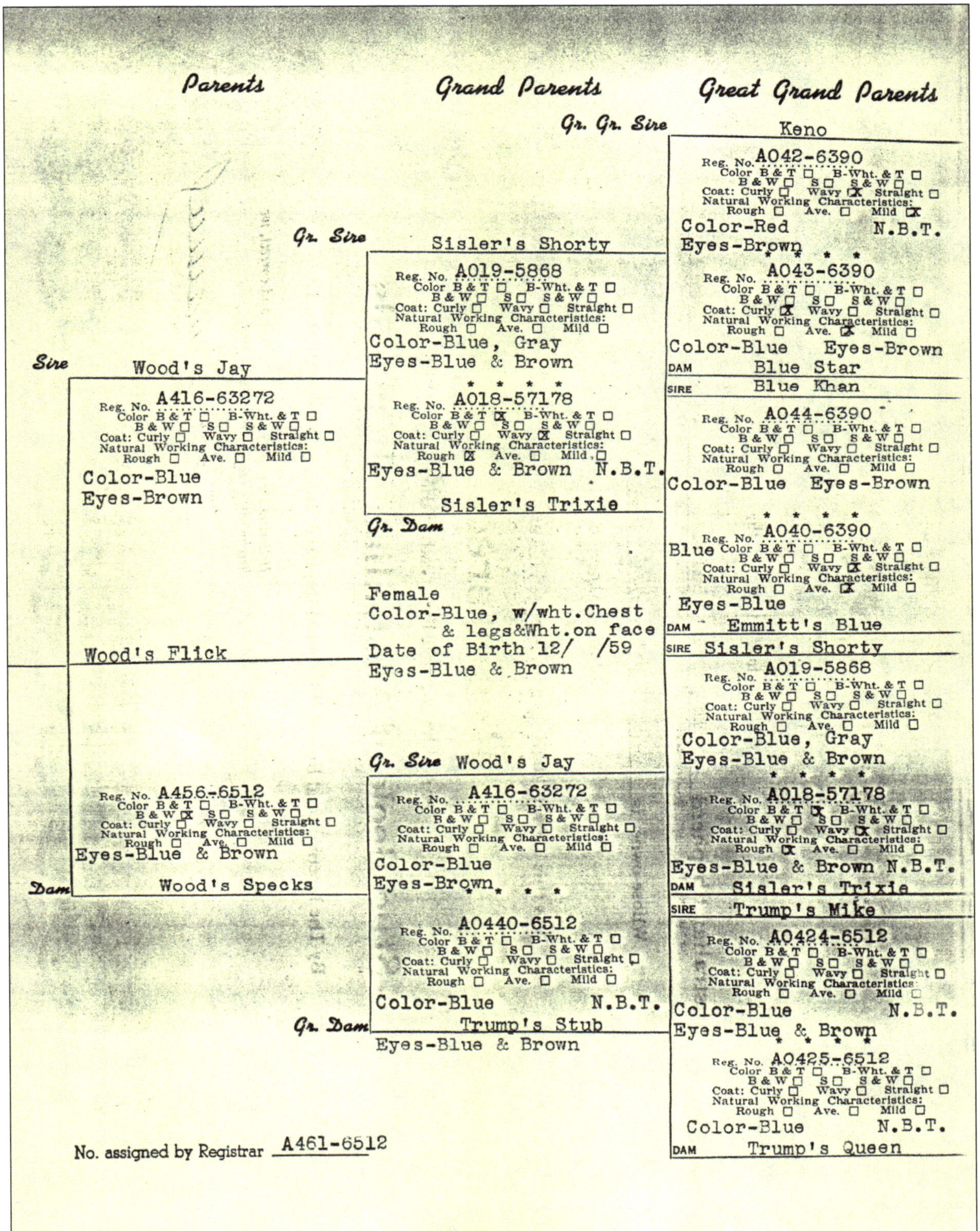

Parents | **Grand Parents** | **Great Grand Parents**

Sire: Wood's Jay
Reg. No. A416-63272
Color B & T ☐ B-Wht. & T ☐ B & W ☐ S ☐ S & W ☐
Coat: Curly ☐ Wavy ☐ Straight ☐
Natural Working Characteristics: Rough ☐ Ave. ☐ Mild ☐
Color-Blue
Eyes-Brown

Gr. Sire: Sisler's Shorty
Reg. No. A019-5868
Color B & T ☐ B-Wht. & T ☐ B & W ☐ S ☐ S & W ☐
Coat: Curly ☐ Wavy ☐ Straight ☐
Natural Working Characteristics: Rough ☐ Ave. ☐ Mild ☐
Color-Blue, Gray
Eyes-Blue & Brown

Gr. Gr. Sire: Keno
Reg. No. A042-6390
Color B & T ☐ B-Wht. & T ☐ B & W ☐ S ☐ S & W ☐
Coat: Curly ☐ Wavy ☒ Straight ☐
Natural Working Characteristics: Rough ☐ Ave. ☐ Mild ☒
Color-Red N.B.T.
Eyes-Brown

Reg. No. A043-6390
Color B & T ☐ B-Wht. & T ☐ B & W ☐ S ☐ S & W ☐
Coat: Curly ☒ Wavy ☐ Straight ☐
Natural Working Characteristics: Rough ☐ Ave. ☒ Mild ☐
Color-Blue Eyes-Brown
DAM Blue Star

Gr. Dam: Sisler's Trixie
Reg. No. A018-57178
Color B & T ☒ B-Wht. & T ☐ B & W ☐ S ☐ S & W ☐
Coat: Curly ☐ Wavy ☒ Straight ☐
Natural Working Characteristics: Rough ☒ Ave. ☐ Mild ☐
Eyes-Blue & Brown N.B.T.

SIRE Blue Khan
Reg. No. A044-6390
Color B & T ☐ B-Wht. & T ☐ B & W ☐ S ☐ S & W ☐
Coat: Curly ☐ Wavy ☐ Straight ☐
Natural Working Characteristics: Rough ☐ Ave. ☐ Mild ☐
Color-Blue Eyes-Brown

Reg. No. A040-6390
Blue
Color B & T ☐ B-Wht. & T ☐ B & W ☐ S ☐ S & W ☐
Coat: Curly ☐ Wavy ☒ Straight ☐
Natural Working Characteristics: Rough ☐ Ave. ☒ Mild ☐
Eyes-Blue
DAM Emmitt's Blue

Wood's Flick
Female
Color-Blue, w/wht.Chest & legs&Wht.on face
Date of Birth 12/ /59
Eyes-Blue & Brown

Dam: Wood's Specks
Reg. No. A456-6512
Color B & T ☐ B-Wht. & T ☐ B & W ☒ S ☐ S & W ☐
Coat: Curly ☐ Wavy ☐ Straight ☐
Natural Working Characteristics: Rough ☐ Ave. ☐ Mild ☐
Eyes-Blue & Brown

Gr. Sire: Wood's Jay
Reg. No. A416-63272
Color B & T ☐ B-Wht. & T ☐ B & W ☐ S ☐ S & W ☐
Coat: Curly ☐ Wavy ☐ Straight ☐
Natural Working Characteristics: Rough ☐ Ave. ☐ Mild ☐
Color-Blue
Eyes-Brown

SIRE Sisler's Shorty
Reg. No. A019-5868
Color B & T ☐ B-Wht. & T ☐ B & W ☐ S ☐ S & W ☐
Coat: Curly ☐ Wavy ☐ Straight ☐
Natural Working Characteristics: Rough ☐ Ave. ☐ Mild ☐
Color-Blue, Gray
Eyes-Blue & Brown

Reg. No. A018-57178
Color B & T ☒ B-Wht. & T ☐ B & W ☐ S ☐ S & W ☐
Coat: Curly ☐ Wavy ☒ Straight ☐
Natural Working Characteristics: Rough ☒ Ave. ☐ Mild ☐
Eyes-Blue & Brown N.B.T.
DAM Sisler's Trixie

Gr. Dam: Trump's Stub
Reg. No. A0440-6512
Color B & T ☐ B-Wht. & T ☐ B & W ☐ S ☐ S & W ☐
Coat: Curly ☐ Wavy ☐ Straight ☐
Natural Working Characteristics: Rough ☐ Ave. ☐ Mild ☐
Color-Blue N.B.T.
Eyes-Blue & Brown

SIRE Trump's Mike
Reg. No. A0424-6512
Color B & T ☐ B-Wht. & T ☐ B & W ☐ S ☐ S & W ☐
Coat: Curly ☐ Wavy ☐ Straight ☐
Natural Working Characteristics: Rough ☐ Ave. ☐ Mild ☐
Color-Blue N.B.T.
Eyes-Blue & Brown

Reg. No. A0425-6512
Color B & T ☐ B-Wht. & T ☐ B & W ☐ S ☐ S & W ☐
Coat: Curly ☐ Wavy ☐ Straight ☐
Natural Working Characteristics: Rough ☐ Ave. ☐ Mild ☐
Color-Blue N.B.T.
DAM Trump's Queen

No. assigned by Registrar A461-6512

National Stock Dog Registry pedigree for Wood's Flick—full sister to Wood's Slate. Courtesy of Mary Wood via Sorensen

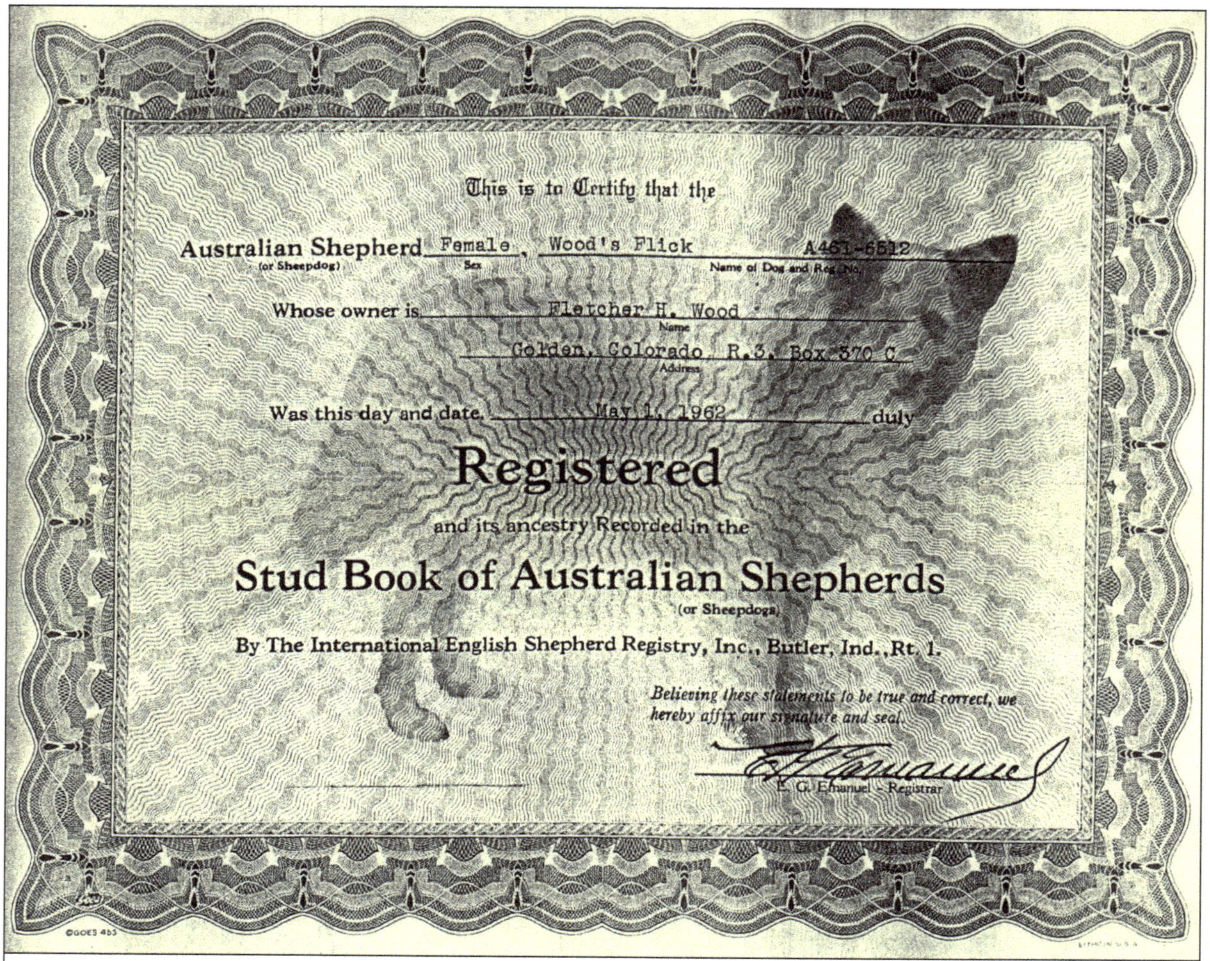

This is to Certify that the

Australian Shepherd (or Sheepdog) Female (Sex), Wood's Flick A461-6512 (Name of Dog and Reg. No.)

Whose owner is Fletcher H. Wood (Name)

Golden, Colorado R.3. Box 370 C (Address)

Was this day and date May 1, 1962 duly

Registered

and its ancestry Recorded in the

Stud Book of Australian Shepherds
(or Sheepdogs)

By The International English Shepherd Registry, Inc., Butler, Ind., Rt. 1.

Believing these statements to be true and correct, we hereby affix our signature and seal.

E. G. Emanuel - Registrar

National Stock Dog registration for Wood's Flick. Courtesy of Mary Wood via Sorensen

time so Stubbie hated Kent for awhile. He obviously outgrew this phase as he and Kent were eventually best of friends.

Dick Sorensen remembers Stubbie as a substantial dog with the broad, wedge-shaped head typical of his ancestors. His descriptions echo those of Les Burlile when describing Keno and Sisler's Shorty. A deep, dark, clear blue merle with only a spot of white on his chest, Stubbie is probably the dog most responsible for stamping the line for color.

It is as a stud dog that Wood's Stubbie will always be well remembered. He is the fountainhead for a number of kennels and was especially good at producing excellent sires in his offspring. A number of his sons became superb foundation studs for many, many kennels. Their stories live on today and follow in these pages.

The very dark Wood's Stubbie, a near clone of his sire, grandsire and great-grandsire, Wood's Jay. Courtesy of Mary Wood

Wood's Stubbie. Photo by Danite of Fort Collins, courtesy of Sorensen.

NOTE: THE SKETCH OF THE DOG ON THIS REGISTRY CERTIFICATE, WAS CHOSEN AND APPROVED BY THE BOARD OF DIRECTORS OF THE AUSTRALIAN SHEPHERD CLUB OF AMERICA, INC., TO BE REPRESENTATIVE OF THIS BREED, IN GENERAL, AND USED IN COMPLIANCE WITH THEIR REQUEST. AUGUST, 1962

Gr. Gr. Sire Keno
Reg. No. A042-6390 Unreg. No.
Natural Bob Tail ☒ Docked ☐
Color Eyes Blue ☐ Brown ☒ Left ..... Right.....
Color Blk. & White ☐ & Copper ☐ RED
Blue & White ☐ & Copper ☐
Blue Merle ☐ Red Merle ☐ Spotted ☐
Nat. Working Char. - Rough ☐ Average ☐ Mild ☐

Reg. No. A043-6390 Unreg. No.
Natural Bob Tail ☐ Docked ☐
Color Eyes Blue ☐ Brown ☒ Left ..... Right.....
Color Blk. & White ☐ & Copper ☐ BLUE
Blue & White ☐ & Copper ☐
Blue Merle ☐ Red Merle ☐ Spotted ☐
Nat. Working Char. - Rough ☐ Average ☒ Mild ☐

DAM Blue Star

SIRE Blue Khan
Reg. No. A044-6390 Unreg. No.
Natural Bob Tail ☐ Docked ☐
Color Eyes Blue ☐ Brown ☒ Left ..... Right.....
Color Blk. & White ☐ & Copper ☐ BLUE
Blue & White ☐ & Copper ☐
Blue Merle ☐ Red Merle ☐ Spotted ☐
Nat. Working Char. - Rough ☐ Average ☐ Mild ☐

Reg. No. A040-6390 Unreg. No.
Natural Bob Tail ☐ Docked ☐
Color Eyes Blue ☐ Brown ☐ Left ..... Right.....
Color Blk. & White ☐ & Copper ☐
Blue & White ☐ & Copper ☐
Blue Merle ☐ Red Merle ☐ Spotted ☐
Nat. Working Char. - Rough ☐ Average ☒ Mild ☐

DAM Emmett's Blue

SIRE Sisler's Shorty
Reg. No. A019-5868 Unreg. No.
Natural Bob Tail ☐ Docked ☐
Color Eyes Blue ☒ Brown ☒ Left ..... Right.....
Color Blk. & White ☐ & Copper ☐
Blue & White ☐ & Copper ☐ Blue Gray
Blue Merle ☐ Red Merle ☐ Spotted ☐
Nat. Working Char. - Rough ☒ Average ☐ Mild ☐

Reg. No. A018-57178 Unreg. No.
Natural Bob Tail ☒ Docked ☐
Color Eyes Blue ☒ Brown ☒ Left ..... Right.....
Color Blk. & White ☐ & Copper ☐
Blue & White ☐ & Copper ☒ & Blue
Blue Merle ☐ Red Merle ☐ Spotted ☐
Nat. Working Char. - Rough ☒ Average ☐ Mild ☐
Sisler's

DAM Australian Shepherd (Trixie)

SIRE Wood's Jay
Reg. No. A416-63272
Natural Bob Tail ☐ Docked ☐
Color Eyes Blue ☐ Brown ☒ Left ..... Right.....
Color Blk. & White ☐ & Copper ☐ BLUE
Blue & White ☐ & Copper ☐
Blue Merle ☐ Red Merle ☐ Spotted ☐
Nat. Working Char. - Rough ☒ Average ☐ Mild ☐

Reg. No. A0440-6512 Unreg. No.
Natural Bob Tail ☒ Docked ☐
Color Eyes Blue ☒ Brown ☒ Left ..... Right.....
Color Blk. & White ☐ & Copper ☐ BLUE
Blue & White ☐ & Copper ☐
Blue Merle ☐ Red Merle ☐ Spotted ☐
Nat. Working Char. - Rough ☐ Average ☐ Mild ☐

DAM Trump's Stub

Gr. Sire Sisler's Shorty
Reg. No. A019-5868 Unreg. No.
Natural Bob Tail ☐ Docked ☐
Color Eyes Blue ☒ Brown ☒ Left ..... Right.....
Color Blk. & White ☐ & Copper ☐ Blue Gray
Blue & White ☐ & Copper ☐
Blue Merle ☐ Red Merle ☐ Spotted ☐
Nat. Working Char. - Rough ☒ Average ☐ Mild ☐

Reg. No. A018-57178 Unreg. No.
Natural Bob Tail ☒ Docked ☐
Color Eyes Blue ☒ Brown ☒ Left ..... Right.....
Color Blk. & White ☐ & Copper ☐
Blue & White ☐ & Copper ☒ & Blue
Blue Merle ☐ Red Merle ☐ Spotted ☐
Nat. Working Char. - Rough ☒ Average ☐ Mild ☐
Sisler's
Australian Shepherd (Trixie)

Gr. Dam

Sire Wood's Jay
Reg. No. A416-63272
Natural Bob Tail ☐ Docked ☒
Color Eyes Blue ☐ Brown ☒ Left ..... Right.....
Color Blk. & White ☐ & Copper ☐ BLUE
Blue & White ☐ & Copper ☐
Blue Merle ☐ Red Merle ☐ Spotted ☐
Nat. Working Char. - Rough ☒ Average ☐ Mild ☐

Wood's Stubbie

Male
Color-Blue, White Chest,
Right Front Foot White
Date of Birth 3/16/63
Color - Brown
Not natural bob tail

Gr. Sire Wood's Jay
Reg. No. A416-63272
Natural Bob Tail ☐ Docked ☒
Color Eyes Blue ☐ Brown ☒ Left ..... Right.....
Color Blk. & White ☐ & Copper ☐
Blue & White ☐ & Copper ☐ BLUE
Blue Merle ☐ Red Merle ☐ Spotted ☐
Nat. Working Char. - Rough ☒ Average ☐ Mild ☐

Reg. No. A456-6512
Natural Bob Tail ☐ Docked ☐
Color Eyes Blue ☒ Brown ☒ Left ..... Right.....
Color Blk. & White ☐ & Copper ☐
Blue & White ☒ & Copper ☐
Blue Merle ☐ Red Merle ☐ Spotted ☐
Nat. Working Char. - Rough ☐ Average ☐ Mild ☐

Gr. Dam Wood's Specks

Reg. No. A457-63272
Natural Bob Tail ☐ Docked ☒
Color Eyes Blue ☒ Brown ☒ Left ..... Right.....
Color Blk. & White ☐ & Copper ☐ BLUE
Blue & White ☐ & Copper ☐
Blue Merle ☐ Red Merle ☐ Spotted ☐
Nat. Working Char. - Rough ☐ Average ☐ Mild ☐

Wood's Slate

Dam

No. assigned by Registrar A2688-61095

National Stock Dog Registry pedigree of Wood's Stubbie. Courtesy of Mary Wood via Sorensen

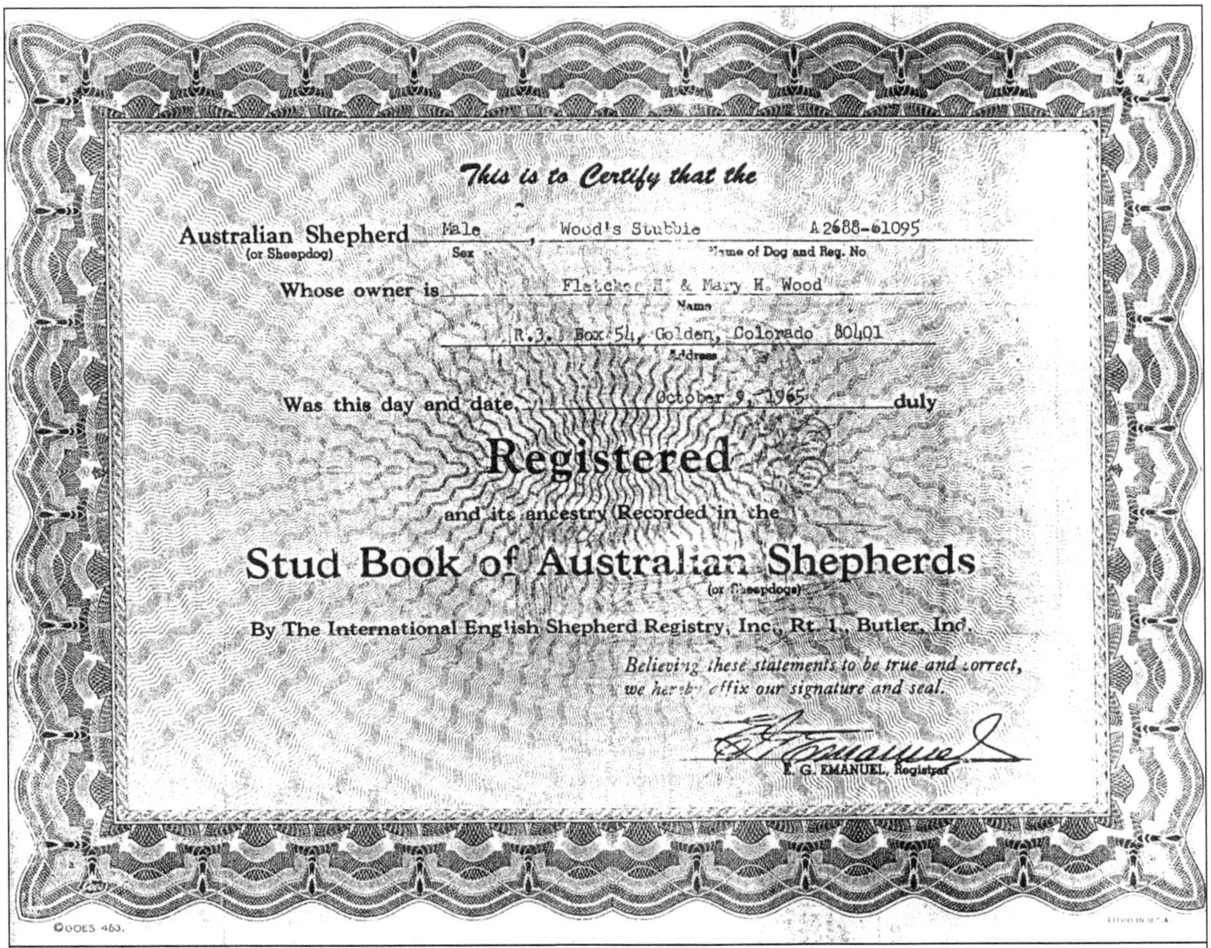

This is to Certify that the

Australian Shepherd Male , Wood's Stubbie A2688-61095
(or Sheepdog) Sex Name of Dog and Reg. No

Whose owner is Fletcher H. & Mary H. Wood
Name

R.3. Box 54, Golden, Colorado 80401
Address

Was this day and date October 9, 1965 **duly**

Registered

and its ancestry Recorded in the

Stud Book of Australian Shepherds
(or Sheepdogs)

By The International English Shepherd Registry, Inc., Rt. 1., Butler, Ind.

Believing these statements to be true and correct, we hereby affix our signature and seal.

E. G. EMANUEL, Registrar

National Stock Dog registration of Wood's Stubbie. Courtesy of Mary Wood via Sorensen

Hard working ranch dogs can also be beloved house pets. Kent Wood with (left) Sisler's Wood's Panda and (right) Wood's Stubbie. Courtesy of Mary Wood

Wood's Stubbie with his best pal, Kent Wood. Courtesy of Mary Wood

This dog is thought to be either Wood's Stubbie or another Wood-bred dog playing with one of the Wood daughters. Courtesy of Mary Wood

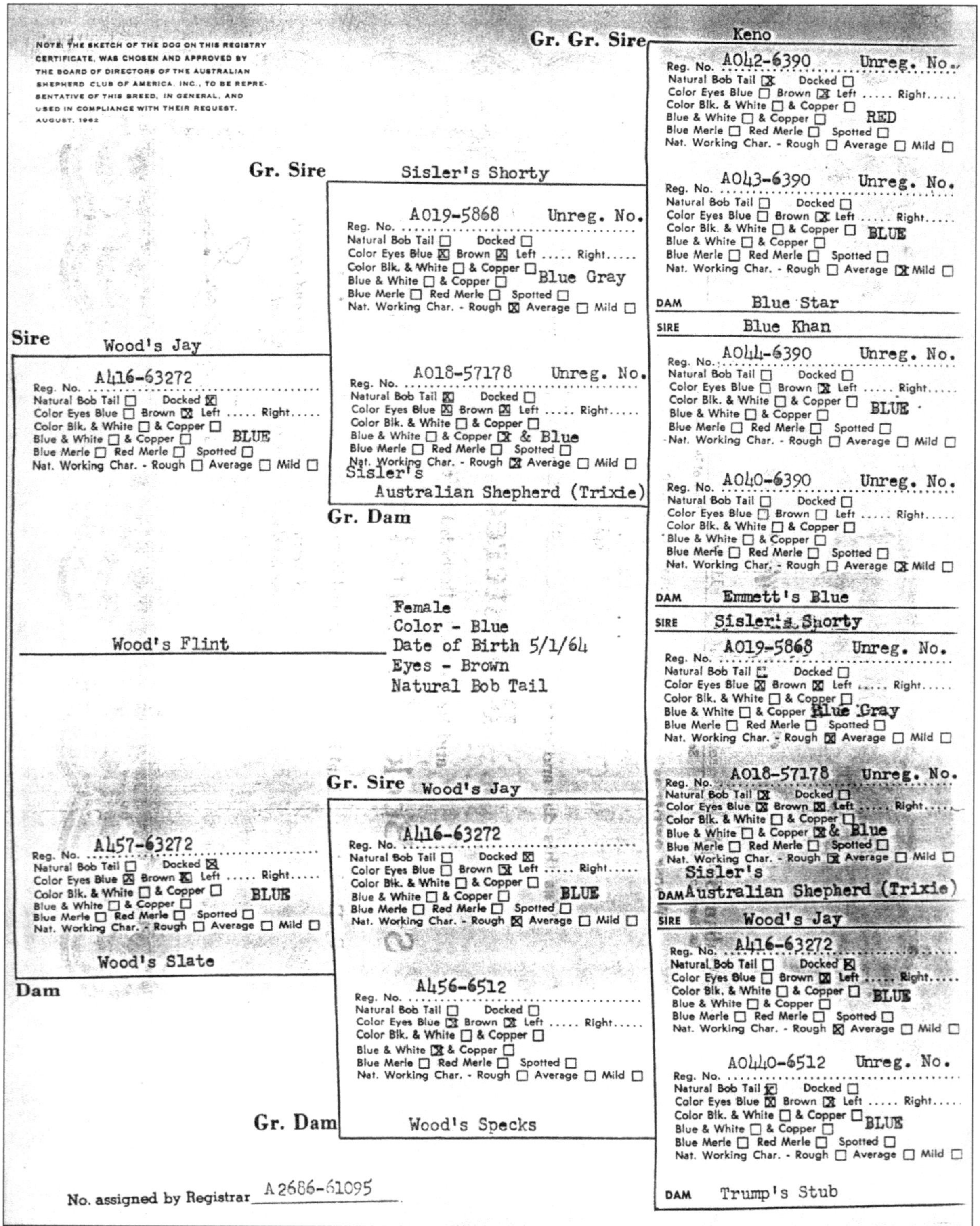

NOTE: THE SKETCH OF THE DOG ON THIS REGISTRY CERTIFICATE, WAS CHOSEN AND APPROVED BY THE BOARD OF DIRECTORS OF THE AUSTRALIAN SHEPHERD CLUB OF AMERICA, INC., TO BE REPRESENTATIVE OF THIS BREED, IN GENERAL, AND USED IN COMPLIANCE WITH THEIR REQUEST. AUGUST, 1962

Sire Wood's Jay
Reg. No. A416-63272
Natural Bob Tail ☐ Docked ☒
Color Eyes Blue ☐ Brown ☒ Left ..... Right.....
Color Blk. & White ☐ & Copper ☐
Blue & White ☐ & Copper ☐ BLUE
Blue Merle ☐ Red Merle ☐ Spotted ☐
Nat. Working Char. - Rough ☐ Average ☐ Mild ☐

Gr. Sire Sisler's Shorty
Reg. No. A019-5868 Unreg. No.
Natural Bob Tail ☐ Docked ☐
Color Eyes Blue ☒ Brown ☒ Left ..... Right.....
Color Blk. & White ☐ & Copper ☐ Blue Gray
Blue & White ☐ & Copper ☐
Blue Merle ☐ Red Merle ☐ Spotted ☐
Nat. Working Char. - Rough ☒ Average ☐ Mild ☐

Reg. No. A018-57178 Unreg. No.
Natural Bob Tail ☒ Docked ☐
Color Eyes Blue ☒ Brown ☒ Left ..... Right.....
Color Blk. & White ☐ & Copper ☐
Blue & White ☐ & Copper ☒ & Blue
Blue Merle ☐ Red Merle ☐ Spotted ☐
Nat. Working Char. - Rough ☒ Average ☐ Mild ☐
Sisler's Australian Shepherd (Trixie)
Gr. Dam

Gr. Gr. Sire Keno
Reg. No. A042-6390 Unreg. No.
Natural Bob Tail ☒ Docked ☐
Color Eyes Blue ☐ Brown ☒ Left ..... Right.....
Color Blk. & White ☐ & Copper ☐
Blue & White ☐ & Copper ☐ RED
Blue Merle ☐ Red Merle ☐ Spotted ☐
Nat. Working Char. - Rough ☐ Average ☐ Mild ☐

Reg. No. A043-6390 Unreg. No.
Natural Bob Tail ☐ Docked ☐
Color Eyes Blue ☐ Brown ☒ Left ..... Right.....
Color Blk. & White ☐ & Copper ☐ BLUE
Blue & White ☐ & Copper ☐
Blue Merle ☐ Red Merle ☐ Spotted ☐
Nat. Working Char. - Rough ☐ Average ☒ Mild ☐
DAM Blue Star

SIRE Blue Khan
Reg. No. A044-6390 Unreg. No.
Natural Bob Tail ☐ Docked ☐
Color Eyes Blue ☐ Brown ☒ Left ..... Right.....
Color Blk. & White ☐ & Copper ☐ BLUE
Blue & White ☐ & Copper ☐
Blue Merle ☐ Red Merle ☐ Spotted ☐
Nat. Working Char. - Rough ☐ Average ☐ Mild ☐

Reg. No. A040-6390 Unreg. No.
Natural Bob Tail ☐ Docked ☐
Color Eyes Blue ☐ Brown ☐ Left ..... Right.....
Color Blk. & White ☐ & Copper ☐
Blue & White ☐ & Copper ☐
Blue Merle ☐ Red Merle ☐ Spotted ☐
Nat. Working Char. - Rough ☐ Average ☒ Mild ☐
DAM Emmett's Blue

Wood's Flint

Female
Color - Blue
Date of Birth 5/1/64
Eyes - Brown
Natural Bob Tail

SIRE Sisler's Shorty
Reg. No. A019-5868 Unreg. No.
Natural Bob Tail ☐ Docked ☐
Color Eyes Blue ☒ Brown ☒ Left ..... Right.....
Color Blk. & White ☐ & Copper ☐
Blue & White ☐ & Copper Blue Gray
Blue Merle ☐ Red Merle ☐ Spotted ☐
Nat. Working Char. - Rough ☒ Average ☐ Mild ☐

Reg. No. A018-57178 Unreg. No.
Natural Bob Tail ☒ Docked ☐
Color Eyes Blue ☒ Brown ☒ Left ..... Right.....
Color Blk. & White ☐ & Copper ☐
Blue & White ☐ & Copper ☒ & Blue
Blue Merle ☐ Red Merle ☐ Spotted ☐
Nat. Working Char. - Rough ☒ Average ☐ Mild ☐
DAM Sisler's Australian Shepherd (Trixie)

Gr. Sire Wood's Jay
Reg. No. A416-63272
Natural Bob Tail ☐ Docked ☒
Color Eyes Blue ☐ Brown ☒ Left ..... Right.....
Color Blk. & White ☐ & Copper ☐
Blue & White ☐ & Copper ☐ BLUE
Blue Merle ☐ Red Merle ☐ Spotted ☐
Nat. Working Char. - Rough ☒ Average ☐ Mild ☐

Reg. No. A457-63272
Natural Bob Tail ☐ Docked ☒
Color Eyes Blue ☒ Brown ☒ Left ..... Right.....
Color Blk. & White ☐ & Copper ☐ BLUE
Blue & White ☐ & Copper ☐
Blue Merle ☐ Red Merle ☐ Spotted ☐
Nat. Working Char. - Rough ☐ Average ☐ Mild ☐
Wood's Slate
Dam

SIRE Wood's Jay
Reg. No. A416-63272
Natural Bob Tail ☐ Docked ☒
Color Eyes Blue ☐ Brown ☒ Left ..... Right.....
Color Blk. & White ☐ & Copper ☐ BLUE
Blue & White ☐ & Copper ☐
Blue Merle ☐ Red Merle ☐ Spotted ☐
Nat. Working Char. - Rough ☒ Average ☐ Mild ☐

Reg. No. A456-6512
Natural Bob Tail ☐ Docked ☐
Color Eyes Blue ☒ Brown ☒ Left ..... Right.....
Color Blk. & White ☐ & Copper ☐
Blue & White ☒ & Copper ☐
Blue Merle ☐ Red Merle ☐ Spotted ☐
Nat. Working Char. - Rough ☐ Average ☐ Mild ☐
Gr. Dam Wood's Specks

Reg. No. A0440-6512 Unreg. No.
Natural Bob Tail ☒ Docked ☐
Color Eyes Blue ☒ Brown ☒ Left ..... Right.....
Color Blk. & White ☐ & Copper ☐ BLUE
Blue & White ☐ & Copper ☐
Blue Merle ☐ Red Merle ☐ Spotted ☐
Nat. Working Char. - Rough ☐ Average ☐ Mild ☐
DAM Trump's Stub

No. assigned by Registrar A 2686-61095

National Stock Dog Registry pedigree for Wood's Flint, full sister to Wood's Stubbie. Courtesy of Mary Wood via Sorensen

This is to Certify that the

Australian Shepherd (or Sheepdog) Female (Sex), Wood's Flint A2686-61095 (Name of Dog and Reg. No.)

Whose owner is Fletcher H. & Mary M. Wood (Name)

R.3. Box 54, Golden, Colorado 80401 (Address)

Was this day and date, October 9, 1965 duly

Registered

and its ancestry Recorded in the

Stud Book of Australian Shepherds
(or Sheepdogs)

By The International English Shepherd Registry, Inc., Rt. 1., Butler, Ind.

Believing these statements to be true and correct, we hereby affix our signature and seal.

E. G. EMANUEL, Registrar

GOES 463. LITHO IN U.S.A.

National Stock Dog registration for Wood's Flint. Courtesy of Mary Wood via Sorensen

Born: ______ Sex: ______
Owned by: ______

PEDIGREE OF Wood's Stubbie

Parents	Grandparents	Great-grandparents	Great-great-grandparents
Sire: Woods Jay	Sire: Sislers Shorty	Sire: Keno	Sire
			Dam
		Dam: Blue Star	Sire
			Dam
	Dam: Sislers Trixie	Sire: Blue Khan	Sire
			Dam
		Dam: Emmetts Blue	Sire
			Dam
Dam: Woods Slate	Sire: Woods Jay	Sire: Sislers Shorty	Sire: Keno
			Dam: Blue Star
		Dam: Sislers Trixie	Sire: Blue Khan
			Dam: Emmetts Blue
	Dam: Woods Specks	Sire: Woods Jay	Sire: Sislers Shorty
			Dam: Sislers Trixie
		Dam: Trumps Stubbie	Sire
			Dam

Bred by: ______ Address: ______ Phone: ______

The details given above are correct to the best of my knowledge. Signed: ______(Breeder)

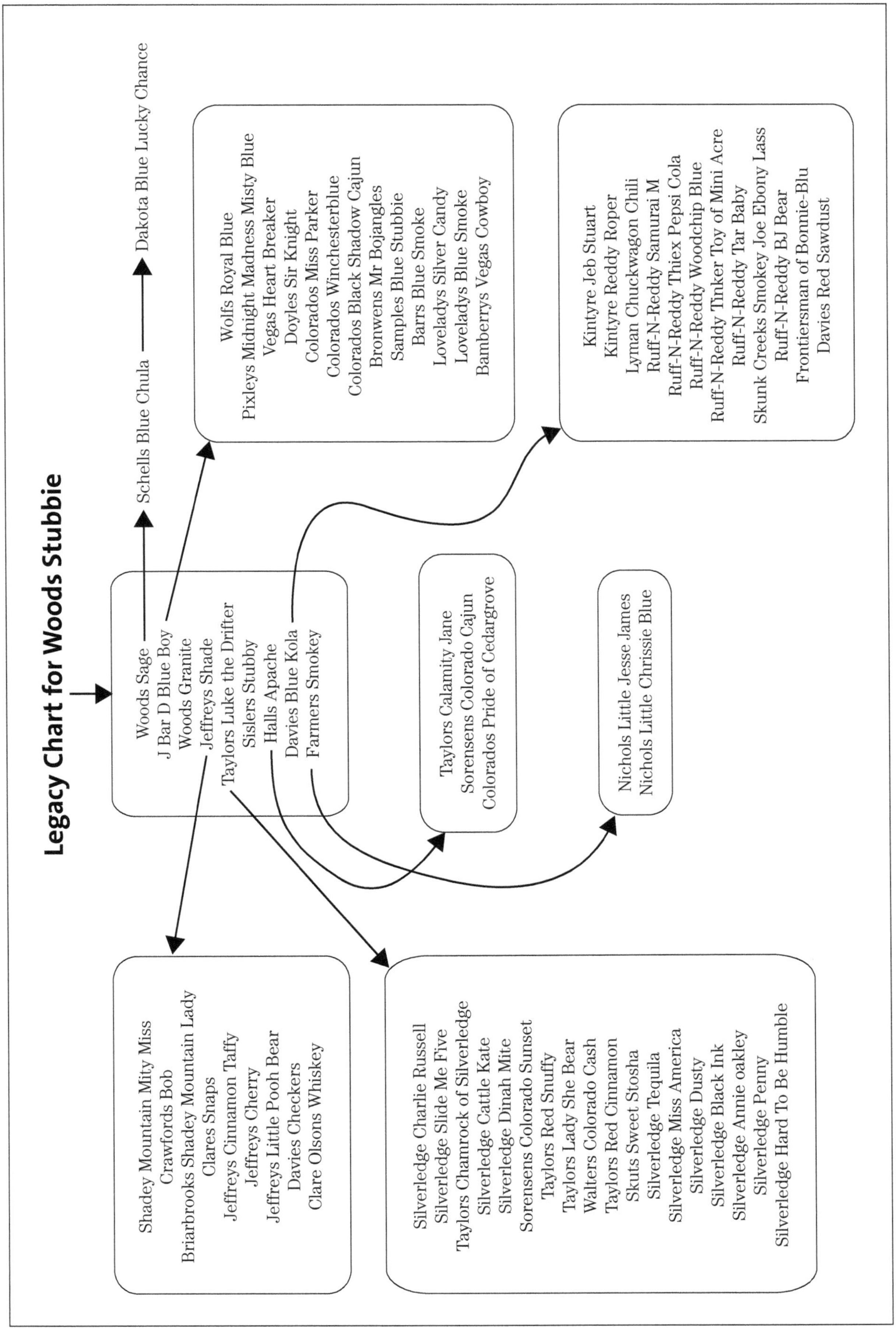
Legacy Chart for Woods Stubbie
Woods Sage
J Bar D Blue Boy
Woods Granite
Jeffreys Shade
Taylors Luke the Drifter
Sislers Stubby
Halls Apache
Davies Blue Kola
Farmers Smokey
Schells Blue Chula
Dakota Blue Lucky Chance
Wolfs Royal Blue
Pixleys Midnight Madness Misty Blue
Vegas Heart Breaker
Doyles Sir Knight
Colorados Miss Parker
Colorados Winchesterblue
Colorados Black Shadow Cajun
Bronwens Mr Bojangles
Samples Blue Stubbie
Barrs Blue Smoke
Loveladys Silver Candy
Loveladys Blue Smoke
Bamberrys Vegas Cowboy
Kintyre Jeb Stuart
Kintyre Reddy Roper
Lyman Chuckwagon Chili
Ruff-N-Reddy Samurai M
Ruff-N-Reddy Thiex Pepsi Cola
Ruff-N-Reddy Woodchip Blue
Ruff-N-Reddy Tinker Toy of Mini Acre
Ruff-N-Reddy Tar Baby
Skunk Creeks Smokey Joe Ebony Lass
Ruff-N-Reddy BJ Bear
Frontiersman of Bonnie-Blu
Davies Red Sawdust
Taylors Calamity Jane
Sorensens Colorado Cajun
Colorados Pride of Cedargrove
Nichols Little Jesse James
Nichols Little Chrissie Blue
Shadey Mountain Mity Miss
Crawfords Bob
Briarbrooks Shadey Mountain Lady
Clares Snaps
Jeffreys Cinnamon Taffy
Jeffreys Cherry
Jeffreys Little Pooh Bear
Davies Checkers
Clare Olsons Whiskey
Silverledge Charlie Russell
Silverledge Slide Me Five
Taylors Chamrock of Silverledge
Silverledge Cattle Kate
Silverledge Dinah Mite
Sorensens Colorado Sunset
Taylors Red Snuffy
Taylors Lady She Bear
Walters Colorado Cash
Taylors Red Cinnamon
Skuts Sweet Stosha
Silverledge Tequila
Silverledge Miss America
Silverledge Dusty
Silverledge Black Ink
Silverledge Annie oakley
Silverledge Penny
Silverledge Hard To Be Humble

Davies' Blue Kola, blue dark blue merle and white female by Wood's Stubbie x Davies' Checkers (Davies' Checkers: Jeffreys' Shade x Wood's Little Kola), Dam of Ch Kintyre Jeb Stuart and a legion of other great dogs. Courtesy of Sorensen

The legacy of a top breeder—consistency, prepotency, and lasting quality. From L to R: Charlene Jeffreys with Jeffreys' Shade, Cathy Jones with Davies' Blue Kola, Patty Ennen with Piz's Sheba, and Sue Foster with Clare Olson's Blue Joe. Courtesy of Sorensen

chapter 12

Tom and Charlene Jeffreys
SHADY MOUNTAIN

An influential branch of the Wood's Jay family—through Stubbie—was produced by Tom and Charlene Jeffreys who, in the beginning, used their last name and later used the kennel name, Shady Mountain, in naming their dogs. Their Jeffreys' Shade was a son of Wood's Stubbie out of Jeffreys' Missy Blue—a Williams' Laddie daughter and half sister to Sorensen's Gunsmoke. The legacy of Gunsmoke as a potent and treasured cross with the Wood's dogs will be covered in greater detail in the chapter on the J Bar D, Sorensen and Colorado dogs.

Tom Jeffreys with the Shady Mountain foundation dogs (L to R: Jeffreys' Sheila, Jeffreys' Missy Blue, and Jeffreys' Shade). Courtesy of Sorensen

The first Aussie owned by Tom and Charlene was named Jeffreys' Dead Eye Jr. and called "Ted" by his family. This blue merle male was obtained from Jeanne Farned and was by her dog Farned's Dead Eye. Farned's Dead Eye was the constant companion of Jeanne and knew hundreds of words and tricks. A favorite story is of a time Jeanne was putting new shingles on her roof. Dead Eye would wait on the ground for Jeanne's instructions on

Jeffreys' Dead Eye Jr. riding Tom Jeffreys' good mule, Rusty. Courtesy of Sorensen

Jeffreys' Sheila, foundation matriarch of the Jeffreys' Shady Mountain kennel. Courtesy of Melissa Couch

what tools she needed. Dead Eye would pick up the correct tool, climb the ladder and deliver it to Jeanne. He would climb down the ladder again and wait for more instructions from Jeanne! Dead Eye was also influential in the J Bar D foundations as the grandsire (through his daughter McGuire's Palojami Gold) of the Rowe's foundation bitch, Rowe's Dingo.

The second good Aussie acquired by the Jeffreys' was a bitch named Jeffreys' Sheila. Dick remembers Sheila as a fairly large bitch with blue eyes. She was a blue merle with a white head, although she was colored over her eyes and ears. Sheila's greatest legacy is through her son, Sorensen's Gunsmoke. She also produced a litter by the magnificent Jeffreys' Shade but little is known about these puppies and their fate.

Before his partnership with Jerry and Sharon Rowe and his marriage to Leslie, Dick Sorensen was a close friend of the Jeffreys. His founding sire, Sorensen's Gunsmoke, was bred by the Jeffreys and he remembers their dogs as highly intelligent. Sheila was no exception to this rule. In addition to their brains, the Jeffreys' dogs were also very athletic. Dick remembers watching Sheila as she rode in the back of Tom's pickup while he was driving. She was so nimble she could walk along the edge of the moving truck's bed like an Olympic gymnast—never losing her balance or falling off.

Sheila was assigned the task of ranch watchdog during the day. She took the job very seriously and developed many effective tactics for carrying out her duties. The REA meter reader once arrived after dark to take a reading. At that time the meters were up on a pole. This public servant found himself in a difficult situation when he tried to climb down from his perch. Sheila had decided he was up to no good. During his six to eight hour captivity he tried many ploys to escape. However, Sheila had other ideas and spent the time toying with the man. She would pretend to leave to lure him down the pole then wait for the perfect moment to send him back to a safe height. He was only freed when someone returned to the ranch to confine her! When the whole pack of dogs was loose on guard duty, intruders rarely made it out of their vehicles. Several would jump onto the hood and bark through the windshield while one waited underneath to grab the foot of anyone foolish enough to try to exit.

Another use of the Jeffreys' dogs involved removing vermin from high country barns. Hippies were invading the Colorado mountains and would find a hay barn in the mountains and move in. Unfortunately their activities often resulted in barn fires and dead cattle. When a new colony was discovered, Jeffreys, Dick, and sometimes joined by Jerry, would wait until late at night, then load up their dogs. They would drive to the infested barn and release their dogs into the darkness. The dogs apparently worked silently since the first sounds to pierce the darkness would be screams of pain when the dogs found their targets. Following a successful "hippie hunt," a visit to the local bar was made in honor of the canine heroes—who were allowed inside to accept their praise!

A third foundation bitch owned by the Jeffreys was called "Dee" but no record exists of any breedings from her.

Jeffreys' Sheila. Courtesy of Melissa Couch

L to R: Jeffreys' Sheila, Jeffreys' "Dee," Porter's Yakima Blue Squaw, Rowe's Dingo. Courtesy of Rowe

L to R: Jeffreys' Sheila, Jeffreys' "Dee," Porter's Yakima Blue Squaw. Courtesy of Rowe

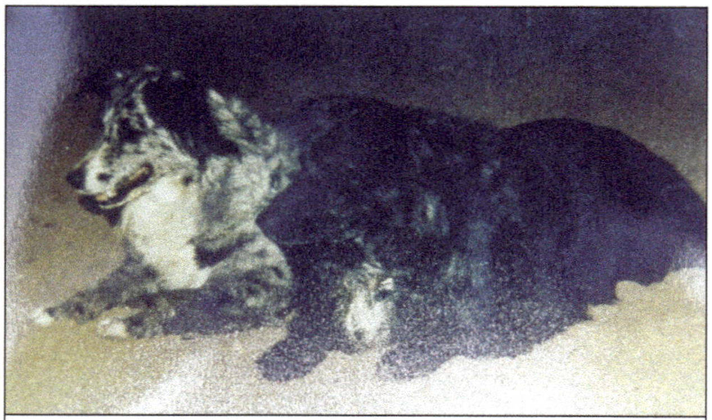

The beautiful Jeffreys' Missy Blue (left) with her most famous son, Jeffreys' Shade (right). Courtesy of Melissa Couch

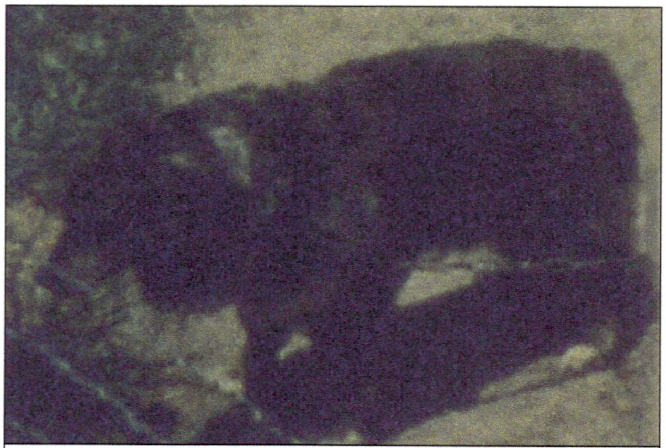

Williams' Laddie, influential sire of this era. Laddie was typical of the mountain-type dogs, highly intelligent, a prepotent producer and exhibiting the look common in these dogs. Courtesy of Melissa Couch

She was probably purchased from the Thorntons—early breeders whose dogs are behind many of the original mountain-type Aussies.

The next important female in the Shady Mountain line is the lovely blue merle bitch, Jeffreys' Missy Blue. As history would later show in hindsight, her bloodlines were impeccable. Her dam was Picraux's Juliet, a puppy from the Jeffreys' first male, Dead Eye Jr and a half sister to Sheila through their dam Williams Georgina. Georgina traced back to Jay Sisler's original dogs through Ely's Blue and was also a granddaughter of the Green's male, Greens (Manskers) Kim. Missy was a half sister to Sorensen's Gunsmoke through their sire, Williams Laddie.

Laddie was a well known sire in the area in the early 60s and is behind many of the original dogs of this era. Missy and Gunsmoke are probably the two most significant dogs he produced, but he is also the sire of Shady Mountain Flirt, Van Pelts Tillie and he was a full brother to the strikingly beautiful Malone's Lotsa Freckles. Flirt was an especially valuable female since she was from the cross of Laddie back to his granddaughter Shady Mountain Lady. Van Pelts Tillie produced a number of sons for the Ely kennel which in turn had descendants such as Cornwell's Cody. Malone's Lotsa Freckles lived in Arizona and was influential in founding numerous lines in that state, including Cattle Call, Hazelwood, Ironwood, and others.

Laddie reproduced himself in Sorensen's Gunsmoke and Dick remembers many similarities between the two. In appearance, Laddie was a typical dark brown eyed, silver blue merle without many white markings—even less than Gunsmoke. He had a very intelligent and easily trained mind and could be trained by anyone. He also displayed a quirky trait found in many dogs of these lines. Very friendly to nearly every person he met, he was a persistent beggar for petting. However, if his benefactor decided enough was enough and withdrew their affections,

Laddie would make a quick nip to tell them he thought he needed a little more! This sometimes startling trait is still seen in his distant descendants today.

Jeffreys' Shade, there is no mistaking his Wood background—color, type, size, temperament, intelligence, and sound genetics. This dog had it all. Courtesy of Sorensen

Jeffreys' Missy Blue was bred to the extremely inbred Wood's Stubbie to produce Jeffreys' Shade. Jeffreys' Shade would likely qualify as the Jeffreys' foundation sire. He was the typical Wood type dog—large, very dark blue merle with little white and no copper markings. Missy was also bred to Gunsmoke to produce the Sorensen's treasured bitch J Bar D Flirt—who later produced a dynasty of outstanding dogs for the Colorado kennel. Not much else is known about Missy. By virtue of Shade and Flirt she has secured her place in the history of the mountain Aussie.

Jeffreys' Shade, 1978. Courtesy of Sorensen

By the time Jeffreys' Shade was born, the growth of the Aussie as a purebred breed was beginning to take off. Through the popularity of Jay Sisler's acts and the breeding acumen of Fletcher Wood, a solid foundation for the future had been laid. Everybody taught their dogs tricks during this time due to the wide influence of Jay Sisler and his brilliant dogs. Dogs were man's constant companion on the ranches and in the community with ample time and opportunity to pick up all kinds of clever stunts. The Jeffreys' dogs would stay with a coat laid on the ground until they were told they could leave. All of the dogs were taught not to take food from anyone. Parlor tricks abounded and served to attract the public when infant Aussie clubs set up booths at community events.

Shade was a stunning example of the soundness of his ancestors. A large dog in the tradition of the Sisler and Wood dogs, he was around 22-23 inches tall with the large, square head which stamped this line all the way back to the Sisler foundation sire, Keno. He had the midnight blue color which trademarked Fletcher Wood's dogs and genetics which

PEDIGREE OF Jeffreys Shade

Born: ____________ Sex: ________
Owned by: ____________________

<table>
<tr><td rowspan="8">Sire
Woods Stubbie</td><td rowspan="4">Sire
Woods Jay</td><td rowspan="2">Sire
Sislers Shorty</td><td>Sire Keno</td></tr>
<tr><td>Dam Blue Star</td></tr>
<tr><td rowspan="2">Dam
Sislers Trixie</td><td>Sire Blue Khan</td></tr>
<tr><td>Dam Emmetts Blue</td></tr>
<tr><td rowspan="4">Dam
Woods Slate</td><td rowspan="2">Sire
Woods Jay</td><td>Sire Sislers Shorty</td></tr>
<tr><td>Dam Sislers Trixie</td></tr>
<tr><td rowspan="2">Dam
Woods Specks</td><td>Sire Woods Jay</td></tr>
<tr><td>Dam Trumps Stubbie</td></tr>
<tr><td rowspan="8">Dam
Jeffreys Missy Blue</td><td rowspan="4">Sire
Williams Laddie</td><td rowspan="2">Sire
Uhleudorfs Tippy</td><td>Sire Parks Joe</td></tr>
<tr><td>Dam Uhlendorfs Linda</td></tr>
<tr><td rowspan="2">Dam
Thorntons Chrissie</td><td>Sire Watsons Old Blue</td></tr>
<tr><td>Dam Johnsons Sue</td></tr>
<tr><td rowspan="4">Dam
Picrauxs Juliette</td><td rowspan="2">Sire
Jeffreys Dead Eye Jr</td><td>Sire Farneds Dead Eye</td></tr>
<tr><td>Dam Hoopers Blue Keno</td></tr>
<tr><td rowspan="2">Dam
Williams Georgina</td><td>Sire Elys Joker</td></tr>
<tr><td>Dam Barretts Blue Belle</td></tr>
</table>

Bred by: ____________ Address: ____________________ Phone: ____________

The details given above are correct to the best of my knowledge. Signed: ____________________(Breeder)

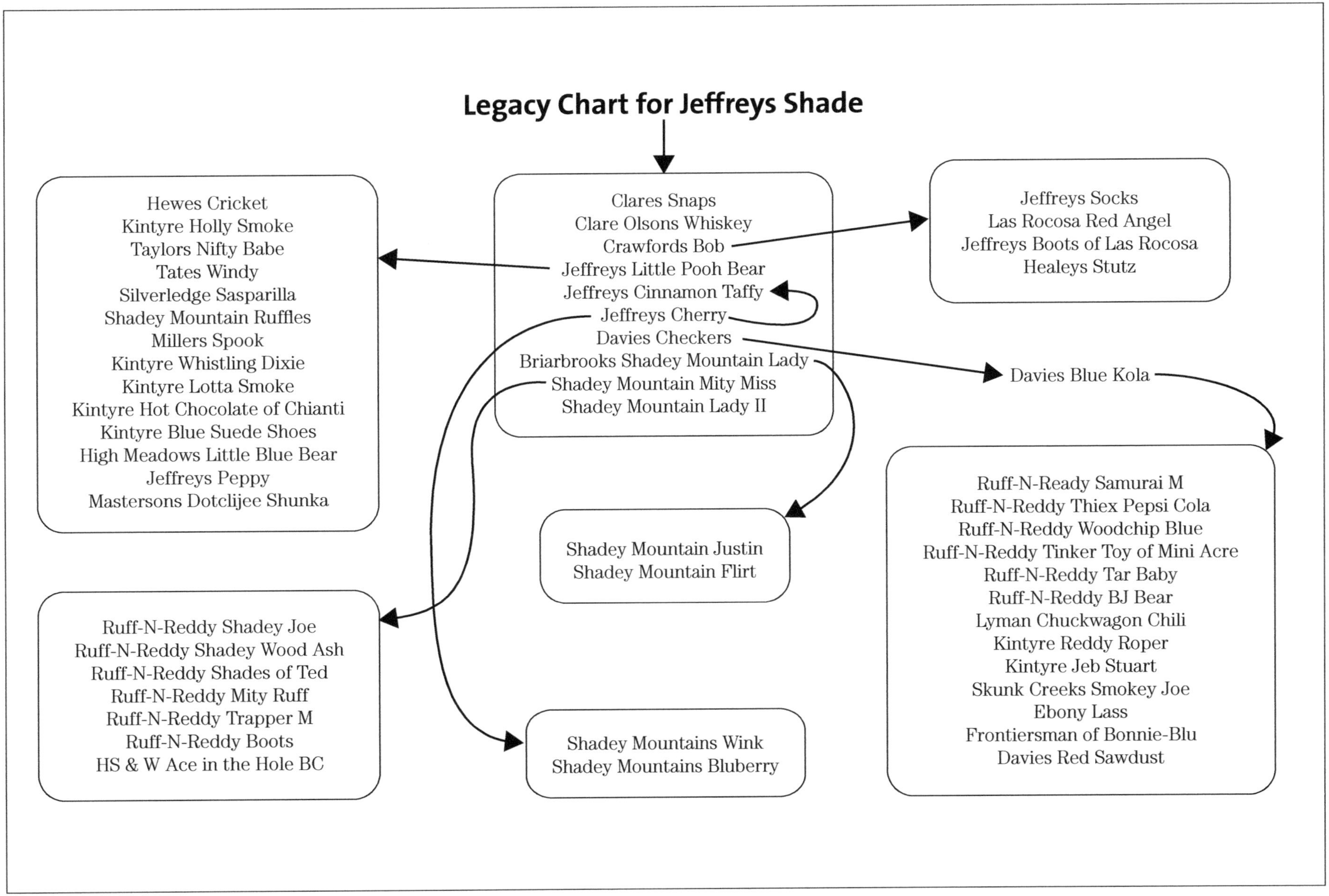
Legacy Chart for Jeffreys Shade
Clares Snaps
Clare Olsons Whiskey
Crawfords Bob
Jeffreys Little Pooh Bear
Jeffreys Cinnamon Taffy
Jeffreys Cherry
Davies Checkers
Briarbrooks Shadey Mountain Lady
Shadey Mountain Mity Miss
Shadey Mountain Lady II
Hewes Cricket
Kintyre Holly Smoke
Taylors Nifty Babe
Tates Windy
Silverledge Sasparilla
Shadey Mountain Ruffles
Millers Spook
Kintyre Whistling Dixie
Kintyre Lotta Smoke
Kintyre Hot Chocolate of Chianti
Kintyre Blue Suede Shoes
High Meadows Little Blue Bear
Jeffreys Peppy
Mastersons Dotclijee Shunka
Jeffreys Socks
Las Rocosa Red Angel
Jeffreys Boots of Las Rocosa
Healeys Stutz
Davies Blue Kola
Ruff-N-Ready Samurai M
Ruff-N-Reddy Thiex Pepsi Cola
Ruff-N-Reddy Woodchip Blue
Ruff-N-Reddy Tinker Toy of Mini Acre
Ruff-N-Reddy Tar Baby
Ruff-N-Reddy BJ Bear
Lyman Chuckwagon Chili
Kintyre Reddy Roper
Kintyre Jeb Stuart
Skunk Creeks Smokey Joe
Ebony Lass
Frontiersman of Bonnie-Blu
Davies Red Sawdust
Shadey Mountain Justin
Shadey Mountain Flirt
Ruff-N-Reddy Shadey Joe
Ruff-N-Reddy Shadey Wood Ash
Ruff-N-Reddy Shades of Ted
Ruff-N-Reddy Mity Ruff
Ruff-N-Reddy Trapper M
Ruff-N-Reddy Boots
HS & W Ace in the Hole BC
Shadey Mountains Wink
Shadey Mountains Bluberry

Jeffreys' Shade at a gathering of Aussie lovers.
Courtesy of Melissa Couch

Jeffreys' Shade on a day in the mountains.
Courtesy of Raymie Jones

had been forged by necessity and skillful selection. Shade was used on the ranch just as all the other dogs were but his skills garnered him at least one accolade. He was one of only a few dogs invited to the very first Aussie stockdog trial organized by Bob Carrillo in Santa Rosa, California. He joined his relatives Colorado's Cherry Brandy and J Bar D Comanche Warrior at this historical event.

Shade's descendants are a Who's Who of the best dogs of that era and today. If Shade had only produced his son out of Sorensen's Blue Mist named Jeffreys' Little Pooh Bear, his place in history would have been secure. But like all outstanding sires, he left an even larger legacy. His outstanding daughter Shady Mountain Mity Miss left an incredible legacy of her own—especially for a bitch. His daughter, Shady Mountain Lady was mentioned earlier as the granddaughter of Williams Laddie who was bred back to Laddie to produce the exquisite female, Shady Mountain Flirt. Daughter Jeffreys' Cherry was a key female in the red Wood's bloodline developed by Joe Piz and Limon Davies. A blue daughter, Davies Checkers, was another important link in the Piz/Davies program—bringing back a close cross to Wood's Jay through her dam, the Jay daughter Wood's Little Kola. Like many of the dogs of that time, it is likely many other dogs produced by Shade were used without any record of the event. We will never know exactly how far-reaching the effect of this grand monarch really was. The heritage we have recorded stands second to none in quality, importance and influence.

The pedigrees of the Shady Mountain dogs reflect the small, tight-knit community that was the Colorado Aussie world during the 1960s and '70s. The very close breeding on Wood's Jay was continued through Shade crossed on dogs of Joe Piz's breeding (which were all closely line and inbred on Wood's Jay) sprinkled liberally with crosses to Sorensen's Gunsmoke and the early J Bar D dogs.

Jeffreys' Little Pooh Bear was the image of his sire. Not only pick of his litter but also the only survivor of this cross, Pooh Bear carried on for Shade the way Stubbie carried on for Jay. The strength of the Wood breeding continued to come through and was, perhaps, strengthened even more by the combination of Shade with Pooh Bear's dam Sorensen's Blue Mist. Misty was a daughter of Sorensen's Gunsmoke and Rowe's Dingo—the first dogs of the J Bar D legacy. Gunsmoke and Shade's dam, Missy, were half brother and sister by Williams' Laddie. This was another of the magic cross of Gunsmoke with Wood's line dogs which has proven over the decades to be one of the best for the mountain-type Aussies. Pooh Bear was an outstanding sire who contributed greatly to Tom and Jean Taylor's Silverledge line, Casey Masterson Newman's HS & W line. and many others.

Jeffreys' Shade in 1978. Courtesy of Sorensen

Jeffreys' Cinnamon Bear was not well known in her own right but contributed greatly to the background of many influential dogs through her red merle son Jeffreys' Cinnamon Man and her blue merle great-granddaughter Shady Mountain Mity Miss. Cinnamon Bear was from the cross which combined the inbreeding of Joe Piz's red dogs with an inbred Sorensen's Gunsmoke sire. Another instance of the golden combination which produced so many top dogs of the mountain-type, Montoya's Rowdy, inbred on a Sorensen's Gunsmoke son, was crossed to Montoya's Snipper, inbred on Piz Joseph. Snipper was a full sister to Davies' Ginger Red, the dam of Ch. Jones Reddy Teddy CDX OTD. Cinnamon Bear's most influential son was the red merle Jeffreys' Cinnamon Man. He was sired by the inbred Piz Joseph male Beien's Red Tuffy (Piz Little Joe x Piz Lady II—both Joseph puppies). This cross again concentrated the Wood's background.

The Sorensens recall another of the excellent red dogs produced from Shady Mountain dogs named Johnson's Quiller. Owned by Richard and Robin Johnson she is described by Leslie as a "dark Hershey bar red." This intense liver pigment is a hallmark of the Wood bred reds. Shady Mountain (Jeffreys) Hobo Joe was Quiller's sire, a red merle dog closely bred to the Piz dogs and tracing back to Jeffreys' Shade. Her dam was Johnson's Gretchen, another good daughter of Jeffreys' Cinnamon

Jeffreys' Shade and Tom Jeffreys. Courtesy of Sorensen

The finest Shade son, Jeffreys' Pooh Bear. Courtesy of Raymie Jones

Bear and a half sister to Hobo Joe by their sire Beiens Red Tuffy (inbred Piz).

Shady Mountain Mity Miss is possibly the most important and unheralded female produced by the Jeffreys'. A beautiful silvery blue merle with a sweet, kind temperament, she was bred to a number of different males in her lifetime and produced exceptional individuals every time. Missy was a daughter of Jeffreys' Shade. Her dam was owned by Cathy Jones and named Jones Mity Bite. Mity Bite was a daughter of Jeffreys' Cinnamon Man and from another Piz dog named Sheba. The Cinnamon Man x Sheba cross again re-concentrated the Piz breeding as Sheba was inbred on Piz Joseph, being by him out of his daughter Piz Lady II. Sheba was a full sister to another well known sire named Olson's (Clare's) Blue Joe.

Mity Miss had the concentrated Wood's breeding which made so many of these dogs such excellent producers. One daughter, Ruff N Reddy Boots was the result of Missy bred back to her grandson Jones Little Redwood. Boots was the dam of the Sorensen's Ruff N Reddy Ross, who was a striking silver blue merle like his granddam. Boots' full sister Ruff N Reddy Shadey Wood Ash was a top producer for dairyman Al Combee in Florida when bred to the Reddy Teddy son Ruff N Reddy Joe Willy Wang. Two sons from this cross have been top sires for Roger Steven's Pincie Creek kennel. Another sister from the cross of Mity Miss to her grandson Redwood produced another top bitch, Ruff N Reddy Shades of Ted.

Mity Miss was bred to Jones Reddy Teddy and produced another great sire in the red merle, Ruff N Reddy Shadey Joe, known as "Clank." Clank was bred to Red Oliver's bitch, Slash V Slide Me Sweet—a heavily linebred Wood's bitch—to produce a litter of nice working dogs.

Jeffreys' Shade (far left), Cathy Jones, Davies' Blue Kola, Patty Ennen, Piz's Sheba, Sue Foster with Clare Olson's Blue Joe. Courtesy of Sorensen

Missy's last home was with Casey Masterson Newman. Casey was able to breed Missy to her beloved Newman's Blue Sprog for one final litter. Only one puppy survived to adulthood, the solid black male, HS & W Ace in the Hole BC, CD. Ace has produced several nice litters and his progeny are producing the mountain-type dogs today.

Another Shady Mountain female who found a final home with Casey Newman was Shady Mountain Mity Blu Goblin. The very dark blue Goblin was a half sister to Mity Miss out of the exceptional Jones Mity Bite and sired by the great Jones Reddy Teddy. A large bitch with a congenial temperament, Goblin was bred only once in her life to Newman's Blue Sprog and produced only one puppy named Masterson's Gitel. Gitel was a dark blue like her parents and has served as the link from past to present for both HS & W and Kristone kennels through her daughters, HS & W Blazing Star and Kristone Mity Black Medicine. Both of these daughters were from outcross sires but both have been bred back to heavily linebred Wood's type dogs to carry this type into the future.

The breathtaking Jeffreys' Pooh Bear, a typical dark blue merle and foundation of a dynasty of Wood dogs which spread their influence throughout the breed. Photo by Sue Noll, courtesy of Melissa Couch.

Shady Mountain Flirt was the last of the Shady Mountain dogs to find a final home with Casey Newman. An attempt was made to breed her to Sprog but her advanced age defeated all attempts at this cross. Flirt was another exceptional female who had produced a number of excellent dogs for other breeders. She was bred at least twice to the Jones Reddy Teddy son, Kintyre Jeb Stuart. She was one of the last Williams' Laddie daughters out of Shady Mountain Lady II—a Jeffreys' Shade daughter from the beautiful Poudre Condesa. Condesa was closely related to the great McGuire's SeMeGo Hemi and also dam of Casey Masterson Newman's foundation bitch by Jeffreys' Pooh Bear, Masterson's Dotclijee Shunka.

Jeffreys' (Little) Pooh Bear as a young dog. Courtesy of Sorensen

Briarbrooks Crafty Lady, blue merle and white female (Malone's Lotsa Freckles x Jeffreys' Peppy), a show ring winner and low heeling driving dog. Courtesy of Sorensen

Briarbrooks Crafty Lady. Sired by Sorensen's Gunsmoke's full brother and out of a littermate to Casey Newman's foundation female Masterson's Dotclijee Shunka, Lady was pure mountain breeding. Courtesy of Sorensen

Crawfords Bob (Jeffreys' Shade x Davies' Chaw). Courtesy of Melissa Couch

The magnificent Jeffreys' Pooh Bear (also called Jeffreys' Little Pooh Bear). Photo by Casey Masterson Newman

Pooh Bear had the massive head typical of his line dating back to Sisler's Shorty and his sire Keno. Photo by Sue Noll, courtesy of Jean Taylor.

Jeffreys' Pooh Bear. Photo by Sue Noll, courtesy of Jean Taylor.

The Sisler legacy of excellent head type is easily seen in Jeffreys' Pooh Bear. Courtesy of Melissa Couch

Jeffreys' Pooh Bear. Courtesy of Raymie Jones

Legacy Chart for Jeffreys Little Pooh Bear

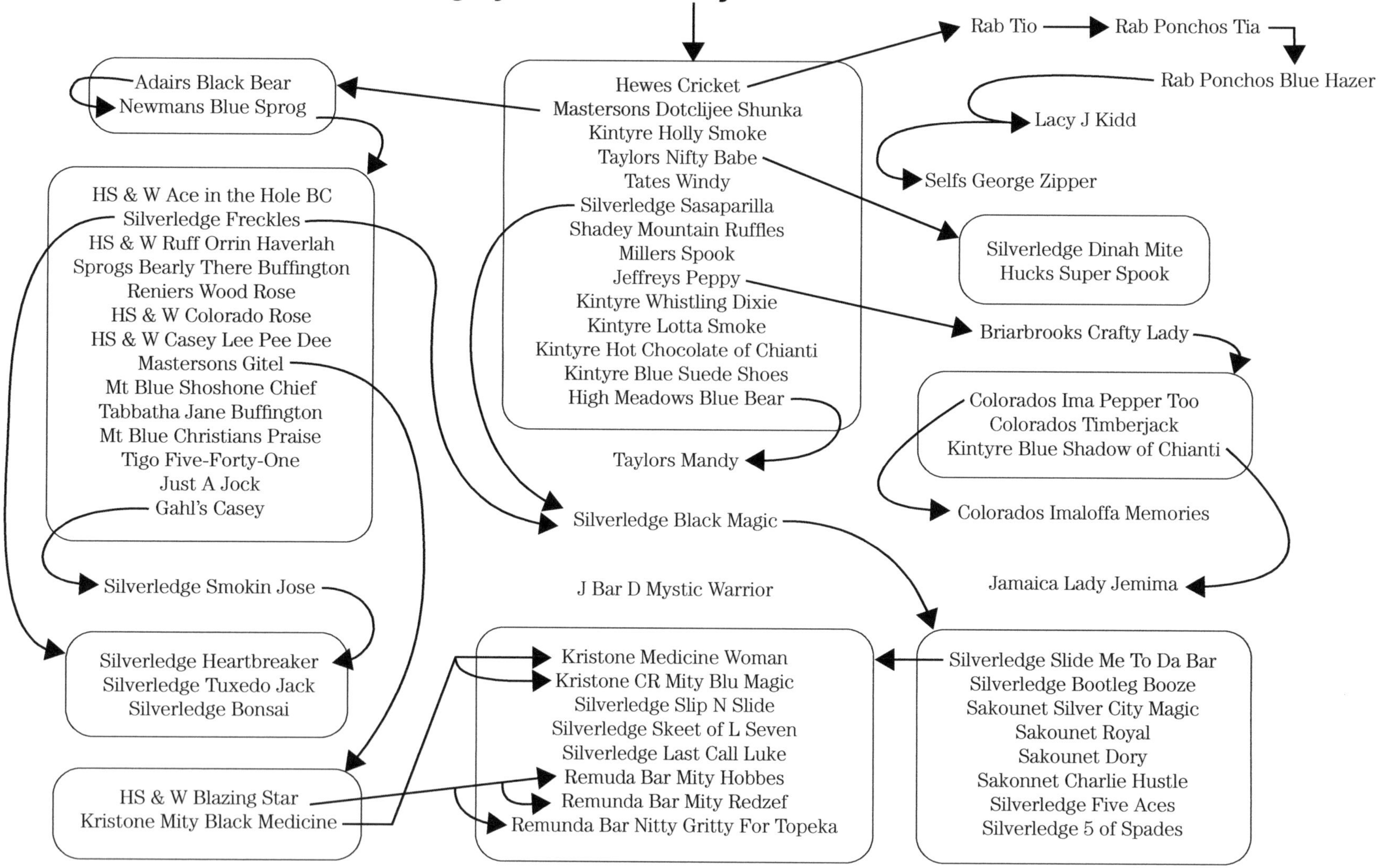

The solid dark red female Jeffreys' Cinnamon Bear, an important link to the red Wood-type dogs. Courtesy of Raymie Jones

Jeffreys' Cinnamon Bear. Courtesy of Raymie Jones

The red merle male, Beien's Red Tuffy, sire of Jeffreys' Cinnamon Man. Courtesy of Melissa Couch

Beien's Red Tuffy introduced the red gene to the Jeffreys' bloodline. Courtesy of Melissa Couch

Although it has not been completely confirmed, this red merle dog is likely Jeffreys' Cinnamon Man. This dog's call name was Pawnee which was also the known call name of Cinnamon Man. Pawnee was the sire of Cathy Jones' foundation female, Jones' Mity Bite. Courtesy of Mary Wood

Born: ______________ Sex: ________
Owned by: ______________________

PEDIGREE OF Jeffreys Cinnamon Bear

Parents	Grandparents	Great-grandparents	Great-great-grandparents
Sire: Montoyas Rowdy	Sire: Clingers Royal Blue Merle	Sire: Sorensens Gunsmoke	Sire: Williams Laddie
			Dam: Jeffreys Sheila
		Dam: Van Gordons Blue Asta	Sire: Elys Major
			Dam: Lighters Asta
	Dam: Kalbs Princess	Sire: Clingers Royal Blue Merle	Sire: Sorensens Gunsmoke
			Dam: Van Gordons Blue Asta
		Dam: Shockleys Blue Samantha	Sire: Clare Olsons Blue Chips
			Dam: Clares Troubles
Dam: Montoyas Snipper	Sire: Piz Joseph	Sire: Woods Jay	Sire: Sislers Shorty
			Dam: Sislers Trixie
		Dam: Woods Chili	Sire: Woods Jay
			Dam: Greens Dingo
	Dam: Pizs Lady II	Sire: Piz Joseph	Sire: Woods Jay
			Dam: Woods Chili
		Dam: Piz Lady I	Sire: Woods Jay
			Dam: Woods Chili

Bred by: ______________ Address: ______________ Phone: ______________

The details given above are correct to the best of my knowledge. Signed: ______________ (Breeder)

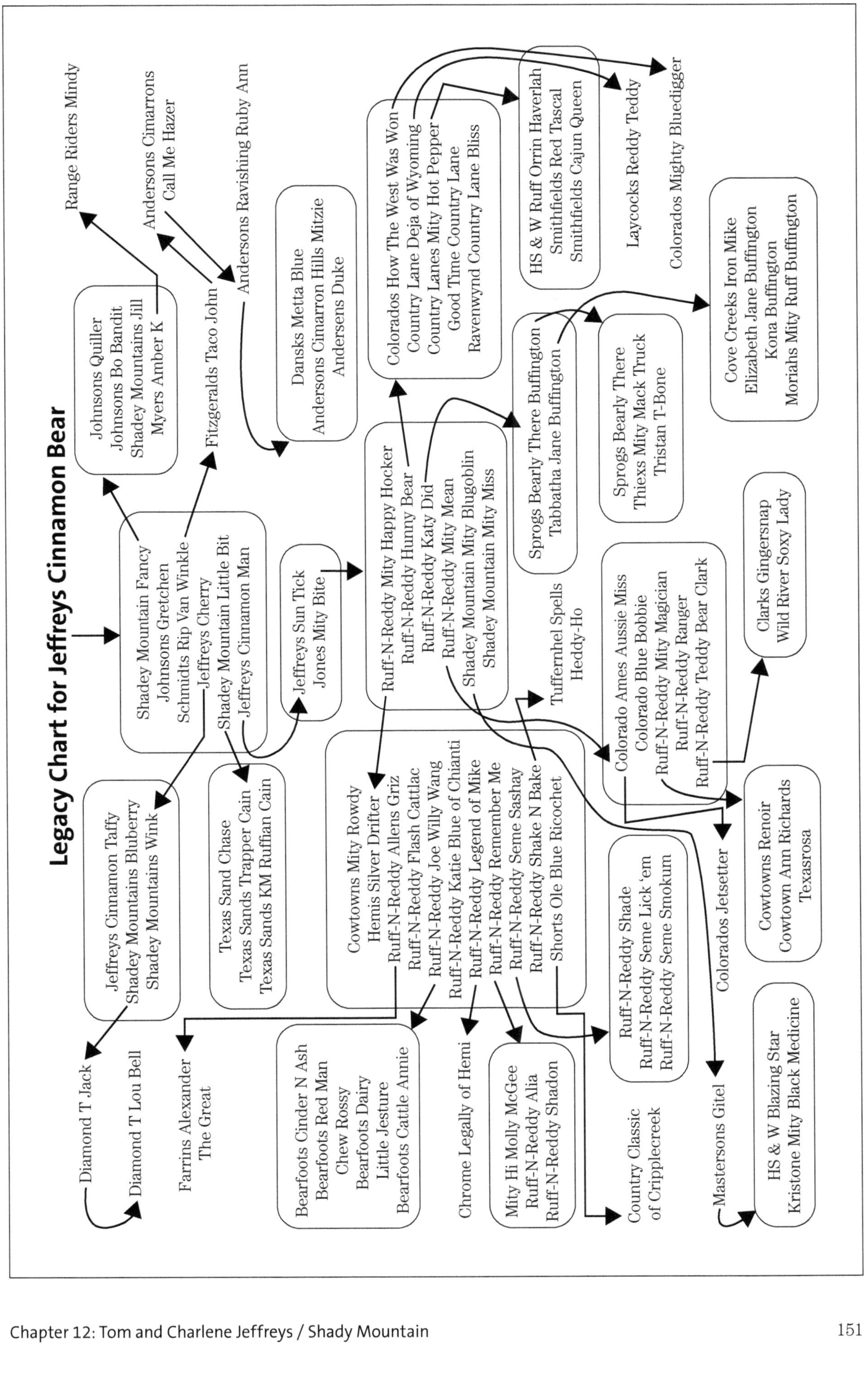
Legacy Chart for Jeffreys Cinnamon Bear
Shadey Mountain Fancy
Johnsons Gretchen
Schmidts Rip Van Winkle
Jeffreys Cherry
Shadey Mountain Little Bit
Jeffreys Cinnamon Man
Johnsons Quiller
Johnsons Bo Bandit
Shadey Mountains Jill
Myers Amber K
Range Riders Mindy
Andersons Cimarrons
Call Me Hazer
Fitzgeralds Taco John
Andersons Ravishing Ruby Ann
Dansks Metta Blue
Andersons Cimarron Hills Mitzie
Andersens Duke
Jeffreys Cinnamon Taffy
Shadey Mountains Bluberry
Shadey Mountains Wink
Diamond T Jack
Diamond T Lou Bell
Texas Sand Chase
Texas Sands Trapper Cain
Texas Sands KM Ruffian Cain
Farrins Alexander
The Great
Jeffreys Sun Tick
Jones Mity Bite
Ruff-N-Reddy Mity Happy Hocker
Ruff-N-Reddy Hunny Bear
Ruff-N-Reddy Katy Did
Ruff-N-Reddy Mity Mean
Shadey Mountain Mity Blugoblin
Shadey Mountain Mity Miss
Colorados How The West Was Won
Country Lane Deja of Wyoming
Country Lanes Mity Hot Pepper
Good Time Country Lane
Ravenwynd Country Lane Bliss
HS & W Ruff Orrin Haverlah
Smithfields Red Tascal
Smithfields Cajun Queen
Laycocks Reddy Teddy
Colorados Mighty Bluedigger
Cowtowns Mity Rowdy
Hemis Silver Drifter
Ruff-N-Reddy Allens Griz
Ruff-N-Reddy Flash Cattlac
Ruff-N-Reddy Joe Willy Wang
Ruff-N-Reddy Katie Blue of Chianti
Ruff-N-Reddy Legend of Mike
Ruff-N-Reddy Remember Me
Ruff-N-Reddy Seme Sashay
Ruff-N-Reddy Shake N Bake
Shorts Ole Blue Ricochet
Bearfoots Cinder N Ash
Bearfoots Red Man
Chew Rossy
Bearfoots Dairy
Little Jesture
Bearfoots Cattle Annie
Chrome Legally of Hemi
Mity Hi Molly McGee
Ruff-N-Reddy Alia
Ruff-N-Reddy Shadon
Sprogs Bearly There Buffington
Tabbatha Jane Buffington
Sprogs Bearly There
Thiexs Mity Mack Truck
Tristan T-Bone
Cove Creeks Iron Mike
Elizabeth Jane Buffington
Kona Buffington
Moriahs Mity Ruff Buffington
Tuffernhel Spells
Heddy-Ho
Colorado Ames Aussie Miss
Colorado Blue Bobbie
Ruff-N-Reddy Mity Magician
Ruff-N-Reddy Ranger
Ruff-N-Reddy Teddy Bear Clark
Clarks Gingersnap
Wild River Soxy Lady
Ruff-N-Reddy Shade
Ruff-N-Reddy Seme Lick 'em
Ruff-N-Reddy Seme Smokum
Country Classic
of Cripplecreek
Colorados Jetsetter
Cowtowns Renoir
Cowtown Ann Richards
Texasrosa
Mastersons Gitel
HS & W Blazing Star
Kristone Mity Black Medicine

Three generations of superior red dogs. L to R: Beien's Red Tuffy, Jeffreys' Hobo Joe, and Johnson's Quiller. Courtesy of Melissa Couch

The stunning red merle, Jeffreys' Hobo Joe, winning Reserve Winners Dog at the CASA show in 1975. Courtesy of Sorensen

Shady Mountain Mity Miss—foundation dam to many kennels across the country through her excellent sons and daughters. Courtesy of Melissa Couch

A well known sire of good working dogs, Clare Olson's Blue Joe, a dark blue merle. Courtesy of Sorensen

Clare Olson's Blue Joe as a mature dog. Courtesy of Jean Taylor

Shady Mountain Flirt in her prime. Courtesy of Sorensen

The dark blue merle female, Shady Mountain Mity Blu Goblin as an older dog. Courtesy of Casey Masterson Newman

Shady Mountain Flirt, beautifully correct blue merle female. Courtesy of Casey Masterson Newman

Born: ____________ Sex: ________
Owned by: ______________________

PEDIGREE OF Shadey Mountain Mity Miss

Parents	Grandparents	Great-grandparents	Great-great-grandparents
Sire: Jeffreys Shade	Sire: Woods Stubbie	Sire: Woods Jay	Sire: Sislers Shorty
			Dam: Sislers Trixie
		Dam: Woods Slate	Sire: Woods Jay
			Dam: Woods Specks
	Dam: Jeffreys Missy Blue	Sire: Williams Laddie	Sire: Uhlendorfs Tippy
			Dam: Thorntons Chrissie
		Dam: Picrauxs Juliette	Sire: Jeffreys Dead Eye Jr
			Dam: Williams Georgina
Dam: Jones Mity Bite	Sire: Jeffreys Cinnamon Man	Sire: Beiens Red Tuffy	Sire: Pizs Little Joe
			Dam: Pizs Lady II
		Dam: Jeffreys Cinnamon Bear	Sire: Montoyas Rowdy
			Dam: Montoyas Snipper
	Dam: Piza Sheba	Sire: Piz Joseph	Sire: Woods Jay
			Dam: Woods Chili
		Dam: Pizs Lady II	Sire: Piz Joseph
			Dam: Piz Lady I

Bred by: ____________ Address: ____________ Phone: ____________

The details given above are correct to the best of my knowledge. Signed: ____________ (Breeder)

Legacy Chart for Shadey Mountain Mity Miss

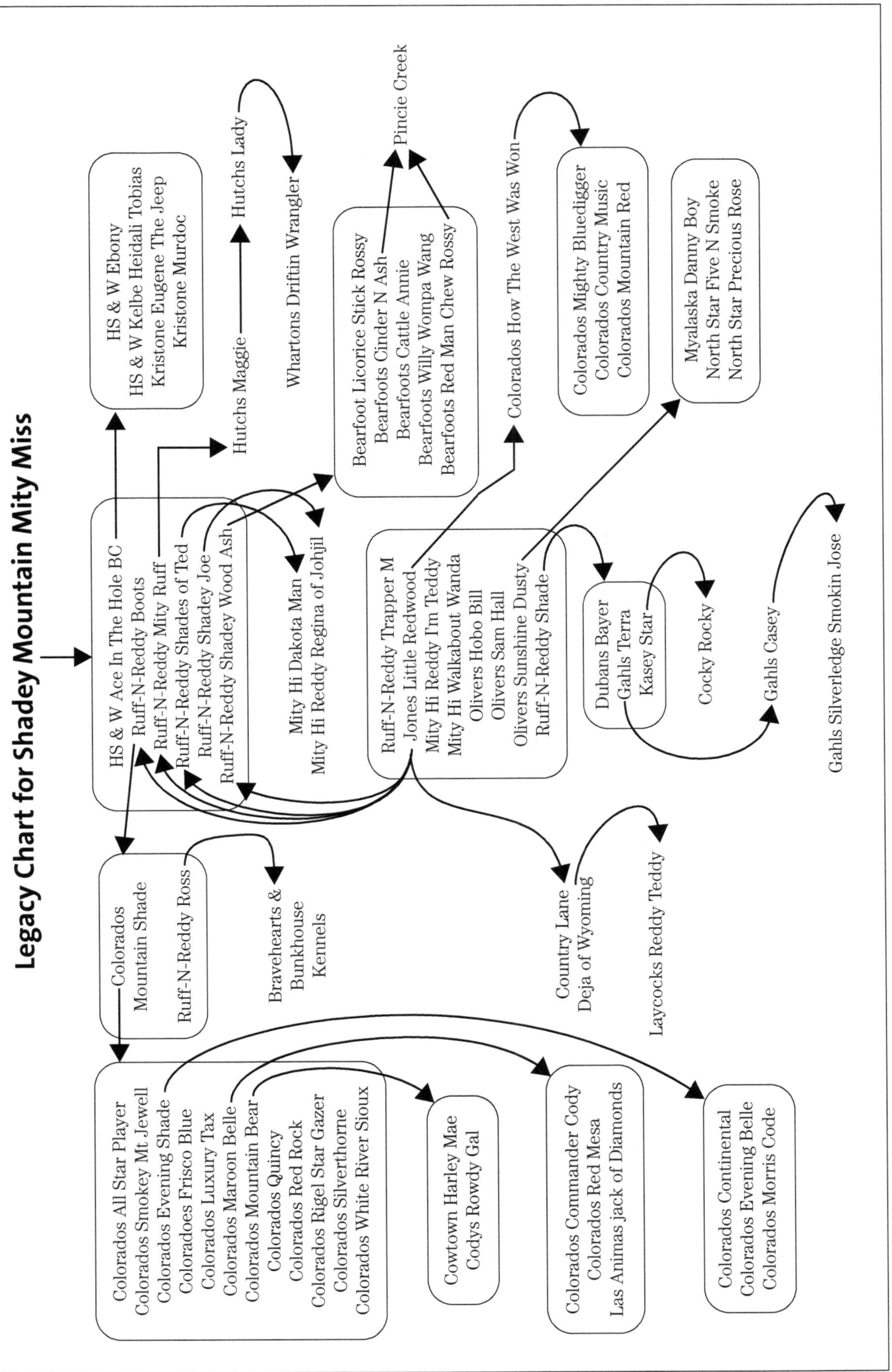

Shady Mountain Flirt in old age. Courtesy of Casey Masterson Newman

The Shady Mountain dogs influenced many working kennels across the country. Shady Mountain Little Bit, a black tri female, was the foundation for Rick Dill's kennel in Texas. Courtesy of Melissa Couch

Shady Mountain Sumac, owned at one time by Cathy Jones. Blue merle female. Courtesy of Melissa Couch

chapter 13

Other Wood Dogs and Influence in Other Lines

Wood's Jay and Wood's Stubbie were the most well known and influential dogs Fletcher Wood owned and used. However Jay produced a number of other dogs of note and Mr. Wood owned and used other dogs which have influenced the breed and merit mention in this book.

One of Mary Wood's favorite bitches was a dark blue merle named Wood's Little Blue. Over the years a great deal of confusion has arisen regarding the lineage of this female. It is unclear what has prompted this confusion but an early breed reference, The ASCA Yearbook 1957-1977, contains a statement from Fletcher Wood stating the lineage of this influential female. Wood clearly states that Wood's Little Blue was a daughter of the Sisler bred Ely's Blue. In an interview with Mary Wood by Leslie Sorensen, she also noted Wood's Little Blue was obtained from Jay Sisler. Knowing Sisler's penchant for trading dogs this is a logical source for this female. It would not have been unusual for him to have gotten a dog from the good producing Ely's Blue—especially since she was a dog of his and Gene's own breeding.

The National Stock Dog Registry registration papers for Wood's Little Blue show her to be sired by Blue Boy and out of Blue Shadow—the most widely seen pedigree for her. The confusing habit of the National Stock Dog Registry of changing dog names when they changed owners raises questions about the identity of Blue Shadow. This is one possible source of confusion about Little Blue's lineage.

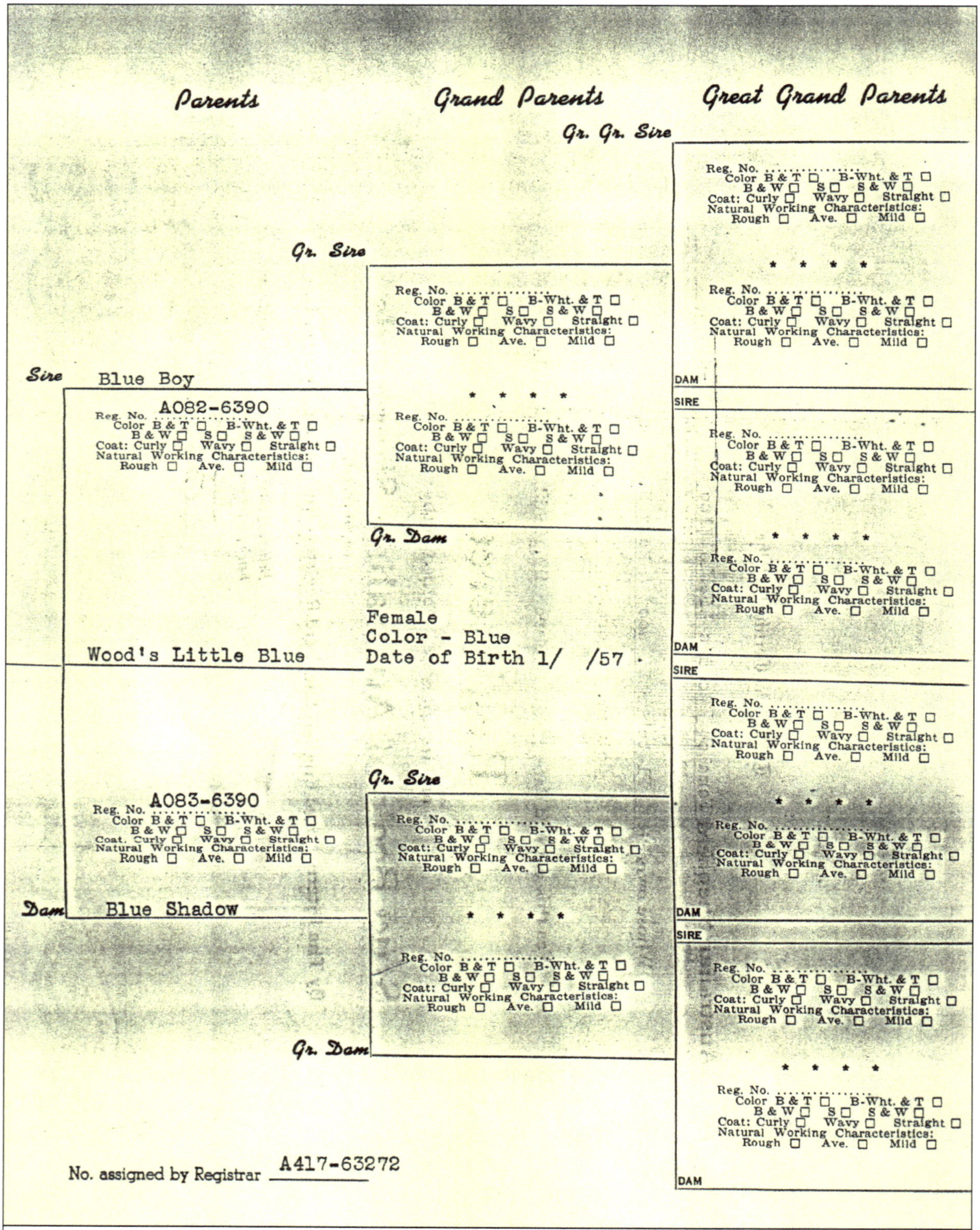
Parents
Grand Parents
Great Grand Parents
Gr. Gr. Sire
Gr. Sire
Sire Blue Boy
Reg. No. A082-6390
Color B & T ☐ B-Wht. & T ☐ B & W ☐ S ☐ S & W ☐
Coat: Curly ☐ Wavy ☐ Straight ☐
Natural Working Characteristics: Rough ☐ Ave. ☐ Mild ☐
Gr. Dam
Wood's Little Blue
Female
Color - Blue
Date of Birth 1/ /57
Gr. Sire
Reg. No. A083-6390
Color B & T ☐ B-Wht. & T ☐ B & W ☐ S ☐ S & W ☐
Coat: Curly ☐ Wavy ☐ Straight ☐
Natural Working Characteristics: Rough ☐ Ave. ☐ Mild ☐
Dam Blue Shadow
Gr. Dam
DAM
SIRE
No. assigned by Registrar A417-63272

National Stock Dog Registry pedigree of Wood's Little Blue. Courtesy of Mary Wood via Sorensen

This is to Certify that the

Australian Shepherd Female , Wood's Little Blue A417-63272
(or Sheepdog) Sex Name of Dog and Reg. No.

Whose owner is Fletcher H. Wood & Mary H. Wood
Name
Golden, Colorado R.3. Box 370 C
Address

Was this day and date, March 27, 1962 duly

Registered

and its ancestry Recorded in the

Stud Book of Australian Shepherds
(or Sheepdogs)

By The International English Shepherd Registry, Inc., Butler, Ind., Rt. 1.

Believing these statements to be true and correct, we hereby affix our signature and seal.

E. G. Emanuel - Registrar

LITHO IN U.S.A.

National Stock Dog registration papers of Wood's Little Blue. Courtesy of Mary Wood via Sorensen

In various sources Wood's Little Blue is claimed to also have been known as a daughter of Hartnagle's Goody (a bitch said to also have been known as Wood's Blue Shadow—a dog unknown to Mary Wood). However the National Stock Dog Registry habit of attaching owner names to the names of dogs belies this claim also. If, indeed, Blue Shadow was a dog owned by Fletcher Wood (as implied by the name Wood's Blue Shadow) then it would have appeared that way on Wood's Little Blue's registration papers. But it was not recorded this way. In fact, Fletcher Wood said to more than one person that he acquired Wood's Little Blue from Jay Sisler.

Were Blue Shadow and Ely's Blue the same dog? This possibility exists if Ely's Blue was known to Jay Sisler as Blue Shadow before she was sold or traded to Juanita Ely. In this case, Fletcher Wood would have registered Wood's Little Blue as being out of Blue Shadow since that would have been the name of her dam at the time Wood got Little Blue from Jay Sisler. It would also account for Fletcher Wood's later statement that Little Blue was a daughter of Ely's Blue if her name was changed after Juanita Ely became her owner. Little Blue proved to be a top producer and this fact would also have encouraged the breedings Fletcher Wood did between his own Wood's Jay and Ely's Blue.

It is unlikely Fletcher Wood looked outside the Sisler bred dogs he knew and valued for breeding stock. The few times he did (such as when he tried to introduce the blood of Wood's Dandy) were unsatisfactory and Wood eliminated such lines from his breeding program. Mary Wood stated to this author that she had no memory of her husband obtaining dogs from the Hartnagles at any time when he was breeding dogs. This statement is supported by the registrations of these two dogs. Blue Shadow is registered with the National Stock Dog registry under the number A083-6390. Hartnagle's Goody is registered with the National Stock Dog registry under the number A098-63260. Although it was common practice for the National Stock Dog registry to alter the name of a dog when it changed owners, the dog retained the same registration number throughout its life. The numbers for Blue Shadow and Goody are not the same. In addition, the registration numbers issued by NSDR were coded to indicate the date on which the dog was registered by them. The numbers following the hyphen are this code. The first number and last number of the series indicate the year the dog was registered. In the case of both Blue Shadow and Goody, that year is 1960. In addition, the numbers between the first and last indicate the day and month the dog was registered. Blue Shadow and Goody were registered in the same month of 1960 as well (March 9, 1960, for Blue Shadow and March 26, 1960, for Goody). If these were two names for the same dog and registration was

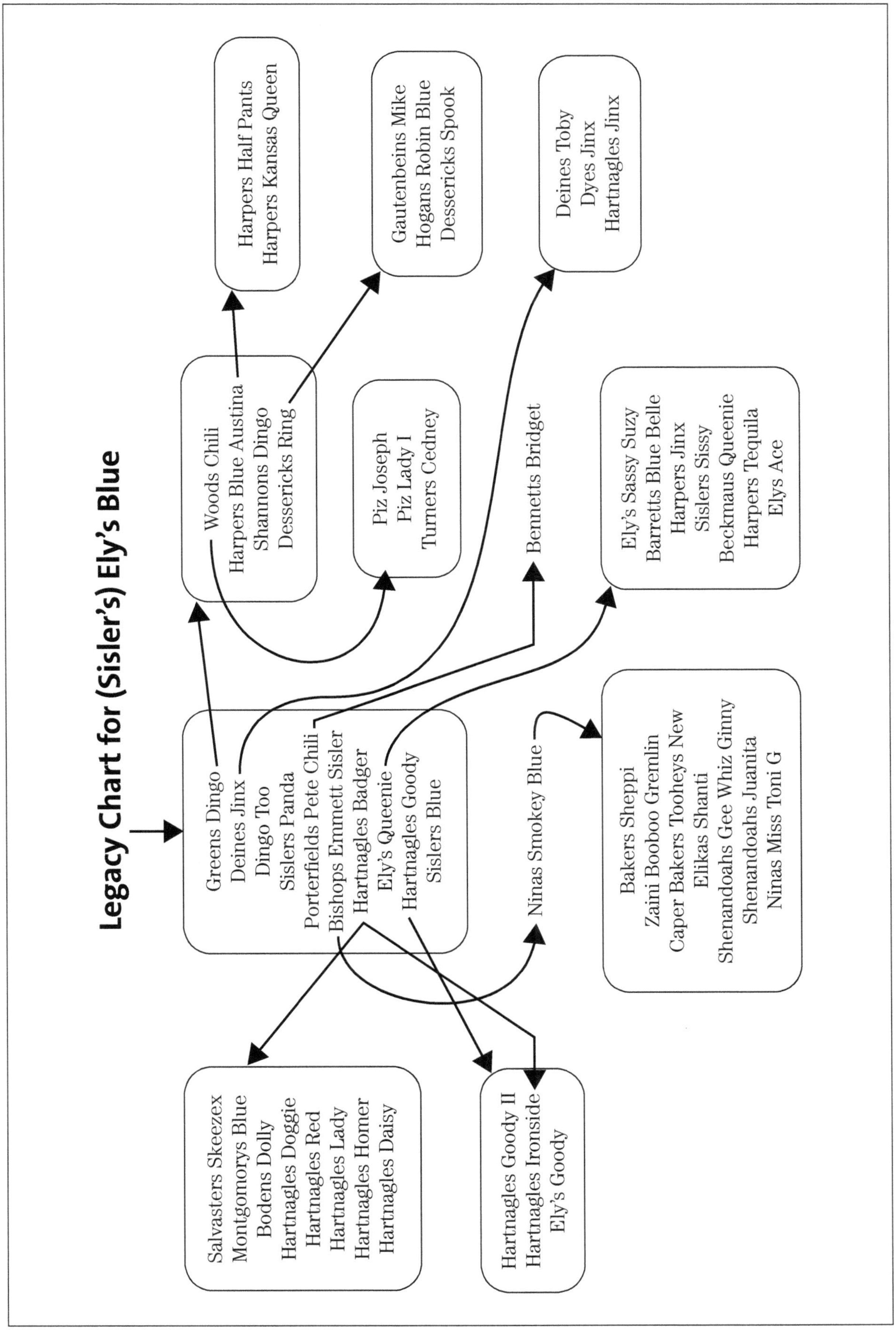
Legacy Chart for (Sisler's) Ely's Blue
Greens Dingo
Deines Jinx
Dingo Too
Sislers Panda
Porterfields Pete Chili
Bishops Emmett Sisler
Hartnagles Badger
Ely's Queenie
Hartnagles Goody
Sislers Blue
Woods Chili
Harpers Blue Austina
Shannons Dingo
Dessericks Ring
Harpers Half Pants
Harpers Kansas Queen
Gautenbeins Mike
Hogans Robin Blue
Dessericks Spook
Deines Toby
Dyes Jinx
Hartnagles Jinx
Piz Joseph
Piz Lady I
Turners Cedney
Bennetts Bridget
Ely's Sassy Suzy
Barretts Blue Belle
Harpers Jinx
Sislers Sissy
Beckmaus Queenie
Harpers Tequila
Elys Ace
Ninas Smokey Blue
Bakers Sheppi
Zaini Booboo Gremlin
Caper Bakers Tooheys New
Elikas Shanti
Shenandoahs Gee Whiz Ginny
Shenandoahs Juanita
Ninas Miss Toni G
Salvasters Skeezex
Montgomorys Blue
Bodens Dolly
Hartnagles Doggie
Hartnagles Red
Hartnagles Lady
Hartnagles Homer
Hartnagles Daisy
Hartnagles Goody II
Hartnagles Ironside
Ely's Goody

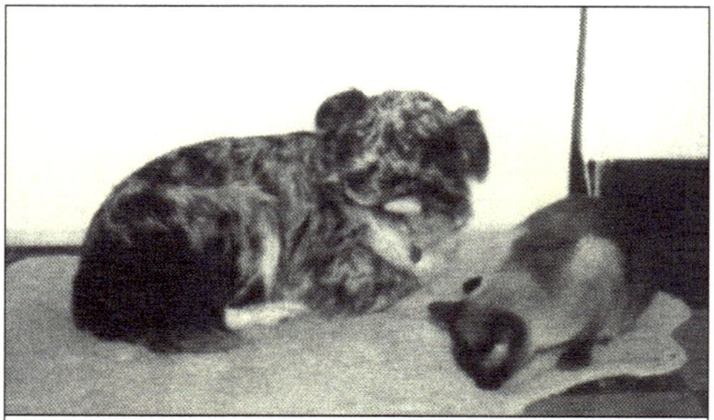

Wood's Little Blue, blue merle and white female, nearly as influential as her housemate Wood's Jay in perpetuating the Wood and mountain-type Australian Shepherd. Courtesy of Sorensen

sought for both in the same month and year, why would there be two different numbers for them?

Goody was a daughter of Ely's (Sisler's) Blue. If Fletcher Wood's claim that Little Blue was a daughter of Ely's (Sisler's) Blue is true, it is possible that this is the source of the confusion and Goody and Little Blue were, in fact, half-sisters and not mother and daughter. It was not uncommon in this era for many pedigrees to be hand written. Many times the transcriber would get off a generation when copying pedigrees and once the mistake was circulated widely it tended to be taken as fact. The accuracy of actual registration papers and the knowledge of the dog's owner and breeder is a much more reliable source of true information than unofficial sources of pedigree.

To further confuse the issue, Wood's Little Blue is shown on the National Stock Dog registry pedigree of Thornton's Blue Blue as having been sired by Blue Khan and out of "Sisler's Blue Female" (see

Mary Wood's favorite house dog, Wood's Little Blue with a litter of puppies. Courtesy of Mary Wood

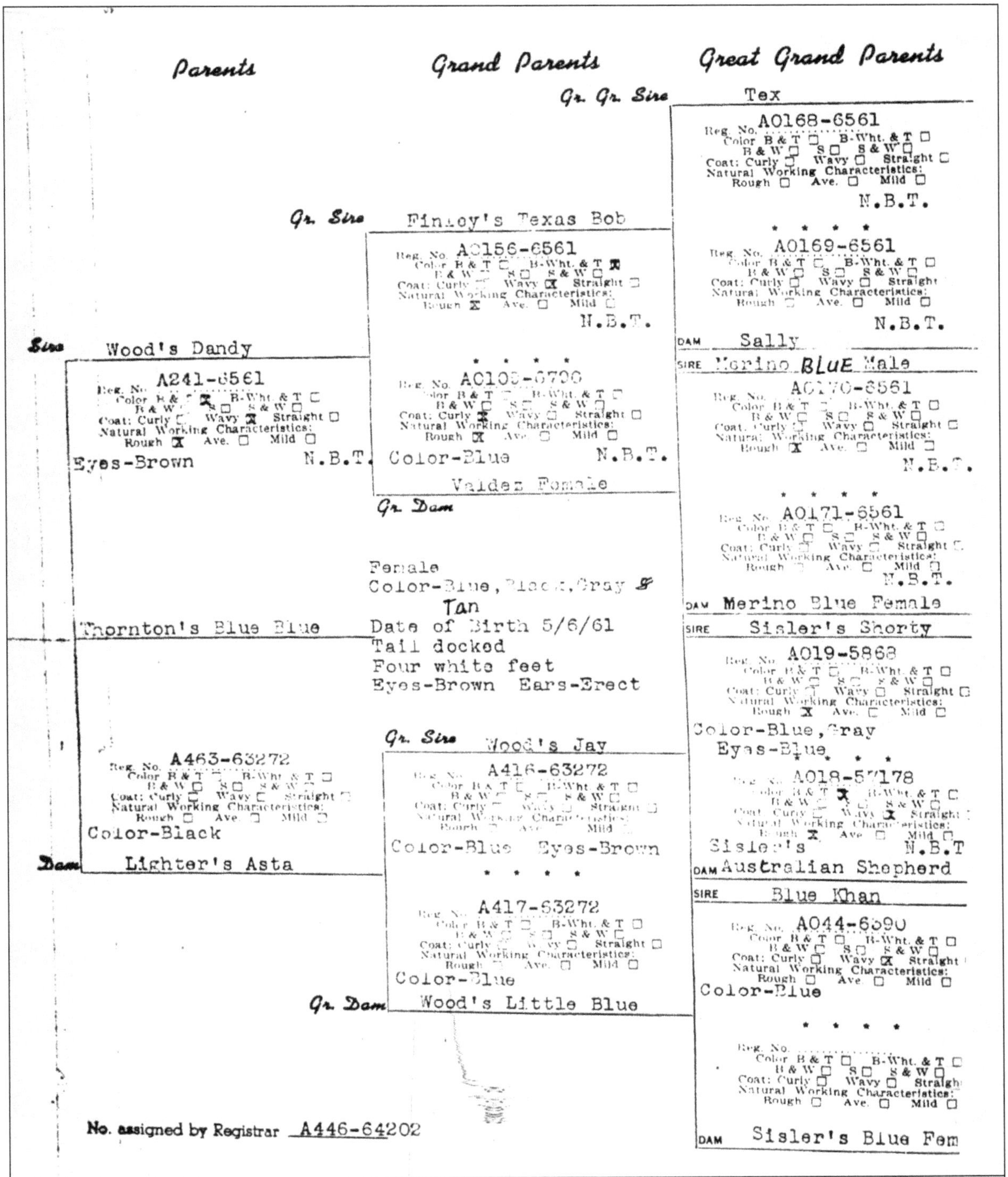

Parents

Grand Parents

Great Grand Parents

Gr. Gr. Sire Tex
Reg. No. A0168-6561
N.B.T.

Gr. Sire Finley's Texas Bob
Reg. No. A0156-6561
N.B.T.

Reg. No. A0169-6561
N.B.T.
DAM Sally
SIRE Merino BLUE Male
Reg. No. A0170-6561
N.B.T.

Sire Wood's Dandy
Reg. No. A241-6561
Eyes-Brown N.B.T.

Reg. No. A0103-6790
Color-Blue N.B.T.
Valdez Female
Gr. Dam

Reg. No. A0171-6561
N.B.T.
DAM Merino Blue Female

Female
Color-Blue, Black, Gray & Tan
Date of Birth 5/6/61
Tail docked
Four white feet
Eyes-Brown Ears-Erect

Thornton's Blue Blue

SIRE Sisler's Shorty
Reg. No. A019-5868
Color-Blue, Gray
Eyes-Blue

Reg. No. A463-63272
Color-Black
Dam Lighter's Asta

Gr. Sire Wood's Jay
Reg. No. A416-63272
Color-Blue Eyes-Brown

Reg. No. A018-57178
Sisler's N.B.T
DAM Australian Shepherd
SIRE Blue Khan
Reg. No. A044-6590
Color-Blue

Reg. No. A417-63272
Color-Blue
Gr. Dam Wood's Little Blue

DAM Sisler's Blue Fem

No. assigned by Registrar A446-64202

National Stock Dog Registry pedigree for Thornton's Blue Blue showing information on Wood's Little Blue—clearly stating her dam was a Sisler dog. Courtesy of Mandy Lees

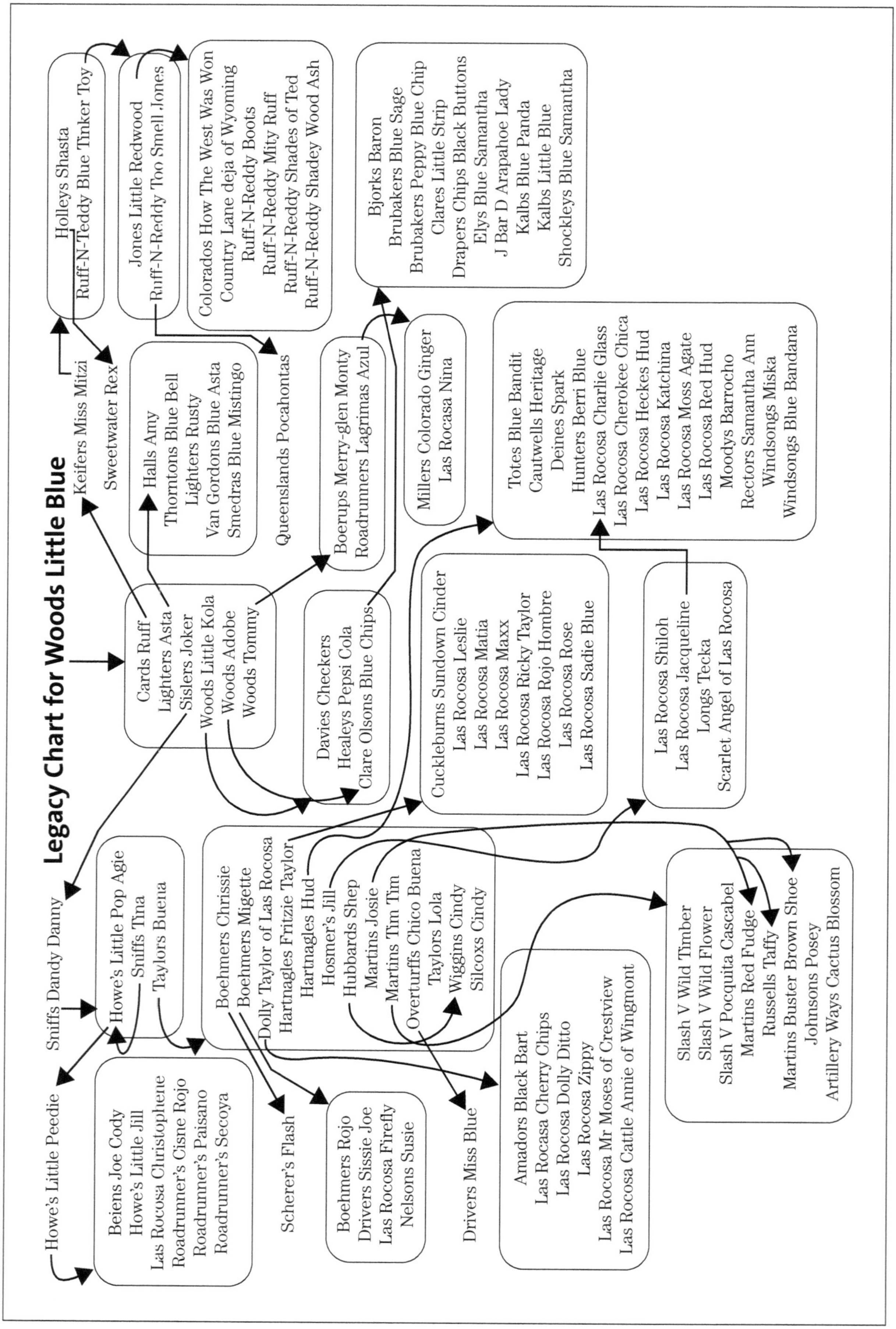
Legacy Chart for Woods Little Blue
Cards Ruff
Lighters Asta
Sislers Joker
Woods Little Kola
Woods Adobe
Woods Tommy
Keifers Miss Mitzi
Sweetwater Rex
Holleys Shasta
Ruff-N-Teddy Blue Tinker Toy
Jones Little Redwood
Ruff-N-Reddy Too Smell Jones
Colorados How The West Was Won
Country Lane deja of Wyoming
Ruff-N-Reddy Boots
Ruff-N-Reddy Mity Ruff
Ruff-N-Reddy Shades of Ted
Ruff-N-Reddy Shadey Wood Ash
Halls Amy
Thorntons Blue Bell
Lighters Rusty
Van Gordons Blue Asta
Smedras Blue Mistingo
Queenslands Pocahontas
Boerups Merry-glen Monty
Roadrunners Lagrimas Azul
Bjorks Baron
Brubakers Blue Sage
Brubakers Peppy Blue Chip
Clares Little Strip
Drapers Chips Black Buttons
Elys Blue Samantha
J Bar D Arapahoe Lady
Kalbs Blue Panda
Kalbs Little Blue
Shockleys Blue Samantha
Millers Colorado Ginger
Las Rocasa Nina
Davies Checkers
Healeys Pepsi Cola
Clare Olsons Blue Chips
Totes Blue Bandit
Cautwells Heritage
Deines Spark
Hunters Berri Blue
Las Rocosa Charlie Glass
Las Rocosa Cherokee Chica
Las Rocosa Heckes Hud
Las Rocosa Katchina
Las Rocosa Moss Agate
Las Rocosa Red Hud
Moodys Barrocho
Rectors Samantha Ann
Windsongs Miska
Windsongs Blue Bandana
Cuckleburrs Sundown Cinder
Las Rocosa Leslie
Las Rocosa Matia
Las Rocosa Maxx
Las Rocosa Ricky Taylor
Las Rocosa Rojo Hombre
Las Rocosa Rose
Las Rocosa Sadie Blue
Las Rocosa Shiloh
Las Rocosa Jacqueline
Longs Tecka
Scarlet Angel of Las Rocosa
Sniffs Dandy Danny
Howe's Little Pop Agie
Sniffs Tina
Taylors Buena
Howe's Little Peedie
Beiens Joe Cody
Howe's Little Jill
Las Rocosa Christophene
Roadrunner's Cisne Rojo
Roadrunner's Paisano
Roadrunner's Secoya
Boehmers Chrissie
Boehmers Migette
Dolly Taylor of Las Rocosa
Hartnagles Fritzie Taylor
Hartnagles Hud
Hosmer's Jill
Hubbards Shep
Martins Josie
Martins Tim Tim
Overturffs Chico Buena
Taylors Lola
Wiggins Cindy
Silcoxs Cindy
Scherer's Flash
Boehmers Rojo
Drivers Sissie Joe
Las Rocosa Firefly
Nelsons Susie
Drivers Miss Blue
Amadors Black Bart
Las Rocasa Cherry Chips
Las Rocosa Dolly Ditto
Las Rocosa Zippy
Las Rocosa Mr Moses of Crestview
Las Rocosa Cattle Annie of Wingmont
Slash V Wild Timber
Slash V Wild Flower
Slash V Pocquita Cascabel
Martins Red Fudge
Russells Taffy
Martins Buster Brown Shoe
Johnsons Posey
Artillery Ways Cactus Blossom

accompanying photo). The Thornton dog was registered in 1962. However, in an interesting twist, Blue Khan was registered on exactly the same day as Blue Boy and Blue Shadow! No number or specific name is given for "Sisler's Blue Female" but this reference to Sisler could indicate either Ely's (Sisler's) Blue or the possibility that Blue Shadow was another Sisler dog. In an interesting twist, if this pedigree is true, it would mean Little Blue was at least a half-sister to Wood's Jay's dam, Sisler's Trixie, through the sire, Blue Khan. If "Sisler's Blue Female" on this version of Little's Blue's pedigree is Emmett's Blue (Trixie's dam), it would mean Little Blue and Trixie were full sisters! If either of these scenarios is true, they would still fit in well with the Wood philosophy of close breeding to set type and retain the good qualities he sought.

In summary, it is very unlikely to be true that Hartnagle's Goody was dam of Wood's Little Blue. She may have been known under the name "Wood's Blue Shadow" at some time but it is doubtful she was owned or used by Fletcher Wood or Jay Sisler. It is also doubtful "Wood's Blue Shadow" and the Blue Shadow shown on Little Blue's registration papers were the same female. Available records do not support this claim nor do the facts surrounding Fletcher Wood's personal breeding philosophy. Fletcher Wood and Jay Sisler were close friends and Wood recognized the qualities he desired in his own dogs were coming from the dogs Jay had access to in Idaho. Wood had seen the excellent results from the cross of Wood's Jay on Ely's (Sisler's) Blue and he would logically have looked to the Sisler brothers to find females to cross with Jay. It is unlikely the true background of Wood's Little Blue or Blue Shadow will ever be known for certain. Fletcher Wood and Jay Sisler were probably the only two people who could clear up this confusion and they are both gone. We have only Fletcher's word and these documents to use to try to solve this mystery.

The true background of Wood's Little Blue is important because she produced many of the most influential dogs in the breed. Although Jay Sisler never sought the credit he deserved in the history of the Australian Shepherd, credit should still be given where it belongs. Fletcher Wood's wise use of the dogs that Jay and Gene produced assured the breed a strong foundation and his contribution should not be minimized, diluted, or claimed by others.

As usual, Mr. Wood's instincts were correct as Little Blue became a top producer when bred to, Jay and his sire, Sisler's Shorty. By Shorty, Little Blue produced one of Jay Sisler's performing dogs, Sisler's Joker. Joker later founded his own strain of dogs which is detailed in the section on Jay Sisler in this book.

By Wood's Jay, Little Blue produced the incomparable Lighter's Asta. Asta produced a dynasty of great producing dogs, including Thornton's

Born: ____________ Sex: ________
Owned by: ________________________

PEDIGREE OF Clare Olsons Blue Chips

Parents	Grandparents	Great-grandparents	Great-great-grandparents
Sire: Woods Adobe	Sire: Woods Dandy	Sire: Finleys Texas Bob	Sire
			Dam
		Dam: Merinos Blue Bitch	Sire
			Dam
	Dam: Woods Little Blue	Sire: (Sislers) Blue Boy	Sire
			Dam
		Dam: (Sislers) Blue Shadow	Sire
			Dam
Dam: Woods Little Kola	Sire: Woods Jay	Sire: Sislers Shorty	Sire: Keno
			Dam: Blue Star
		Dam: Sislers Trixie	Sire: Blue Khan
			Dam: Emmetts Blue
	Dam: Woods Little Blue	Sire: (Sislers) Blue Boy	Sire
			Dam
		Dam: (Sislers) Blue Shadow	Sire
			Dam

Bred by: ____________ Address: ____________ Phone: ____________

The details given above are correct to the best of my knowledge. Signed: ____________ (Breeder)

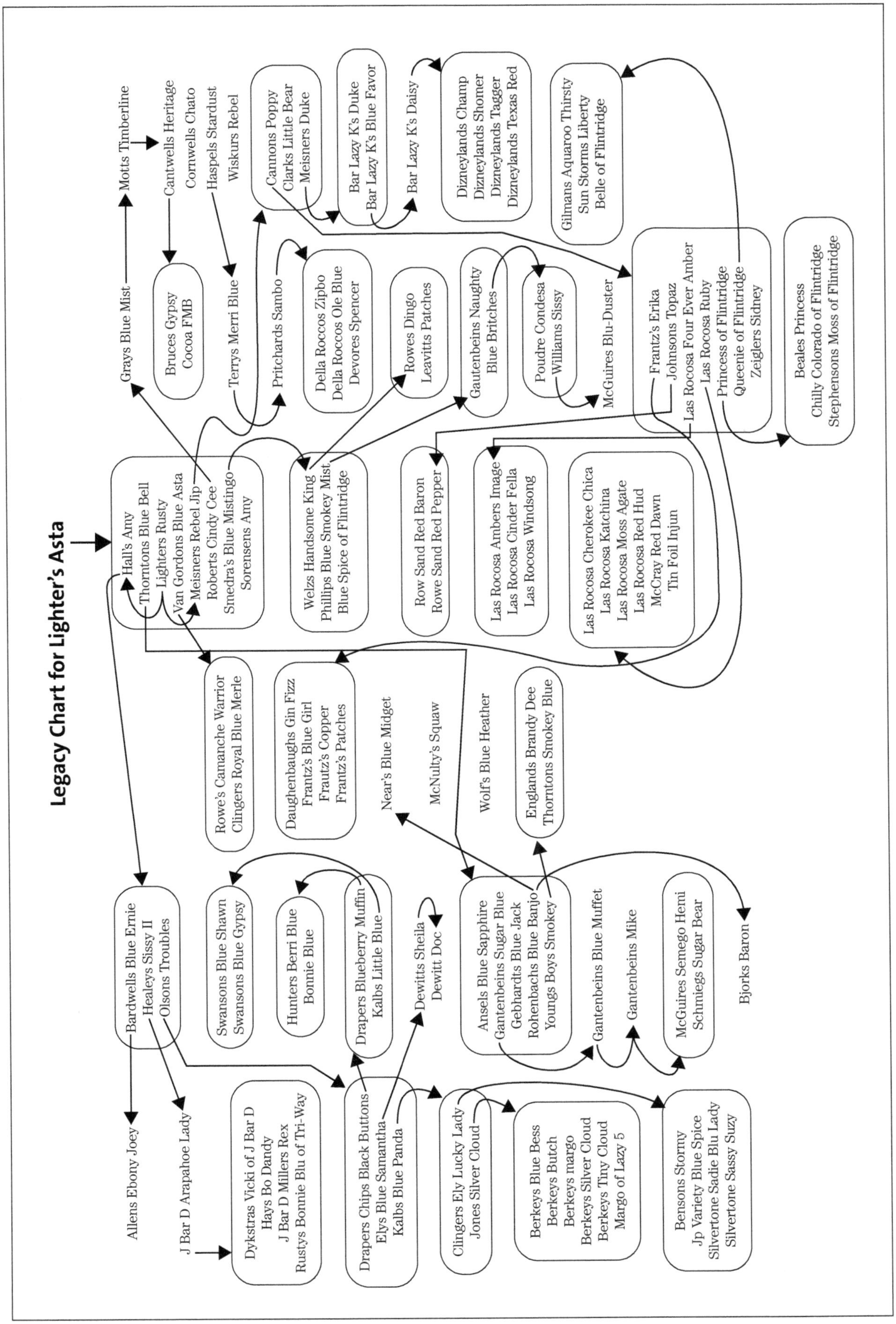
Legacy Chart for Lighter's Asta
Hall's Amy
Thorntons Blue Bell
Lighters Rusty
Van Gordons Blue Asta
Meisners Rebel Jip
Roberts Cindy Cee
Smedra's Blue Mistingo
Sorensens Amy
Grays Blue Mist
Motts Timberline
Cantwells Heritage
Cornwells Chato
Haspels Stardust
Wiskurs Rebel
Bruces Gypsy
Cocoa FMB
Terrys Merri Blue
Pritchards Sambo
Cannons Poppy
Clarks Little Bear
Meisners Duke
Bar Lazy K's Duke
Bar Lazy K's Blue Favor
Bar Lazy K's Daisy
Della Roccos Zipbo
Della Roccos Ole Blue
Devores Spencer
Dizneylands Champ
Dizneylands Shomer
Dizneylands Tagger
Dizneylands Texas Red
Welzs Handsome King
Phillips Blue Smokey Mist
Blue Spice of Flintridge
Rowes Dingo
Leavitts Patches
Gautenbeins Naughty
Blue Britches
Poudre Condesa
Williams Sissy
McGuires Blu-Duster
Gilmans Aquaroo Thirsty
Sun Storms Liberty
Belle of Flintridge
Frantz's Erika
Johnsons Topaz
Las Rocosa Four Ever Amber
Las Rocosa Ruby
Princess of Flintridge
Queenie of Flintridge
Zeiglers Sidney
Beales Princess
Chilly Colorado of Flintridge
Stephensons Moss of Flintridge
Row Sand Red Baron
Rowe Sand Red Pepper
Las Rocosa Ambers Image
Las Rocosa Cinder Fella
Las Rocosa Windsong
Las Rocosa Cherokee Chica
Las Rocosa Katchina
Las Rocosa Moss Agate
Las Rocosa Red Hud
McCray Red Dawn
Tin Foil Injun
Rowe's Camanche Warrior
Clingers Royal Blue Merle
Daughenbaughs Gin Fizz
Frantz's Blue Girl
Frautz's Copper
Frantz's Patches
Near's Blue Midget
McNulty's Squaw
Wolf's Blue Heather
Englands Brandy Dee
Thorntons Smokey Blue
Bardwells Blue Ernie
Healeys Sissy II
Olsons Troubles
Swansons Blue Shawn
Swansons Blue Gypsy
Hunters Berri Blue
Bonnie Blue
Drapers Blueberry Muffin
Kalbs Little Blue
Dewitts Sheila
Dewitt Doc
Ansels Blue Sapphire
Gantenbeins Sugar Blue
Gebhardts Blue Jack
Rohenbachs Blue Banjo
Youngs Boys Smokey
Gantenbeins Blue Muffet
Gantenbeins Mike
McGuires Semego Herni
Schmiegs Sugar Bear
Bjorks Baron
Allens Ebony Joey
J Bar D Arapahoe Lady
Dykstras Vicki of J Bar D
Hays Bo Dandy
J Bar D Millers Rex
Rustys Bonnie Blu of Tri-Way
Drapers Chips Black Buttons
Elys Blue Samantha
Kalbs Blue Panda
Clingers Ely Lucky Lady
Jones Silver Cloud
Berkeys Blue Bess
Berkeys Butch
Berkeys margo
Berkeys Silver Cloud
Berkeys Tiny Cloud
Margo of Lazy 5
Bensons Stormy
Jp Variety Blue Spice
Silvertone Sadie Blu Lady
Silvertone Sassy Suzy

Olson's Clare Blue Chips, blue merle and white male (Wood's Adobe x Wood's Little Kola). Both Adobe and Little Kola were progeny of Olson's Clare Blue Chips, who was a double bred Wood's Little Blue grandson. Sire of great note behind a variety of bloodlines which can be found from coast to coast. Courtesy of Sorensen

Olson's Clare Blue Chips (1977). Courtesy of Sorensen

Blue Blue, Smedra's Blue Mistingo, Van Gordon's Blue Asta (dam of Rowe's Comanche Warrior), Hall's Amy, Lightner's (or Lighter's) Rusty, and many others. Another Little Blue daughter by Wood's Jay was Wood's Little Kola. Kola was another dam of great note, producing Davies' Checkers, Olson's Clare's Whiskey, Olson's Clare Blue Chips, and Healy's Pepsi Cola—all dogs which continued the great Wood bloodline in the breed.

A treasured house pet of Mary Wood, Wood's Little Blue also had great character. A good mother, she was protective of her puppies and would often stand guard during tail docking, warning the vet with a quick nip if she thought he wasn't treating her puppies well.

Her daughter Wood's Little Kola took the nipping game to a new level. Most visitors to the Wood ranch were cowboys who wore high topped boots. Kola learned her nips at their heels were ineffective and she found she could get her desired reaction by reaching just above the boot top to find unprotected flesh. A Wood family friend once wore a pair of boots with a higher eighteen inch top, thinking he would fool Kola and thwart her game. She wasn't fooled and took her nips on this particular visitor on the back of his knee!

Another little known Wood dog was the male, Wood's Teddy Blue Boy. Fletcher Wood bought Teddy Blue Boy from Warren Levitt. Levitt's brother had brought this dog to Colorado from Idaho. He was a long tailed dog (likely blue merle, judging by his name) with a shaggy coat. Fletcher Wood decided to dock his tail after buying him but soon regretted the decision as the dog lost his balance along with his tail.

There are few records of Wood's Teddy Blue Boy. The only remaining connection between him and the modern Australian Shepherd is through his daughter, Porter's Yakima Blue Squaw. Yakima was the only puppy in her litter and possibly the only puppy ever produced by Teddy Blue Boy.

Yakima was one of the foundation females for the Rowe's J Bar D kennel and the main reason Fletcher Wood did not breed further from Teddy Blue Boy. In an effort to learn more about the breed and her own lines, Sharon Rowe questioned Fletcher about Yakima's sire. He stated plainly that he did not like the dog and cited Yakima as a big reason for his dislike. In this one dog (Yakima) were all the traits Fletcher disliked and did not want to perpetuate in his dogs—blue eyes, sunburst pupils, white trim, butterfly nose, flat feet, and a short, squatty build.

Olson's Clare Blue Chips (1977). Courtesy of Sorensen

Lest confusion erupt regarding the identity of two dogs named "Blue Boy," another clarification is in order. Sharon Rowe personally verified during a conversation with Fletcher Wood that the dog appearing on the pedigree of Wood's Little Blue as her sire "Blue Boy" and the dog "Wood's Teddy Blue Boy" were not the same dog. Originality in naming dogs in those days was not common! (Although it added to the confusion when discussing lineage, where confirmation can be made in identity, it will be noted in this book.)

Davies' Cholla, blue merle female (Jeffreys' Shade x Wood's Little Kola—a top producing daughter of Wood's Little Blue). Courtesy of Rowe

A last Wood dog deserves mention due to the belief of Fletcher Wood in the "Basque" heritage of Australian Shepherds. His connections in the horse world included Tom Finley of Arizona, who was the AQHA Hall of Fame breeder, judge and 17th President of the American Quarter Horse Association. Finley was a cutting horse legend in the breed and the breeder of the Quarter Horse, Doc Bar. He was also a casual dog breeder and it was in this capacity that he enters the history of the Australian Shepherd.

Fletcher Wood was seeking to improve his own dogs and thought the mythical Basque breeding was the route to this end. Tom Finley apparently had some dogs believed to be of this background and Wood decided to try one of them with his Wood's Jay foundation dogs. Wood's Dandy was a dog sent to him by Finley. Dandy was a one-man dog who was stubborn and independent. He was a destructive house dog and continued these ways out of doors. A favorite occupation was tearing clothes off

PEDIGREE OF Woods Little Kola

Born: ______ Sex: ______
Owned by: ______

Sire	Sire	Sire	Sire
Woods Jay	Sislers Shorty	Keno	Dam
		Dam	Sire
		Blue Star	Dam
	Dam	Sire	Sire
	Sislers Trixie	Blue Khan	Dam
		Dam	Sire
		Emmetts Blue	Dam
Dam	Sire	Sire	Sire
Woods Little Blue	(Sislers) Blue Boy		Dam
		Dam	Sire
			Dam
	Dam	Sire	Sire
	(Sislers) Blue Shadow		Dam
		Dam	Sire
			Dam

Bred by: ______ Address: ______ Phone: ______

The details given above are correct to the best of my knowledge. Signed: ______(Breeder)

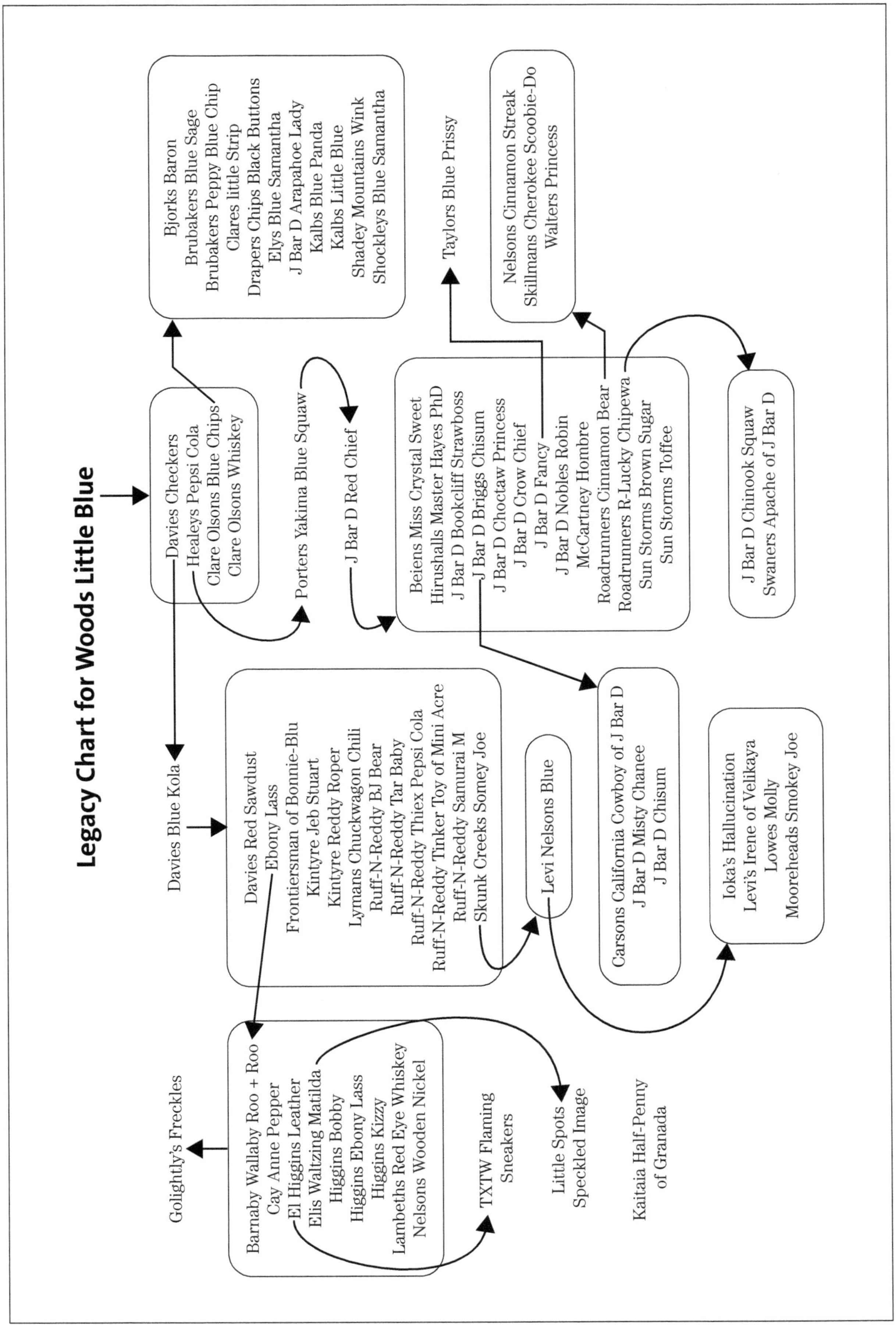
Legacy Chart for Woods Little Blue
Davies Checkers
Healeys Pepsi Cola
Clare Olsons Blue Chips
Clare Olsons Whiskey
Bjorks Baron
Brubakers Blue Sage
Brubakers Peppy Blue Chip
Clares little Strip
Drapers Chips Black Buttons
Elys Blue Samantha
J Bar D Arapahoe Lady
Kalbs Blue Panda
Kalbs Little Blue
Shadey Mountains Wink
Shockleys Blue Samantha
Porters Yakima Blue Squaw
J Bar D Red Chief
Taylors Blue Prissy
Beiens Miss Crystal Sweet
Hirushalls Master Hayes PhD
J Bar D Bookcliff Strawboss
J Bar D Briggs Chisum
J Bar D Choctaw Princess
J Bar D Crow Chief
J Bar D Fancy
J Bar D Nobles Robin
McCartney Hombre
Roadrunners Cinnamon Bear
Roadrunners R-Lucky Chipewa
Sun Storms Brown Sugar
Sun Storms Toffee
Nelsons Cinnamon Streak
Skillmans Cherokee Scoobie-Do
Walters Princess
J Bar D Chinook Squaw
Swaners Apache of J Bar D
Davies Blue Kola
Davies Red Sawdust
Ebony Lass
Frontiersman of Bonnie-Blu
Kintyre Jeb Stuart
Kintyre Reddy Roper
Lymans Chuckwagon Chili
Ruff-N-Reddy BJ Bear
Ruff-N-Reddy Tar Baby
Ruff-N-Reddy Thiex Pepsi Cola
Ruff-N-Reddy Tinker Toy of Mini Acre
Ruff-N-Reddy Samurai M
Skunk Creeks Somey Joe
Carsons California Cowboy of J Bar D
J Bar D Misty Chanee
J Bar D Chisum
Levi Nelsons Blue
Ioka's Hallucination
Levi's Irene of Velikaya
Lowes Molly
Mooreheads Smokey Joe
Golightly's Freckles
Barnaby Wallaby Roo + Roo
Cay Anne Pepper
El Higgins Leather
Elis Waltzing Matilda
Higgins Bobby
Higgins Ebony Lass
Higgins Kizzy
Lambeths Red Eye Whiskey
Nelsons Wooden Nickel
TXTW Flaming
Sneakers
Little Spots
Speckled Image
Kaitaia Half-Penny
of Granada

DeWitt Doc, a top working dog for Calvin Tupps and double great grandson of Olson's Clare Blue Chips. His sire was a full brother to the dam of Taylor's Luke the Drifter. Courtesy of Rowe

Mary Wood's clothesline before moving on to doing the same damage at all the neighbors' homes. In the end, Mary put down her foot and told her husband either the dog went or she did. His disappointment with the produce of this dog made the decision an easy one.

Fletcher's belief in Basque heritage of the Aussie, a common misconception of the time, led him to seek out this dog of supposed Basque background. He makes the following statement in the ASCA Yearbook—"I learned of a family named Moreno from California who brought some larger heavy coated dogs from Spain that were supposed to be good working stock. I crossed into these dogs to reintroduce some of the Basque character, but found that they were poor at wearing and gathering. One of these dogs was known as Finley's Texas Bob who produced a curly-coated tricolor male named (Wood's) Dandy. This dog was a real tough worker with a lot of stubbornness; he had good structure, though. I finally sold him to a dairyman in Wray, Colorado. With a lot of culling, I finally selected this blood out of my line."

Dakota Blue Lucky Chance, red merle and white male (Las Rocosa Comanches Chance x Schells Blue Chula). His sire traced many times to Wood's Jay and Sisler's Shorty through Lamar's Turk and through Las Rocosa Ruby—a linebred Wood's Jay female. His dam was a Wood's Stubbie granddaughter by Wood's Sage out of the famed Bjorks Chula. Courtesy of Rowe

This statement implies Wood did not, in fact, believe his own dogs were of Basque breeding or he would not have felt a need to "reintroduce" this type of dog into dogs he had bred from Wood's Jay. In addition, Basques are not Spanish people so the connection between them and a

California family with Spanish dogs is also highly questionable. Les Burlile speculates it is very unlikely the Aussie is bred from dogs of Basque foundation nor does he believe any of Jay Sisler's dogs were of Basque origin. It is also significant to note from Fletcher Wood's quote above that if Wood's Dandy—whom he believed to be "Basque"—was really the correct type of dog, he would not have made the efforts he did to "weed out" all traces of the crosses he made with Dandy. Fletcher Wood recognized that what he had in Wood's Jay and his descendants was the type of dog he wished to own, work, and breed. There is no evidence he ever again tried to introduce any dog on the basis of assumed Basque heritage into his line.

If any doubt remains, Mr. Wood made a number of statements in his interview for the *Australian Shepherd Quarterly* which leave no question in the reader's mind as to his stance on the supposed "Basque" dog, Wood's Dandy:

> "I did have one dog I weeded out of the program. We called him Dandy. His problem was that he would latch onto a cow's leg and wouldn't quit. We had a hard time breaking him loose. Also, he would tail a cow quite a bit. I learned the easiest way to break a dog from tailing cows is to run the cows where there are a bunch of cockleburs. The cockleburs get in their tail, and when the dogs grab onto it, they quit very quickly and go back to the heels."

> "The Dandy dog, whose blood I weeded completely out of the program because he was too mean, would grab onto a calf on its hind leg. He wasn't a head dog at all; he was strictly a heeling type of dog. He was just mean and rough all the way through."

> "...if they are rough with cattle, they're probably rough with sheep. Although it sounds crazy, Dandy made a good sheepdog. A man named George Miller ran a bunch of sheep on the wheat ground. I let him use Dandy. George said, 'He's the best dog I've ever had. Those sheep actually put on more weight than any bunch of sheep I had that year.' I said, 'Why? He's kind of rough with them.' He said,

PEDIGREE OF Clare Olsons Blue Chips

Born: ____ Sex: ____
Owned by: ____

Parents	Grandparents	Great-grandparents	Great-great-grandparents
Sire: Woods Adobe	Sire: Woods Dandy	Sire: Finleys Texas Bob	Sire
			Dam
		Dam: Merinos Blue Bitch	Sire
			Dam
	Dam: Woods Little Blue	Sire: (Sislers) Blue Boy	Sire
			Dam
		Dam: (Sislers) Blue Shadow	Sire
			Dam
Dam: Woods Little Kola	Sire: Woods Jay	Sire: Sislers Shorty	Sire: Keno
			Dam: Blue Star
		Dam: Sislers Trixie	Sire: Blue Khan
			Dam: Emmetts Blue
	Dam: Woods Little Blue	Sire: (Sislers) Blue Boy	Sire
			Dam
		Dam: (Sislers) Blue Shadow	Sire
			Dam

Bred by: ____ Address: ____ Phone: ____

The details given above are correct to the best of my knowledge. Signed: ____ (Breeder)

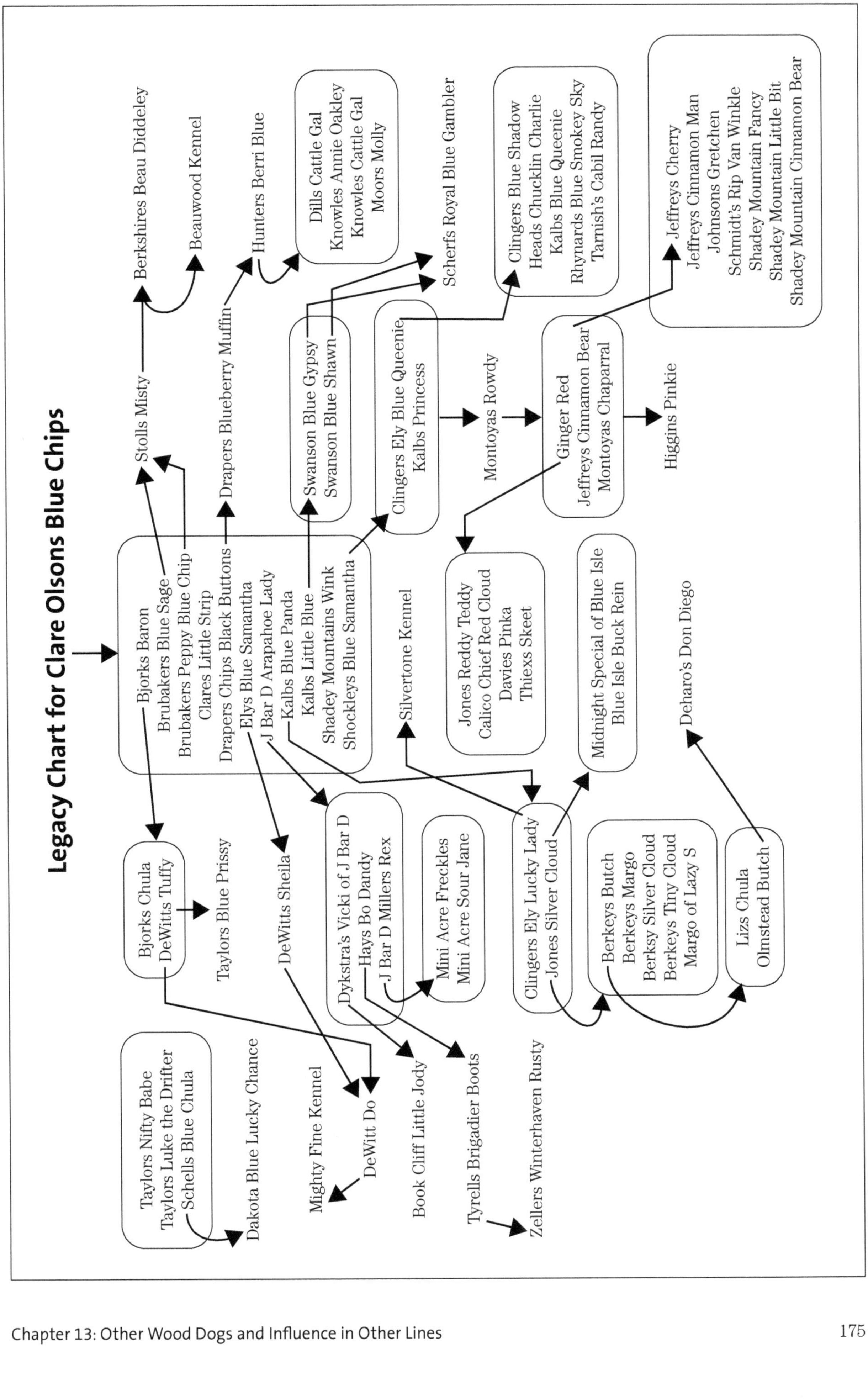
Legacy Chart for Clare Olsons Blue Chips
Bjorks Baron
Brubakers Blue Sage
Brubakers Peppy Blue Chip
Clares Little Strip
Drapers Chips Black Buttons
Elys Blue Samantha
J Bar D Arapahoe Lady
Kalbs Blue Panda
Kalbs Little Blue
Shadey Mountains Wink
Shockleys Blue Samantha
Stolls Misty
Berkshires Beau Diddeley
Beauwood Kennel
Drapers Blueberry Muffin
Hunters Berri Blue
Dills Cattle Gal
Knowles Annie Oakley
Knowles Cattle Gal
Moors Molly
Swanson Blue Gypsy
Swanson Blue Shawn
Scherfs Royal Blue Gambler
Clingers Ely Blue Queenie
Kalbs Princess
Clingers Blue Shadow
Heads Chucklin Charlie
Kalbs Blue Queenie
Rhynards Blue Smokey Sky
Tarnish's Cabil Randy
Montoyas Rowdy
Ginger Red
Jeffreys Cinnamon Bear
Montoyas Chaparral
Jeffreys Cherry
Jeffreys Cinnamon Man
Johnsons Gretchen
Schmidt's Rip Van Winkle
Shadey Mountain Fancy
Shadey Mountain Little Bit
Shadey Mountain Cinnamon Bear
Higgins Pinkie
Silvertone Kennel
Jones Reddy Teddy
Calico Chief Red Cloud
Davies Pinka
Thiexs Skeet
Midnight Special of Blue Isle
Blue Isle Buck Rein
Deharo's Don Diego
Bjorks Chula
DeWitts Tuffy
Taylors Blue Prissy
DeWitts Sheila
Dykstra's Vicki of J Bar D
Hays Bo Dandy
J Bar D Millers Rex
Mini Acre Freckles
Mini Acre Sour Jane
Clingers Ely Lucky Lady
Jones Silver Cloud
Berkeys Butch
Berkeys Margo
Berksy Silver Cloud
Berkeys Tiny Cloud
Margo of Lazy S
Lizs Chula
Olmstead Butch
Taylors Nifty Babe
Taylors Luke the Drifter
Schells Blue Chula
Dakota Blue Lucky Chance
Mighty Fine Kennel
DeWitt Do
Book Cliff Little Jody
Tyrells Brigadier Boots
Zellers Winterhaven Rusty

PEDIGREE OF Van Gordons Blue Asta

Born: ______ Sex: ______
Owned by: ______

Parents	Grandparents	Great-grandparents	Great-great-grandparents
Sire Elys Major	Sire Breazeales Shiek	Sire Tucsons Joe	Sire Taylors Idaho Lad
			Dam Taylors Idaho Lil
		Dam Tucsons Sis	Sire Ayers Shep
			Dam Ayers Bell
	Dam Breazeales Sandy	Sire Sam Boy	Sire
			Dam
		Dam Sambo	Sire
			Dam
Dam Lighters Asta	Sire Woods Jay	Sire Sislers Shorty	Sire Keno
			Dam Blue Star
		Dam Sislers Trixie	Sire Blue Khan
			Dam Emmetts Blue
	Dam Woods Little Blue	Sire (Sislers) Blue Boy	Sire
			Dam
		Dam (Sislers) Blue Shadow	Sire
			Dam

Bred by: ______ Address: ______ Phone: ______

The details given above are correct to the best of my knowledge. Signed: ______ (Breeder)

Legacy Chart for Van Gordons Blue Asta

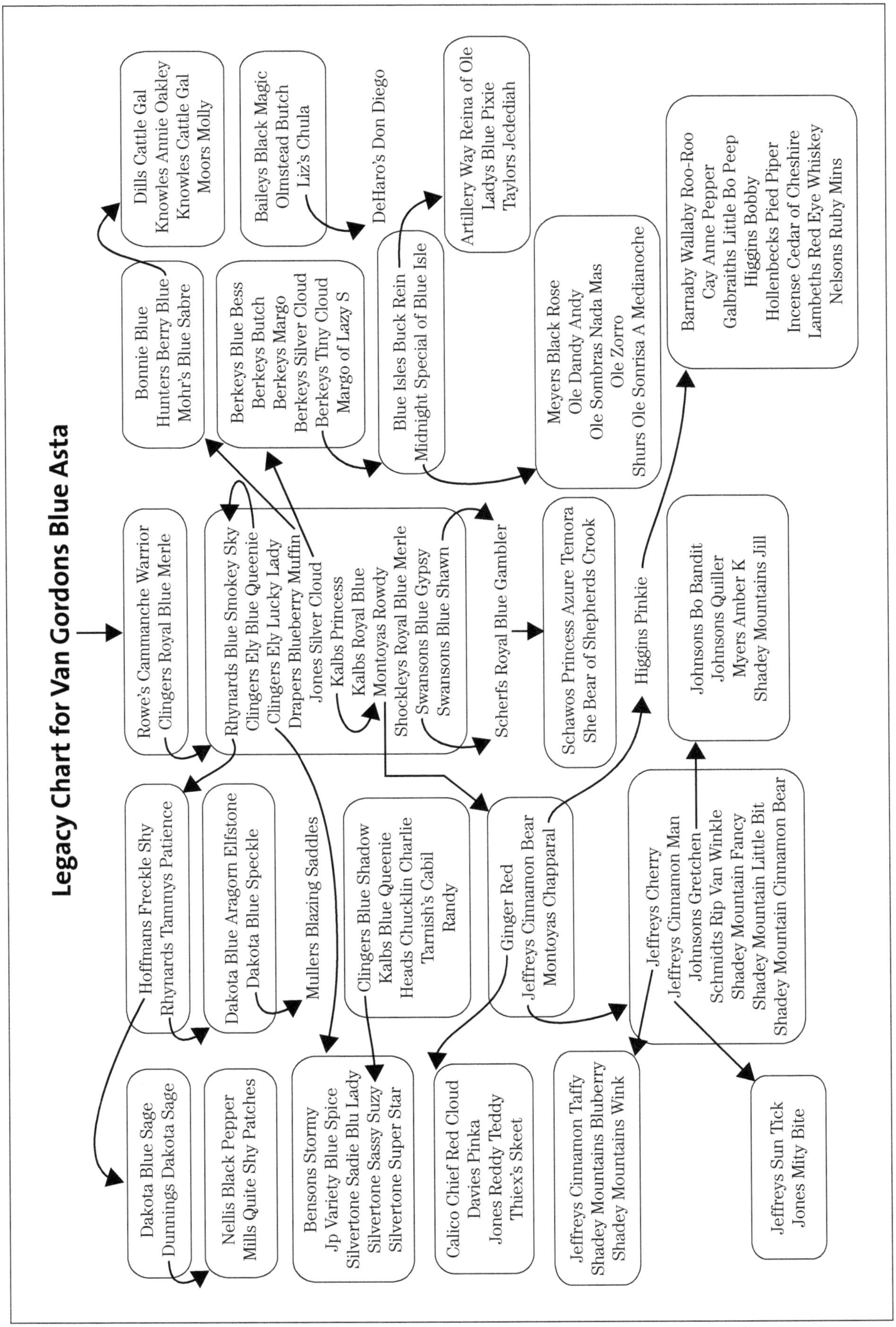

'They kept their mind on their grazing and weren't up and roaming around. The minute they'd break and start moving too quickly, he was in front of them and stopping them.' He would grab a nasty ewe and actually flop her on the ground. He was very tough."

"A bear came in among the sheep, and he challenged the bear. The bear slapped him on the head and actually scalped him. Part of the scalp fell back over the eyes. That didn't slow him down at all. He whipped the bear out of the herd of sheep. Then they took him in and sewed him up. He was a super tough dog. He bit a little bit too much. I finally let him go to some people around Rye, Colorado, on a cow outfit down there. He worked out well for them down there. They had mostly mature cows."

Wood's Dandy was used in the Wood breeding program for a brief time. A few dogs from these crosses did survive and breed on and Dandy's name can still be found at the far right hand side of some pedigrees. His name is rarely, however, found in the pedigrees of what we now consider "Woods" type dogs. His value to the Australian Shepherd is mostly as an interesting footnote. He served to negate any notions Mr. Wood might have had about looking outside his own superior foundations for breeding stock. He also serves as an example of the many mistaken beliefs which are held regarding the assumed "Basque" foundations for the Australian Shepherd. The dogs which founded the breed were a varied group of dogs which were chosen on the basis of performance. They did not come exclusively from the Pyrenees mountains nor were they brought here by Basque people. The unknown background of the breed is just that—unknown. Wherever Basque or "Spanish" influence can be verified, it is known to have been either weeded out, as with Wood's Dandy, used only lightly or not found in the pedigrees of dogs which formed the breed at all.

Another interesting footnote regarding Wood's Dandy comes from Fletcher's wife, Mary. Tom Finley was breeding Australian Cattle Dogs during the same era he was experimenting with the "Moreno" dogs behind Wood's Dandy. In fact, Fletcher also obtained one of these Australian Cattle Dogs from Finley and also named it Wood's Dandy! The only known photo of a dog named Wood's Dandy is found in the ASCA Yearbook 1957-1977. The photo shows a prick eared, short coated dog which may be the Australian Cattle Dog Dandy and not the Dandy behind dogs in the Australian Shepherd breed. It is easy to see where confusion

Thornton's Blue Blue (Wood's Dandy x Lighters Asta). Blue, copper, and white female which did not meet Fletcher Wood's idea of a good Aussie. However she did leave her mark on the breed by producing a number of prolific offspring, among them were Ansel's Blue Sapphire, Gantenbein's Sugar Blue, Gebhardt's Blue Jack, Rohenbach's Blue Banjo, and Young's Boys Smokey. Courtesy of Rowe

may have occurred and how later research often uncovers errors in prior sources.

When the term "Woods" type is used today, it refers to dogs of two origins. First were those dogs which were bred by Fletcher Wood and others who obtained his dogs and continued his breeding practices. These dogs were bred with extreme concentration on Wood's Jay. Second were those dogs which resembled the type of dogs Fletcher Wood bred—what I am calling the "mountain-type" dog in this book. These were dogs of similar look and possibly background (since many were of unknown origin) and which were often crossed with dogs of Fletcher Wood's breeding. They were dogs selected for preservation of this type regardless of pedigree, creating a relatively uniform foundation for the Australian Shepherd.

A photo identified in the ASCA Yearbook 1957-1975 as being of "Wood's Dandy." Is this the Wood's Dandy behind some Australian Shepherds or the later acquisition by Fletcher Wood of an Australian Cattle Dog from Tom Finley? The dog in this photo more resembles an Australian Cattle Dog than the written description of the Aussie, Wood's Dandy. The Aussie Dandy is described as a curly coated tricolor dog—traits which can not be seen in the dog in this photo. Wood was familiar with the Cattle Dog since Dr. Kenneth and Lela Green began importing them for their ranch use after losing Green's Dingo. Regardless of the identity of the dog in this photo, he was not important to the foundation of the Wood line. Courtesy of Sorensen

A quick mention of another Aussie bloodline is called for at this point, as odd claims are made concerning "Woods" background in this line. While, on paper, the pedigrees of Flintridge dogs, like many others, unarguably trace to a few bitches with some Wood's background, Flintridge dogs were intensively inbred on a dog named Harper's Old Smokey. Harper's Old Smokey was a heavily Irish marked, light blue merle dog of unknown lineage and different type and character from the mountain-type Aussie. Harper's Old Smokey was the antithesis of the type of dog Fletcher Wood attempted to produce—no more acceptable to him than the Rowe's Yakima. The founder of the Flintridge bloodline, Weldon Heard DVM, inbred, culled and selected (using methods reminiscent of Fletcher Wood) on Harper's Old Smokey. His culling of dogs with Wood's characteristics and his selection of breeding stock was entirely opposite of Fletcher Wood's—choosing dogs of light color with heavy white markings to represent his bloodline. While color alone is not the sole indication of mountain-type breeding, the selection and culling of dogs to preserve heavy coat, maximum white trim and copper markings, softer temperament, and large bone eliminated both phenotypic and genotypic Woods or mountain-type characteristics from this line in a very short time. Exclusive selection and continued close breeding to Harper's Old

Smokey for show ring traits has virtually eliminated any genetic tie to the Wood's type dog. The only way to bring it back to this line now is by crossing back to todays mountain-type dogs.

chapter 14

Dick and Leslie Sorensen

J BAR D / COLORADO

Much of Australian Shepherd history has been written from the singular perspective of each writer. As stated in the introduction to this book, this sort of vision is deliberate and unapologetic in this volume. The light treatment of the influence of Jay Sisler and Fletcher Wood has only been matched by the dismissal of the significance of the J Bar D and Colorado Australian Shepherd bloodlines and the impact of Jerry and Sharon Rowe and Dick and Leslie Sorensen on this breed.

The groundwork for much of the Australian Shepherd's history was laid by the quiet and hard working families who first comprised the J Bar D name. The fortunate alliance of the Rowe's and Sorensen's happened in 1964 when they met at a community obedience class held at Fletcher Wood's home-away-from-home, the Jefferson County Fairgrounds. The Rowe's had recently acquired their first Australian Shepherd, Rowe's Dingo, and decided she needed some training. Dick Sorensen had made the same decision about his young male, Sorensen's Gunsmoke. By the end of the class, both dogs were in competition for top dog, with Dingo narrowly beating her future mate. The friendship grew steadily and soon Dick and the Rowes formed the partnership which became the J Bar D kennel (created using the first letters of Jerry and Dick's names).

The friendship and the Rowe's involvement with the breed grew from this point, but Dick Sorensen was merely continuing his lifelong involvement with the Australian Shepherd. Dick Sorensen grew up with a yellow Aussie female named Queenie. In a manner reminiscent of so many Aussies of that day, the family obtained her from a neighbor while they were living on a farm west of Arvada, Colorado. Queenie was born on a cattle ranch before she moved in with the Sorensen's in 1947. Although Dick says she would work the calves and goats around the farm, she wasn't as good as his later dogs. She did, however, have one unique talent. At that time, neighbors would run flocks of turkeys in a common area. Queenie could be sent out in the evening to collect the six to eight Sorensen turkeys and she never failed to bring home the right ones!

Dick Sorensen's family had a long association with the Australian Shepherd as it developed. Pictured are his great uncle Gilbert Poppe and Gilbert's mother, Ada, with their blue merle Aussie in Kiowa, Nebraska. Circa 1920. Courtesy of Sorensen

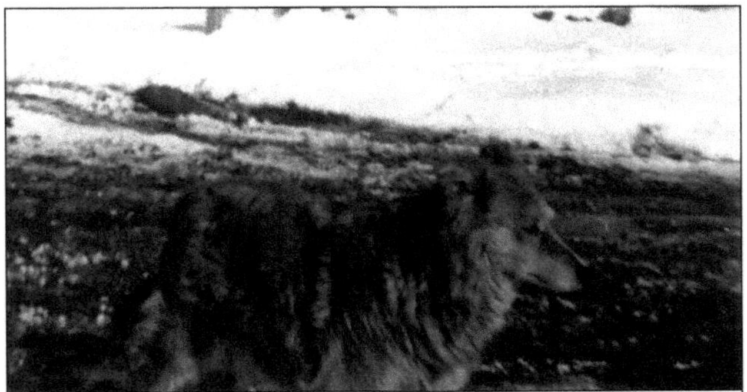

Dick Sorensen's first Australian Shepherd, Queenie, pictured in 1959. Courtesy of Sorensen

Through Junior and Senior High School in the mid-1950s, Dick belonged to FFA and raised a series of calves and lambs. He owned and trained a number of dogs, but none became the dog which had what he was looking for. He soon looked up a friend from his elementary school days, Tom Jeffreys, who had started in Aussies with a dog named Jeffreys' Deadeye Jr.—known as Ted. Tom was completely sold on Australian Shepherds and was beginning to breed from two superior bitches named Jeffreys' Sheila and Jeffreys' Missy Blue. Dick's life with Australian Shepherds changed forever when he got Sorensen's Gunsmoke from Sheila's first litter sired by Williams Laddie. As Gunsmoke grew, he displayed his talents for working stock on Dick's sheep. Since Dick had no knowledge of training stock dogs, Gunsmoke was on his own in figuring out his job. Dick decided he needed to learn how to train dogs if he was going to make the most of this remarkable puppy.

Tom Jeffreys was well acquainted with Fletcher Wood and Dick was often invited to help the two gather cattle off the mountain ranches Tom worked. It was on these drives Dick began to learn about working dogs on livestock, to absorb the lessons about the breed's history from one

Dick Sorensen with his two sisters riding a sled pulled by Dick's first Aussie, Queenie. Circa 1950. Courtesy of Sorensen

Malone's Lotsa Freckles, blue merle male, brown eyes, 21", 62 pounds. Full brother to Gunsmoke's sire, Williams Laddie. An influential sire in the southwestern United States for the Cattle Call and Hazelwood kennels. Later used by the Sorensens to bring back some of Gunsmoke's lines into their breeding program. Courtesy of Sorensen

of it's major influences and to become familiar with the founding dogs firsthand. Gunsmoke, too, gained priceless experience on these drives and continued to develop his genetic potential as a working dog to the fullest. It was also during this time Dick was fortunate enough to meet the great Jay Sisler. It was, of course, at the National Western Stock Show in Denver where he remembers meeting Jay and seeing his second group of dogs including Joker and the Greyhound Sparky.

For awhile, Dick worked in Wyoming at the Highlands Sheep Ranch. It was during this time he finally learned how to really use a good working dog. A border collie trainer was working dogs on the ranch and Dick was able to see how it was done. Although he attempted to learn from this trainer, the man would not teach him what was involved in training a stock dog.

When Dick returned to Colorado he joined Tom Jeffreys and wife Charlene in leasing the Flagstaff Ranch west of Boulder. The three partners spent many long days working cattle and dogs in the mountain pastures. Dick now had Gunsmoke and three blue merle bitches—Sorensen's Blue Mist (pick puppy from Gunsmoke and Rowe's Dingo's first litter), Sorensen's Dolly, and Sorensen's (Nightingales) Tammy—that came to the ranch with him. He had plans to do some breeding when he was drafted and sent to Vietnam from 1967-1969. While he was gone, his parents and friends kept his dogs. By the time he left, he had acquired another female named (Hall's) Sorensen's Amy. Amy, left at home, became particularly attached to Dick's mother and was a wonderful watch dog for her. Always happy to allow people into the home, Amy required permission from a family member to allow them back out. She kept the neighbor dogs off the Sorensen's half acre of land and let Dick's mother know when visitors arrived. Dick had gotten Amy from Dennis Hall who, Dick thought, got her from Jim Healey. In Jim's hands Amy had been hand fed and refused to eat regular dog food. The family finally discovered that putting her dog food in a skillet fooled her into thinking she was getting human food and she would eat. However when it was her turn for a piece of toast she always refused unless it was buttered. Amy's breeding was familiar—she was by Wood's Jay out of a Wood's Jay daughter, Lighter's Asta. A long lived dog, she was still around late into the '70s holding court at home with Dick's mom.

Sorensen's Gunsmoke in 1964 as a very young puppy. Courtesy of Sorensen

Dick reclaimed most of his dogs after his service was done and joined the local Aussie fanciers continuing to learn how to work their dogs. Working with the Rowe's, Fletcher Wood, the Gantenbein's, Jeanne Farned, Floyd Leavitt, and others, the group formed one of the early Aussie clubs in Colorado, ASC of Colorado. They purchased a small band of cull sheep and used their imaginations to come up with ways to try to train the dogs. The early attempts to understand working instinct and how to use it often ended up comically. Nearly everyone had a little bit of knowledge but the group was not able to make sense of the entire picture and good trainers were rare or reluctant to help. Unless the dogs were working their own stock on familiar ground, chances were they were locked up when any out-of-the-ordinary stock work was being done.

Sorensen's Gunsmoke, the iconic anchor of the Colorado bloodline. Courtesy of Sorensen

Sorensen's Gunsmoke. His intelligence and quality speak across the years. Courtesy of Sorensen

Briarbrook's Crafty Lady, blue and white female (Malones Lotsa Freckles x Jeffreys' Peppy. Jeffreys' Peppy was a littermate to Masterson's Dotclijee Shunks by Jeffreys' Pooh Bear x Poudre Condesa), Granddam of Colorados Ima Lotta Memories. Courtesy of Sorensen

Gunsmoke was Dick's main working dog in those days. In 1971-1972 he had some steers on a ranch in Toponas and used Gunsmoke on them from time to time. He remembers Gunsmoke as having nice wear behind the stock but not much bite. In this same time period, Dick and his wife, Leslie, moved to Toponas, Colorado, on the western slope of the Rockies, leaving the J Bar D name to the Rowe's. Although many of the Sorensen's early dogs bear their last name, it was soon after their move they adopted the name "Colorado" for their kennel. It is under this name the genetic worth of Gunsmoke became most famous.

Gunsmoke had not gone unused prior to the move and formation of the Colorado kennel. When the Rowe's decided to breed their Dingo she was bred to Gunsmoke and produced the beautiful dark blue merle bitch, Sorensen's Blue Mist. She is described as having small black spots all over with a silvery blue base color. Dick says she was "cute headed" and remembers many of Gunsmoke's females had very good heads. Blue Mist, typically, had few white markings, just a little on her feet. She was 20 to 20-1/2 inches tall and about 40 pounds with the same streamlined body type of her sire. Blue Mist was one of the bitches Dick took to Toponas and it was there she was bred to Jeffreys' Shade to produce the excellent sire, Jeffreys' Pooh Bear. Gunsmoke crossed with Sorensen's (Nightingale's) Tammy produced Sorensen's Queenie, granddam of Piz's Little Joe through her daughter J Bar D Choctaw Chili.

One of Gunsmoke's most well known sons was the Rowe's own J Bar D Comanche Warrior, a stud fee puppy from the Wood's Jay granddaughter, Van Gordon's Blue Asta. From that same cross came Kalb's Royal Blue who, when bred back to his daughter Kalb's Princess, produced Montoya's Rowdy, who sired the dam of Jones Reddy Teddy, Ginger Red. If Gunsmoke had never sired another litter after this group of dogs, his place in Aussie history would have been secure. However, he was to continue producing top quality Aussies for many years to come.

Williams' Laddie, dark blue merle and white male. Courtesy of Rowe

It was these early breedings which gave the Sorensen's their first indication Gunsmoke was the perfect cross with the predominantly Wood's dogs which were abundant in this era. Fletcher Wood's breeding program produced a top quality cattle dog which was ideal for remote areas and tough terrain. This type dog could be sent into rough country and would bring back any livestock he could find. These dogs were valuable for protection and would do much of their own thinking for themselves. At times these traits could be a disadvantage, especially with the growing interest in formal stock training and trialing. A slightly more biddable dog was needed to work stock in closer quarters and the growing popularity of the breed required a dog with a little more social personality. Gunsmoke was proving to be that perfect additive to the good Wood breeding. Without sacrificing the good qualities of Fletcher Wood's dogs, Gunsmoke could contribute needed traits which nicked extremely well.

Dick Sorensen in 1965 with his original dogs. L to R: Sorensen's Gunsmoke, Sorensen's Blue Mist, and Sorensen's Hall's Amy. Courtesy of Sorensen

Gunsmoke, like many "first" dogs, was the one who went everywhere and did everything with Dick. A famous photo of Gunsmoke shows him jumping through a hoop held above the ground by Dick. His repertoire of parlor tricks was extensive. He developed a liking for beer and would drool copiously whenever cold ones were brought out. After Dick drained a bottle, Gunsmoke would hold it between his paws

Colorado's Cody of Cattle Call, blue and white male (Malone's Lotsa Freckles x Chilton's Matilda). This very correct male was an early cross of old working lines with foundation show lines. The prepotency of the working dogs shines through easily in this superior male. Courtesy of Sorensen

to lick it dry. He would ride on horseback or on a motorcycle. Dick would tell him to "stick 'em up" and Gunsmoke would hold his paws in the air. Dick would shoot him and he would "die." A cluck of the tongue would send him and his housemates to bed. A whistle was the signal to either get in the truck or out. Gunsmoke could easily find his own truck in a parking lot full of them.

Gunsmoke's attachment to Dick cost him first place in the obedience class he attended with Sharon Rowe. Gunsmoke and Dingo were tied for first in the class and during a runoff were required to down-stay across the ring from their owners. Gunsmoke couldn't stand being separated this far from Dick. Although Gunsmoke remained "down" for the entire time, he slithered across the floor on his belly in order to be at Dick's feet for the duration.

Gunsmoke went everywhere with Dick, including to work. The rule of the truck was "no dogs on the seat." Although Dick never caught him there, the seat was always unusually warm when he returned. During the summer, Dick left the windows of the truck open and Gunsmoke spent the day sleeping underneath it. He could jump in and out of the window as he pleased. Gunsmoke and Tammy often went to the feed store with Dick. They enjoyed jumping onto the roof of the cab to watch the feed being loaded. One day, Dick was inside paying for his purchases and visiting with the locals when the young man who was loading his feed came into the store with a stricken look on his face. It seems he had made a mistake and loaded the wrong feed on the truck. When he discovered his mistake, he attempted to unload the feed. The dogs, however, had already determined this was their feed and promptly stopped the boy from taking it back. A streak of mischief once prompted Dick and his friends to set up an inebriated bar patron to try the same thing. From their vantage point inside the bar, the group enjoyed a great show from the dogs and the would-be thief.

Leslie describes Gunsmoke as a top sire of outstanding females and the record confirms this. Important dams of lines which became legendary can be traced back to Gunsmoke. His best known son was Rowe's

Comanche Warrior, who became the first working titled dog in the breed. Comanche, too, was a top sire and very long lived. Leslie describes Gunsmoke in an interview for the *Australian Shepherd Quarterly*:

> "...his puppies were an absolute joy. They had this head and coat and expression, this do or die attitude, in every single one of them. Little, fat tumble bugs—they all were so cobby, so nice and just so together, had such neat little personalities, every one of them. He stamped his puppies with a look that I've never seen in anybody else's bloodline except through Gunsmoke. He was one of those rare individuals that could produce himself, not only his temperament but his individual type, his longevity, just everything about him. He was exceptional, kind of a Man O' War among dogs, so to speak." Leslie continues, "The Gunsmoke pup was the pup that had an incredible head, an incredible expression, the ability to look straight into your eyes; he never left your side even at five, six, or seven weeks of age; he followed you right at your heels wherever you went. He did his best to do everything that you did. He bent over backwards to try and understand everything you were saying. He had incredible concentration. The face was complete with a set of eyes that you could just swim in, the depth of character that was in those faces. It's an uncanny thing, hard to put your finger on it, but we would get those particular puppies in a litter out of Gunsmoke, and they were just incredible."

Colorado's Ima Lotta Memories, blue and white female (Colorado's Talk of the Town x Colorados Ima Pepper Too). (Colorado's Ima Pepper Too was a daughter of Colorado's Clint x Briarbrook's Crafty Lady.) The Sorensen's continued to concentrate the Gunsmoke lines for many years, preserving their original type in beautiful bitches such as Lottie. Courtesy of Sorensen

Physically, Gunsmoke was the ideal the breed standard describes. A moderate 21 inches in height and 50 pounds in weight, he was big and strong enough for any kind of job without any excess. He was a steel blue merle with minimal white markings and carried a rock solid topline. He would work any type of livestock including horses. In the method of the time he was used to wear behind livestock

Legacy Chart for Malones Lotsa Freckles

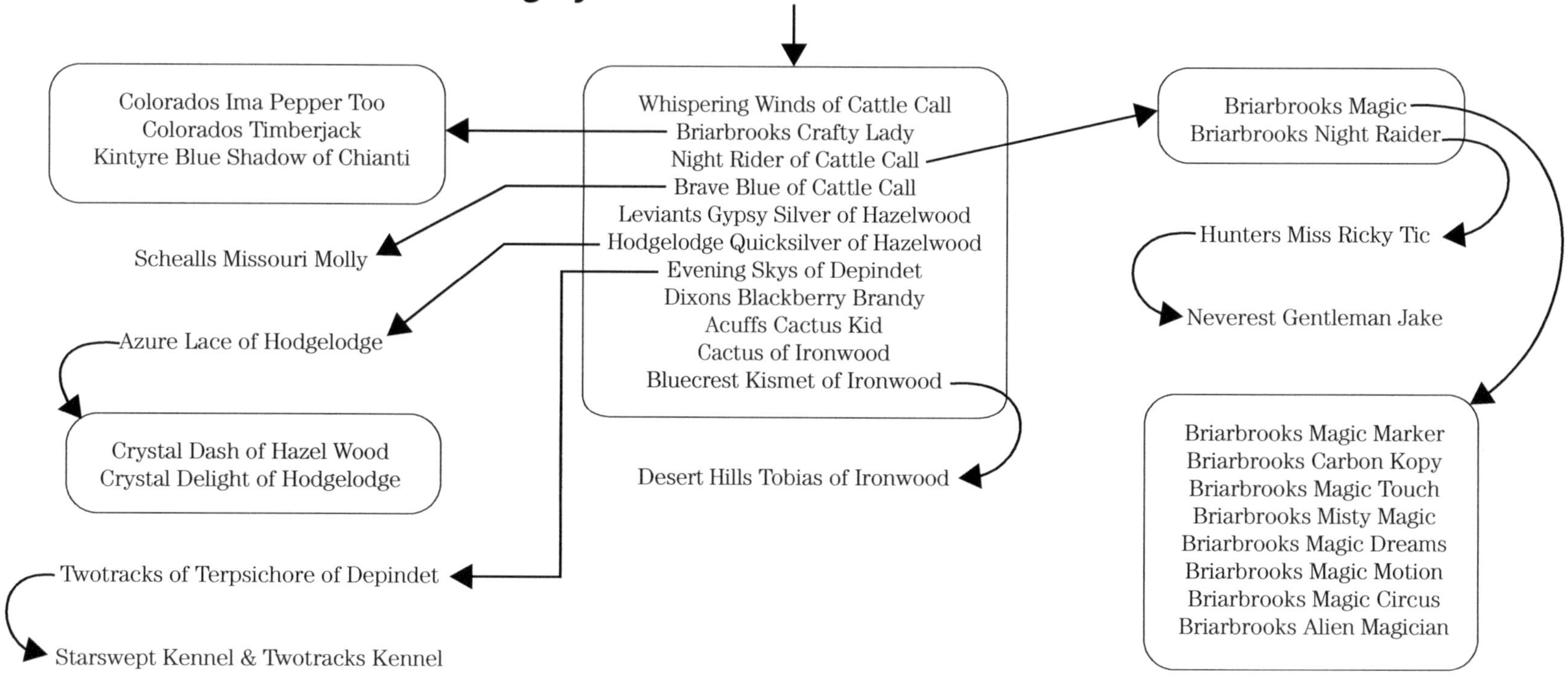

and push them. People who had and used Aussies had not yet learned to appreciate and use heading ability in a dog and discouraged or got rid of dogs which showed this trait. Dick recalls Gunsmoke working in wide natural 180 degree sweeps behind his stock, never losing or missing any individuals in his group.

A couple of interesting minor points about Gunsmoke's genetics involve tail lengths and dewclaws. Gunsmoke's natural bob tail was about 4 inches long with a "spike" on the end marking the tip of his tail. Like many Aussies, he grew an impressive flag noticeable when he carried his tail erect. This trait is typical of many natural bob tailed dogs and he produced quite a few natural bob tails, reinforcing the frequency of the NBT in this line. He also shared another trait common to many of the Wood and mountain-type dogs. He did not produce many puppies which had rear dewclaws. This is a trait which also is still frequently found in dogs of this line today.

Besides Amy and Blue Mist, Dick was able to acquire several other top quality bitches. The Wood's Jay granddaughter, Sorensen's (Nightingales) Tammy, was acquired from the Nightingale family as an older bitch. Tammy was a prolific producer with 50 puppies on the ground already from 5 litters. Tammy was an easy going blue bitch with a fair amount of white markings. She was thick bodied but smaller than Gunsmoke and had been worked by the Nightingales as well as being used for breeding. She had already proven her worth in the whelping box as dam of Nettesheims Twinkle (a Wood's Jay daughter and foundation female of the Flintridge line) so she was an excellent prospect to

Colorado's Ima Lotta Memories as an older dog. Soundness equates to long and healthy lives. Courtesy of Melissa Couch

The Flagstaff Ranch leased by Dick Sorensen with Tom and Charlene Jeffreys. In the foreground are 3 of Dick's dog—Gunsmoke, Blue Mist, and an unidentified third dog. Courtesy of Sorensen

Dick Sorensen with Sorensen's Gunsmoke (jumping through hoop) and Nightingale's Tammy (to Dick's right). Circa 1965. Courtesy of Sorensen

J Bar D Deka (Gunsmoke x Weimers Deesa). Courtesy of Rowe

cross with Gunsmoke. This cross was as good as expected, producing Sorensen's Queenie and a number of puppies which were sold to local dairies. Queenie was a very dark blue with a lot of black and no white. She was a small bitch, standing about 18-19 inches tall and slightly built. Queenie became the mate of Rowe's Warrior Bow but only one puppy survived in this litter, J Bar D Choctaw Chilli—a key female behind a number of great dogs. Chilli was bred to Piz's Joseph to produce the legendary Piz's Little Joe—sire of Jones Reddy Teddy.

Chilli was later bred back to her sire to produce J Bar D Brandy. Brandy bred to Piz's Joseph produced Sorensen's Cherry Brandy—the first conformation champion and working titled dog in the breed. Cherry was also one of a handful of dogs from across the United States invited to the first Aussie only trial in the country, organized by Bob Carrillo as a half time show at the Santa Rosa, California, rodeo. The trial was run on cattle with a configuration similar to today's ASCA A Course.

Carrillo traveled the country selecting dogs for his invitational trial based on their ability to complete a trial course in a controlled manner. He had visited the Rowe's to watch the first all breed pasture trials the new Colorado club was holding. After this trial, Carrillo encouraged the club members to focus more on arena trials due to the greater availability of these venues. At this time Aussie stockdog training was still in its infancy and the invitational trial was intended as a showcase for the breed and a model for future development of a trial program. Carrillo wanted only the best to show to the livestock owning public. To the everlasting credit of the efforts of the Coloradans, out of the 8 dogs chosen for the invitational, 4 were from their club and 3 of those were of the old mountain-type/Wood's bloodlines—Sorensen's Cherry Brandy, Rowe's Comanche Warrior, and Jeffreys' Shade!

Cherry's preparation for the trial consisted of helping Dick work the 300 cows on the L & H Land and Cattle Ranch in McCoy, Colorado, south of Toponas every day. By now, the Sorensens had learned quite a lot about training and working dogs and were refining their own techniques based on their own real needs. Cherry was involved in every aspect of this work, from branding and chute work to checking cows, gathering and calving on the 10,000 acre ranch. Cherry knew her job and was a good choice for Carrillo's showcase trial.

Dick Sorensen with his riding mule, Rusty and Sorensen's Gunsmoke. Circa 1966. Courtesy of Sorensen

When Leslie arrived at Santa Rosa, she didn't quite know what to expect. After all, trialing in front of an audience was a new experience for all the Aussie handlers that day. And this was no ordinary audience. The rodeo halftime performance was taking place in front of knowledgable stockmen and women who knew how they wanted to see livestock handled. Not only were the trial judges marking their performance, but so were the experienced spectators. The Australian Shepherd was having it's first formal audition as a working ranch dog!

Sorensen's Nightingales Tammy (foreground) and Sorensen's Gunsmoke in October 1965. Courtesy of Sorensen

So Leslie was a bundle of nerves when she walked through the gate into the arena. But, as many trialers can now relate to, her partner just saw another day's work, handling the cattle and the course largely on her own. Leslie doesn't remember much about the run other than opening and closing the take pen gate. Cherry apparently did the majority of the

PEDIGREE OF Sorensen's Gunsmoke

Born: ______ Sex: ______
Owned by: ______

Parents	Grandparents	Great-grandparents	Great-great-grandparents
Sire: Williams Laddie	Sire: Uhlendorf's Tippy	Sire: Parks Joe	Sire: Ceuters Teddy
			Dam: Parks Cookie
		Dam: Uhlendorfs Londa	Sire: Ceuters Teddy
			Dam: Centers Buff
	Dam: Thornton's Chrissie	Sire: Watsons Old Blue	Sire: Smiths Joe
			Dam: Ronalds Caro
		Dam: Johnsons Sue	Sire: Harts George
			Dam: Meyers Missey
Dam: Jeffrey's Sheila	Sire: Ely's major	Sire: Breazeales Sheik	Sire: Joe
			Dam: Sistery
		Dam: Breazeales Sandy	Sire: Sam Boy
			Dam: Sambo
	Dam: Williams Georgina	Sire: Ely's Joker (Herpers)	Sire: Breazeales Bob
			Dam: (Breazeales) Dons Judy
		Dam: Barrett's Blue Belle	Sire: Greens Kim
			Dam: Elys Queenie

Bred by: ______ Address: ______ Phone: ______

The details given above are correct to the best of my knowledge. Signed: ______ (Breeder)

job that day, although Leslie must surely have offered some guidance to Cherry about what obstacles to take and in what order! The end result was a first place for Leslie and Cherry to add to their many other honors.

Cherry left a rich legacy of dogs to the Australian Shepherd breed. Bred to Taylor's Whiskey she produced Colorado Lace Leggins, a chocolate and white female who was sold to Melvin Townsend of the Kintyre Cattle Co. as a working dog and brood bitch. Lace Leggins was about 19-1/2 inches tall and weighed about 40 pounds. She had a wonderful temperament toward people but was a tough cow dog when Melvin worked her. She preferred to hit heads on cattle but wouldn't hit the heels. Lace Leggins was bred to a number of Colorado, J Bar D and Wood dogs, producing working ranch dogs which were instrumental in keeping the working heritage alive in many other kennels.

Dick Sorensen with Sorensen's Gunsmoke (left) and Nightingale's Tammy (right) in 1965. Courtesy of Sorensen

Sorensen's Blue Mist. Courtesy of Rowe

Cherry's cross to Slash V Little Rock produced the wonderful sire Colorado's Applejack of Mini Acre. Applejack was owned by a cattle showman named Ken Claussen in Illinois whose kennel name was Mini Acre. A beautiful red merle with copper and white, he was shown to his championship and earned a number of working titles. He stood about 21-1/2 inches tall and was an excellent working dog, exhibiting the low heeling style common in his bloodline. Claussen bred Applejack to a number of his own females and those of others, including the Sorensens own Nichols Little Bonnie. A female from this cross named Colorado Mini Orphan Annie was the pick of this litter and kept by Ken for his breeding program and to work his ranch. A dark red and white, Annie was bred to several straight Wood bred males to produce another generation of top

Sorensen's Gunsmoke as a young dog in 1965. Timeless beauty, correct breed type, superb clear blue color. Courtesy of Sorensen

A superior foundation in any age. Dick Sorensen with his special group of blue dogs. L to R: Sorensen's Gunsmoke, Sorensen's Blue Mist, Sorensen's Dolly, Sorensen's Nightingale's Tammy, and Sorensen's Queenie. Courtesy of Sorensen

working dogs. An Applejack brother, Colorado's Red Kando sired one of the first litters produced at the Talkook kennel in Nevada. Other puppies from this cross made their mark as outcrosses for kennels in Texas and on the east coast.

George's Buckeye Bobby was another of Cherry's mates. A daughter of this cross, Slash V Colorado Flame, was owned by Terry Martin of Slash V kennel. Flame produced the two breed champions Slash V On the Rocks and Slash V Brimstone when bred to George's Rocky Top. Flame's sister, Colorado Buckeye Barbie, was bred to Jones Reddy Teddy to produce several foundation females for other kennels. The first of these was Ruff N Reddy Magic Sparkle, a multi-titled red female who produced another generation of top cattle dogs for Kintyre Cattle Co. when bred to the heavily linebred Wood dog, Olson's Blue Joe. Another Ted and Barbie daughter, Ruff N Reddy Teddys Tuffit became the foundation for Sue Noll's Tuffernhel kennel. Tuffit was also bred to the foundation Wood male, Olson's Blue Joe and then back to the classic Jeffreys' Pooh Bear. The Jeffreys retained a female named Shady Mountain Ruffles from this cross.

J.T. Walters in Oklahoma obtained his first good dogs from the Sorensens from Cherry's cross with Sorensen's Red Man. The chocolate red merle champion Sorensen's Colorado Royal became an outstanding foundation sire for Walters. Royal stood a mere 20 inches and weighed only about 45 pounds but he was all dog. He had good bone and muscle, nice movement and the good temperament the Colorado dogs were famous for. He was used extensively at stud and worked the ranch on a daily basis. His red full sister, Walters Colorado Brandy, joined him in Oklahoma and became a pillar of the Walters breeding program. Through these many good offspring, Cherry's excellent qualities were preserved for the future.

Dick also owned and trained a Comanche son named Brown's Coon when he and Leslie lived in Toponas. He was purchased from his breeders, Marvin and Paula Brown, because of his resemblance to his sire. Coon was a grown dog when he came to Sorensens and Dick describes him as leggy with good color. Coon took awhile to decide to become a stock dog. While he loved to chase cats, rabbits, deer, and squirrels around the ranch he had little interest in cattle in the beginning. Dick remembers Coon tagging along with him and Cherry while he made his rounds on horseback for 3 months without showing any interest in the cattle. One day Dick found some cows on the wrong side of a BLM fence. He was a long ride from the nearest gate and needed to get those cattle back quickly. Cherry and Coon were sent over the fence to bring the cows along while Dick rode ahead to open the gate. As the dogs approached the gate with the cattle, Coon somehow managed to get kicked so severely by one of the cows that Dick was certain he had been killed. He emptied a canteen of water on him in an attempt to revive him with no success. After about half an hour, Coon staggered to his feet and seemed to be able to walk, so Dick and Cherry picked up the cattle and resumed their journey home. Whether it was a vendetta against the beast who nearly killed him or his constant exposure to Cherry's heeling talents, later that day Coon suddenly realized what he was supposed to do. By the time the herd reached home Coon had expertly heeled three or four cows and was a reliable worker for the Sorensens from that day forward.

Sorensen's Blue Mist winning in obedience with Sharon Rowe. Courtesy of Rowe

Coon had other talents, high jumping being a particularly valuable one. He could jump seven feet into the air over Dick's head to get into the truck. Dick could ask him to clear any fence merely by throwing his jacket over the top wire. This skill came in handy when he found cattle on the opposite side of the fence while horseback. Instead of having to ride quite a distance to find a gate, then trying to find the cattle again, he would send Coon over the fence to bring the cattle along until they came to a gate. Dick would open the gate and Coon brought the cattle through. Jumping ability, whether over fences, into

PEDIGREE OF Sorensen's Blue Mist

Born: ______________ Sex: __________
Owned by: ______________________________

Parents	Grandparents	Great-grandparents	Great-great-grandparents
Sire Sorensens Gunsmoke	Sire Williams Laddie	Sire Uhlendorfs Tippy	Sire Parks Joe
			Dam Uhlendorfs Linda
		Dam Thorntons Chrissie	Sire Watsons Old Blue
			Dam Johnsons Sue
	Dam Jeffreys Sheila	Sire Elys Major	Sire Breazeales Shiek
			Dam Breazeales Sandy
		Dam Williams Georgina	Sire Elys Joker
			Dam Barretts Blue Belle
Dam Rowes Dingo	Sire Welz's Handsome King	Sire Harpers Old Smokey	Sire Tucsons Joe
			Dam Tucsons Sis
		Dam Smedras Blue Mistingo	Sire Woods Dandy
			Dam Lighters Asta
	Dam McGuires Palojami Gold	Sire Farneds Dead Eye	Sire Gregg
			Dam Whiskey Bell
		Dam Hoopers Blue Keno	Sire Lillys Mr. Blue Chips
			Dam Lillys Pecos Jane

Bred by: ______________ Address: ______________________ Phone: ______________

The details given above are correct to the best of my knowledge. Signed: ______________________(Breeder)

trucks, onto horseback or into and on a thousand other obstacles was such a valuable skill, it was originally written into the very first breed standard! Not long before the Sorensens moved back to the eastern slopes of the Rockies, Coon was sold to a Midwestern buyer. When the man came to pick him up, Dick told him about Coon's remarkable jumping abilities. The unbelieving new owner decided to test this claim by asking Coon to jump onto the top of his horse trailer, which Coon promptly did. He was soon to regret his boldness as Coon refused to come down!

The first dog show for Australian Shepherds ever held in Colorado (1969). L to R: Jerry Rowe with Rowe's Comanche Warrior, Dick Sorensen with Sorensen's Gunsmoke, and Sharon Rowe with Sorensen's Blue Mist. Courtesy of Sorensen

Another Gunsmoke son was returned to Sorensens following a stint in a pet home. Being a dog who needed a job and being unemployed in this household, he began to create "jobs" for himself. Excavating the family's backyard, dragging their child around at the end of a leash and using his inborn jumping ability to make a mockery of their fenced yard, this dog soon overwhelmed his owner's abilities. He even seemed to have spent time climbing trees as Sorensens found he had been speared by a branch through the skin and brought home part of a tree when he returned.

Another of Dick's earliest breeding bitches was a Gunsmoke daughter named J Bar D Flirt. Flirt's dam was the Jeffreys' foundation bitch, Jeffreys' Missy Blue and Flirt's breeding brought together half brother and sister through their sire, Williams Laddie. Her chain of ownership is somewhat unclear but she was likely obtained by Dick prior to his Army service. Jim Healey had her while Dick was in Vietnam and worked her on his livestock during that time. Dick got her back when he returned home and she remained with the Sorensens for the rest of her life.

Leslie remembers her as a dark self merle with tiny bits of white on her toes and chest. She had excellent conformation and fluid, free movement. Her size was about 19 inches and 37 pounds and she was a natural bob tail. Although she was worked on the Healey ranch, the Sorensens didn't use her at their place. She was their loved companion and raised several good litters for them. Her wonderful temperament made her ideal for both jobs.

Her daughter, Hall's (Sorensen's) Apache, was an early influence on both the Colorado and Silverledge breeding programs. She is described by the Sorensens as a smaller blue with lots of black with a nice earset,

A young Sorensen's Gunsmoke. Courtesy of Rowe

natural bob tail and was similar to Flirt overall. Like Dick's earlier female Amy, Apache was acquired from Dennis Hall after years of trying to purchase her. Hall worked her on the cattle at his ranch where she was a valuable asset. The Sorensens were interested in adding her excellent breeding to their kennel.

Sired by the incomparable Wood's Stubbie, Apache produced several important progeny by the outcross sire, Ch George's Buckeye Bobby. One of these was another foundation bitch for the Silverledge kennel, Taylor's Calamity Jane CD. Her story will be told in greater detail in the Silverledge chapter but one of her sons, Ch Taylor's Shamrock of Silverledge CDX, later returned to Colorado to contribute to the Sorensens—and the breed's—history. Another Buckeye Bobby x Apache daughter was Ch Sorensen's Colorado Cajun, a blue merle, copper and white female influential in several east coast kennels, the Blue Lad kennel and the early Briarbrook dogs.

The Flirt daughter J Bar D Fancy was another very important female bred by the Sorensens. Sired by Ch J Bar D Red Chief, Fancy produced her own dynasty of great dogs. She is described by Leslie as a quiet, soft tempered girl. On the small side, she measured only 18-1/2 inches tall and weighed just under 40 pounds. Her coat was a brilliant silver blue, she had high-set ears that just tipped over and a wonderful flying trot.

Another well known Fancy daughter is Taylor's Blue Prissy. Prissy was one of the first dogs Tom and Jean Taylor acquired for the fledgling Silverledge kennel. Sired by Dewitt's Tuffy, Prissy was linebred on Wood's Jay with the Gunsmoke cross, which nicked so well, plus a dash of the California McConkey breeding. The Rowe's were also adding the McConkey lines to their breeding program and proving this cross to be a good one.

Prissy traveled to the east coast to the Taylors horse training headquarters. There she was bred to the Wood line Taylor's Luke the Drifter and to Taylor's Red Wilbur "Willy" aka Sorensen's Red Man. The mountain working dog of the west was first introduced to many eastern dog lovers by the puppies produced from these crosses. The Luke x Prissy daughter, Taylor's Lady She Bear, was bred to Willy to produce the foundation bitch of Southwynd kennel, Boice's Black Pepper. She Bear's sister, Zipper Lady, produced the foundation dogs of the Good Time kennel of Al and Joan Della Rocco. The Luke x Prissy son, Taylor's Red Snuffy, eventually became the grandsire of the Canadian Briergreen foundation bitch Briergreen's Blue Bobbie.

Weimer's Deesa, blue merle and white female. One of Gunsmoke's early mates. (Jeffreys' Shade x McCallum's Blue Heather). Courtesy of Sorensen

Female by Gunsmoke x Weimer's Deesa. Courtesy of Sorensen

Prissy's first breeding to Willy produced a sire the Taylors would use several times after they were established on some of their later females, Circle K's Big Blue. His influence will be discussed in more detail in the chapter on Silverledge kennel. Prissy's final breeding while in the hands of the Taylor's was back to Willy. She was sold while in whelp to Linda Gray of Sakonnet kennel and became the foundation female for this program. From this litter came Ch Sakonnet American Revolution CD, the founding sire behind most Sakonnet dogs today. Prissy was later bred to an outcross Flintridge male to produce other dogs whose descendants still appear in the Sakonnet line.

While Prissy was establishing her dynasty in the eastern United States, J Bar D Fancy continued to produce important progeny for the Sorensens. A daughter by Sorensen's Colorado Royal, Colorados Ginger, was used extensively by JT Walters and produced a number of dogs for the Diamond J kennel. She was bred to both Slash V and Las Rocosa males, whose descendants founded the early Briarbrook kennel and the P Bar J line in the Midwest and Texas. Fancy's cross to the Farned's Dead Eye grandson Roadrunners Picacho also brought the McConkey lines back in once again.

When the Sorensens returned to the eastern slopes of the Rockies in June 1973 the Colorado breeding program began to really take off. Both Dick and Leslie now had a wealth of success and experience in both breeding and using the Australian Shepherd. From this depth of

J Bar D Deka, blue merle female (Gunsmoke x Weimers Deesa). Courtesy of Rowe

understanding they knew what they wanted in the breed and also how and where to get it. Jean Taylor states, "Leslie had a LOT of good dogs at the time...and she had a good eye for a dog." By leasing and buying a number of dogs and bitches they began to seriously develop a recognizable bloodline for the Colorado kennel.

Gunsmoke had always been, and was still, the most important part of the Colorado breeding plan. His proven worth when crossed with the Wood dogs was already established and would be retained as time went on. But other high quality dogs were to play a part in the development of the bloodline and the mountain-type Aussie. It takes a number of top flight females to prove a sire and the Sorensens were able to obtain quite a few of the very best.

Higgins Bobbie was a Wood bred red merle, highly versatile female obtained from Lloyd Higgins. Her Wood breeding came through the Piz and Montoya dogs who were well known for producing excellent cattle dogs. She was about 20 inches tall and 45 pounds with a somewhat long body. Leslie remembers her as a dog with a highly developed sense of humor and she loved people. But when it came to working cattle she became a hard headed machine. She was a low heeler which made her valuable as a deluxe chute dog. She was so good she could put "5 head of steers down a 3 head wide alley"! One thing she did not tolerate was being kicked by a cow. This crime switched on Bobbie's "search and destroy" button and her vengeance would be delivered swiftly. The Sorensens tried to trial her in stockdog trials but she was unstoppable once a cow took a swipe at her. She was truly an all around dog and earned a CDX in obedience, excelled at Schutzhund work, had a great nose for tracking and would retrieve anything. She was the type of dog everyone looked for and she had the breeding to reproduce it.

Weimer's Deesa, blue and white female. Courtesy of Sorensen

Bobbie produced a couple of litters in California before coming to Colorado. Puppies by outcross sires from her days in California found their way into the breeding programs of Betty Nelson, Pat Lambeth (Little Spot), and Myra Fourwinds. After arriving at Colorado kennels, she was bred only to linebred Wood/Gunsmoke males. Ruff-N-Reddy Mity Mean, Taylors Shamrock of Silverledge, J Bar D Blue Boy, and Colorados Talk of the Town were all crossed with Bobbie and each litter produced good dogs typical of the line.

This blue merle and white female is J Bar D Deka. Courtesy of Sorensen

Nichols Little Bonnie was a striking black and white female who came to the Sorensens sometime around the age of 3 or 4 years. She was by the double Sisler's Shorty sire Arington's Stub (Wood's Jay x Harper's Jinx) out of a ranch dog named McLean's Lady, of unknown breeding. Bonnie was a breeder's dream with a flat blue-black coat, flying trot and a superb friendly temperament. She excelled as a brood bitch, mothering not only her own puppies but frequently caring for puppies from other litters. Before the Sorensens found her, she spent the first few years of life tied near the office in a stable. The barn had a phone at the office with a bell wired to it so it could be heard throughout the area. When the phone rang, people ran from all directions to answer it, startling Bonnie and creating a number of habits in her. For the rest of her life she would bark uncontrollably anytime she heard a phone. While tied in the barn she learned to jump at the phone, knocking the receiver off the hook and barking as people ran to pick it up. One wonders about the impression left with the caller who heard this mayhem on the other end of the line! As a stockdog she never lost a battle. In a 1987, *Australian Shepherd Quarterly*, interview, Leslie describes her this way, "She was a fighter, a ruthless heeler on stock; I mean no stock could stand her down, they could not back her down. She'd wail into them and move them no matter what, she was so tough. She was aggressive, but had a pleasant disposition to be around, maybe a little bit

Sorensen's Gunsmoke, age 10. Courtesy of Sorensen

Gunsmoke puppy out of J Bar D Cree Squaw. Courtesy of Sorensen

The Grand Old Man, Sorensen's Gunsmoke, at age 12. Courtesy of Sorensen

Ch Nichols Little Bonnie working cattle on the Sorensen's ranch. Courtesy of Sorensen

on the high strung, quick side, but very interesting."

Bonnie was used for breeding both before and after the Sorensens obtained her. She was always used with dogs of Wood or Gunsmoke breeding and proved her worth producing top quality dogs from each litter. Her mates included Gunsmoke, Colorado's Applejack of Mini Acre, Wolf's Royal Blue, Farmers Smokey and Scherf's Royal Blue Gambler. From Applejack came Colorados Mini Orphan Annie, a red and white female who produced some excellent cattle dogs for the Kintyre Cattle Co. From Gunsmoke came one of the finest bitches ever produced by the Sorensens, Colorado Perky Miss.

Perky Miss was the epitome of the near perfection that could be achieved with the Gunsmoke/Wood cross. With nearly flawless conformation and type, she could easily have become a Champion if Sorensens had shown her. Her gun metal self blue merle color was an unmistakable stamp of her breeding. Her size of 20 inches and 45 pounds was the ideal for an Australian Shepherd female. Her temperament matched the breed standard word for word, reserved with strangers but a loving companion to those she knew. Although she was never used for working livestock her record for producing top working dogs speaks for itself.

Her prolific breeding record is second to none. Bred to a variety of males with widely differing bloodlines, she never failed to produce top quality offspring. Crossed to a grandson of Gunsmoke, Colorado's Sabbath, she produced Colorado's Smooth Smoke, a female who made a significant impact producing top working dogs for the Crown Point breeding program. When bred to the Gunsmoke grandson Dorns Poudre Lakohta, her son Ch Colorado's Clint became Leslie's Gunsmoke "gene bank" to take her breeding program into the future. Clint, too, was a typical dark blue self merle who strongly resembled Gunsmoke. He had the same excellent conformation with good balance all around and an outstanding head. He was a moderate 21-1/2 inches tall and weighed around 50 pounds. His temperament was superb, exhibiting the loyal, upbeat and friendly but reserved ideal called for in the standard. Another high energy dog, he was in constant motion waiting for direction from his people. He was the sire of Ch Colorados Talk of the Town, another heavily used sire at Colorado kennel, who brought in the Flintridge outcross Leslie has used to the present day. Crown Point bred Colorado's Smooth Smoke back to her half brother Clint to produce two females which contributed a number of top working and breeding dogs to their program. Clint was eventually sold to California where he was used extensively by Betty Nelson to bring the old mountain-type breeding back up close in her program.

Kelly's Madi, dark blue merle and white female of exquisite mountain-type and unknown pedigree. Leslie describes her temperament as "fabulous" and says she had good conformation and maternal instincts. Courtesy of Sorensen

Kelly's Madi exhibiting the low heeling style typical of the mountain-type dogs. According to Leslie she was "the hardest low-heeler ever...right on the ground like Ted (Jones' Reddy Teddy)." She was used to work the hot walk at the race track. Courtesy of Sorensen

The Perky Miss daughter by the heavily linebred Wood dog Olson's Blue Joe, Ch Colorado's Columbine CD, was a wonderful all around dog. Another typical dark blue, she was the image of her sire with the more correct head type of her dam. Standing 19-1/2 inches and weighing about 40 pounds, she embodied the perfection described in the breed standard. Shown in conformation, obedience and Junior Showmanship, she achieved in every field. She was worked at home and proved to be a sensible, strong, easily trained working dog. Leslie states she was a high energy dog who was always busily moving around looking for something to do.

Gunsmoke puppy out of Kelly's Madi. Courtesy of Sorensen

Gunsmoke puppy out of Kelly's Madi. Courtesy of Sorensen

Her cross to Ch Taylor's Shamrock of Silverledge produced a striking dark blue merle dog named Colorado's Ketchum. A very intuitive dog, Ketchum would always go to the barn with anyone heading out to do chores. Showing the natural trainability of these dogs, he picked up a number of commands almost on his own. One of his most memorable was his willingness to stay when told no matter what went on around him. Ketchum was another dog who eventually was sold to a ranch in Oklahoma as a working and breeding dog.

The list of great offspring from Perky Miss goes on and on. Her daughter by Ch Wolfs Royal Blue, Colorado's Blue Majesty, was another superior working and breeding dog for the Kintyre Cattle Co. Another daughter by Olson's Blue Joe, Shady Mountain Su-Mac became a breed Champion and trial dog. When bred back to her grandson Ch Colorados Talk of the Town, she produced yet another outstanding litter of beautiful dogs. Her influence throughout the breed was important and lasting.

Wolf's Blue Heather was another lovely female leased by the Sorensens for breeding. Heather's dam was a half sister to Nichols Little Bonnie through their sire Arington's Stub, bringing the Sisler/Wood blood back into the lines. Heather produced at least two litters while at Colorado kennel. One litter was by the outstanding sire J Bar D Blue Boy, a son of Wood's Stubbie out of the Gunsmoke daughter Sorensen's Blue Mist. This cross to Blue Boy produced an amazing dog named Ch. Wolf's Royal Blue. This stunning heavy coated blue merle dog, NBT, stood 22 inches tall and weighed approximately 60 pounds. Leslie was able to finish his breed championship and stand him at stud for a short period of time. She remembers him as an extremely intelligent but gentle dog who was strongly devoted to his people. The couple who owned him evidently shared that level of devotion to him. When they divorced several years later they spent 2 years in court fighting over the custody of Royal Blue. Blue Heather produced at least one other litter for the Sorensens by Sorensen's Red Man. One puppy from this litter was later to produce another great dog in breed history, Ch. Powder River City Slicker CD, ATDcsd, AD.

Sorensen's Red Man was a dog of many names. At this time in Aussie history, the National Stock Dog Registry was the main registry. Their policy regarding names was that a new owner had the right to name their dog even if it was already registered. It was very common to change a dog's name every time the dog changed hands. Often the new owner would merely change the former owner's name to their own and leave the remainder of the name the same. Sometimes the new owner would create an entirely new name. This was the case with Sorensen's Red Man.

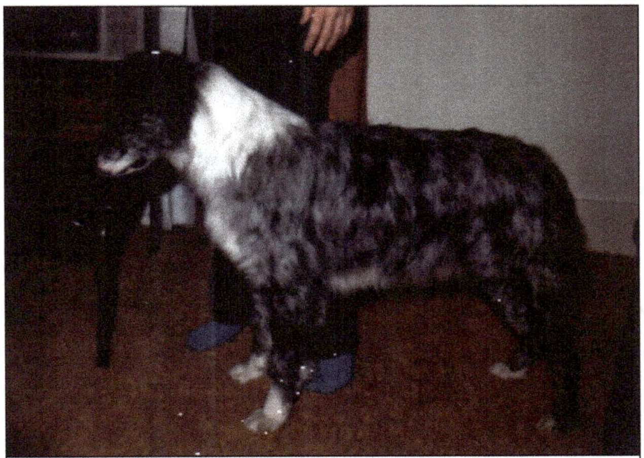

Colorado's Joshua, a beautiful clear blue merle and white male by Gunsmoke x Kelly's Madi. Sire of Mini Acre Wooden Nickel. Courtesy of Sorensen

Mini Acre Wooden Nickel, dark blue merle female (Colorado's Joshua x Taylor's Silver Buffie). Another good working dog for Ken Claussen. Courtesy of Sorensen

This lovely red merle male was bred by the parents of barrel racer Ardith Bruce on their Utah ranch. He was obtained in 1971 by Tom Taylor (of the Silverledge kennel) on one of his horse show circuits as a one year old, unregistered youngster. Tom's wife, Jean, did not travel with Tom on the circuit since she was managing the couple's eastern Pennsylvania 50 horse stable. She doesn't know exactly how Tom was able to get the handsome pup but says, "Tom could talk the stars out of heaven if he wanted something" and she speculates he may have traded Ardith some fancy show gear for the dog.

So Tom arrived home with, in Jean's words, "the cute little devil" and they got him registered as Taylor's Red Wilbur with Bruce's Blue and Bruce's Blue Susie listed as his parents. Jean states the Bruce's were likely ranchers and Willie's parents were likely their ranch dogs. In later years, Jean bred one of her females to Taylor's Whiskey and noted an uncanny resemblance between Whiskey and Willie. She describes the similiarities like this "Size—fairly small, about 35-40 lbs, type—real tight

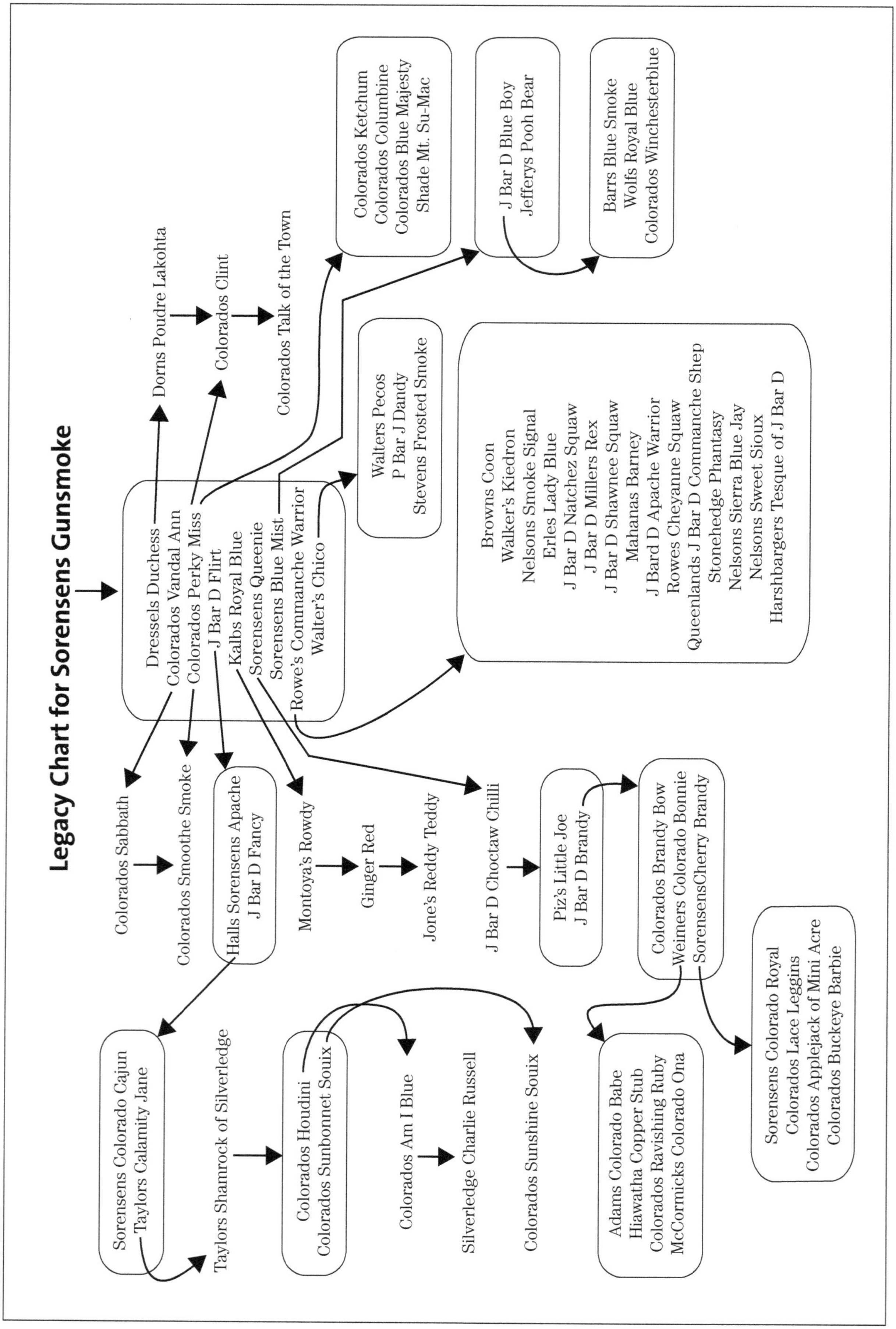

Legacy Chart for Sorensens Gunsmoke
Dressels Duchess
Colorados Vandal Ann
Colorados Perky Miss
J Bar D Flirt
Kalbs Royal Blue
Sorensens Queenie
Sorensens Blue Mist
Rowe's Commanche Warrior
Walter's Chico
Dorns Poudre Lakohta
Colorados Clint
Colorados Talk of the Town
Colorados Ketchum
Colorados Columbine
Colorados Blue Majesty
Shade Mt. Su-Mac
J Bar D Blue Boy
Jefferys Pooh Bear
Barrs Blue Smoke
Wolfs Royal Blue
Colorados Winchesterblue
Walters Pecos
P Bar J Dandy
Stevens Frosted Smoke
Browns Coon
Walker's Kiedron
Nelsons Smoke Signal
Erles Lady Blue
J Bar D Natchez Squaw
J Bar D Millers Rex
J Bar D Shawnee Squaw
Mahanas Barney
J Bard D Apache Warrior
Rowes Cheyanne Squaw
Queenlands J Bar D Commanche Shep
Stonehedge Phantasy
Nelsons Sierra Blue Jay
Nelsons Sweet Sioux
Harshbargers Tesque of J Bar D
Colorados Sabbath
Colorados Smoothe Smoke
Halls Sorensens Apache
J Bar D Fancy
Montoya's Rowdy
Ginger Red
Jone's Reddy Teddy
J Bar D Choctaw Chilli
Piz's Little Joe
J Bar D Brandy
Colorados Brandy Bow
Weimers Colorado Bonnie
SorensensCherry Brandy
Sorensens Colorado Royal
Colorados Lace Leggins
Colorados Applejack of Mini Acre
Colorados Buckeye Barbie
Sorensens Colorado Cajun
Taylors Calamity Jane
Taylors Shamrock of Silverledge
Colorados Houdini
Colorados Sunbonnet Souix
Colorados Am I Blue
Silverledge Charlie Russell
Colorados Sunshine Souix
Adams Colorado Babe
Hiawatha Copper Stub
Colorados Ravishing Ruby
McCormicks Colorado Ona

topline, good angulation, tight feet, pretty head, and there was something about the head…squint of the eyes, wide topskull, expression…could have been twins. AND the distinctive sound of the bark. And the dark liver color. Too much to be an accident." Because so many ranchers traded puppies on a whim and the Aussie's ancestors were widely scattered because of this practice, it is not unlikely the two were somehow related.

The beautiful and important Sorensen's Nightingale's Tammy. Courtesy of Rowe

Willie lived at Silverledge for 4-5 years and was bred twice to a striking dark blue female obtained from Leslie Sorensen named Taylor's Blue Prissy (the Taylor dogs are discussed in greater depth in the section titled "Silverledge"). Prissy was by Bjork's Baron out of J Bar D Fancy and Jean remembers the pups as outstanding. They were all sold to local ranchers and many dogs of this type still exist on farms and ranches in the east. Prissy was eventually sold to Linda Gray in whelp to Willie and this breeding produced her Ch. Sakonnet American Revolution.

Sorensen's Nightingale's Tammy. Courtesy of Rowe

Jean's memory isn't clear about the deal that brought Willie to the Sorensens but it was likely a trade for some of Leslie's dogs which no longer fit into her program. When Willie came to Colorado, he was registered with the new moniker, Sorensen's Red Man and it is under this name that he became most well known. Willie had obviously been a well loved and well trained dog while with the Taylors. Leslie describes his manners as impeccable and he was a joy to have in the house. She could leave food on the stove or the counters and Willie would not touch it. He only had one habit which was forbidden in the Sorensen household—getting on the couch. It only took one quick lesson for Willie to learn the couch was loaded with set mouse traps and he needed to find a comfy sleeping spot elsewhere.

In Leslie's hands, Willie contributed to producing another pillar of the Australian Shepherd breed, Ch Powder River City Slicker CD, ATDcsd,

Born: ____________ Sex: ________
Owned by: ____________________

PEDIGREE OF Nightingales Tammy

Parents	Grandparents	Great-grandparents	Great-great-grandparents
Sire: Elys Chico	Sire: Byes Rusty II	Sire: Shermans Casey	Sire: Davey
			Dam: Anza
		Dam: Byes Snooks II	Sire: Byes Taffy
			Dam: Byes Snooks
	Dam: Byes Grypsie	Sire: Bob	Sire: unknown
			Dam: unknown
		Dam: Gyp	Sire: unknown
			Dam: unknown
Dam: Hollands Nikki	Sire: Woods Jay	Sire: Sislers Shorty	Sire: Keeno
			Dam: Blue Star
		Dam: Sislers Trixie	Sire: Blue Khan
			Dam: Emmetts Blue
	Dam: Knolls Kiki	Sire: unknown	Sire: unknown
			Dam: unknown
		Dam: unknown	Sire: unknown
			Dam: unknown

Bred by: ____________ Address: ____________ Phone: ____________

The details given above are correct to the best of my knowledge. Signed: ____________(Breeder)

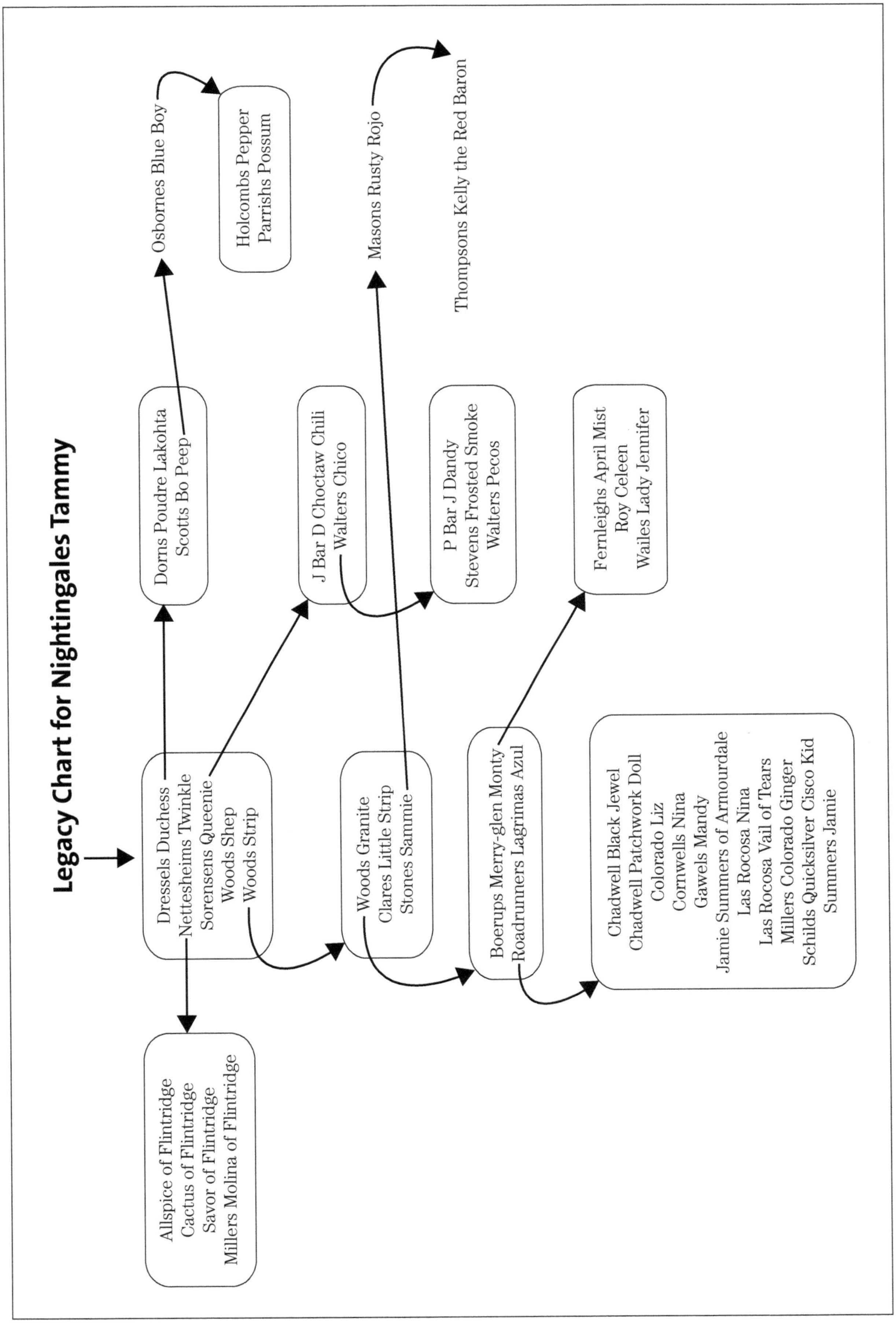
Legacy Chart for Nightingales Tammy
Dressels Duchess
Nettesheims Twinkle
Sorensens Queenie
Woods Shep
Woods Strip
Allspice of Flintridge
Cactus of Flintridge
Savor of Flintridge
Millers Molina of Flintridge
Dorns Poudre Lakohta
Scotts Bo Peep
Osbornes Blue Boy
Holcombs Pepper
Parrishs Possum
J Bar D Choctaw Chili
Walters Chico
P Bar J Dandy
Stevens Frosted Smoke
Walters Pecos
Woods Granite
Clares Little Strip
Stones Sammie
Masons Rusty Rojo
Thompsons Kelly the Red Baron
Boerups Merry-glen Monty
Roadrunners Lagrimas Azul
Fernleighs April Mist
Roy Celeen
Wailes Lady Jennifer
Chadwell Black Jewel
Chadwell Patchwork Doll
Colorado Liz
Cornwells Nina
Gawels Mandy
Jamie Summers of Armourdale
Las Rocosa Nina
Las Rocosa Vail of Tears
Millers Colorado Ginger
Schilds Quicksilver Cisco Kid
Summers Jamie

Ch Sorensen's Cherry Brandy CWD, red merle female (Piz's Joseph x J Bar D Brandy). First working titled Champion female in the history of the Ausralian Shepherd. Winner of the first Australian Shepherd Invitational Trial in Santa Rosa, California. Courtesy of Sorensen

Ch Sorensen's Cherry Brandy and Leslie Sorensen negotiating an obstacle at the Santa Rosa Invitational Trial. Photo by Stratton Photography, courtesy of Sorensen.

AD. She leased Slick's dam, Doyles Blue Buttons for the litter because she was a daughter of Sorensen's Red Man out of Wolf's Blue Heather. Buttons was a dark blue merle with copper and white trim about 19 inches tall and 40 pounds. Leslie remembers her as being pretty but stocky and well built. She was a good family dog with a superior temperament.

Leslie owned the great Slash V Little Rock CD and chose him to sire Buttons' litter. City Slicker was a solidly built black tri who matured to 22-1/2 inches and 60 pounds. He was purchased by Jim and Sue Foster as a foundation sire for their Powder River kennel. Jim Foster was a Meat Inspector for the Colorado Department of Agriculture and found Slick a useful partner on his rounds. Foster used Slick on all kinds of livestock and in a variety of circumstances. Much of his work involved inspecting herds and herd health and Slick excelled at handling stock in the pens and chutes required for this close work. The Fosters valued the old Wood's bloodlines, too, and owned a number of females from this line through Jones Reddy Teddy which were bred to Slick. Ruth McGuire remembers Slick as the best working dog she ever saw. He could work any type of livestock, had style, training and the all around ability that all the early breeders strove for.

Another good trade from the Taylors to the Sorensens was Taylor's Silver Buffie, a red merle female of unknown background but typical old mountain-type. Buffie's story is told in more detail in the Silverledge chapter, but a reminder of her influence in the hands of the Sorensens is appropriate. Her most notable sons and daughters were Walters Colorado Cash (sire of Ch Colorados Ravishing Ruby CDX, OTDcs, ATDd), Colorado Sunset, Mini Acre Wooden Nickel (great granddam of Ch Zellers Winterhaven Rusty), Colorado Link, Piney Wood's Colorado Valor (sire of Marys Rocky

Born: ______________ Sex: ________

Owned by: ______________________

PEDIGREE OF Sorensen's Cherry Brandy

Generation 1	Generation 2	Generation 3	Generation 4
Sire: Piz's Joseph	Sire: Woods Jay	Sire: Sislers Shorty	Sire: Kono
			Dam: Blue Star
		Dam: Sislers Trixie	Sire: Blue Khan
			Dam: Emmett's Blue
	Dam: Wood's Chili	Sire: Woods Jay	Sire: Sislers Shorty
			Dam: Sislers Trixie
		Dam: Greens Dingo	Sire: Woods Jay
			Dam: Ely's Blue
Dam: J Bar D Brandy	Sire: Rowe's Warrior Bow	Sire: Ronsley's Charcoal Smoke	Sire: Rifter's Streak
			Dam: Ritters Lass
		Dam: Murphy's Cochina	Sire: Ronsleys Charcoal Smoke
			Dam: Ronsley's Mitzi Buck
	Dam: J Bar D Choctaw Chilli	Sire: Rowe's Warrior Bow	Sire: Ronsley's Charcoal Smoke
			Dam: Murphy's Cochina
		Dam: Sorensens Queenie	Sire: Sorensens Gunsmoke
			Dam: Nightingales Tammy

Bred by: ______________ Address: ______________________ Phone: ______________

The details given above are correct to the best of my knowledge. Signed: ______________________(Breeder)

A historic moment in Australian Shepherd history. The awards ceremony at the Santa Rosa Invitational Trial, Petaluma, California. R to L: First Place—Leslie Sorensen (Ch Sorensen's Cherry Brandy), Second Place—unidentified person (Hartnagle's Leslie), Third Place—Jerry Rowe with Rowe's Comanche Warrior, Fourth Place—David King (King's Troy), Fifth Place—Bob Carrillo (Carrillo's Jenny), Sixth Place—Charlene Jeffreys (Jeffreys' Shade), Seventh Place—Colleen Gage with Gage's Cookie B. Photo by Stratton Photography, courtesy of Sorensen.

Ch Sorensen's Cherry Brandy at the Santa Rosa Invitational Trial. This was a cattle only trial. Photo by Stratton Photography, courtesy of Sorensen.

Mountain Babe), Colorados Charisma of Little Sioux and Taylors Red Cinnamon. All of these dogs appear in the pedigrees of countless Australian Shepherds today, proving the worth of this little dog of unknown background.

Of course, the Sorensen's old friends, Jerry and Sharon Rowe, were not forgotten. Over the years, Colorado kennel was able to acquire a number of J Bar D dogs which soon proved their worth. J Bar D Blue Boy was previously mentioned as a sire of superior quality. He came by this quality honestly. Sired by the incomparable Wood's Stubbie and out of the Sorensens beautiful Gunsmoke daughter, Blue Mist, he epitomized the quality possible from the Woods/Gunsmoke blend. In the tradition of the Wood dogs, Blue Boy was a large, impressive male. He stood 23 inches tall and and weighed about 65 pounds. He was a dark, gunmetal blue merle with the broad square head which was a trademark of the line tracing all the way back to Sisler's Shorty. Leslie remembers him as a short bodied dog with good movement. Blue Boy excelled working steers in the feedlots. He produced a number of key dogs which influenced the Colorado bloodline and other good working bloodlines around the country.

Ch Sorensen's Cherry Brandy CSD with the trophy she won at the ASCA Invitational Herding Trial held at Santa Rosa, CA. July 22, 1974. Trophy presenters and sponsors are Robert and Jean Carrillo. Handler is Leslie Sorensen. Courtesy of Sorensen

Cherry and Leslie working cattle through an obstacle at the Santa Rosa Invitational Trial. Photo by Stratton Photography, courtesy of Sorensen.

His previously mentioned son, Wolf's Royal Blue, produced a number of excellent dogs. The breathtaking Gunsmoke daughter, Colorado's Perky Miss, produced a litter which included the Kintyre Cattle Co.'s top working female, Colorado's Blue Majesty. The silver blue merle Blue Majesty worked everything on the ranch and produced a wealth of top working dogs when bred to their Jones Reddy Teddy son, Kintyre Jeb Stuart. She was moderate in every way, standing about 20-1/2 inches tall and weighing around 45 pounds. Royal Blue was also crossed with Nichols Little Bonnie to produce Colorado's Razzamatazz—

Ch Colorado's Applejack of Mini Acre, red merle male. The foundation sire for Ken Claussen's Mini Acre kennel and producer of a number of top working dogs. Courtesy of Sorensen

Colorado Mini Orphan Annie, red and white female who produced many good working cattle dogs for Ken Claussen (Ch Colorados Applejack of Mini Acre x Ch Nichols' Little Bonnie). Courtesy of Sorensen

a female later used to outcross with Slash V Wild Timber. A daughter of Nichols Little Bonnie by a linebred Gunsmoke sire named She Bear of Shepherds Crook was another of Royal Blue's mates. When bred to the intensely Wood bred Davies' Blue Kola (Wood's Stubbie x Davies' Checkers—Jeffreys' Shade x Wood's Little Kola) he produced a dog obtained by Phil Wildhagen named Frontiersman of Bonnie-Blu. Frontiersman was a dark blue with white trim who worked Arab horses at the Wolf stable. Royal Blue was also bred to a Wildhagen bred Flintridge female named Herzogs Smokey of Bonnie-Blu to produce Herzogs Heidi of Chadilane. Lastly, he was crossed to the McGuire's SeMeGo Hemi daughter, Hemis Spittin Image, to produce the jet black Colorado Angus for the Sorensens.

A full brother to Royal Blue was a dog named Barrs Blue Smoke, owned by Kenneth Barr of Texas. This dog was one of K-Barr kennel's foundation sires and produced many fine working dogs in Texas including Reaves Subaru Brat K and Reaves Stubbette K—who left an astounding legacy for the Black Powder kennel when bred to the legendary Slash V Little Rock.

Doyles Blue Buttons was another superior female bred to J Bar D Blue Boy. A daughter named Colorados Miss Parker produced a prolific female named Diamond J Checkers. Checkers was sired by the Sorensen's Red Man x Sorensen's Cherry Brandy son Sorensen's Colorado Royal. Colorado Royal was owned by J.T. Walters of Muldrow, Oklahoma where he was used extensively to produce top of the line working dogs which were valued all over the southern United States.

Another important link to J Bar D Blue Boy was through his cross to Higgin's Bobby. This Wood linebreeding produced the important sire Colorado's Winchesterblue, owned by Joe LaCount. Winchesterblue was a truly top dog in all respects. As a working dog he was second to none, showing impeccable judgment in handling cattle in the tradition of many Wood dogs. He was a hard, low heeler when necessary yet he would sleep with baby calves in the pasture in the wintertime. As a ranch dog he handled all the normal chores expected of him. A true companion, he loved children as much as he loved working cattle and was friendly to people who were acceptable to his boss. Agile and powerful, Winchesterblue stood about 22 inches and weighed about 60 pounds. He was a typical superior example of the Wood bloodline. His sister Colorado's Black Shadow was bred to a full brother of Powder River City Slicker at least once.

Ch Sorensen's Cherry Brandy. Courtesy of Sorensen

Colorado's Buckeye Barbie, red merle and white female. Photo taken at 2-1/2 months of age. Courtesy of Sorensen

From the cross of Winchesterblue to Colorado's Lace Leggins (Taylor's Whiskey x Ch Sorensen's Cherry Brandy) came the red merle female, Colorado's Cherry Punch. This speedy low heeler was a moderate 18-1/2 inches and weighed around 35 pounds. Her athletic build was topped by an attractive head. She was bred several times to good working males and eventually joined many good Colorado bred dogs at the cattle ranch of Melvin Townsend in the southeastern United States. Townsend's manager, Colville Jackson (Kintyre kennel) preferred quiet working dogs and Cherry Punch fit the bill perfectly.

J Bar D Brandy was another important female produced by Jerry and Sharon Rowe. Inbred on their Rowe's Warrior Bow—he was Brandy's sire and sire of her dam J Bar D Choctaw Chili—Brandy provided an important genetic outcross to the Gunsmoke and Wood dogs while still retaining the mountain-type of this family of dogs. (Warrior Bow and Choctaw Chili will be discussed in greater depth under the section "J Bar D/Rowe"). Brandy was a red merle with a wonderful temperament and excellent working ability. Leslie remembers she was a natural on sheep—working them from the start with excellent style and absolutely no training.

Thiex's Griz, red merle, copper and white male (Ch Sorensen's Colorado Royal x Walter's Princess) shown winning Best of Breed over 40 dogs on February 4, 1978. Courtesy of ' Sorensen

Colorado's Red Moon at one year of age. Red merle, copper and white male (Ch Sorensen's Colorado Royal x Colorado Charisma—Ch Slash V Little Rock x Taylor's Silver Buffie). Courtesy of Sorensen

Brandy was bred to a number of different Wood bred sires and produced winners every time. Her daughter, Sorensen's Cherry Brandy, sired by Piz' Joseph, was one of the Sorensen's first big winners. She was later bred to Joseph's most famous grandson, Jones Reddy Teddy, to produce the great Colorado's Brandy Bow. Brandy Bow was a red female of good temperament who would work anything she faced. Brandy Bow was sold to Melvin Townsend of the Kintyre Cattle Co. and bred to Jeffreys' Little Pooh Bear. This litter was a treasure chest of consistent high quality working dogs. J Bar D Brandy was also bred to Ch Georges Buckeye Bobby for an outcross litter. Buckeye Bobby was a popular outcross for the Wood type dogs and was also used by Silverledge to produce a slightly different type of working dog.

Another successful cross with Brandy was her mating to Sorensen's Red Man. From this cross came the influential female, (Weimers) Colorado Bonnie. Bonnie was a pretty headed, small red merle, standing about 18-1/2 inches tall. She was sold as a puppy to Dawn Weimer who used her for working their horses. She had a superb family temperament but she had one bad habit which caused her return to the Sorensens—she had a hobby of killing cats. In the hands of these eminent breeders, Bonnie began to prove her real worth. Her most prolific and influential daughter was Ch Colorados Ravishing Ruby CDX OTDcs ATDd, owned by Jim and Sue Foster. Ruby was a dark chocolate solid red female standing about 19-1/2 inches tall and weighing about 40 pounds. As her pedigree would predict she was a top of the line obedience and working dog and a superb companion. When raising puppies she was an outstanding mother.

The Fosters already had a good mate for Ruby in their other Sorensen bred dog, Ch Powder River City Slicker CD, ATDcsd, AD. This cross

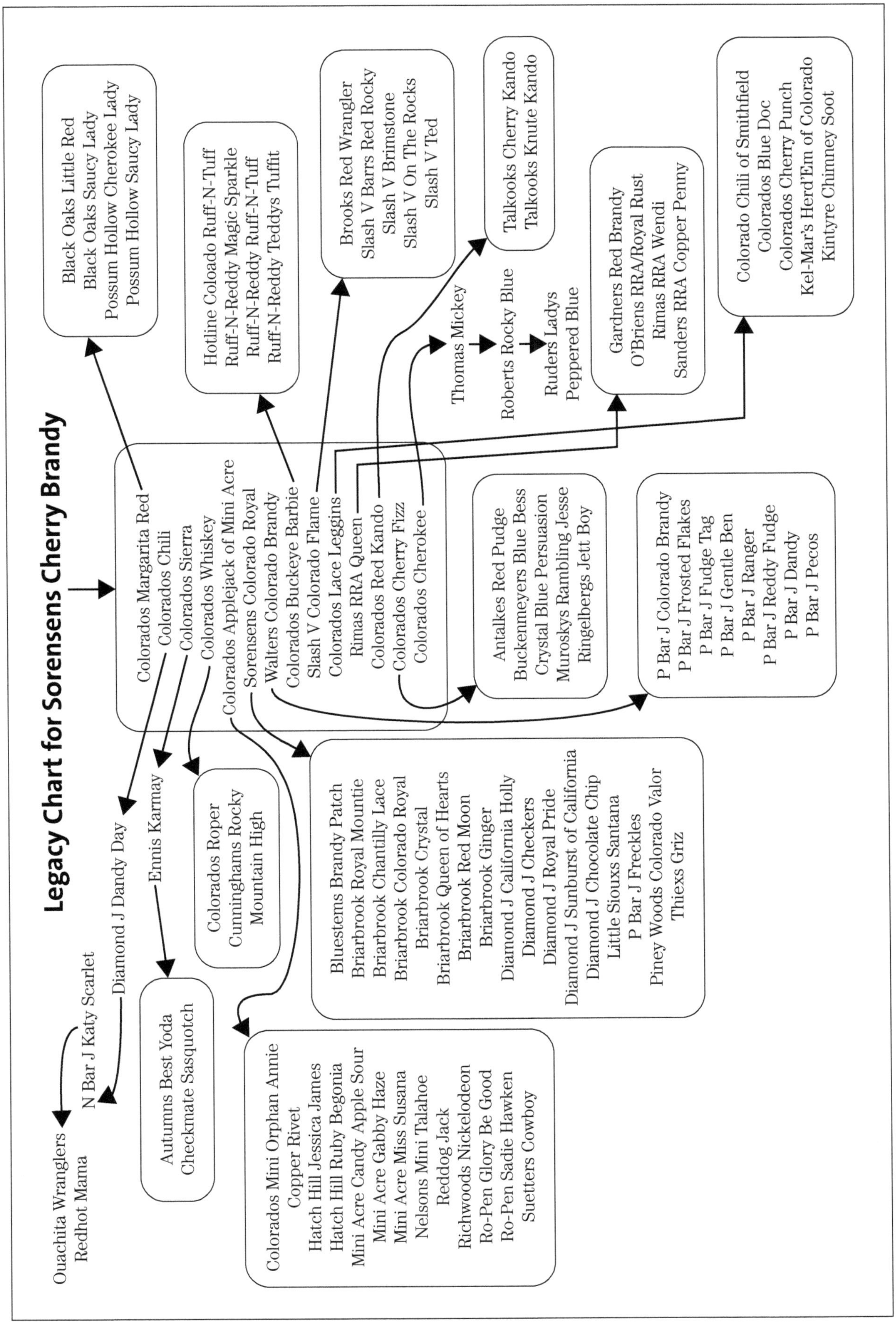
Legacy Chart for Sorensens Cherry Brandy
Colorados Margarita Red
Colorados Chili
Colorados Sierra
Colorados Whiskey
Colorados Applejack of Mini Acre
Sorensens Colorado Royal
Walters Colorado Brandy
Colorados Buckeye Barbie
Slash V Colorado Flame
Colorados Lace Leggins
Rimas RRA Queen
Colorados Red Kando
Colorados Cherry Fizz
Colorados Cherokee
Black Oaks Little Red
Black Oaks Saucy Lady
Possum Hollow Cherokee Lady
Possum Hollow Saucy Lady
Hotline Coloado Ruff-N-Tuff
Ruff-N-Reddy Magic Sparkle
Ruff-N-Reddy Ruff-N-Tuff
Ruff-N-Reddy Teddys Tuffit
Brooks Red Wrangler
Slash V Barrs Red Rocky
Slash V Brimstone
Slash V On The Rocks
Slash V Ted
Talkooks Cherry Kando
Talkooks Knute Kando
Thomas Mickey
Roberts Rocky Blue
Ruders Ladys Peppered Blue
Gardners Red Brandy
O'Briens RRA/Royal Rust
Rimas RRA Wendi
Sanders RRA Copper Penny
Colorado Chili of Smithfield
Colorados Blue Doc
Colorados Cherry Punch
Kel-Mar's Herd'Em of Colorado
Kintyre Chimney Soot
Antalkes Red Pudge
Buckenmeyers Blue Bess
Crystal Blue Persuasion
Muroskys Rambling Jesse
Ringelbergs Jett Boy
P Bar J Colorado Brandy
P Bar J Frosted Flakes
P Bar J Fudge Tag
P Bar J Gentle Ben
P Bar J Ranger
P Bar J Reddy Fudge
P Bar J Dandy
P Bar J Pecos
Diamond J Dandy Day
Ennis Karmay
Colorados Roper
Cunninghams Rocky Mountain High
Bluestems Brandy Patch
Briarbrook Royal Mountie
Briarbrook Chantilly Lace
Briarbrook Colorado Royal
Briarbrook Crystal
Briarbrook Queen of Hearts
Briarbrook Red Moon
Briarbrook Ginger
Diamond J California Holly
Diamond J Checkers
Diamond J Royal Pride
Diamond J Sunburst of California
Diamond J Chocolate Chip
Little Sioux Santana
P Bar J Freckles
Piney Woods Colorado Valor
Thiexs Griz
N Bar J Katy Scarlet
Ouachita Wranglers Redhot Mama
Autumns Best Yoda
Checkmate Sasquotch
Colorados Mini Orphan Annie
Copper Rivet
Hatch Hill Jessica James
Hatch Hill Ruby Begonia
Mini Acre Candy Apple Sour
Mini Acre Gabby Haze
Mini Acre Miss Susana
Nelsons Mini Talahoe
Reddog Jack
Richwoods Nickelodeon
Ro-Pen Glory Be Good
Ro-Pen Sadie Hawken
Suetters Cowboy

(Sorensen's) Halls Apache, very dark blue merle with white female. Genetic powerhouse behind both the Colorado and Silverledge bloodlines. Courtesy of Sorensen

Ch Taylor's Shamrock of Silverledge CDX, red merle male obtained from Jean Taylor. Courtesy of Sorensen

produced dogs which made their mark all over the United States. Powder River Ginger Snap went to Cutter Hill kennel in Texas and produced a number of good working dogs when crossed with Zephyr's Crimson King. Another daughter, Colorados Rave Review, also went to Texas to produce working dogs for Dizneyland kennel. Several other progeny were scattered throughout the Midwest and their descendants are behind many successful dogs today.

Another excellent cross with Ruby was to Jones Reddy Teddy. From this union a daughter named Powder River Outlaw Annie went to the Midwest and produced a number of nice litters from both related and unrelated sires. A son, Ch Powder River Tuffernhel CD STDd, was the pride and joy of Sue Noll and became a well known sire in her hands. Many other siblings were scattered across the country and used as outcrosses for other bloodlines. In addition, Ruby was bred to the outcross sire, Windsongs Raisin Cain and produced several progeny used in breeding programs across the country. Through Ruby's many contributions, Bonnie's influence was felt in kennels producing many types of Aussies during a period of great growth in the breed.

Bonnie's place in history would have been secure through her daughter Ruby, but she continued to produce dogs whose influence rippled through the breed. A son by Ch George's Buckeye Bobby, Ch Talkooks Colorado Clay, went to Nevada to become an important sire for Jonna Cole. He produced Ch Miss Nutmeg of Talkook and other puppies which helped found new kennels in California and other western states. Another Buckeye

Bobby daughter, Adams Colorado Babe, produced Marys Rocky Mountain Babe (by Piney Wood's Colorado Valor, a son of Ch Sorensen's Colorado Royal x Taylors Silver Buffie). Mary's Rocky Mountain Babe produced many dogs for Betty Nelson, most notably a son named Walkers Joshua STDs, sired by J Bar D Cimmeron Warrior. Joshua was an important foundation sire for the Nelson kennel and is found behind many dogs of her lineage today.

A day of triumph for Ch Taylor's Shamrock of Silverledge CDX. Pictured at the May 1981 show where he won Best of Breed, Best Junior Handler with Becky Sorensen, High Score in Obedience, and he was sire of the Winners Bitch and Futurity Winner, Colorado's Miss Behavin and sire of the Reserve Winners dog, Colorado's Ketchum. Courtesy of Jean Taylor

Bonnie's daughter by Walters Colorado Cash (Taylors Luke the Drifter x Taylors Silver Buffie), named McCormicks Colorado Ona, produced a foundation sire for Pincie Creek kennel in Alabama—McCormick's Smokey Seminole. Her son Hiawatha Copper Stub (by Samples Blue Stubbie, by J Bar D Blue Boy x Wolfs Blue Heather) sired Loida's Pepsi Will Do, dam of the well known working dog, Loida's Just Plain Charlie. These many branches emanating from Weimers Colorado Bonnie were the foundation for a variety of breeders spanning the entire United States during a time of great growth in the breed. Her quality and influence assured a solid genetic base for many generations.

A fortunate discovery by Leslie Sorensen was the Gunsmoke grandson, Dorns Poudre Lakohta. This stunning blue merle, copper and white dog was the well loved pet of Kathy Dorn, who lived in Greeley. His dam, Dressel's Duchess was a Gunsmoke x Sorensen's Nightingale's Tammy daughter which the Sorensens had sold years before to Carol Sue Dressel. She had been bred to Ch Iacovetta's Buck who traced through his dam to many early Sisler dogs. He was about 21 inches tall, very compact and muscular with an attractive head. He had an excellent temperament. At the time of his discovery he was not registered and papers could not be obtained for him. Leslie applied to ASCA for registration and he was accepted under the early inspection program ASCA had for unregistered

Ch Taylor's Shamrock of Silverledge CDX with Leslie Sorensen capturing yet another win in his successful show career. Courtesy of Sorensen

dogs. Lakohta was used with several of Colorado's very best females, producing Ch Colorados Clint out of Colorados Perky Miss, and good litters from Doyles Blue Buttons and Colorados Sharade.

Colorados Am I Blue was another pivotal female used by Silverledge from the Sorensens. Her dam, Colorados Houdini, was a granddaughter of Taylor's Luke the Drifter through her sire Taylor's Shamrock of Silverledge. Houdini was sold as a puppy but later bred to Ch Slash V Wild Timber while he was a resident of Colorado kennel. Am I Blue was a stud fee puppy who was named for her unusual color. Although she appeared to be a black tri, she had one small patch of merling on her face which revealed her color genetics. Leslie remembers her as a good tempered dog with show dog looks. She was sold to Liz Winn and later leased by Jean Taylor, who bred her back to her great grandfather Taylor's Luke the Drifter to produce her most famous son, Silverledge Charlie Russell.

A little known female who has had a lasting influence was Colorado's Vandal Ann. From the proven superior cross of Gunsmoke to a Wood line female named Holt's Misty Blue (by J Bar D Blue Boy), Vandal Ann was a very dark blue with a black mask—hence her name. She was somewhat tall like her Wood ancestors, standing 21 inches and weighing around 50 pounds. Leslie showed her in the conformation ring and remembers her incredible movement. So flawlessly and rapidly could she gait around a show ring, she would insult Leslie by glancing over her right shoulder to see if this doddering human was going to even try to keep up with her! The Sorensens never worked Vandal Ann on livestock but her legacy speaks well of her genetics.

Crossed to a Flintridge line male, Ann produced a blue-eyed black male named Colorado Sabbath (Sabbath was later sold and spent his life working on a sheep ranch). Although he was not a major contributor to the Colorado bloodline he did produce at least one litter with the awe-inspiring Gunsmoke/Wood bred dam Colorado Perky Miss. From this cross came Colorado's Smooth Smoke, a female who would eventually become a major contributor to the Crown Point working kennel of Wayne and Kaye Harris. Leslie states Smooth Smoke was a superior producer, as one would expect from her pedigree, sound, trainable and exceedingly beautiful.

Ch Taylor's Shamrock of Silverledge CDX showing his top obedience ability. Courtesy of Sorensen

Colorado's Hotline Paige was not bred by the Sorensens but obtained from Brenda Myrup of Utah. She was an important outcross for the Colorado dogs bringing in some old California breeding which had been used by both Sorensens and Rowes with success. She produced Colorados Sharade for the Sorensens and was later sold to Diamond J where she joined other Colorado bred dogs, producing several litters with Slash V bred males.

Ch Sorensen's Colorado Cajun, very correct blue merle, copper and white female who was the foundation for a number of eastern kennels. Courtesy of Sorensen

Colorados Sharade was a mostly black blue merle with copper and white markings. Her diminutive 18 inch height and 35 pound weight gave her a "cute" look. She was not fine boned but nicely moderate and well balanced. Sharade was Becky Sorensens first dog and granddam of Becky's favorite Junior Showmanship dog in later years. She was used very little for working livestock but made her mark as dam of a number of significant breeding females.

Colorados Sharade produced well with a number of different males. She is best known as the dam of one of Briarbrook kennels first Champions who was an excellent dam for this kennel, Ch Colorados Sizzlin Sioux. Sizzlin Sioux not only produced several champions for Briarbrook but also whelped foundation females for Aristocrat and Southwynd kennels.

Sharade's other significant daughter by Ch Taylor's Shamrock of Silverledge CDX is Ch Colorado's Sunbonnet Sioux CD STDs. As evidenced by her titles, Sioux was an all around performer and earned many All

Taylor's Calamity Jane, blue merle and white female who helped found the Silverledge kennel and provided the perfect outcross for Jean Taylor to her intensely bred Wood foundations. Courtesy of Sorensen

Colorado Pride of Cedar Grove, blue merle and white. (Slash V Buckeye Bobby x Halls Apache) winning Best of Opposite Sex Puppy at the 1973 Phoenix Specialty. Courtesy of Sorensen

Around Awards with Becky Sorensen during her years as a Junior Handler. She was a muscular red merle standing 19-1/2 inches with a nice head and earset. Her love of working livestock, especially sheep, and her natural biddability made her the perfect teammate for Becky Sorensen. Sunny had a knack for handling difficult sheep with a non-threatening manner and saved many a touchy situation with her good judgment. Becky was 10 years old at the time and did all the training of Sunny. Her devotion was repaid by Sunny, who was reserved around strangers and rather protective of Becky. When showing in obedience, Sunny kept an eye on the judges at all times. This alertness rarely affected their scores as the team won the Reserve Champion Junior in Obedience at the Colorado State Fair and many awards together at the ASCA National Specialties including being the youngest team to ever win ASCA's All Around Junior award.

Sunny bred to Ch Powder Rivers City Slicker CD, ATDcsd, AD produced a male named Colorado's Sun City Slick, who sired several good working dogs for Ken Claussen at Mini Acre. Sunny bred to an outcross Flintridge male produced a significant breeding bitch for Colorado named Colorado's Sunshine Sioux. Sunshine Sioux was used with both working sires and Flintridge sires to form the foundation of Leslie's current breeding program through her son, AKC/ASCA Ch Colorados Terry Blue OTDc. He carried lines back to old Colorado breeding through his sire's dam, French Toast, as well as his own dam, Sunny. Owned by Chris and Vicki Lembcke of Applecross kennel, he was siring litters until a few months before his death at age 14.

The influence of the Sorensen's Colorado kennel has been far reaching and vital to the formation of the Australian Shepherd. Many dogs produced by their breeding program have been scattered throughout the United States and still bring forth top quality dogs today. This overview is not exhaustive and does not include every dog produced by the Sorensens. It does provide a foundation for understanding the genetic worth of the dogs bearing their name.

J Bar D Fancy, blue merle female of beautiful foundation head type and excellent conformation. Courtesy of Sorensen

The wide range of experience of the Sorensens has made them in high demand as mentors and judges in the breed. Dick became the first ASCA stock dog judge to come from the Apprentice program. His knowledge and experience not only qualified him to eventually become one of ASCA's most well known and popular judges but also benefited the breed through his generosity with that knowledge. Today he shares that knowledge and experience through the many training clinics and lessons he conducts near home.

A priceless collection of Aussie genetics. circa 1969. L to R: J Bar D Blue Boy, J Bar D Flirt, Rowe's Commanche Warrior, Sorensen's Blue Mist, Sorensen's Gunsmoke. Courtesy of Rowe

Leslie married into Australian Shepherds from a lifelong background in dogs inherited from her mother. Dick was well into his breeding program when they met and married and Leslie was immediately smitten with the breed. She has become an ASCA Senior Breeder Judge and an AKC Licensed judge for several breeds and Junior Showmanship. The Colorado breeding program continues on a smaller scale than in the past and a new generation of working dogs is in residence, including J Bar D Colorado Blacksmith. The Sorensens devotion to the breed continues and talking dogs and answering questions about Aussies remain one of their favorite activities.

Born: ____________ Sex: ________
Owned by: ____________________

PEDIGREE OF J Bar D Fancy

Parents	Grandparents	Great-grandparents	Great-great-grandparents
Sire: J Bar D Red Chief	Sire: Rowe's Warrior Bow	Sire: Ronsleys Charcoal Smoke	Sire: Ritters Streak
			Dam: Ritters Lass
		Dam: Murphys Cochina	Sire: Ronsleys Charcoal Smoke
			Dam: Ronsleys Mitzi Buck
	Dam: Porters Yakima Blue Squaw	Sire: Woods Teddy Blue Boy	Sire: unknown
			Dam: unknown
		Dam: Healys Pepsi Cola	Sire: Woods Jay
			Dam: Woods Little Kola
Dam: J Bar D Flirt	Sire: Sorensens Gunsmoke	Sire: Williams Laddie	Sire: Uhlendorfs Tippy
			Dam: Thorntons Chrissie
		Dam: Jeffreys Sheila	Sire: Elys Major
			Dam: Williams Georgina
	Dam: Jeffreys Missy Blue	Sire: Williams Laddie	Sire: Uhlendorfs Tippy
			Dam: Thorntons
		Dam: Picrauxs Juliette	Sire: Jeffreys Dead Eye Jr
			Dam: Williams Georgina

Bred by: ____________ Address: ____________ Phone: ____________

The details given above are correct to the best of my knowledge. Signed: ____________ (Breeder)

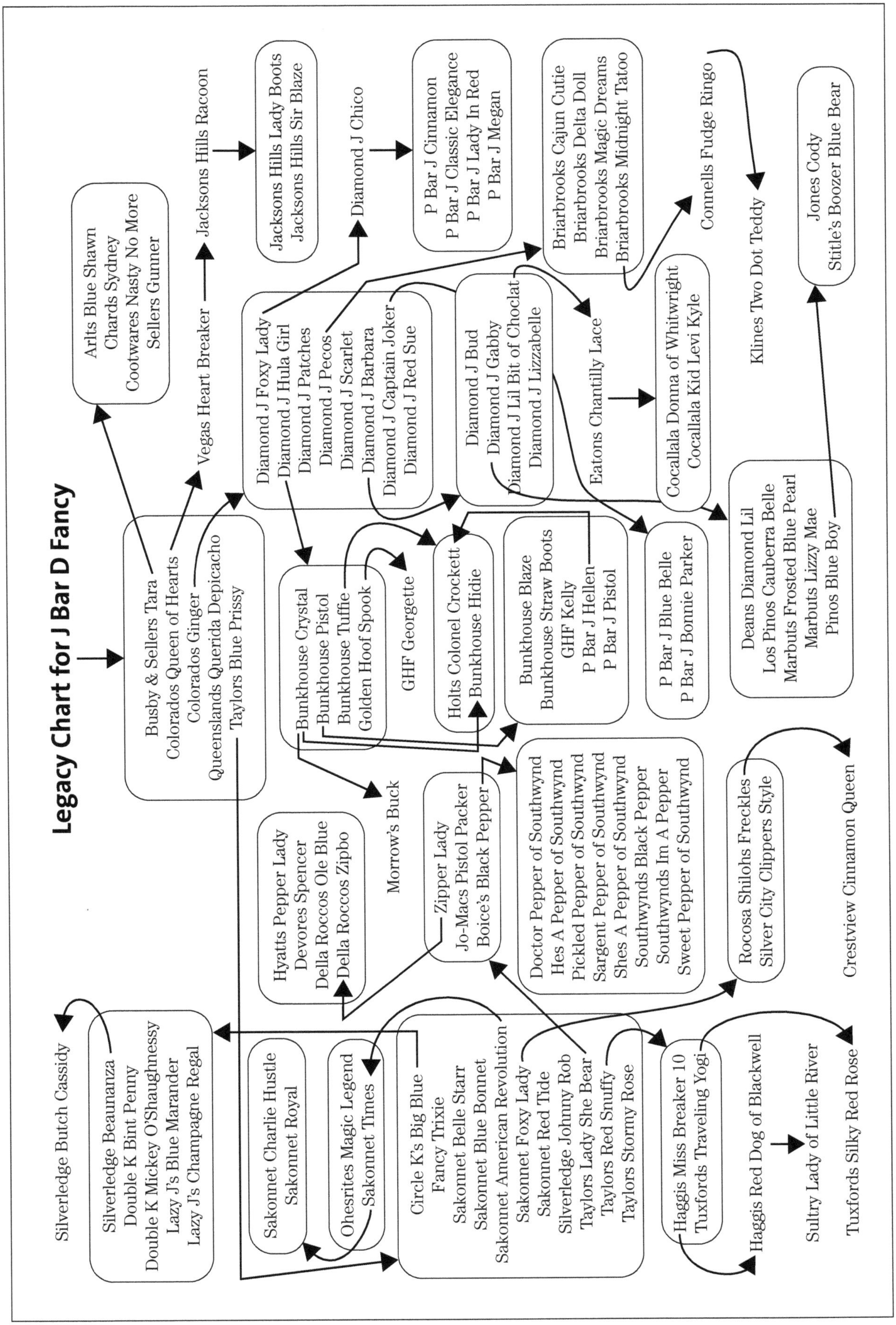

Legacy Chart for J Bar D Fancy
Busby & Sellers Tara
Colorados Queen of Hearts
Colorados Ginger
Queenslands Querida Depicacho
Taylors Blue Prissy
Arlts Blue Shawn
Chards Sydney
Cootwares Nasty No More
Sellers Gunner
Vegas Heart Breaker
Jacksons Hills Racoon
Jacksons Hills Lady Boots
Jacksons Hills Sir Blaze
Diamond J Chico
P Bar J Cinnamon
P Bar J Classic Elegance
P Bar J Lady In Red
P Bar J Megan
Briarbrooks Cajun Cutie
Briarbrooks Delta Doll
Briarbrooks Magic Dreams
Briarbrooks Midnight Tatoo
Connells Fudge Ringo
Jones Cody
Stitle's Boozer Blue Bear
Diamond J Foxy Lady
Diamond J Hula Girl
Diamond J Patches
Diamond J Pecos
Diamond J Scarlet
Diamond J Barbara
Diamond J Captain Joker
Diamond J Red Sue
Diamond J Bud
Diamond J Gabby
Diamond J Lil Bit of Choclat
Diamond J Lizzabelle
Eatons Chantilly Lace
Cocallala Donna of Whitwright
Cocallala Kid Levi Kyle
Klines Two Dot Teddy
Bunkhouse Crystal
Bunkhouse Pistol
Bunkhouse Tuffie
Golden Hoof Spook
GHF Georgette
Holts Colonel Crockett
Bunkhouse Hidie
Bunkhouse Blaze
Bunkhouse Straw Boots
GHF Kelly
P Bar J Hellen
P Bar J Pistol
P Bar J Blue Belle
P Bar J Bonnie Parker
Deans Diamond Lil
Los Pinos Cauberra Belle
Marbuts Frosted Blue Pearl
Marbuts Lizzy Mae
Pinos Blue Boy
Hyatts Pepper Lady
Devores Spencer
Della Roccos Ole Blue
Della Roccos Zipbo
Morrow's Buck
Zipper Lady
Jo-Macs Pistol Packer
Boice's Black Pepper
Doctor Pepper of Southwynd
Hes A Pepper of Southwynd
Pickled Pepper of Southwynd
Sargent Pepper of Southwynd
Shes A Pepper of Southwynd
Southwynds Black Pepper
Southwynds Im A Pepper
Sweet Pepper of Southwynd
Rocosa Shilohs Freckles
Silver City Clippers Style
Crestview Cinnamon Queen
Silverledge Butch Cassidy
Silverledge Beaunanza
Double K Bint Penny
Double K Mickey O'Shaughnessy
Lazy J's Blue Marander
Lazy J's Champagne Regal
Sakonnet Charlie Hustle
Sakonnet Royal
Ohesrites Magic Legend
Sakonnet Times
Circle K's Big Blue
Fancy Trixie
Sakonnet Belle Starr
Sakonnet Blue Bonnet
Sakonnet American Revolution
Sakonnet Foxy Lady
Sakonnet Red Tide
Silverledge Johnny Rob
Taylors Lady She Bear
Taylors Red Snuffy
Taylors Stormy Rose
Haggis Miss Breaker 10
Tuxfords Traveling Yogi
Haggis Red Dog of Blackwell
Sultry Lady of Little River
Tuxfords Silky Red Rose

The multi-talented Higgins Bobbie CDX with Leslie Sorensen working on obedience at home. Courtesy of Sorensen

Higgins Bobbie CDX enjoyed everything she was asked to do. Here she is working the sleeve at a Schutzhund training session. Courtesy of Sorensen

First place in the Breeder's Class at an early Colorado show. Left to right: Rowe's Commanche Warrior with Jerry Rowe, Sorensen's Gunsmoke with Dick Sorensen and Sorensen's Blue Mist with Sharon Rowe. Courtesy of Rowe

Higgins Bobbie, red merle female, a powerhouse on livestock. Courtesy of Sorensen

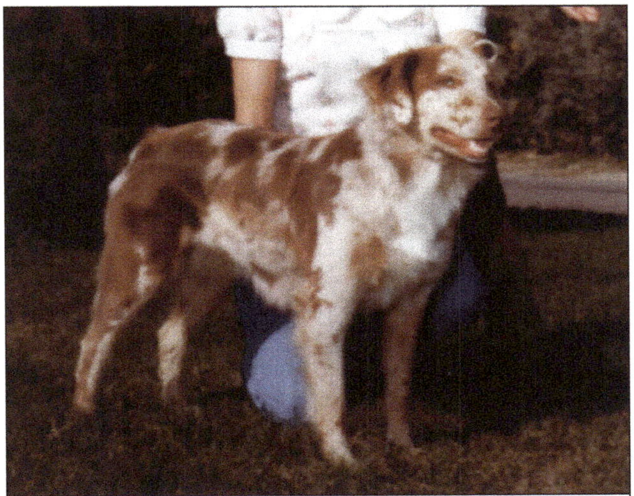

Higgins Bobbie CDX winning again at a 1977 CASA Match. Courtesy of Sorensen

Higgins Bobbie CDX moving cattle at the ranch. Courtesy of Sorensen

PEDIGREE OF Higgins Bobby

Born: ______ Sex: ______
Owned by: ______

Parents	Grandparents	Great-grandparents	Great-great-grandparents
Sire: Higgins Pinkie	Sire: Montoyas Chapparel	Sire: Montoyas Rowdy	Sire: Clingers Royal Blue Merle
			Dam: Kalbs Princess
		Dam: Montoyas Snipper	Sire: Piz Joseph
			Dam: Piz's Lady II
	Dam: Montoyas Snipper	Sire: Pix Joseph	Sire: Woods Jay
			Dam: Woods Chili
		Dam: Pizs Lady II	Sire: Piz Joseph
			Dam: Piz Lady I
Dam: Ebony Lass	Sire: Piz's Little Joe	Sire: Piz Joseph	Sire: Woods Jay
			Dam: Woods Chili
		Dam: J Bar D Choctaw Chili	Sire: Rowes Warrior Bow
			Dam: Sorensens Queenie
	Dam: Davies Blue Kola	Sire: Woods Stubbie	Sire: Woods Jay
			Dam: Woods Slate
		Dam: Davies Checkers	Sire: Jeffreys Shade
			Dam: Woods Little Kola

Bred by: ______ Address: ______ Phone: ______

The details given above are correct to the best of my knowledge. Signed: ______(Breeder)

Colorado's Macho Azo of Fourwinds, blue and white male foundation sire of Fourwinds kennel (Higgins Fast Buck of Berrycreek x Higgins Bobbie.) Courtesy of Rowe

Higgins Bobbie CDX at an early Schutzhund training day. Courtesy of Sorensen

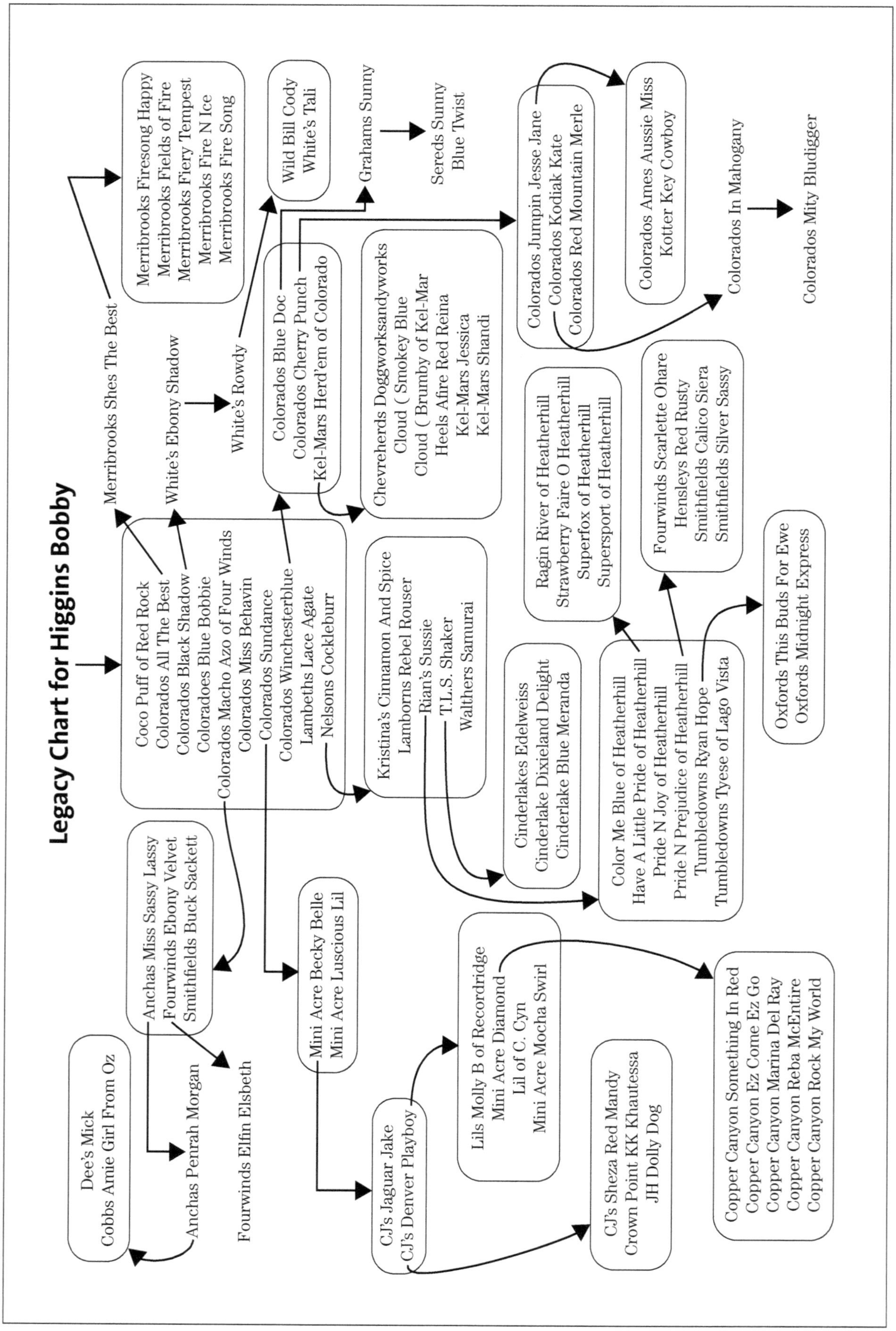

Legacy Chart for Higgins Bobby
Coco Puff of Red Rock
Colorados All The Best
Colorados Black Shadow
Coloradoes Blue Bobbie
Colorados Macho Azo of Four Winds
Colorados Miss Behavin
Colorados Sundance
Colorados Winchesterblue
Lambeths Lace Agate
Nelsons Cockleburr
Merribrooks Shes The Best
Merribrooks Firesong Happy
Merribrooks Fields of Fire
Merribrooks Fiery Tempest
Merribrooks Fire N Ice
Merribrooks Fire Song
White's Ebony Shadow
White's Rowdy
Wild Bill Cody
White's Tali
Grahams Sunny
Sereds Sunny
Blue Twist
Colorados Blue Doc
Colorados Cherry Punch
Kel-Mars Herd'em of Colorado
Chevreherds Doggworksandyworks
Cloud (Smokey Blue
Cloud (Brumby of Kel-Mar
Heels Afire Red Reina
Kel-Mars Jessica
Kel-Mars Shandi
Colorados Jumpin Jesse Jane
Colorados Kodiak Kate
Colorados Red Mountain Merle
Colorados Ames Aussie Miss
Kotter Key Cowboy
Colorados In Mahogany
Colorados Mity Bludigger
Kristina's Cinnamon And Spice
Lamborns Rebel Rouser
Rian's Sussie
T.L.S. Shaker
Walthers Samurai
Ragin River of Heatherhill
Strawberry Faire O Heatherhill
Superfox of Heatherhill
Supersport of Heatherhill
Fourwinds Scarlette Ohare
Hensleys Red Rusty
Smithfields Calico Siera
Smithfields Silver Sassy
Cinderlakes Edelweiss
Cinderlake Dixieland Delight
Cinderlake Blue Meranda
Color Me Blue of Heatherhill
Have A Little Pride of Heatherhill
Pride N Joy of Heatherhill
Pride N Prejudice of Heatherhill
Tumbledowns Ryan Hope
Tumbledowns Tyese of Lago Vista
Oxfords This Buds For Ewe
Oxfords Midnight Express
Dee's Mick
Cobbs Amie Girl From Oz
Anchas Miss Sassy Lassy
Fourwinds Ebony Velvet
Smithfields Buck Sackett
Anchas Penrah Morgan
Fourwinds Elfin Elsbeth
Mini Acre Becky Belle
Mini Acre Luscious Lil
CJ's Jaguar Jake
CJ's Denver Playboy
Lils Molly B of Recordridge
Mini Acre Diamond
Lil of C. Cyn
Mini Acre Mocha Swirl
CJ's Sheza Red Mandy
Crown Point KK Khautessa
JH Dolly Dog
Copper Canyon Something In Red
Copper Canyon Ez Come Ez Go
Copper Canyon Marina Del Ray
Copper Canyon Reba McEntire
Copper Canyon Rock My World

Ch Nichols' Little Bonnie, black and white female and a treasure of good mountain-type genes, including close up crosses to Sisler's Shorty. Courtesy of Sorensen

Ch Nichols Little Bonnie winning Best of Breed over 2 specials for a 5 point major on July 29, 1973. Courtesy of Sorensen

Born: ____________ Sex: ________

Owned by: ____________________

PEDIGREE OF Nichols Little Bonnie

Parents	Grandparents	Great-grandparents	Great-great-grandparents
Sire Arington's Stub	Sire Wood's Jay	Sire Sisler's Shorty	Sire Keno
			Dam Blue Star
		Dam Sisler's Trixie	Sire Blue Khan
			Dam Emmetts Blue
	Dam Harper's Jinx	Sire Sisler's Shorty	Sire Keno
			Dam Blue Star
		Dam Elys Queenie	Sire Sislers Spike
			Dam Blue Star
Dam McLean's Lady	Sire unknown	Sire	Sire
			Dam
		Dam	Sire
			Dam
	Dam unknown	Sire	Sire
			Dam
		Dam	Sire
			Dam

Bred by: ____________ Address: ____________ Phone: ____________

The details given above are correct to the best of my knowledge. Signed: ____________(Breeder)

Ch Nichols' Little Bonnie showing her strong style working cattle. Courtesy of Sorensen

Ch Nichols' Little Bonnie taking another win with Leslie Sorensen. Courtesy of Sorensen

The nearly perfect Colorado's Perky Miss, blue merle and white female. Her type, conformation and intelligence personified the breed standard like no other. Courtesy of Sorensen

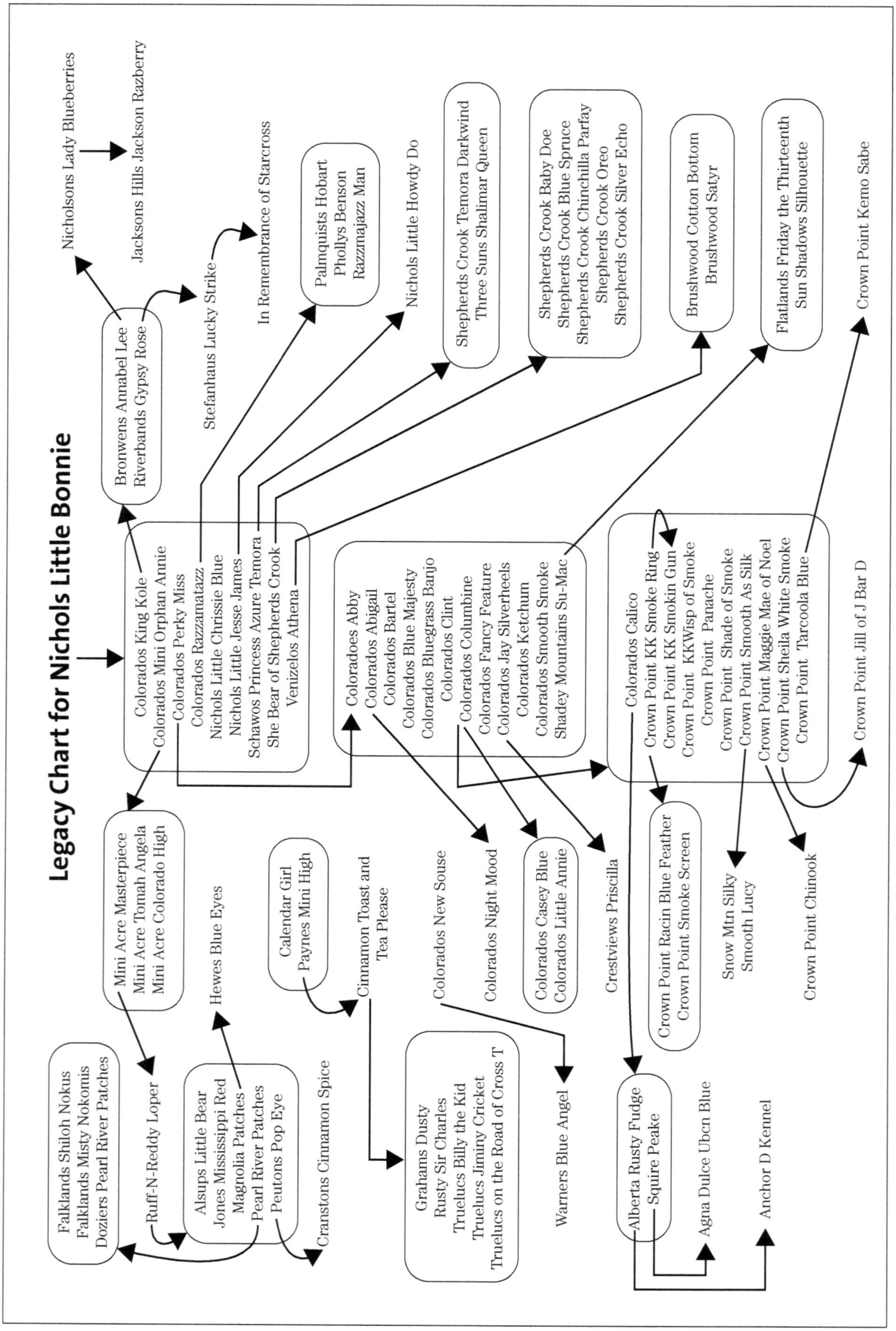
Legacy Chart for Nichols Little Bonnie
Nicholsons Lady Blueberries
Jacksons Hills Jackson Razberry
Bronwens Annabel Lee
Riverbands Gypsy Rose
Stefanhaus Lucky Strike
In Remembrance of Starcross
Palmquists Hobart
Phollys Benson
Razzmajazz Man
Nichols Little Howdy Do
Shepherds Crook Temora Darkwind
Three Suns Shalimar Queen
Shepherds Crook Baby Doe
Shepherds Crook Blue Spruce
Shepherds Crook Chinchilla Parfay
Shepherds Crook Oreo
Shepherds Crook Silver Echo
Brushwood Cotton Bottom
Brushwood Satyr
Flatlands Friday the Thirteenth
Sun Shadows Silhouette
Crown Point Kemo Sabe
Colorados King Kole
Colorados Mini Orphan Annie
Colorados Perky Miss
Colorados Razzamatazz
Nichols Little Chrissie Blue
Nichols Little Jesse James
Schawos Princess Azure Temora
She Bear of Shepherds Crook
Venizelos Athena
Coloradoes Abby
Colorados Abigail
Colorados Bartel
Colorados Blue Majesty
Colorados Bluegrass Banjo
Colorados Clint
Colorados Columbine
Colorados Fancy Feature
Colorados Jay Silverheels
Colorados Ketchum
Colorados Smooth Smoke
Shadey Mountains Su-Mac
Colorados Calico
Crown Point KK Smoke Ring
Crown Point KK Smokin Gun
Crown Point KKWisp of Smoke
Crown Point Panache
Crown Point Shade of Smoke
Crown Point Smooth As Silk
Crown Point Maggie Mae of Noel
Crown Point Sheila White Smoke
Crown Point Tarcoola Blue
Crown Point Jill of J Bar D
Mini Acre Masterpiece
Mini Acre Tomah Angela
Mini Acre Colorado High
Hewes Blue Eyes
Calendar Girl
Paynes Mini High
Cinnamon Toast and
Tea Please
Colorados New Souse
Colorados Night Mood
Colorados Casey Blue
Colorados Little Annie
Crestviews Priscilla
Crown Point Racin Blue Feather
Crown Point Smoke Screen
Snow Mtn Silky
Smooth Lucy
Crown Point Chinook
Falklands Shiloh Nokus
Falklands Misty Nokomis
Doziers Pearl River Patches
Ruff-N-Reddy Loper
Alsups Little Bear
Jones Mississippi Red
Magnolia Patches
Pearl River Patches
Peutons Pop Eye
Cranstons Cinnamon Spice
Grahams Dusty
Rusty Sir Charles
Truelucs Billy the Kid
Truelucs Jiminy Cricket
Truelucs on the Road of Cross T
Warners Blue Angel
Alberta Rusty Fudge
Squire Peake
Agna Dulce Ubcn Blue
Anchor D Kennel

Colorado's Perky Miss in old age. Still sound, still beautiful. Courtesy of Melissa Couch

Ch Colorado's Clint, blue merle and white male, the Gunsmoke gene bank for the Colorado kennel and later several west coast kennels. Courtesy of Sorensen

Born: ______________ Sex: ________
Owned by: ______________________

PEDIGREE OF Colorado's Perky Miss

Parents	Grandparents	Great-grandparents	Great-great-grandparents
Sire: Sorensens Gunsmoke	Sire: Williams Laddie	Sire: Uhlendorfs Tippy	Sire: Parks Joe
			Dam: Uhlendorfs Linda
		Dam: Thorntons Chrissie	Sire: Watsons Old Blue
			Dam: Johnsons Sue
	Dam: Jeffreys Sheila	Sire: Elys major	Sire: Breazeales Shiek
			Dam: Breazeales Sandy
		Dam: Williams Georgina	Sire: Elys Joker
			Dam: Barretts Blue Belle
Dam: Nichols Little Bonnie	Sire: Arington's Stub	Sire: Wood's Jay	Sire: Woods Jay
			Dam: Woods Chili
		Dam: Harper's Jinx	Sire: Rowes Warrior Bow
			Dam: Sorensens Queenie
	Dam: McLean's Lady	Sire: unknown	Sire:
			Dam:
		Dam: unknown	Sire:
			Dam:

Bred by: ______________ Address: ______________________ Phone: ____________

The details given above are correct to the best of my knowledge. Signed: ______________________(Breeder)

Colorado's Trail Shadow "Zane," blue merle and white male (Colorado's Talk of the Town x Ruff N Reddy Boots). Courtesy of Sorensen

Ch Colorado's Talk of the Town known as "Manny," blue merle, copper and white male by Ch Colorado's Clint x Ch Posey Patch of Blue Mist. Courtesy of Sorensen

Ch Colorado's All Wheel Drive, blue, copper and white male (Ch Moonstruck's Star Slayer x Ch Colorado's Mountain Shade). An outcross male blending working and show lines. Courtesy of Sorensen

Ch Colorado's Mountain Shade, blue and white female (Colorado's Talk of the Town x Ruff N Reddy Boots). A beautiful, well bred female of largely working blood. Courtesy of Sorensen

Ch Colorado's Toast of the Town known as "Todd," blue merle male sired by Ch Colorado's Talk of the Town and owned by Cathy Jones. Courtesy of Sorensen

Ch Ruff N Reddy Ross STDcsd, blue and white male (Colorado's Toast of the Town x Ruff N Reddy Boots. Boots was linebred on Ch Jones Reddy Teddy and Shady Mountain Mity Miss). Shown winning High in Trial at the USASA National Specialty herding trial with trainer/handler Jerry Rowe. Courtesy of Sorensen

Ch Colorado's Columbine CD as a young dog at an early show handled by Leigh Ann, Leslie's daughter. Courtesy of Sorensen

Ch Colorado's Columbine CD, striking blue merle female and all around performer at a show in 1979, handler, Leigh Ann. Photo by Sue Noll, courtesy of Sorensen.

Ch Colorado's Columbine CD as an aged bitch. Her quality never faded. Courtesy of Melissa Couch

Shadey Mountain Sumac, blue merle and white female, full sister to Ch Colorado's Columbine CD (Olson's Blue Joe x Colorado's Perky Miss) with Cathy Jones. Courtesy of Sorensen

Ch Colorado's Columbine CD on another winning day. Photo by Sue Noll, courtesy of Sorensen

Ch Colorado's Columbine CD in old age as sound and correct as dogs half her age. Courtesy of Melissa Couch

Colorado's Ketchum, dark blue merle male owned by Don and Sherry Sweatt. Courtesy of Sorensen

Ch Wolf's Royal Blue. Courtesy of Sorensen

Born: ____________ Sex: ________
Owned by: ____________

PEDIGREE OF Wolfs Royal Blue

Parents	Grandparents	Great-grandparents	Great-great-grandparents
Sire: J Bar D Blue Boy	Sire: Woods Stubbie	Sire: Woods Jay	Sire: Sislers Shorty
			Dam: Sislers Trixie
		Dam: Woods Slate	Sire: Woods Jay
			Dam: Woods Specks
	Dam: Sorensens Blue Mist	Sire: Sorensens Gunsmoke	Sire: Williams Laddie
			Dam: Jeffreys Sheila
		Dam: Rowe's Dingo	Sire: Welz's Handsome King
			Dam: McGuires Palojami Gold
Dam: Wolfs Blue Heather	Sire: Youngs Tuffy	Sire: Shaws Joker	Sire: Baggerleys Johnnie No. 2
			Dam: Baggerleys Lady
		Dam: Shaws Red Tag	Sire: Gecko
			Dam: Redwood
	Dam: McNulty's Squaw	Sire: Aringtons Stub	Sire: Woods Jay
			Dam: Harpers Jinx
		Dam: Nears Blue Midget	Sire: Lighters Rusty
			Dam: Millers Belon

Bred by: ____________ Address: ____________ Phone: ____________

The details given above are correct to the best of my knowledge. Signed: ____________(Breeder)

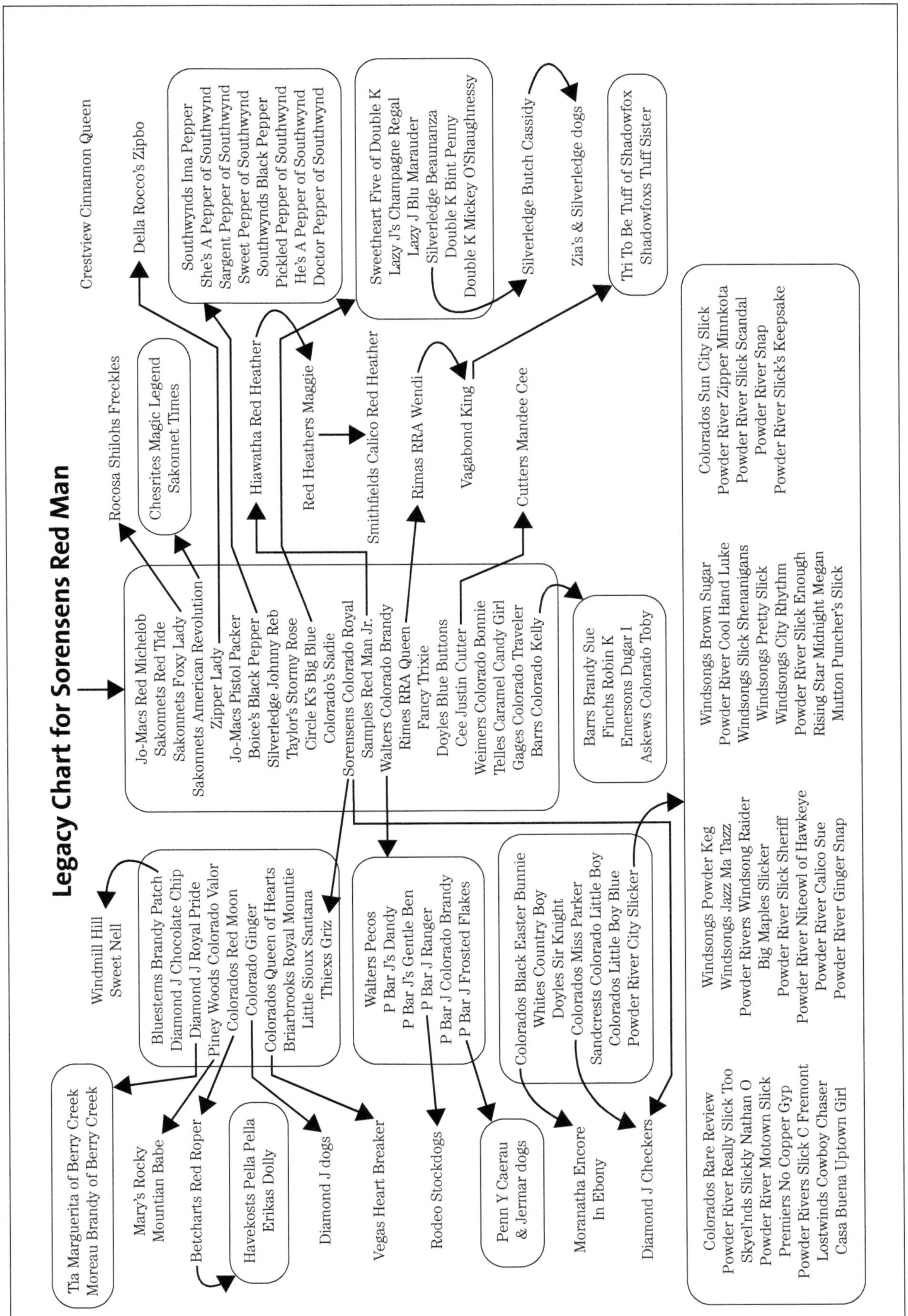
Legacy Chart for Sorensens Red Man
Jo-Macs Red Michelob
Sakonnets Red Tide
Sakonnets Foxy Lady
Sakonnets American Revolution
Zipper Lady
Jo-Macs Pistol Packer
Boice's Black Pepper
Silverledge Johnny Reb
Taylor's Stormy Rose
Circle K's Big Blue
Colorado's Sadie
Sorensens Colorado Royal
Samples Red Man Jr.
Walters Colorado Brandy
Rimes RRA Queen
Fancy Trixie
Doyles Blue Buttons
Cee Justin Cutter
Weimers Colorado Bonnie
Telles Caramel Candy Girl
Gages Colorado Traveler
Barrs Colorado Kelly
Crestview Cinnamon Queen
Della Rocco's Zipbo
Southwynds Ima Pepper
She's A Pepper of Southwynd
Sargent Pepper of Southwynd
Sweet Pepper of Southwynd
Southwynds Black Pepper
Pickled Pepper of Southwynd
He's A Pepper of Southwynd
Doctor Pepper of Southwynd
Sweetheart Five of Double K
Lazy J's Champagne Regal
Lazy J Blu Marauder
Silverledge Beaunanza
Double K Bint Penny
Double K Mickey O'Shaughnessy
Silverledge Butch Cassidy
Zia's & Silverledge dogs
Tri To Be Tuff of Shadowfox
Shadowfoxs Tuff Sister
Rocosa Shilohs Freckles
Chesrites Magic Legend
Sakonnet Times
Hiawatha Red Heather
Red Heathers Maggie
Smithfields Calico Red Heather
Rimas RRA Wendi
Vagabond King
Cutters Mandee Cee
Barrs Brandy Sue
Finchs Robin K
Emersons Dugar I
Askews Colorado Toby
Colorados Sun City Slick
Powder River Zipper Minnkota
Powder River Slick Scandal
Powder River Snap
Powder River Slick's Keepsake
Windsongs Brown Sugar
Powder River Cool Hand Luke
Windsongs Slick Shenanigans
Windsongs Pretty Slick
Windsongs City Rhythm
Powder River Slick Enough
Rising Star Midnight Megan
Mutton Puncher's Slick
Windmill Hill
Sweet Nell
Bluestems Brandy Patch
Diamond J Chocolate Chip
Diamond J Royal Pride
Piney Woods Colorado Valor
Colorados Red Moon
Colorado Ginger
Colorados Queen of Hearts
Briarbrooks Royal Mountie
Little Sioux Santana
Thiexs Griz
Walters Pecos
P Bar J's Dandy
P Bar J's Gentle Ben
P Bar J Ranger
P Bar J Colorado Brandy
P Bar J Frosted Flakes
Colorados Black Easter Bunnie
Whites Country Boy
Doyles Sir Knight
Colorados Miss Parker
Sandercrests Colorado Little Boy
Colorados Little Boy Blue
Powder River City Slicker
Windsongs Powder Keg
Windsongs Jazz Ma Tazz
Powder Rivers Windsong Raider
Big Maples Slicker
Powder River Slick Sheriff
Powder River Niteowl of Hawkeye
Powder River Calico Sue
Powder River Ginger Snap
Tia Marguerita of Berry Creek
Moreau Brandy of Berry Creek
Mary's Rocky
Mountian Babe
Betcharts Red Roper
Havekosts Pella Pella
Erikas Dolly
Diamond J dogs
Vegas Heart Breaker
Rodeo Stockdogs
Penn Y Caerau
& Jermar dogs
Moranatha Encore
In Ebony
Diamond J Checkers
Colorados Rare Review
Powder River Really Slick Too
Skyel'nds Slickly Nathan O
Powder River Motown Slick
Premiers No Copper Gyp
Powder Rivers Slick C Fremont
Lostwinds Cowboy Chaser
Casa Buena Uptown Girl

Ch Wolf's Royal Blue, a stunning dark blue and white male who epitomized the quality inherent in the Wood dogs and the quality of the Wood lines when crossed with Gunsmoke. Courtesy of Sorensen

J Bar D Blue Boy (Wood's Stubbie x Sorensen's Blue Mist), the dark blue merle genetic powerhouse. Courtesy of Sorensen

Ch Wolf's Royal Blue (J Bar D Blue Boy x Wolf's Blue Heather) Courtesy of Sorensen

Ch Powder River City Slicker CD, ATDcsd, AD trialing on cattle. Courtesy of Sorensen

Doyle's Blue Buttons, half sister to Ch Wolf's Royal Blue and dam of Ch Powder River City Slicker CD, ATDcsd, AD. Courtesy of Sorensen

Ch Powder River City Slicker CD, ATDcsd, AD winning Best Opposite Sex at the 1979 Utah Specialty. Shown by his owner Jim Foster. Courtesy of Sorensen

Ch Wolf's Royal Blue. Courtesy of Sorensen

Doyle's Blue Buttons, blue merle, copper and white female (Sorensen's Red Man x Wolf's Blue Heather). Dam of Ch Powder River City Slicker CD, ATDcsd, AD. Courtesy of Sorensen

Ch Powder Rivers City Slicker CD, ATDcsd, AD at a Colorado trial showing his good low heeling style on cattle. Courtesy of Sorensen

Born: ______	Sex: ______
Owned by: ______	

PEDIGREE OF Powder River City Slicker

Parents	Grandparents	Great-grandparents	Great-great-grandparents
Sire: Slash V Little Rock	Sire: Georges Rocky Top	Sire: Georges Dago Rojo	Sire: Tartaglia's Chocolate
			Dam: Caligaris Lady
		Dam: Keeps Waltzing Matilda	Sire: Bells Bandit
			Dam: Van Matres Riata
	Dam: Myrups Gabby	Sire: Moodys Burrocho	Sire: Hartnagles Hud
			Dam: Las Rocosa Mandaville
		Dam: Martins Red Fudge	Sire: Martins Tim Tim
			Dam: Martins Josie
Dam: Doyles Blue Buttons	Sire: Sorensens Red Man	Sire: Bruce's Blue	Sire: unknown
			Dam: unknown
		Dam: Bruce's Blue Susie	Sire: unknown
			Dam: unknown
	Dam: Wolfs Blue Heather	Sire: Youngs Tuffy	Sire: Shaws Joker
			Dam: Shaws Red Tag
		Dam: McNultys Squaw	Sire: Aringtons Stub
			Dam: Nears Blue Midget

Bred by: ______ Address: ______ Phone: ______

The details given above are correct to the best of my knowledge. Signed: ______ (Breeder)

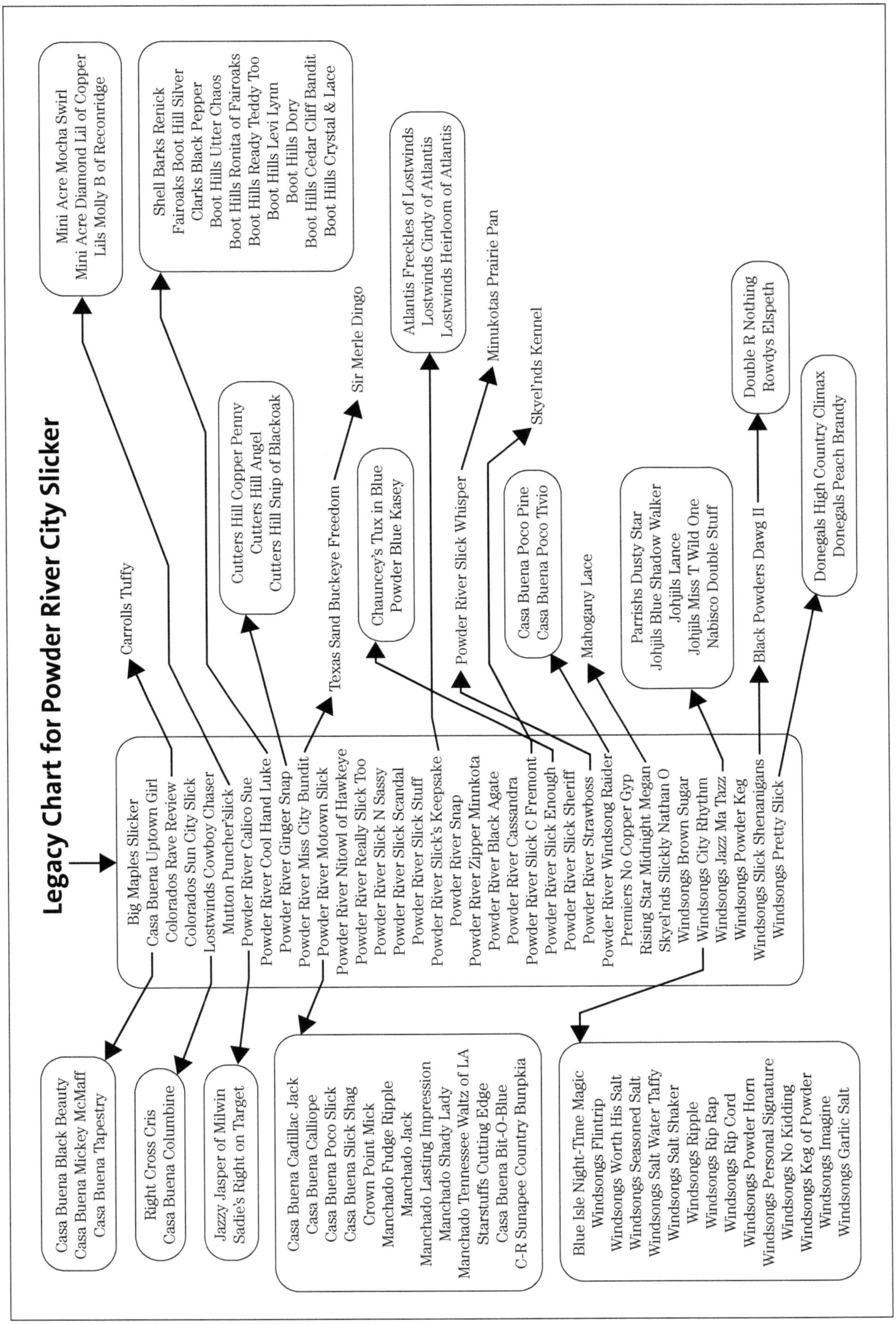
Legacy Chart for Powder River City Slicker
Big Maples Slicker
Casa Buena Uptown Girl
Colorados Rave Review
Colorados Sun City Slick
Lostwinds Cowboy Chaser
Mutton Puncher'slick
Powder River Calico Sue
Powder River Cool Hand Luke
Powder River Ginger Snap
Powder River Miss City Bundit
Powder River Motown Slick
Powder River Nitowl of Hawkeye
Powder River Really Slick Too
Powder River Slick N Sassy
Powder River Slick Scandal
Powder River Slick Stuff
Powder River Slick's Keepsake
Powder River Snap
Powder River Zipper Minnkota
Powder River Black Agate
Powder River Cassandra
Powder River Slick C Fremont
Powder River Slick Enough
Powder River Slick Sheriff
Powder River Strawboss
Powder River Windsong Raider
Premiers No Copper Gyp
Rising Star Midnight Megan
Skyel'nds Slickly Nathan O
Windsongs Brown Sugar
Windsongs City Rhythm
Windsongs Jazz Ma Tazz
Windsongs Powder Keg
Windsongs Slick Shenanigans
Windsongs Pretty Slick
Mini Acre Mocha Swirl
Mini Acre Diamond Lil of Copper
Lils Molly B of Reconridge
Shell Barks Renick
Fairoaks Boot Hill Silver
Clarks Black Pepper
Boot Hills Utter Chaos
Boot Hills Ronita of Fairoaks
Boot Hills Ready Teddy Too
Boot Hills Levi Lynn
Boot Hills Dory
Boot Hills Cedar Cliff Bandit
Boot Hills Crystal & Lace
Carrolls Tuffy
Cutters Hill Copper Penny
Cutters Hill Angel
Cutters Hill Snip of Blackoak
Texas Sand Buckeye Freedom
Sir Merle Dingo
Chauncey's Tux in Blue
Powder Blue Kasey
Atlantis Freckles of Lostwinds
Lostwinds Cindy of Atlantis
Lostwinds Heirloom of Atlantis
Powder River Slick Whisper
Minukotas Prairie Pan
Skyel'nds Kennel
Casa Buena Poco Pine
Casa Buena Poco Tivio
Mahogany Lace
Parrishs Dusty Star
Johjils Blue Shadow Walker
Johjils Lance
Johjils Miss T Wild One
Nabisco Double Stuff
Black Powders Dawg II
Double R Nothing
Rowdys Elspeth
Donegals High Country Climax
Donegals Peach Brandy
Casa Buena Black Beauty
Casa Buena Mickey McMaff
Casa Buena Tapestry
Right Cross Cris
Casa Buena Columbine
Jazzy Jasper of Milwin
Sadie's Right on Target
Casa Buena Cadillac Jack
Casa Buena Calliope
Casa Buena Poco Slick
Casa Buena Slick Shag
Crown Point Mick
Manchado Fudge Ripple
Manchado Jack
Manchado Lasting Impression
Manchado Shady Lady
Manchado Tennessee Waltz of LA
Starstuffs Cutting Edge
Casa Buena Bit-O-Blue
C-R Sunapee Country Bunpkia
Blue Isle Night-Time Magic
Windsongs Flintrip
Windsongs Worth His Salt
Windsongs Seasoned Salt
Windsongs Salt Water Taffy
Windsongs Salt Shaker
Windsongs Ripple
Windsongs Rip Rap
Windsongs Rip Cord
Windsongs Powder Horn
Windsongs Personal Signature
Windsongs No Kidding
Windsongs Keg of Powder
Windsongs Imagine
Windsongs Garlic Salt

Taylor's Silver Buffie showing her natural ability on cattle. This was the trait most highly valued during the creation of the Australian Shepherd. After the Sorensens obtained Buffie she turned into a good heeling dog on cattle. Courtesy of Sorensen

Ch Powder Rivers City Slicker CD, ATDcsd, AD, black tri male. One of the best working dogs of his day and sire of innumerable top dogs. Courtesy of Sorensen

Ch Powder Rivers City Slicker CD, ATDcsd, AD was a versatile dog capable of working any type of livestock. He had plenty of grip on cattle but showed authoritative eye on sheep and ducks. Courtesy of Sorensen

Ch Powder Rivers City Slicker CD, ATDcsd, AD—the complete Australian Shepherd. Well trained to use his instincts as required for the job at hand. He passed on his good qualities in abundance and influenced working bloodlines from coast to coast. Courtesy of Sorensen

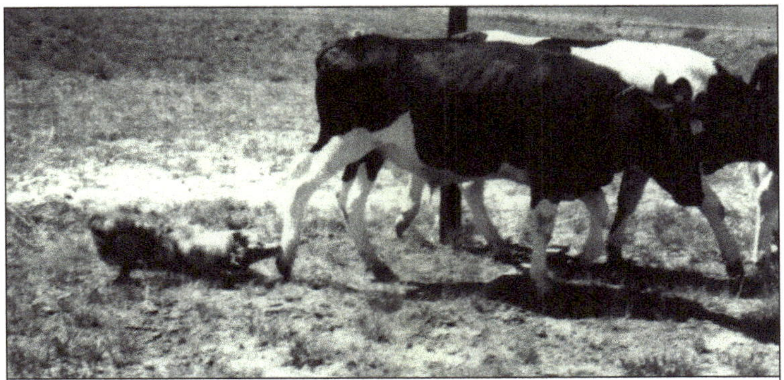

Taylor's Silver Buffie was well known for her low heeling style. Courtesy of Sorensen

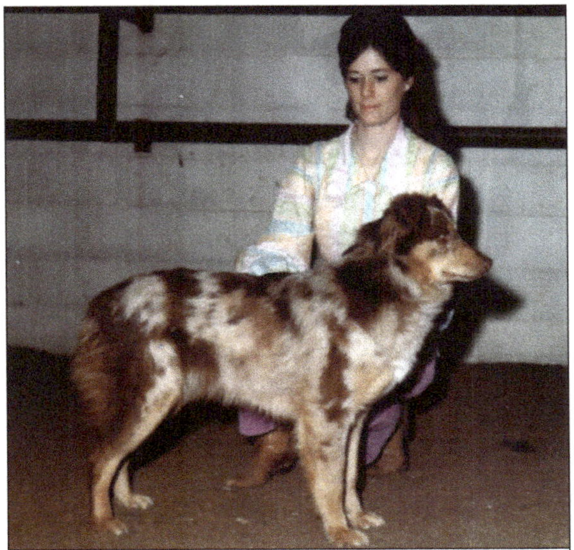

Ch Colorado's Sunset, red merle female and daughter of Taylor's Luke the Drifter x Taylor's Silver Buffie. Courtesy of Jean Taylor

Ch (Sorensen's) Colorado Sunset winning first place in a 6-9 month puppy bitch class at about 8 months of age. Courtesy of Sorensen

Taylor's Silver Buffie, red merle female of unknown background but beautiful type, temperament, working ability and structure. When the Taylors got Buffie she had been beaten for chasing horses and wouldn't look at them if someone was watching. To register her, a film was submitted to ASCA of Buffie heeling a staked out calf! Courtesy of Sorensen

Mary's Rocky Mountain Babe, solid red female who later had a profound influence on Betty Nelson's breeding program on the west coast (Piney Wood's Colorado Valor x Piney Wood's Colorado Babe). A strong cattle working dog who had great style. Pictured here in January 1976, at the Rowe Ranch in Broomfield, Colorado, at 10 months of age. Courtesy of Sorensen

J Bar D Blue Boy, blue merle male, sired by Wood's Stubbie, a typical Wood's type dog with the proven cross of Gunsmoke, his grandsire. Courtesy of Sorensen

Jody Blue, a J Bar D Blue Boy daughter. Consistent quality was the hallmark of Blue Boy. Courtesy of Sorensen

Blue Boy puppy out of Weimers Deesa. Typical very dark blue merle of excellent type. Courtesy of Sorensen

Blue Boy puppy out of Sorensen's Hall's Apache. Blue merle with the obvious quality of a linebred Wood's dog. Courtesy of Sorensen

Silver Candy, a J Bar D Blue Boy puppy. Courtesy of Sorensen

Blue Boy litter out of Sorensen's Hall's Apache. Consistent quality was the legacy of Fletcher Wood's lines. Courtesy of Sorensen

Silver Belle by J Bar D Blue Boy. Courtesy of Sorensen

Born: ____________ Sex: ______
Owned by: ____________________

PEDIGREE OF J Bar D Blue Boy

Parents	Grandparents	Great-grandparents	Great-great-grandparents
Sire: Woods Stubbie	Sire: Woods Jay	Sire: Sislers Shorty	Sire: Keeno
			Dam: Blue Star
		Dam: Sislers Trixie	Sire: Blue Khan
			Dam: Emmetts Blue
	Dam: Woods Slate	Sire: Woods Jay	Sire: Sislers Shorty
			Dam: Sislers Trixie
		Dam: Woods Specks	Sire: Woods Jay
			Dam: Trumps Stubbie
Dam: Sorensens Blue Mist	Sire: Sorensens Gunsmoke	Sire: Williams Laddie	Sire: Uhlendorfs Tippy
			Dam: Thorntons Chrissie
		Dam: Jeffreys Sheila	Sire: Elys Major
			Dam: Williams Georgina
	Dam: Rowes Dingo	Sire: Welz's Handsome King	Sire: Harpers Old Smokey
			Dam: Smedras Blue Mistingo
		Dam: McGuires Palojami Gold	Sire: Farneds Dead Eye
			Dam: Hoopers Blue Keno

Bred by: ____________ Address: ____________ Phone: ____________

The details given above are correct to the best of my knowledge. Signed: ____________(Breeder)

Colorado's Blue Majesty RD, STD, blue and white female who was a top working dog and producer for the Kintyre Cattle Co. A natural low heeler learning to head in this photo. Courtesy of Sorensen

The classic Colorado's Blue Majesty. Courtesy of Sorensen

Colorado's Puffa Smoke, blue merle and white 5 month old puppy (J Bar D Blue Boy x Colorado's Perky Miss), Brown eyed 1/4 NBT. Courtesy of Sorensen

Colorado's Blue Majesty at about 10 months of age. Described by her owner as a good worker with an excellent temperament who would hit on heads and heels. Courtesy of Sorensen

The quiet heeler, Colorado's Blue Majesty, at the Deep South ASC show and trial on November 13, 1977. She won Reserve Winners Bitch and qualified on Started Cattle at this event. Courtesy of Sorensen

Colorado's Angus. Courtesy of Sorensen

Colorado's Winchesterblue. Courtesy of Sorensen

Colorado's Winchesterblue, blue merle male of strong Wood heritage. A top working dog and sire. Courtesy of Sorensen

Colorado Angus, solid black male by Wolf's Royal Blue x Hemis Spittin Image shows the quality that was achieved when crossing the Wood dogs with dogs of that general type such as the renowned Hemi. Courtesy of Sorensen

Colorado's Cherry Punch, red merle female reminiscent of her granddam Cherry in type and working ability. Courtesy of Sorensen

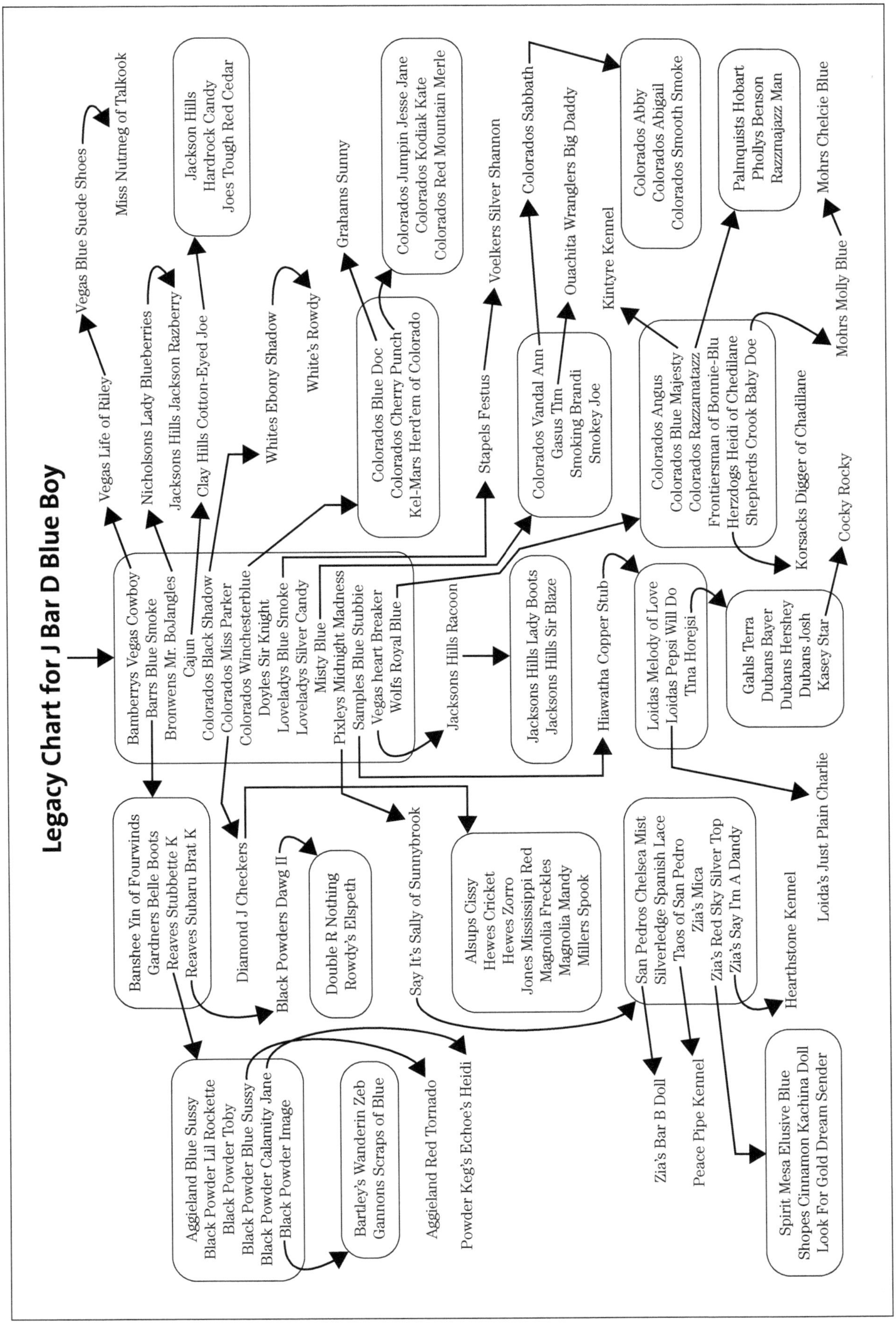
Legacy Chart for J Bar D Blue Boy
Bamberrys Vegas Cowboy
Barrs Blue Smoke
Bronwens Mr. BoJangles
Cajun
Colorados Black Shadow
Colorados Miss Parker
Colorados Winchesterblue
Doyles Sir Knight
Loveladys Blue Smoke
Loveladys Silver Candy
Misty Blue
Pixleys Midnight Madness
Samples Blue Stubbie
Vegas heart Breaker
Wolfs Royal Blue
Vegas Life of Riley
Vegas Blue Suede Shoes
Miss Nutmeg of Talkook
Nicholsons Lady Blueberries
Jacksons Hills Jackson Razberry
Clay Hills Cotton-Eyed Joe
Jackson Hills
Hardrock Candy
Joes Tough Red Cedar
Whites Ebony Shadow
White's Rowdy
Colorados Blue Doc
Colorados Cherry Punch
Kel-Mars Herd'em of Colorado
Grahams Sunny
Colorados Jumpin Jesse Jane
Colorados Kodiak Kate
Colorados Red Mountain Merle
Stapels Festus
Voelkers Silver Shannon
Colorados Vandal Ann
Gasus Tim
Smoking Brandi
Smokey Joe
Colorados Sabbath
Ouachita Wranglers Big Daddy
Colorados Abby
Colorados Abigail
Colorados Smooth Smoke
Colorados Angus
Colorados Blue Majesty
Colorados Razzamatazz
Frontiersman of Bonnie-Blu
Herzdogs Heidi of Chedilane
Shepherds Crook Baby Doe
Kintyre Kennel
Palmquists Hobart
Phollys Benson
Razzmajazz Man
Mohrs Molly Blue
Mohrs Chelcie Blue
Korsacks Digger of Chadilane
Cocky Rocky
Jacksons Hills Racoon
Jacksons Hills Lady Boots
Jacksons Hills Sir Blaze
Hiawatha Copper Stub
Loidas Melody of Love
Loidas Pepsi Will Do
Tina Horejsi
Gahls Terra
Dubans Bayer
Dubans Hershey
Dubans Josh
Kasey Star
Loida's Just Plain Charlie
Banshee Yin of Fourwinds
Gardners Belle Boots
Reaves Stubbette K
Reaves Subaru Brat K
Diamond J Checkers
Black Powders Dawg II
Double R Nothing
Rowdy's Elspeth
Say It's Sally of Sunnybrook
Alsups Cissy
Hewes Cricket
Hewes Zorro
Jones Mississippi Red
Magnolia Freckles
Magnolia Mandy
Millers Spook
San Pedros Chelsea Mist
Silverledge Spanish Lace
Taos of San Pedro
Zia's Mica
Zia's Red Sky Silver Top
Zia's Say I'm A Dandy
Hearthstone Kennel
Aggieland Blue Sussy
Black Powder Lil Rockette
Black Powder Toby
Black Powder Blue Sussy
Black Powder Calamity Jane
Black Powder Image
Bartley's Wanderin Zeb
Gannons Scraps of Blue
Aggieland Red Tornado
Powder Keg's Echoe's Heidi
Zia's Bar B Doll
Peace Pipe Kennel
Spirit Mesa Elusive Blue
Shopes Cinnamon Kachina Doll
Look For Gold Dream Sender

J Bar D Brandy, dark red merle female, perhaps one of the most important dams in the history of the Australian Shepherd. Courtesy of Sorensen

Colorado's Brandy Bow, red merle female owned by Kintyre Cattle Co. A working dog deluxe as all the Kintyre dogs had to be. Courtesy of Sorensen

Colorado's Brandy Bow. Courtesy of Sorensen

Colorado's Reddy Raider, Red merle and white male (Ch Jones Reddy Teddy CDX CWD x J Bar D Brandy) pictured at about 7 months of age. Another of the Kintyre Cattle Company's treasured males. Courtesy of Sorensen

Born: ______________ Sex: ________
Owned by: ______________________

PEDIGREE OF J Bar D Brandy

Parents	Grandparents	Great-grandparents	Great-great-grandparents
Sire: Rowe's Warrior Bow	Sire: Ronsleys Charcoal Smoke	Sire: Ritters Streak	Sire: Ritters Duke
			Dam: Ritters Babe
		Dam: Ritters Lass	Sire: Ritters Duke
			Dam: Ritters Nickie
	Dam: Murphys Cochina	Sire: Ronsleys Charcoal Smoke	Sire: Ritters Streak
			Dam: Ritters Lass
		Dam: Ronsleys Mitzi Buck	Sire: Byes Rusty II
			Dam: Byes Penny
Dam: J Bar D Choctaw Chilli	Sire: Rowe's Warrior Bow	Sire: Ronsleys Charcoal Smoke	Sire: Ritters Streak
			Dam: Ritters Lass
		Dam: Murphys Cochina	Sire: Ronsleys Charcoal Smoke
			Dam: Ronsleys Mitzi Buck
	Dam: Sorensens Queenie	Sire: Sorensens Gunsmoke	Sire: Williams Laddie
			Dam: Jeffreys Sheila
		Dam: Nightingales Tammy	Sire: Elys Chico
			Dam: Hollands Nikki

Bred by: ______________ Address: ______________________ Phone: ____________

The details given above are correct to the best of my knowledge. Signed: ______________________(Breeder)

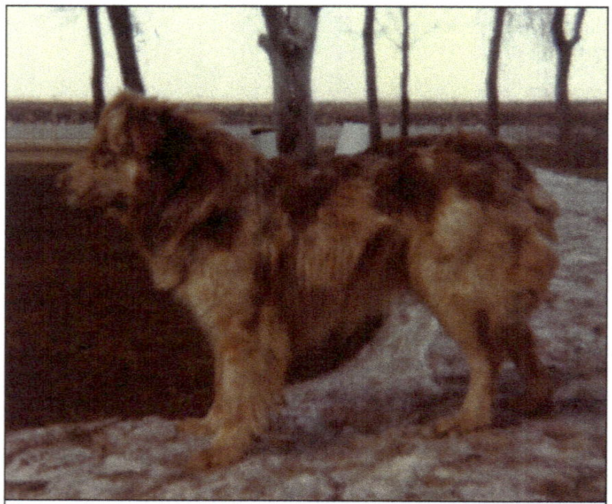

(Weimer's) Colorado Bonnie, red merle female of impeccable pedigree and wide influence throughout the breed. Courtesy of Sorensen

Ch Colorado's Ravishing Ruby CDX OTDcs ATDd, solid red female. The female foundation of Powder River kennel and producer of working dogs which influenced breeding programs from coast to coast. Courtesy of Sorensen

Walker's Joshua, blue merle male who was influential in many west coast breeding programs and a top working dog in his own right. Courtesy of Sorensen

Dorns Poudre Lakohta, blue merle, copper and white male, (Ch Iacovetta's Buck x Dressel's Duchess), an important sire for Colorado, carrying on the tradition of the Gunsmoke line well into the future. Courtesy of Sorensen

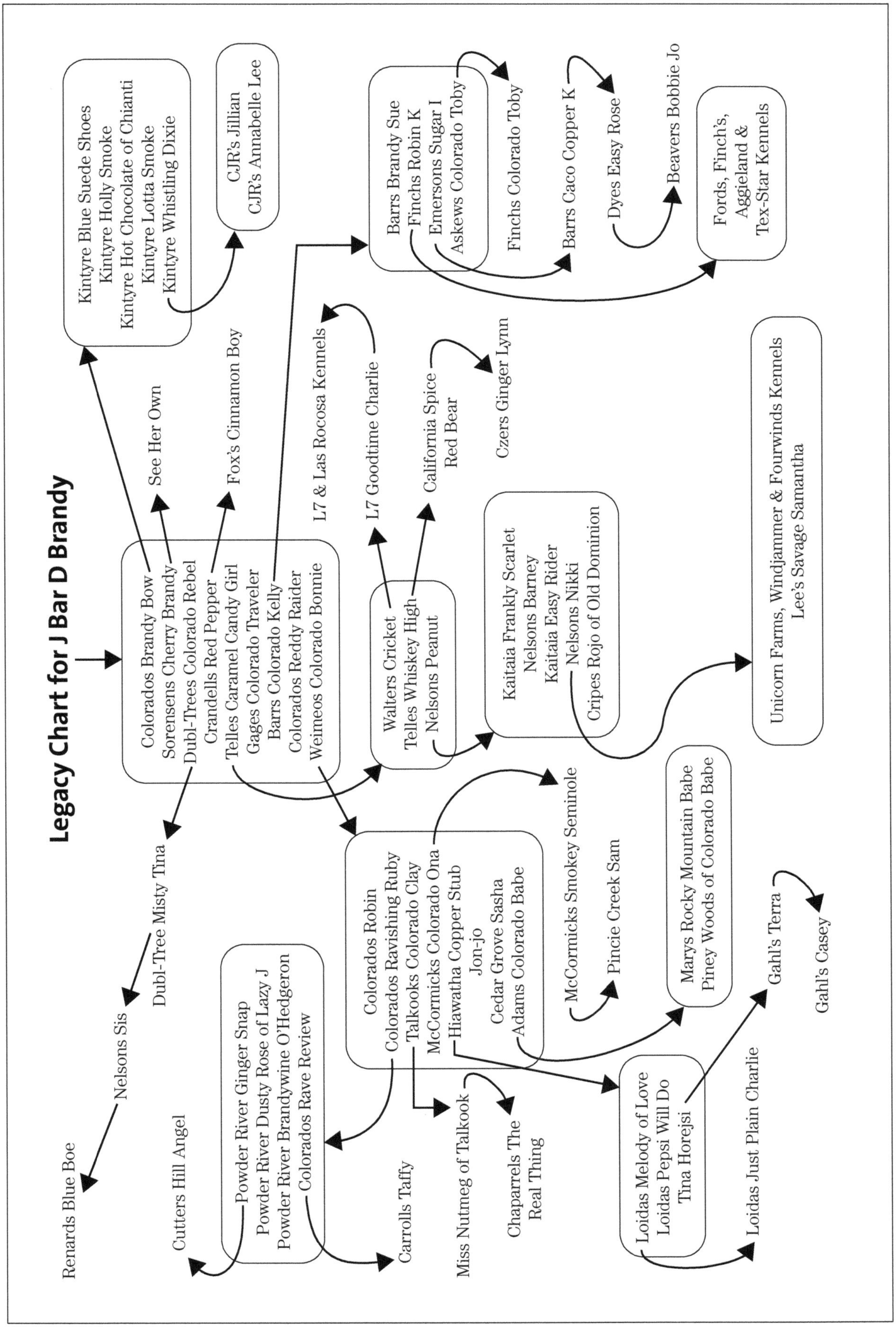
Legacy Chart for J Bar D Brandy
Colorados Brandy Bow
Sorensens Cherry Brandy
Dubl-Trees Colorado Rebel
Crandells Red Pepper
Telles Caramel Candy Girl
Gages Colorado Traveler
Barrs Colorado Kelly
Colorados Reddy Raider
Weimeos Colorado Bonnie
Kintyre Blue Suede Shoes
Kintyre Holly Smoke
Kintyre Hot Chocolate of Chianti
Kintyre Lotta Smoke
Kintyre Whistling Dixie
CJR's Jillian
CJR's Annabelle Lee
See Her Own
Fox's Cinnamon Boy
Barrs Brandy Sue
Finchs Robin K
Emersons Sugar I
Askews Colorado Toby
Finchs Colorado Toby
Barrs Caco Copper K
Dyes Easy Rose
Beavers Bobbie Jo
Fords, Finch's, Aggieland & Tex-Star Kennels
L7 & Las Rocosa Kennels
L7 Goodtime Charlie
California Spice
Red Bear
Czers Ginger Lynn
Walters Cricket
Telles Whiskey High
Nelsons Peanut
Kaitaia Frankly Scarlet
Nelsons Barney
Kaitaia Easy Rider
Nelsons Nikki
Cripes Rojo of Old Dominion
Unicorn Farms, Windjammer & Fourwinds Kennels
Lee's Savage Samantha
Dubl-Tree Misty Tina
Nelsons Sis
Renards Blue Boe
Colorados Robin
Colorados Ravishing Ruby
Talkooks Colorado Clay
McCormicks Colorado Ona
Hiawatha Copper Stub
Jon-jo
Cedar Grove Sasha
Adams Colorado Babe
McCormicks Smokey Seminole
Pincie Creek Sam
Marys Rocky Mountain Babe
Piney Woods of Colorado Babe
Gahl's Terra
Gahl's Casey
Powder River Ginger Snap
Powder River Dusty Rose of Lazy J
Powder River Brandywine O'Hedgeron
Colorados Rave Review
Cutters Hill Angel
Carrolls Taffy
Miss Nutmeg of Talkook
Chaparrels The Real Thing
Loidas Melody of Love
Loidas Pepsi Will Do
Tina Horejsi
Loidas Just Plain Charlie

PEDIGREE OF Nightingales Tammy

Born: ______ Sex: ______
Owned by: ______

Generation 1	Generation 2	Generation 3	Generation 4
Sire: Elys Chico	Sire: Byes Rusty II	Sire: Shermans Casey	Sire: Davey
			Dam: Anza
		Dam: Byes Snooks II	Sire: Byes Taffy
			Dam: Byes Snooks
	Dam: Byes Grypsie	Sire: Bob	Sire: unknown
			Dam: unknown
		Dam: Gyp	Sire: unknown
			Dam: unknown
Dam: Hollands Nikki	Sire: Woods Jay	Sire: Sislers Shorty	Sire: Keeno
			Dam: Blue Star
		Dam: Sislers Trixie	Sire: Blue Khan
			Dam: Emmetts Blue
	Dam: Knolls Kiki	Sire: unknown	Sire: unknown
			Dam: unknown
		Dam: unknown	Sire: unknown
			Dam: unknown

Bred by: ______ Address: ______ Phone: ______

The details given above are correct to the best of my knowledge. Signed: ______(Breeder)

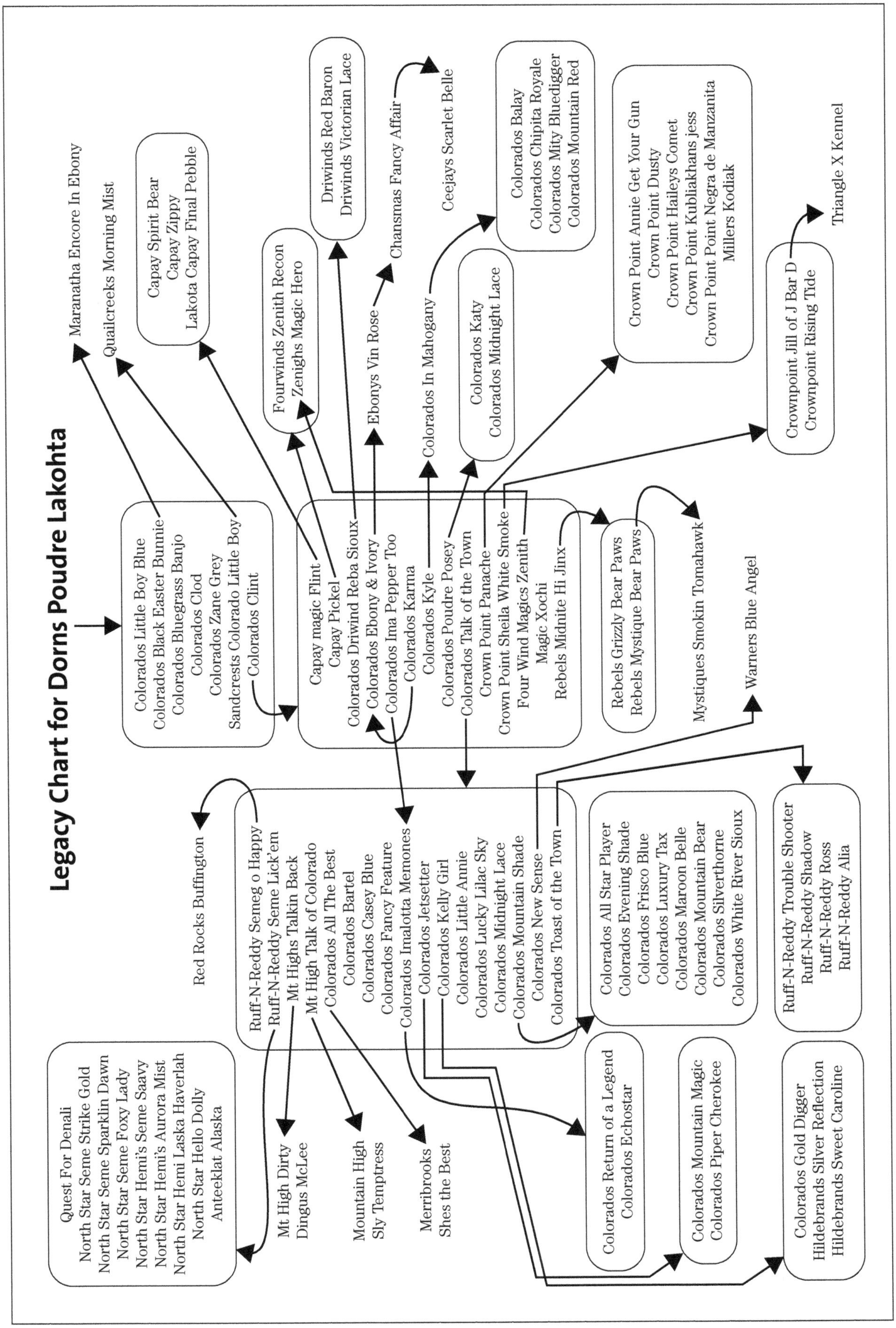
Legacy Chart for Dorns Poudre Lakohta
Colorados Little Boy Blue
Colorados Black Easter Bunnie
Colorados Bluegrass Banjo
Colorados Clod
Colorados Zane Grey
Sandcrests Colorado Little Boy
Colorados Clint
Maranatha Encore In Ebony
Quailcreeks Morning Mist
Capay Spirit Bear
Capay Zippy
Lakota Capay Final Pebble
Capay magic Flint
Capay Pickel
Colorados Driwind Reba Sioux
Colorados Ebony & Ivory
Colorados Ima Pepper Too
Colorados Karma
Colorados Kyle
Colorados Poudre Posey
Colorados Talk of the Town
Crown Point Panache
Crown Point Sheila White Smoke
Four Wind Magics Zenith
Magic Xochi
Rebels Midnite Hi Jinx
Fourwinds Zenith Recon
Zenighs Magic Hero
Driwinds Red Baron
Driwinds Victorian Lace
Ebonys Vin Rose
Chansmas Fancy Affair
Ceejays Scarlet Belle
Colorados In Mahogany
Colorados Katy
Colorados Midnight Lace
Colorados Balay
Colorados Chipita Royale
Colorados Mity Bluedigger
Colorados Mountain Red
Crown Point Annie Get Your Gun
Crown Point Dusty
Crown Point Haileys Comet
Crown Point Kubliakhans jess
Crown Point Negra de Manzanita
Millers Kodiak
Crownpoint Jill of J Bar D
Crownpoint Rising Tide
Triangle X Kennel
Rebels Grizzly Bear Paws
Rebels Mystique Bear Paws
Mystiques Smokin Tomahawk
Warners Blue Angel
Red Rocks Buffington
Ruff-N-Reddy Semeg o Happy
Ruff-N-Reddy Seme Lick'em
Mt Highs Talkin Back
Mt High Talk of Colorado
Colorados All The Best
Colorados Bartel
Colorados Casey Blue
Colorados Fancy Feature
Colorados Imalotta Memones
Colorados Jetsetter
Colorados Kelly Girl
Colorados Little Annie
Colorados Lucky Lilac Sky
Colorados Midnight Lace
Colorados Mountain Shade
Colorados New Sense
Colorados Toast of the Town
Colorados All Star Player
Colorados Evening Shade
Colorados Frisco Blue
Colorados Luxury Tax
Colorados Maroon Belle
Colorados Mountain Bear
Colorados Silverthorne
Colorados White River Sioux
Ruff-N-Reddy Trouble Shooter
Ruff-N-Reddy Shadow
Ruff-N-Reddy Ross
Ruff-N-Reddy Alia
Quest For Denali
North Star Seme Strike Gold
North Star Seme Sparklin Dawn
North Star Seme Foxy Lady
North Star Hemi's Seme Saavy
North Star Hemi's Aurora Mist
North Star Hemi Laska Haverlah
North Star Hello Dolly
Anteeklat Alaska
Mt High Dirty
Dingus McLee
Mountain High
Sly Temptress
Merribrooks
Shes the Best
Colorados Return of a Legend
Colorados Echostar
Colorados Mountain Magic
Colorados Piper Cherokee
Colorados Gold Digger
Hildebrands Silver Reflection
Hildebrands Sweet Caroline

Colorado's Vandal Ann, blue merle female of superior type. Courtesy of Sorensen

Colorado's Smooth Smoke, blue merle female who brought excellent type, superior pedigree and good working ability to the Crown Point kennel. Courtesy of Sorensen

Colorado's Sharade, blue merle, copper and white female (Ch Slash V Little Rock x Colorado's Hotline Paige), a beautiful companion, show dog, and Junior dog for Becky Sorensen. Courtesy of Sorensen

Ch Colorado's Sizzlin Sioux, red merle female, foundation for the early Briarbrook dogs. Courtesy of Sorensen

Colorado's Sharade, blue merle female. A good working dog even with limited training and opportunity. Courtesy of Sorensen

Born: ____________ Sex: ________
Owned by: ____________

PEDIGREE OF Sorensen's Cherry Brandy

Parents	Grandparents	Great-grandparents	Great-great-grandparents
Sire: Piz's Joseph	Sire: Woods Jay	Sire: Sislers Shorty	Sire: Kono
			Dam: Blue Star
		Dam: Sislers Trixie	Sire: Blue Khan
			Dam: Emmett's Blue
	Dam: Wood's Chili	Sire: Woods Jay	Sire: Sislers Shorty
			Dam: Sislers Trixie
		Dam: Greens Dingo	Sire: Woods Jay
			Dam: Ely's Blue
Dam: J Bar D Brandy	Sire: Rowe's Warrior Bow	Sire: Ronsley's Charcoal Smoke	Sire: Rifter's Streak
			Dam: Ritters Lass
		Dam: Murphy's Cochina	Sire: Ronsleys Charcoal Smoke
			Dam: Ronsley's Mitzi Buck
	Dam: J Bar D Choctaw Chilli	Sire: Rowe's Warrior Bow	Sire: Ronsley's Charcoal Smoke
			Dam: Murphy's Cochina
		Dam: Sorensens Queenie	Sire: Sorensens Gunsmoke
			Dam: Nightingales Tammy

Bred by: ____________ Address: ____________ Phone: ____________

The details given above are correct to the best of my knowledge. Signed: ____________(Breeder)

Ch Colorado's Sunbonnet Sioux CD STDs, red merle female, at one of her first shows as a puppy. Courtesy of Sorensen

Ch Colorado's Sunbonnet Sioux CD STDs, winning as a mature dog. Courtesy of Sorensen

Ch Colorado's Sunbonnet Sioux CD STDs red merle, copper and white female (Ch Taylor's Shamrock of Silverledge CDX x Colorado's Sunshine Sioux). Courtesy of Sorensen

AKC/ASCA Ch Colorado's Terry Blue OTDc, blue merle male. An outcross made by Colorado kennel to incorporate show lines into their excellent working dogs. Courtesy of Sorensen

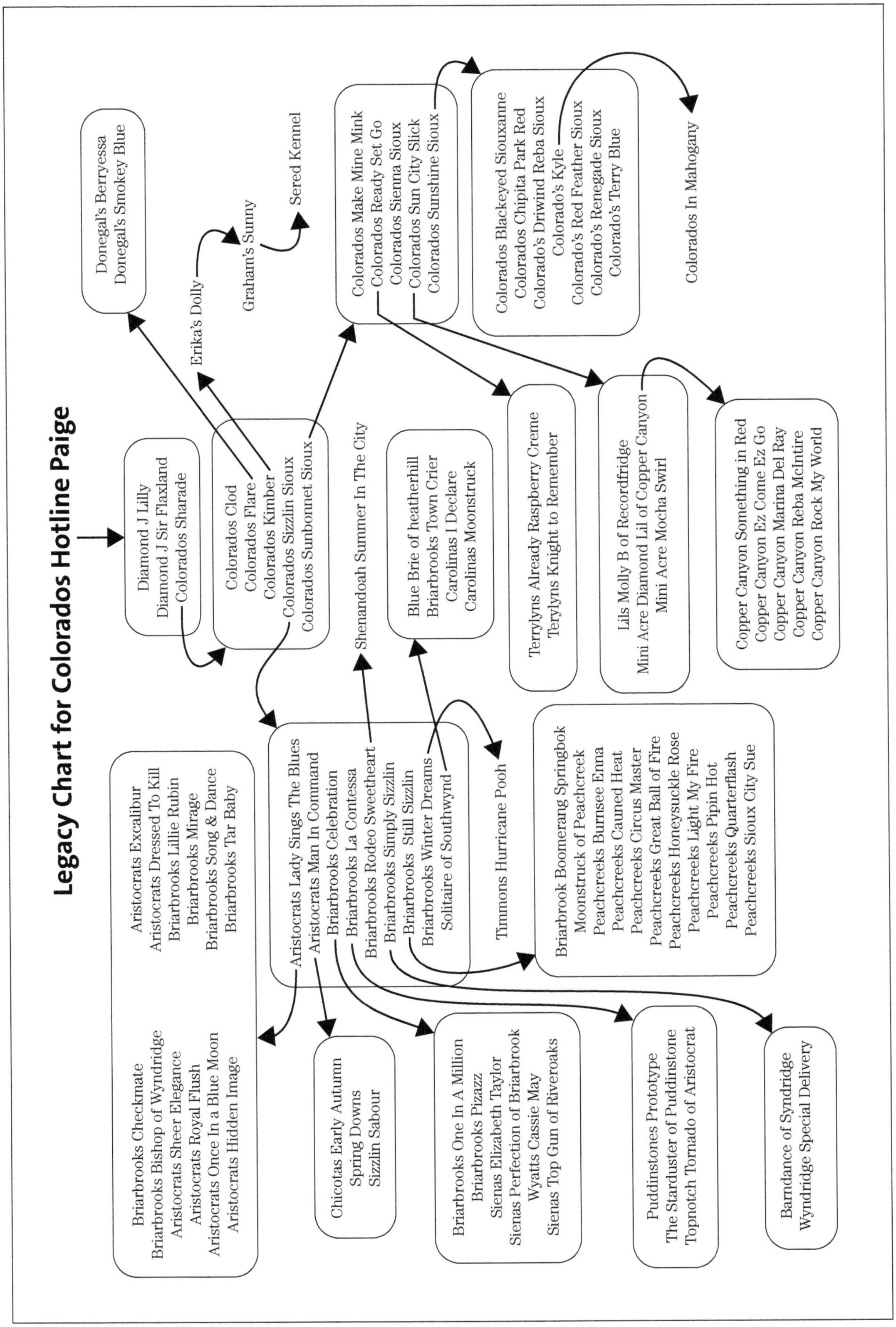
Legacy Chart for Colorados Hotline Paige
Diamond J Lilly
Diamond J Sir Flaxland
Colorados Sharade
Colorados Clod
Colorados Flare
Colorados Kimber
Colorados Sizzlin Sioux
Colorados Sunbonnet Sioux
Donegal's Berryessa
Donegal's Smokey Blue
Erika's Dolly
Graham's Sunny
Sered Kennel
Colorados Make Mine Mink
Colorados Ready Set Go
Colorados Sienna Sioux
Colorados Sun City Slick
Colorados Sunshine Sioux
Colorados Blackeyed Siouxanne
Colorados Chipita Park Red
Colorado's Driwind Reba Sioux
Colorado's Kyle
Colorado's Red Feather Sioux
Colorado's Renegade Sioux
Colorado's Terry Blue
Colorados In Mahogany
Shenandoah Summer In The City
Blue Brie of heatherhill
Briarbrooks Town Crier
Carolinas I Declare
Carolinas Moonstruck
Terrylyns Already Raspberry Creme
Terylyns Knight to Remember
Lils Molly B of Recordfridge
Mini Acre Diamond Lil of Copper Canyon
Mini Acre Mocha Swirl
Copper Canyon Something in Red
Copper Canyon Ez Come Ez Go
Copper Canyon Marina Del Ray
Copper Canyon Reba McIntire
Copper Canyon Rock My World
Aristocrats Excalibur
Aristocrats Dressed To Kill
Briarbrooks Lillie Rubin
Briarbrooks Mirage
Briarbrooks Song & Dance
Briarbrooks Tar Baby
Aristocrats Lady Sings The Blues
Aristocrats Man In Command
Briarbrooks Celebration
Briarbrooks La Contessa
Briarbrooks Rodeo Sweetheart
Briarbrooks Simply Sizzlin
Briarbrooks Still Sizzlin
Briarbrooks Winter Dreams
Solitaire of Southwynd
Timmons Hurricane Pooh
Briarbrook Boomerang Springbok
Moonstruck of Peachcreek
Peachcreeks Burnsee Enna
Peachcreeks Cauned Heat
Peachcreeks Circus Master
Peachcreeks Great Ball of Fire
Peachcreeks Honeysuckle Rose
Peachcreeks Light My Fire
Peachcreeks Pipin Hot
Peachcreeks Quarterflash
Peachcreeks Sioux City Sue
Briarbrooks Checkmate
Briarbrooks Bishop of Wyndridge
Aristocrats Sheer Elegance
Aristocrats Royal Flush
Aristocrats Once In a Blue Moon
Aristocrats Hidden Image
Chicotas Early Autumn
Spring Downs
Sizzlin Sabour
Briarbrooks One In A Million
Briarbrooks Pizazz
Sienas Elizabeth Taylor
Sienas Perfection of Briarbrook
Wyatts Cassie May
Sienas Top Gun of Riveroaks
Puddinstones Prototype
The Starduster of Puddinstone
Topnotch Tornado of Aristocrat
Barndance of Syndridge
Wyndridge Special Delivery

Colorado's Renegade Sioux, red tri female (Ch Bayshores Flapjack x Colorado's Sunshine Sioux. Courtesy of Sorensen

Ch Colorado's Talk of the Town, blue merle, copper and white male (Ch Colorados Clint x Ch Posey Patch of Blue Mist) winning Best of Breed at a 1986 CASA show. Courtesy of Sorensen

Colorado's Kelly Girl, Blue merle, copper and white female (Ch Colorados Talk of the Town x Colorados Christmas Posey). Courtesy of Sorensen

Colorado's Jetsetter. Courtesy of Sorensen

Colorado's Jetsetter (Colorado's Talk of the Town x Colorados Ames Aussie Miss) (Colorado's Ames Aussie Miss was a daughter of Ruff N Reddy Mity Mean x Colorado's Jumpin Jesse Jane) (Colorado's Jumpin Jesse Jane was a daughter of Slash V Wild Timber x Colorado's Cherry Punch). Stunning black female of foundation Colorado lines. Courtesy of Sorensen

Colorado's Jetsetter. Courtesy of Sorensen

Colorado's Jetsetter. Courtesy of Sorensen

Colorado's Country Music (Colorado's How the West Was Won x Colorado's Gold-Digger), blue and white female. Another example of the working and show crosses practiced today at Colorado kennel. Courtesy of Sorensen

Dick and Leslie Sorensen, July 4, 1983. Courtesy of Sorensen

Colorado's Mountain Magic (Colorados Sugar Daddy x Colorados Jetsetter), black and white female representative of the blend of old working lines and new show lines which marks today's Colorado kennel. Magic is dam of Colorado's Cathy Jean, Colorado's Slate and others. Courtesy of Sorensen

chapter 15

Jerry and Sharon Rowe

J BAR D

The J Bar D story has two endings. The first, told in the previous chapter, is the story of the Sorensen family who later left the original kennel name to the Rowes and carried on under the Colorado prefix. The second ending is that of Jerry and Sharon Rowe, who continued breeding under the J Bar D name. The two endings overlap and intertwine, even as they diverge on separate paths. The separation was in name and to some degree in bloodline. However, the Rowe and Sorensen names remained closely tied and the dogs they produced were used often to refresh the lines of the other over the years until the present day.

As children, both Jerry and Sharon had grown up with sheepdog type and shepherd dogs as pets. However, the Rowes were new to the Australian Shepherd breed when they met Dick Sorensen at that first obedience class. They had a young family and were looking for a family pet when they found their first Aussie, Rowe's Dingo, in 1964. Sharon had been searching the newspaper ads for a Norwegian Elkhound but none were available. Then she spied an ad Ron McGuire had placed for a litter of Australian Shepherd puppies he had available. Recalling her own upbringing with dogs known as toy shepherds, she decided to check out this intriguing new breed. What a life altering decision this proved to be!

In keeping with the breed name, the black tri female who changed the direction of their lives was named Dingo after the wild dogs of

Australia. She was a daughter of McGuire's Palojami Gold, a black tri daughter of the multi-talented character, Farned's Dead Eye. Her sire was a blue merle son of Harper's Old Smokey and the Wood's Jay granddaughter, Smedra's Blue Mistingo. Dingo was also closely related to Marie Leavitt's beautiful blue merle bitch, Leavitt's Patches. Patches was also by Dingo's sire, Welz's Handsome King and her dam, Robert's Turning Wing Blue was a full sister to McGuire's Palojami Gold. Patches was later bred to several Flintridge males, producing foundation stock for a number of show kennels. She was also bred to a son of Rowe's Comanche Warrior and a linebred J Bar D male, producing a number of good dogs for other early breeders.

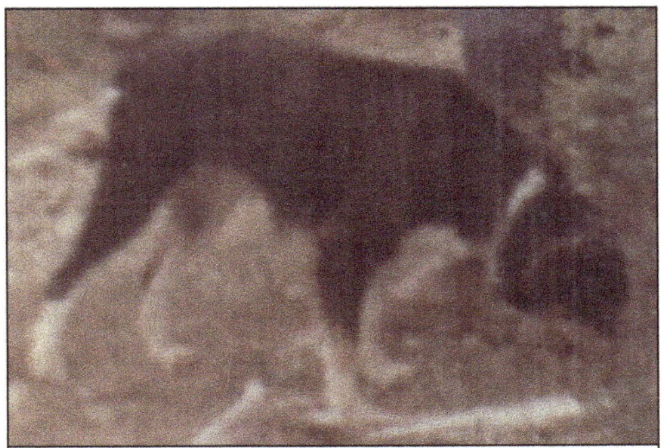

Rowe's Dingo (Welz's Handsome King x McGuire's Palojami Gold) on a trip to the mountains with Rowe family. Courtesy of Rowe

Dingo spent her early days exhibiting her destructive abilities, managing to destroy 70 carefully tended rose bushes and a decorative pond. Sharon realized this active, intelligent puppy was going to need some serious work if she was to remain a happy family member. Sharon knew nothing about training dogs so she again picked up the newspaper to find help. She found it in the person of nationally renowned obedience trainer and competitor Edna Travinek. She signed up for the class which would join the Rowe and Sorensen names forever in Aussie history.

Farned's Dead Eye, (Gregg x Whiskey Bell), blue merle and white male who sired many important foundation dogs is pictured sitting on the back of a horse. Dead Eye was also the ultimate character who was the subject of many adventures with his owner Jeane Farned. Courtesy of Rowe

In this era, most dogs were rural dogs or cowboy's dogs and most people prided themselves on their dogs intelligence, their own abilities as trainers, and their dog's performance at gatherings. When the Rowe's and the other early Colorado breeders joined together to form a club, one of their goals was to bring attention to their great breed as a family dog. There were no shows or other formal competitions for Aussies yet and showing off mainly consisted of trying to outdo one another with the variety and creativity of the tricks each dog knew. Having a booth at festivals, rodeos and other public gatherings where they could show off these skills was one way the club accomplished this goal. Jay Sisler's example served to drive these trainers to greater and greater heights with some truly amazing tricks. Thanks to Fletcher Wood,

Farned's Dead Eye. Courtesy of Rowe

Australian Shepherd Club of Colorado booth at the Arvada Festival, September 1965. Courtesy of Rowe

Sharon, and Jerry had the pleasure of meeting Jay Sisler for the first time during one of his many appearances at the National Western Stock Show. Their memories, like so many others, are of a friendly and generous man who was always willing to share dog stories and training tips with anyone who asked. After this initial meeting, the Rowe's enjoyed many other visits with Jay when he happened to be in town and always made the most of the opportunity to learn more about dog training and the breed.

With their enthusiasm for the breed growing every day, Dick Sorensen and the Rowes decided to seek out other breed enthusiasts in their area and around 1965 the group decided to form a new Colorado club for the breed—the first in the state. They called their group the Australian Shepherd Club of Colorado, which later became the Colorado Australian Shepherd Association (CASA), a club that still exists today. Besides the Rowes and Dick Sorensen, other founding members included Jeanne Farned, Eleanor and Don Gantenbein, Mr. and Mrs. Thornton, Fletcher Wood, Jerry and Joyce Nightingale, Alvie Williams, Jim Healey, Dennis Hall, Bonnie Crowder, Roberta Hooks and Floyd, and Marie Leavitt. The club was attempting to promote the breed and never missed a chance to get the dogs in front of the public. The kids showed in obedience classes in 4-H and the adults manned booths at local fairs. The Arvada Harvest Festival was an annual event and the club kept a permanent booth available for this event and many others, complete with handouts, photos,

A historic day in Aussie history. A gathering of some grand matrons of the breed. Left to right: Jeffreys' Sheila, Porter's Yakima Blue Squaw, Rowe's Dingo, and Jeffreys' "Dee." Courtesy of Rowe

posters and, of course, dogs doing tricks! It was a common practice in those days to raffle a puppy and at least one famed dog passed through the raffle donated by Janeane Harper, Gantenbein's Lady Blue (dam of the legendary Ch. McGuire's SeMeGo Hemi UDT, OTDc, STDsd).

Gunsmoke and Rowe's Dingo were two regular performers at all of these events. Soon their owners decided to cross these foundation dogs to see if they could preserve their much loved qualities. Although Rowe's did not keep any puppies from this litter, Dick Sorensen obtained Sorensen's Blue Mist from this cross. Blue Mist produced a dynasty of important sires, including Jeffreys' Little Pooh Bear and J Bar D Blue Boy and the excellent dam J Bar D Jamie.

Dingo was bred only once because she was unfortunately killed by a car soon after weaning her litter. But the Rowe's success with this cross drove them to look for a replacement for Dingo. The working heritage of the Aussie was important to them and they began searching for good dogs on which to found a bloodline. The most important dog in the founding of the Rowe's J Bar D line was their Sorensen's Gunsmoke son, Rowe's Comanche Warrior. Bum, as he was affectionately known, was out of the Wood's Jay granddaughter, Van Gordon's Blue Asta—a daughter of the legendary Lighters Asta. Dick Sorensen met the Van Gordons through Fletcher Wood when they were looking for a sire for Asta. A deal was made to breed for a puppy back. Dick wasn't looking for another dog at the time so he offered this pick of the litter to Jerry. In 1966, the silver blue male chosen by Dick became the third Australian Shepherd at the Rowes and taking him home was a decision they never regretted.

In this same litter was another blue male who was eventually named Clinger's (Kalbs) Royal Blue. This male, too, became a producer of a multitude of important offspring, among them Jones Silver Cloud (behind such northwest kennels as

Sorensen's Gunsmoke (Williams' Laddie x Jefferys' Sheila). Courtesy of Rowe

Rowe's Dingo at 11 months of age. Courtesy of Rowe

Sorensen's Blue Mist (Sorensen's Gunsmoke x Rowe's Dingo). First place obedience 1969. Courtesy of Rowe

PEDIGREE OF Rowe's Dingo

Born: ____________ Sex: ________
Owned by: ____________________

<table>
<tr><td rowspan="8">Sire
Welz's Handsome King</td><td rowspan="4">Sire
Harpers Old Smokey</td><td rowspan="2">Sire
Tucsons Joe</td><td>Sire Taylor's Idaho Lad</td></tr>
<tr><td>Dam Taylor's Idaho Lil</td></tr>
<tr><td rowspan="2">Dam
Tucsons Sis</td><td>Sire Ayer's Shep</td></tr>
<tr><td>Dam Ayer's Bell</td></tr>
<tr><td rowspan="4">Dam
Smedra's Blue Mist–ingo</td><td rowspan="2">Sire
Woods Dandy</td><td>Sire Finleys Texas Bob</td></tr>
<tr><td>Dam Merino's (Valdez) Blue Bitch</td></tr>
<tr><td rowspan="2">Dam
Lighters Asta</td><td>Sire Woods Jay</td></tr>
<tr><td>Dam Woods Little Blue</td></tr>
<tr><td rowspan="8">Dam
McGuires Palojami Gold</td><td rowspan="4">Sire
Farneds Dead Eye</td><td rowspan="2">Sire
Gregg</td><td>Sire Mr. Arizona</td></tr>
<tr><td>Dam Ty–o–wee</td></tr>
<tr><td rowspan="2">Dam
Whiskey Bell</td><td>Sire Heel Fly</td></tr>
<tr><td>Dam Gree–o</td></tr>
<tr><td rowspan="4">Dam
Hoopers Blue Keno</td><td rowspan="2">Sire
Lillys Mr Blue Chips</td><td>Sire Old Honley</td></tr>
<tr><td>Dam Boltza Lady Blue</td></tr>
<tr><td rowspan="2">Dam
Lillys Pecos Jane</td><td>Sire Madrid Mr Scott</td></tr>
<tr><td>Dam Boltzs Lilly Babe</td></tr>
</table>

Bred by: ____________ Address: ____________________ Phone: ____________

The details given above are correct to the best of my knowledge. Signed: ____________________(Breeder)

Hilltop and Blue Isle), Clinger's Ely Lucky Lady (out of the linebred Wood's bitch Kalb's Blue Panda, Lady founded the Silvertone kennel), Clinger's Elys Blue Queenie (behind the Dakota Blue kennel foundations), Swanson's Blue Shawn and Swanson's Blue Gypsy (full brother and sister who produced the foundation sire of Shepherds Crook kennel), bred to his daughter Kalb's Princess he produced Montoya's Rowdy, sire of Montoya's Chaparal (sire of Higgins' Pinkie), Jeffreys' Cinnamon Bear (behind Jeffreys' and Piz dogs and ultimately, Jones' Reddy Teddy), and Ginger Red (dam of Jones' Reddy Teddy and many others). Over time, the descendants of these two brothers, Clinger's (Kalbs) Royal Blue and Rowe's Commanche Warrior, were brought together many times to produce a rich legacy of great working dogs.

Rowe's Commanche Warrior grew up to fulfill every dream, wish and hope of the Rowes. He was a superior sire, a terrific family dog, a peerless working dog and an individual credit to the breed. He was the perfect first dog to lead Jerry into the world of working livestock dogs. Incredible instinct, high intelligence and a forgiving nature combined to give the Rowes a teacher as well as a partner.

A moderate dog, Commanche was about 23 inches tall but weighed a modest 50-55 pounds. Known affectionately as Bum for his habit of roaming the neighborhood, this rascal grew up teaching the Rowes about not only working livestock with dogs but also teaching them about Australian Shepherds and their nature in general. Tales of their adventures are ones any lover of the true old time Aussie can appreciate! Bum was a family dog of the highest order. The Rowe children grew up with him and treated him as one of their playmates. They often dressed him up and played house with him—the kind of attention this sociable dog thrived on! At the time Jerry worked nights and was also earning money fattening some Angus cows. Sharon came home one night to find his cows loose in the neighborhood. Finding black cows in the dark was not going to be an easy job so she sent Bum out to find them. By the time Sharon had the

Commanche as a mature dog (Sorensen's Gunsmoke x Van Gordon's Blue Asta). Courtesy of Rowe

Kalb's Royal Blue at one year of age, one of Commanche's littermates. Courtesy of Rowe

Shawnee, blue and white female, one of Commanche's littermates. Courtesy of Rowe

Lynn Rowe playing dress up with Ch J Bar D Chinook Squaw. Courtesy of Rowe

Sorensen's Blue Mist, the phenomenal blue female produced by Gunsmoke and Dingo, founder of a dynasty of top dogs. Courtesy of Rowe

Commanche's sire, Sorensen's Gunsmoke as a young dog. Courtesy of Rowe

kids and groceries unloaded, her good working dog had brought all the cows back to the Rowe property.

Jerry also had trouble during this time with theft of gasoline from a tank in his yard. The tank always seemed to be empty and he couldn't catch the thieves. Finally one night he surprised several delinquents draining the tank and watch as they ran into some weeds next to his house. Jerry quietly let Bum out of his kennel and he promptly sent the young criminals on their way. He apparently scared the kids sufficiently that they were overheard the next day threatening to sue for having a dog sic'ed on them. Empty words at best since it would also mean confessing their own transgressions to the authorities.

Bum never forgot a person but he also had the old fashioned good judgment of the traditional Aussie about the intents of anyone who came onto his personal territory. When the Rowes were around and okayed a visitor, he was the perfect hospitable gentleman. But that same visitor had better not return without permission. Bum kept a couple of strangers sitting on their car for nearly an entire day until Jerry came home to rescue them. Another time, Jerry got a ride home from an event with Ernie Hartnagle. It was late and Jerry got out of the car without his hat. When Ernie discovered the hat on his seat it was fairly late but he decided to go back to Jerry's anyway. The Rowes had gone to bed and left Bum in the yard. Bum managed to corner the surprised Mr. Hartnagle for a time until the Rowes were awakened to retrieve the hat and release Bum's prisoner.

Bum, like most dogs at this time, was left loose around the Rowe property all the time. He developed the habit of jumping through the open windows of the Rowe's car, an orange Vega station wagon, to sleep on the front seat. When he wasn't napping he would make the rounds of the neighborhood checking on the news and making sure everything was in order. His wandering habits are what earned him the nickname Bum. One day he didn't return home from his travels. Jerry and Sharon called all the neighbors and the dog pound looking for him. An employee of the dog pound had no knowledge of the Australian Shepherd breed, but they did claim to have an odd looking gray and black Cocker Spaniel in residence. The Rowes hurried down to the shelter and found their nomadic Comanche where a Cocker Spaniel supposedly resided. The employee got an education on his breed identity and the Rowes got an education on the habits of their hobo. The Bum was making his rounds that day and apparently arrived at his nap time. He found a convenient orange Vega station wagon in the driveway of one of the homes in the subdivision he was patrolling and hopped in the front seat for a quick snooze. The Vega's owners came out to go for a drive to find their car had been appropriated as a canine bedroom. Bum was firm about his claim to the Vega so the dogcatcher was called to deal with him. The dog warden knew a little bit about dogs and merely opened the car door and Bum jumped out. He politely hopped into the paddy wagon with the dogcatcher and rode down to the pound to await his bail. Before this incident Bum would disappear every day until noon making his rounds. After his short time in prison he never made his regular rounds again.

This personal recollection of Commanche comes from the files of Betty Nelson and are her observations: "Rowe's Commanche Warrior—this quiet, effective working dog showed considerable eye when working sheep and ducks. He could have competed with the best at this years (97) National. He heeled higher on hind leg than necessary and so was minus some teeth. A hard kick never bothered him.

Rowe's Commanche Warrior, blue merle and white male. Foundation sire for J Bar D Australian Shepherd. Courtesy of Rowe

Rowe's Commanche Warrior as a young dog. Courtesy of Rowe

Rowe's Commanche Warrior. Courtesy of Rowe

Commanche showing his excellent head type. Courtesy of Rowe

Thing that impressed me most besides working (at 1974 Invitational Trial) was his solid good steady temperament. This he threw to pups. I was amazed when I took 2 sons thru obedience and could compare with town raised dogs. Commanche pups were calm, steady—more than any dog of any breed in class no matter what distraction was tried. His pups would work stock independently up to a 1/2 mile away and take quickly any commands they could hear. The friend who did this training liked to show off his dog by sending her out and directing her to cows she couldn't see (owner be on second story balcony of his house) Phil (Walker) let his wife work the dog with only 2 commands—bring cows or bring sheep. Dog would do the rest." Betty Nelson did more line breeding on Commanche than almost anyone and studied her dogs thoroughly. This endorsement is a real recommendation for Commanche and his breeding.

When Jerry began working Bum, their club had rented sheep for an upcoming trial from one of the club members. Jerry was taking the sheep home after the trial with Commanche in the bed of the pickup. Somewhere along the route another dog began chasing the truck and Commanche bailed out to fight this dog unbeknownst to Jerry. By the time Jerry got home, Commanche was lost. The Rowes were distraught and looked for him for days, posting flyers for him all over town and talking to everyone they knew. After several weeks it seemed he was gone forever. Over a month later, Jerry came home to find his best partner waiting on the back step of his home, safe and sound! He never discovered where Bum had been all that time but calculated he must have traveled at least 40 miles and crossed a major interstate highway in order to find his way home.

Walker's Keidron, (Rowe's Commanche Warrior x Nelson's Sierra Jolly Lass), an influential west coast dam. Courtesy of Rowe

The herding club finally decided to purchase some sheep for their own use and Jerry and Dick Sorensen made a deal with a local sheep rancher to trade some chickens for a few sheep. The only way they had to haul the sheep at the time was a pickup with a camper shell on it. The rancher loaded the sheep but when the two proud owners arrived home they were on their own getting them unloaded. They decided to have their novice working dog do the job and Commanche obliged. His skills, however, were not honed as they would be later in life and he had the mistaken belief the job was not finished when the last sheep cleared the tailgate. By the time Jerry and Dick emerged from the orchard, cleared the fence the sheep had taken down and run into the neighboring wheat field they had decided tying up the dog would be a wise decision. The sheep had not stopped to enjoy the wheat but were still shopping for a new home and with a little direction from the wheat farmer from his perch on the combine they finally located the flock and brought them home without any help from a dog.

Commanche, Bum, was a dog who commanded respect from other dogs without ever causing a fight. He had the air of the true alpha dog who never has to prove his right to the top pack position. He assumed his rank through his confidence and manner and by using the universally understood body language of canine nobility throughout the world. He only fought with one dog in his entire 17 year life span, a dog named R Lucky Chipawa. Chipawa had a grudge he couldn't let go and never missed a chance to declare it to Bum. Finally one day Chipawa escaped his kennel, attacked Bum, and the fight was on. Sharon was home alone and not able to separate the two large dogs so she decided to just leave them alone and hope Commanche could finally teach Chipawa a lesson. When she came back from shopping she found the two dogs still locked in combat but she was finally able to separate them. Chipawa never did learn his lesson but he didn't spoil Commanches otherwise good disposition, either.

Rowe's Commanche Warrior showing his quiet, commanding authority working sheep. Courtesy of Rowe

Born: ____________ Sex: ________
Owned by: ____________________

PEDIGREE OF Rowe's Commanche Warrior

Sire: Sorensen's Gunsmoke	Sire: Williams Laddie	Sire: Uhlendorf's Tippy	Sire: Parks Joe
			Dam: Uhlendorfs Linda
		Dam: Thornton's Chrissie	Sire: Watsons Old Blue
			Dam: Johnsons Sue
	Dam: Jeffreys Sheila	Sire: Ely's Major	Sire: Breazeales Sheik
			Dam: Breazeales Sandy
		Dam: Williams Georgina	Sire: (Harpers) Ely's Joker
			Dam: Barretts Blue Belle
Dam: Van Gordon's Blue Asta	Sire: Ely's Major	Sire: Breazeales Sheik	Sire: Joe
			Dam: Sister
		Dam: Breazeales Sandy	Sire: Sam Boy
			Dam: Sambo
	Dam: Lighter's Asta	Sire: Woods Jay	Sire: Sislers Shorty
			Dam: Sislers Trixie
		Dam: Woods Little Blue	Sire: Blue Boy
			Dam: Blue Shadow

Bred by: ____________ Address: ____________ Phone: ____________

The details given above are correct to the best of my knowledge. Signed: ____________(Breeder)

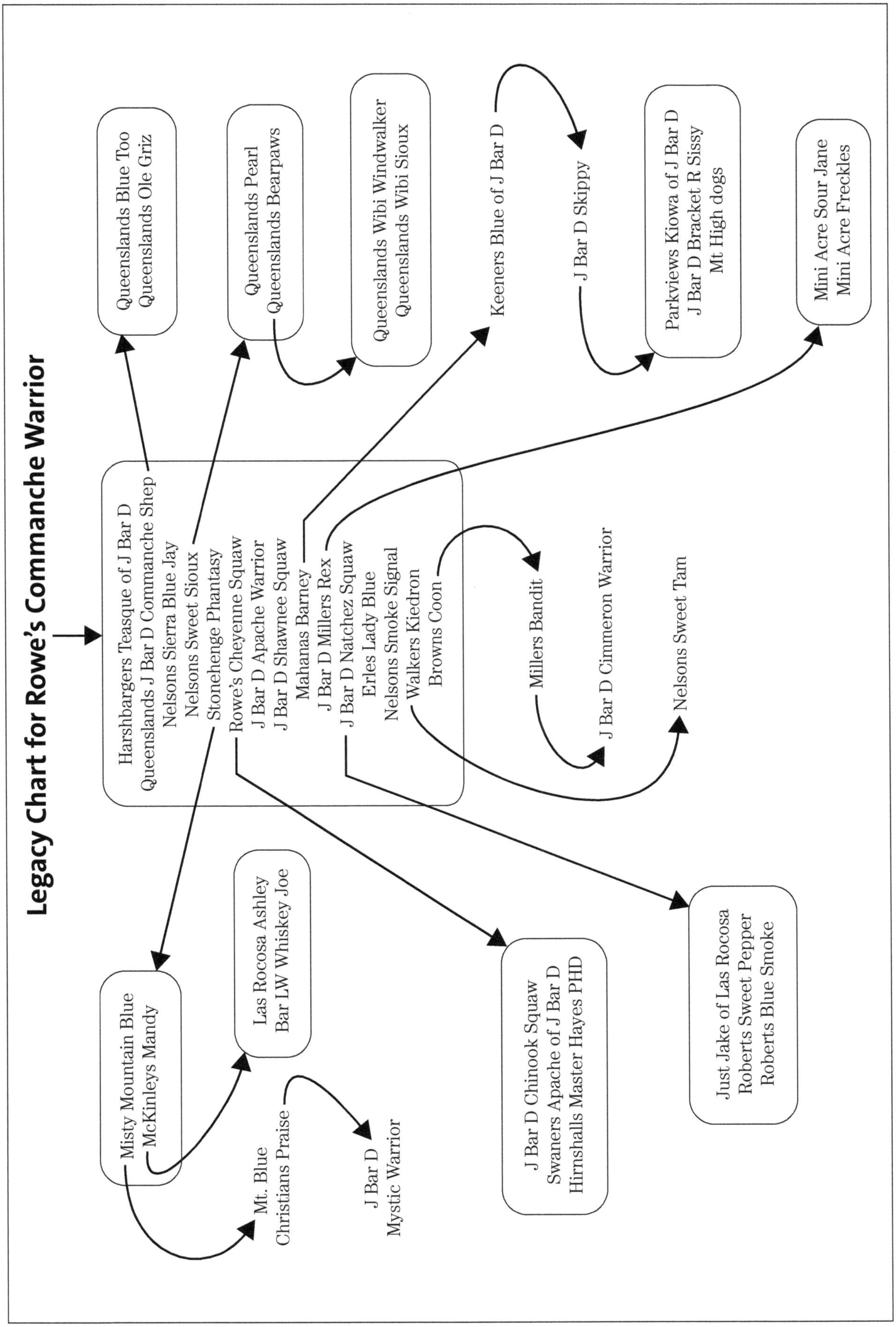
Legacy Chart for Rowe's Commanche Warrior
Harshbargers Teasque of J Bar D
Queenslands J Bar D Commanche Shep
Nelsons Sierra Blue Jay
Nelsons Sweet Sioux
Stonehenge Phantasy
Rowe's Cheyenne Squaw
J Bar D Apache Warrior
J Bar D Shawnee Squaw
Mahanas Barney
J Bar D Millers Rex
J Bar D Natchez Squaw
Erles Lady Blue
Nelsons Smoke Signal
Walkers Kiedron
Browns Coon
Queenslands Blue Too
Queenslands Ole Griz
Queenslands Pearl
Queenslands Bearpaws
Queenslands Wibi Windwalker
Queenslands Wibi Sioux
Keeners Blue of J Bar D
J Bar D Skippy
Parkviews Kiowa of J Bar D
J Bar D Bracket R Sissy
Mt High dogs
Mini Acre Sour Jane
Mini Acre Freckles
Millers Bandit
J Bar D Cimmeron Warrior
Nelsons Sweet Tam
Misty Mountain Blue
McKinleys Mandy
Las Rocosa Ashley
Bar LW Whiskey Joe
Mt. Blue
Christians Praise
J Bar D
Mystic Warrior
J Bar D Chinook Squaw
Swaners Apache of J Bar D
Hirnshalls Master Hayes PHD
Just Jake of Las Rocosa
Roberts Sweet Pepper
Roberts Blue Smoke

At one time, Jerry Rowe was training horses for a living and Commanche helped when needed with the horses. Courtesy of Rowe

Sanger's Smokey of J Bar D drinking with a baby lamb was typical of the J Bar D dogs. Courtesy of Rowe

Commanche was a typical all-around dog. He won awards in conformation, obedience and, of course, working trials long before any titles were given for these achievements. He had a naturally gentle, calm style of working. His own demeanor was picked up by livestock so he could work rank stock in difficult country as well as work very young calves or lambs. He was used as a driving dog which was the only style handlers knew at the time. He had eye which became more developed as he worked. Sharon explains, "We did not realize what eye was in those days. This was a natural trait. In later years we realized how loose eyed dogs used it to control livestock."

When Sharon was raising chicks, Commanche would help her by gingerly catching and picking up the chicks to put them in the chicken house when it rained. When she would raise orphan lambs, Commanche would put two or three lambs under his chin, straddle the tiny flock and bodily push them to their pen. At one time Jerry trained horses for a living. He would occasionally get a horse to train that refused to come to a person to be caught. When the frustrating equine would run off, Commanche would be sent to change it's mind. He was not an alligator when it came to bite but he knew just the right amount of force and how to correctly place it to teach these outlaws they had the wrong idea. When the horse returned to Jerry, he would have Bum lie down. If the horse decided to try freedom again, he received another toothy lesson from Bum. It didn't take too many trips down the pasture for most horses to decide being caught was a desirable alternative. The same philosophy was applied to horses which refused to load into a trailer. As long as they continued with forward movement, all was well. But shifting to reverse was an invitation to Commanche to apply his skills once again. He knew exactly how to train livestock to obey him without harming them or picking a fight. All of these individual traits combined to form a dog who instinctually knew how to handle livestock in any situation.

One tactic tried by Commanche at a young age was barking at the heads of cattle. One rancher who obtained a dog from the Rowes became friends with them and invited them out to help with gathers, branding, vaccinating, sorting, etc. Bum was about nine months old the first time the Rowes took him out for a day of real work. The cattle were being gathered and he would run to their heels and bark at every opportunity. Of course, the cattle would turn back because they had calves and were being protective, which was NOT the direction desired by Jerry and the others. Bum continued barking and the cattle eventually turned around, came after him and ran him off. That lesson and the owner's displeasure about having a dog along that was making extra work taught Bum not to bark on cattle ever again! In later years, Jerry would learn about fetching or heading dogs and appreciate this desire, but in the beginning nobody understood it well or knew how to use it—especially cattlemen!

One of the crowning achievements of Bum's life was the invitation offered by Bob Carrillo to be one of the few dogs to compete in the very first Aussie-only trial in Santa Rosa, California. Carrillo traveled the country looking for well trained Aussies who would give a good showing in a showcase trial held during the halftime of the Santa Rosa rodeo. Commanche was one of 4 dogs invited from the state of Colorado, placing third behind two of the other Colorado invitees. In addition, Commanche

Winners of the First Invitational Trial held in Santa Rose, California. From Right to Left: unknown person, First Place—Leslie Sorensen (Sorensen's Cherry Brandy), Second Place—unknown person (Hartnagle's Leslie), Third Place—Jerry Rowe with Rowe's Commanche Warrior, Fourth Place—David King (King's Troy), Fifth Place—Bob Carrillo (Carrillo's Jenny), Sixth Place—Charlene Jeffreys (Jeffreys' Shade), Seventh Place—Colleen Gage with Gage's Cookie B. Courtesy of Rowe

was pressed into service for a duck herding demonstration during the rodeo. The ducks were locally provided by someone unfamiliar with the finer points of working fowl with dogs. Jerry and Bum entered the arena for their performance and Jerry sent his companion on an outrun to begin the demonstration. Bum quietly and obediently began to circle the ducks, reaching balance just as the flock neatly lifted off and flew over the nearest fence! Nobody had thought to clip their wings in order to make them workable by a dog. While the bewildered Commanche stared after his disappearing flock, the quick thinking Jerry doffed his hat and bowed to the expectant crowd!

In honor of Commanches stock trial achievements and his overall superior working ability, the Rowes decided in the mid-1970s to offer a perpetual trophy in his name. Commanche had won the first Colorado Stockdog Fanciers trials in 1973, the first ASCA National Trial in Arizona in 1974, and was High in Trial at the 1975 and 1976 ASCA trials in Colorado. The 17 inch tall trophy became known as The Commanche Award and had to be won 3 times by the same dog in order to retire it. It was awarded to the High Scoring Aussie at

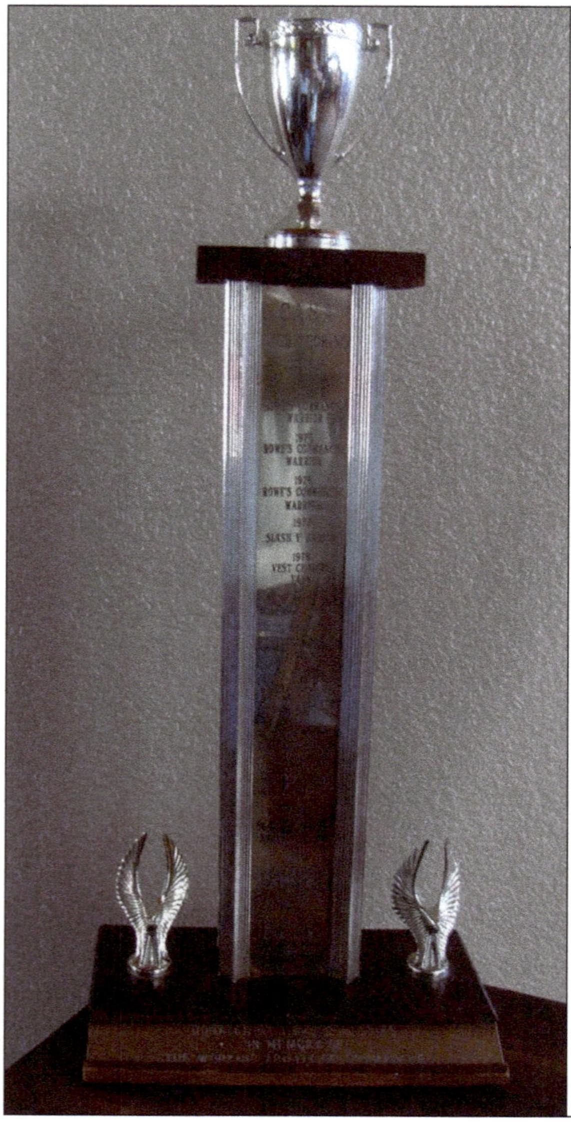

The Commanche Trophy. Inscribed on the plaque at the base is the following "Donated by J Bar D Aussies In Memory of The Working Ability of Commanche" Recorded on the tall trophy on both the front and back are the engraved names of all the winners from 1974 through 1997. Those names are: 1974, 1975 and 1976—Rowe's Commanche Warrior, 1977—Slash V Martini, 1978—Vests Chargin Tank, 1979—Brutons Holtex, 1980—Ch Powder River's City Slicker, 1981—Romers Magnolia Blossom STD, ATD, 1982—Ch Powder Rivers City Slicker ATD, 1983—no winner, 1984—Olivers Moqui CDX TDX ATDcsdh, 1985—no winner, 1986—Slash V Slide Me Sweet STDcsd, 1987—Las Rocosa Kublia Khan of Crown Point ATDcsd, 1988—Las Rocosa Bonny Kyle RDX, ATDcsd, 1989—WTCH Crown Point Inspirator Bar LW RDX, 1990—WTCH Crown Point Inspirator Bar LW RDX, 1991—WTCH Las Rocosa Kublia Khan of Crown Point RDX, 1992—WTCH Bar LW Inspired Blue, 1993—WTCH O'Sages Shinin Times, 1994—"Manchado Jack" WTCH, 1995—WTCH Crownpoint Ace of Spades RTDcs, 1996—WTCH Crownpoint Ace of Spades RTDcs, 1997—WTCH Crownpoint Crimson Tide RTDcs PATDcs. Courtesy of Rowe

Nordstrom's Tammy, an unregistered ranch dog of unknown pedigree but excellent Aussie type, dam of the first litter by Rowe's Commanche Warrior. Courtesy of Rowe

Rowe's Commanche Warrior. Courtesy of Rowe

the annual CASA (Colorado Australian Shepherd Association) stock trial and although it was won at least once by nearly every top working Aussie in the nation over a period of over 20 years, it was never won three times by the same dog. Finally, the Rowes decided to retire it themselves to preserve the history it represented.

Commanche was always a very sociable dog, unlike many of the working dogs of that day, including many of the Wood dogs. While his sire, Gunsmoke, was more a one man dog, Commanche was a friendly, outgoing soul who, nonetheless, retained the judgment to know when a situation required some police action.

Commanche's first litter out of Nordstrom's Tammy in May 1967. When bred to a solid colored female, Commanche always produced only merles. Courtesy of Rowe

Commanche was also a long lived dog who was very healthy and tough until the time of his death. He had done so much jumping in his life he developed arthritis in his front legs. In spite of this, shortly before his death he jumped a fence and bred the neighbor's Labrador! When he was 17 years old, he contracted lung cancer. When the day came to release him from his suffering, Sharon took him to the veterinarian for the lethal injection. The vet weighed him and drew up the necessary dosage and delivered it to Commanche. Nothing happened. He drew up another dose and gave him that. Still Commanche breathed. The vet finally drew

Rowe's Cheyanne Squaw (Rowe's Commanche Warrior x Kelly's Blue Rockette), blue merle and white female, Reserve Winners Bitch, 1968. Courtesy of Rowe

Kelly's Rockette won Grand Champion of the Dog Program (199 points out of a possible 200) at the 1966 Jefferson County 4-H Fair County and District Dog Obedience Trials. Photo courtesy of *The National Stock Dog Magazine* (Spring 1967).

up enough to euthanize a horse and that is what it took to finally ease Commanche from this life.

Like his sire, Commanche sired puppies which had many of the good traits of the Wood dogs. But Commanche added increased biddability and more sensitivity to people than exhibited by many of the Wood dogs. He, like Gunsmoke, was used to cross with the good Wood dogs in order to perpetuate all the good qualities, while toning down some of the hardness and independence of that line. The results of this cross were dogs which fit in better around more suburban settings and out in public when the breed was promoted at fairs and booths. This cross ensured they didn't lose their strong working instincts or fantastic intelligence.

For all his importance in the lives and education of Jerry and Sharon Rowe, Commanche's career as a stud dog had more importance in the bloodlines of many other kennels than in the J Bar D kennel. Several Commanche offspring made their mark in the Rowe's breeding program and many dogs descended from him traced their way back to the J Bar D program. However, much of his most important influence can be found in other well known Australian Shepherd breeding programs.

Commanche sired a good female for J Bar D named Rowe's Cheyenne Squaw. Cheyenne was a dark gunmetal self blue merle with brown eyes and the tiniest amount of white. Of average height—she stood about 19-20 inches—she was of above average intelligence. She was the daughter of a 4-H project puppy named Kelly's Rockette that won the state 4-H Novice Obedience Grand Championship in Colorado in 1966. A Cheyenne Squaw daughter, J Bar D Chinook Squaw was one of the first ASCA breed Champions. Through her daughter, Miller's Sheba, she was behind the

Rowe's J Bar D Cimmeron Warrior who sired many of Betty Nelson's later dogs and, through them, Anita Laycock's foundation female, Laycock's Cheyenne Maurine and Myra Fourwinds' foundation dogs. Commanche also contributed to Cimmeron Warriors background through Brown's Coon. Coon was bred to Miller's Sheba to produce Miller's Bandit, dam of Cimmeron Warrior.

J Bar D Cimmeron Warrior (J Bar D Cherokee Red x Millers Bandit), red merle and white male who taught Jerry as much about working livestock as his other dogs. Courtesy of Rowe

Betty Nelson acquired a number of Commanche offspring to help found her bloodline. These include Nelson's Sierra Bluejay, Nelson's Sweet Sioux, and Nelson's Smoke Signal. Sweet Sioux was especially influential in the founding of the Queensland kennels producing their foundation dogs, Queenslands Pearl and Queensland's Bearpaws. Smoke Signal produced the dogs which founded the Flying X kennel in California as well as Nelson's Coal B Jinxed for Betty. The double Commanche bred dog J Bar D Cimmeron Warrior was also a big influence in the Nelson kennel producing a number of dogs for her and dogs which also went to the Fourwinds kennel.

Commanche also sired a dog named Stonehenge Phantasy who produced two influential females. Misty Mountain Blue was the foundation female for Claudia Nelson (no relation to Betty Nelson). When bred to the Wood and Hemi bred male Newman's Blue Sprog, she produced the female Mountain Blue Christian's Praise. Praise was bred to a Queensland dog tracing many times to Commanche to produce one of the Rowe's later dogs, J Bar D Mystic Warrior. Stonehenge Phantasy also produced a female named McKinley's Mandy who was the dam of two good working dogs, Bar LW Whiskey Joe and Las Rocosa Ashley.

Commanche blood was widely used with great success throughout the breed. Natchez, bred back to her sire Commanche, produced J Bar D Arrow Warrior. Natchez Squaw also produced Jeane Robert's foundation females Roberts' Sweet Pepper and Roberts' Blue Smoke. Natchez Squaw produced for Las Rocosa a nice blue male named Just Jake of Las Rocosa who produced a number of good working dogs for them. Mountain High got another dose of Commanche blood through another J Bar D dog named J Bar D Skippy. Two other Skippy offspring helped found

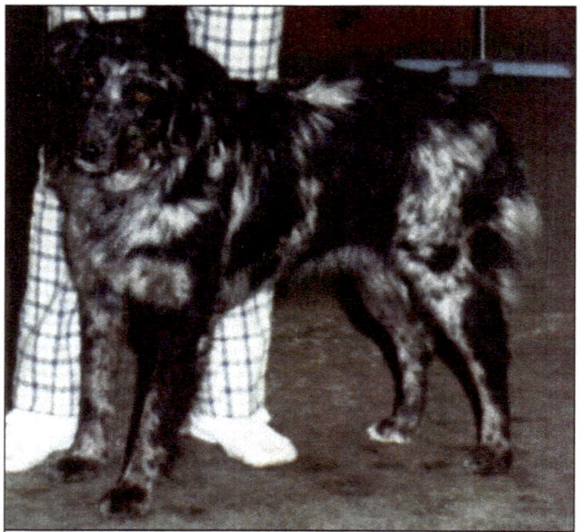

J Bar D Shawnee Squaw, blue female. Courtesy of Rowe

Rowe's Warrior Bow. Courtesy of Rowe

the Bracket R kennel in Iowa and the Parkview kennel in Kansas.

Close breeding was apparently not a problem when using Commanche as the good sire J Bar D Apache Warrior was the product of a father daughter breeding sired by Commanche and out of the Commanche daughter J Bar D Shawnee Squaw. In addition, Betty Nelson, Barrie Abbott breeding the Queensland dogs, and Pat (Clark-Goodwin) Gordon breeding the WIBI dogs, each crossed Commanche descendants many, many times without difficulty. Many foundation dogs would not have withstood this test of time if they had not been healthy genetically.

Another extremely important dog in the development of both the Sorensen and Rowe branches of J Bar D was a dog the Rowes obtained from Reno, Nevada from the Ritter bloodlines named Rowe's Warrior Bow. In 1967 it was still early in the organization of the Australian Shepherd and things were chaotic with the formation of numerous clubs and several registries. One of the clubs was the IASA (International Australian Shepherd Association) which was headquartered in California. ASCA was in it's infancy and had written a breed standard but so had IASA and the two clubs were attempting to unite under one standard. ASCA asked Jerry and Sharon Rowe and Eleanor Gantenbein, another early Colorado breeder, to travel to California to discuss the problems of the two different standards with the IASA officials. Sharon decided to make the trip worthwhile since their route would take them through Reno, Nevada anyway and planned a stop to see the Ritter dogs. She had been corresponding with Stan and Lorena Ronsley for some time and their bloodlines were a proven good cross with the dogs Rowes already had.

When they arrived in Nevada they learned of a dog which was for sale and owned by a neighbor of the Ronsleys. The neighbors had the 9 month old red merle male they called CoCo tied on a spring chain in their barn. Sharon took one look at this poor dog and his awful condition and decided to pay the $35 asking price just to get him out of the situation. She was writing the check when one of the kids volunteered the bad news this dog was a renowned chicken killer. As if the situation wasn't bad enough, the moment the owner took her check she turned and released CoCo who shot out the barn door and headed down the road as fast as he could run! Sharon ran after him hoping she could somehow retrieve him, while the former owner headed for the house offering no help in catching him. Sharon optimistically called him on the off chance that he might obey. Miraculously, he stopped and returned to her, jumped in her car onto the front seat and never looked back. Sharon brought him into their hotel room that night where he slept by the bed and was a perfect gentleman!

Sharon was roundly criticized for bringing a red dog to Colorado. At the time there were few red dogs in the area and not much was known about whether they were acceptable or something to be ashamed of. When Sharon arrived back in Colorado with her new dog, he was the only red dog most people had ever seen other than a few reds produced by Fletcher Wood. Almost everyone who saw him was critical of him, but more than a few eventually tried to buy him. Eleanor Gantenbein tried to buy a part interest in CoCo before she and Sharon arrived home. Another family offered to buy him but the Rowes refused all offers. One day, CoCo, now named Rowe's Warrior Bow, disappeared and the Rowes always believed he had been stolen. Amazingly enough, they received a postcard at Christmas one year from a family who had moved to Georgia and in the photo on the card was a dog that looked exactly like! To this day, the Rowes believe the dog in the photo was theirs but they were never able to follow up on their suspicion or reclaim him.

Rowe's Warrior Bow was one of the last of the old Nevada Ritter dogs. Born July 8, 1966, he was bred by the Ronsleys, who had the last

Sire of Rowe's Warrior Bow, Ronsley's Charcoal Smoke (left) and one of his puppies, Ronsley's Blue Sage (right). Courtesy of Rowe

Ronsley's Charcoal Smoke (standing on hind legs), blue merle male who was sire and grandsire of Rowe's Warrior Bow. In the foreground is Ronsley's Mitzi Buck, granddam of Warrior Bow. Courtesy of Rowe

PEDIGREE OF Rowe's Warrior Bow

Born: ____________ Sex: ________
Owned by: ____________

Parents	Grandparents	Great-grandparents	Great-great-grandparents
Sire Ronsleys Charcoal Smoke	Sire Ritters Streak	Sire Ritters Duke	Sire Ritters Cowboy
			Dam Ritters Lady
		Dam Ritters Babe	Sire unknown
			Dam Ritters Belle
	Dam Ritters Lass	Sire Ritters Duke	Sire Ritters Cowboy
			Dam Ritters Lady
		Dam Ritters Nickie	Sire unknown
			Dam unknown
Dam Murphys Cochina	Sire Ronsleys Charcoal Smoke	Sire Ritters Streak	Sire Ritters Duke
			Dam Ritters Babe
		Dam Ritters Lass	Sire Ritters Duke
			Dam Ritters Nickie
	Dam Ronsleys Mitzi Buck	Sire Byes Rusty II	Sire Shermans Casey
			Dam Byes Snooks II
		Dam Byes Penny	Sire Shermans Casey
			Dam Shermans Lassie

Bred by: ____________ Address: ____________ Phone: ____________

The details given above are correct to the best of my knowledge. Signed: ____________(Breeder)

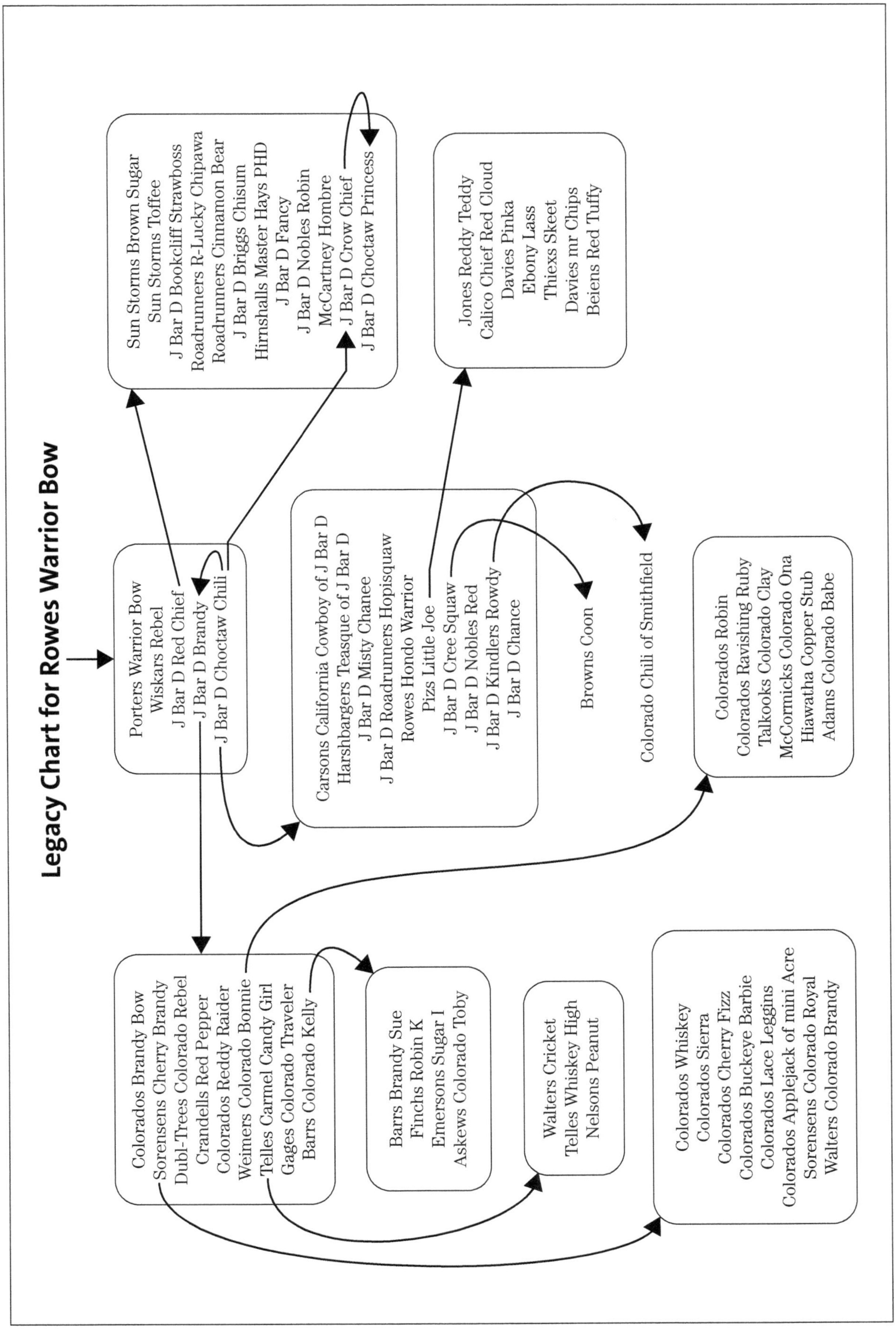
Legacy Chart for Rowes Warrior Bow
Porters Warrior Bow
Wiskars Rebel
J Bar D Red Chief
J Bar D Brandy
J Bar D Choctaw Chili
Sun Storms Brown Sugar
Sun Storms Toffee
J Bar D Bookcliff Strawboss
Roadrunners R-Lucky Chipawa
Roadrunners Cinnamon Bear
J Bar D Briggs Chisum
Hirnshalls Master Hays PHD
J Bar D Fancy
J Bar D Nobles Robin
McCartney Hombre
J Bar D Crow Chief
J Bar D Choctaw Princess
Jones Reddy Teddy
Calico Chief Red Cloud
Davies Pinka
Ebony Lass
Thiexs Skeet
Davies mr Chips
Beiens Red Tuffy
Carsons California Cowboy of J Bar D
Harshbargers Teasque of J Bar D
J Bar D Misty Chanee
J Bar D Roadrunners Hopisquaw
Rowes Hondo Warrior
Pizs Little Joe
J Bar D Cree Squaw
J Bar D Nobles Red
J Bar D Kindlers Rowdy
J Bar D Chance
Browns Coon
Colorado Chili of Smithfield
Colorados Robin
Colorados Ravishing Ruby
Talkooks Colorado Clay
McCormicks Colorado Ona
Hiawatha Copper Stub
Adams Colorado Babe
Colorados Brandy Bow
Sorensens Cherry Brandy
Dubl-Trees Colorado Rebel
Crandells Red Pepper
Colorados Reddy Raider
Weimers Colorado Bonnie
Telles Carmel Candy Girl
Gages Colorado Traveler
Barrs Colorado Kelly
Barrs Brandy Sue
Finchs Robin K
Emersons Sugar I
Askews Colorado Toby
Walters Cricket
Telles Whiskey High
Nelsons Peanut
Colorados Whiskey
Colorados Sierra
Colorados Cherry Fizz
Colorados Buckeye Barbie
Colorados Lace Leggins
Colorados Applejack of mini Acre
Sorensens Colorado Royal
Walters Colorado Brandy

Ritter's Streak, red merle and white male, sire of Ronsley's Charcoal Smoke. Photo taken in 1961. Courtesy of Rowe

J Bar D Choctaw Chilli, red merle female. One of the greatest foundation females in the breed. Courtesy of Rowe

Sorensen's Nightengales Tammy. Courtesy of Rowe

of the Ritter dogs that were still being bred. His sire was Ronsleys Charcoal Smoke and his dam was Murphys Cochina—a daughter of Ronsleys Charcoal Smoke and Ronsleys Mitzi Buck. Rowe's Warrior Bow's genetic strength most likely resulted from the fact that Charcoal Smoke was a son of Ritters Streak and Ritters Lass who were both sired by Ritters Duke. Warrior Bow was from a long line of dogs which had been bred for working ability and type. John Ritter, like Fletcher Wood, used his dogs extensively in his ranching life and needed dogs for work. Ritters Streak is famous for rescuing his owner after he was involved in an accident. The horse he was riding got tangled in barb wire and seriously injured Ritter. He had 3 dogs with him but it was Streak who went 8 miles to find help to save his life.

Rowe's Warrior Bow was a credit to his breeding. The Rowes used him often on a ranch in the mountains sorting and vaccinating cattle. He had the ability to take his cattle back to the same pasture from which they had been gathered all on his own. Nobody had to tell him what to do. He had the same sort of natural ability the Rowes had found in Commanche. His genetic strength made him a good producer of both top working dogs and good conformation dogs. His only downfall was a penchant for biting and jumping fences to roam—also much like his kennel mate Commanche!

Fortunately, before Rowe's Warrior Bow was stolen, the Rowes used him and he produced some of the most influential dogs in the breed. It teases the imagination to wonder what his influence might have been if his breeding career had not been cut short. However, two females he produced made a most profound impact on the Australian Shepherd through their offspring. The first of these was his daughter J Bar D Choctow Chilli. Chilli was out of Sorensen's Queenie, the Sorensen's

Gunsmoke and Nightengales Tammy daughter kept by the Sorensens for their own breeding program. Chilli was the only surviving puppy from this litter but her legacy is legend. She was a 19-1/2 inch tall red merle Sharon describes as not very pretty or friendly but all of her puppies were very sociable and extremely nice. Sharon decided to make sure Chillis puppies were well socialized and taught them at an early age to allow food to be taken away from them. She had tried a similar tactic with Chilli and gotten bitten for her efforts. Sharon taught Chilli not to do this and Sharon never had trouble taking her food dish again, but she was never completely reliable around anyone else. Another Chilli quirk was her part in deciding who she would be bred to. If the male did not meet her approval, she would refuse to stand for him and no amount of force could convince her to change her mind. She must have known what she was doing because her decisions shaped a large part of the future of the Australian Shepherd!

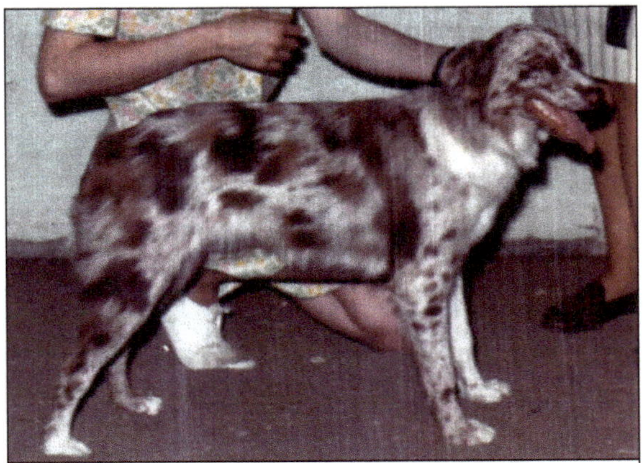

J Bar D Brandy as a young dog. Courtesy of Rowe

Her finest offspring were the result of close inbreeding and line breeding. First among these has to be her daughter J Bar D Brandy, a daughter of Chilli and Chilli's sire, Rowe's Warrior Bow. She was one of the foundation females for the Sorensens and producer of a dynasty of top working dogs. Her daughter, Sorensen's Cherry Brandy, won almost everything available at the time, in addition to being one of the finest ranch hands they ever owned. Cherry's story is told in greater depth in the Sorensen's chapter of this book. However her legacy went very deep and included many dogs not mentioned there. Her most well known offspring were Colorado's Buckeye Barbie, Slash V Colorado Flame, Colorado Lace

Rowe's Commanche Warrior could be as tough as needed on large stock like cattle. Courtesy of Rowe

September 1967 branding at Gypsum, Colorado. Jerry Rowe with Commanche and Yakima. Courtesy of Sorensen

Rowe's Commanche Warrior working goats at a trial at the old Longmont Arena in 1974. Courtesy of Rowe

Leggins, Colorado's Applejack of Mini Acre, Sorensen's Colorado Royal and many more. These dogs carried on the Sorensen's breeding program and were foundations for many other kennels around the country.

Another daughter of J Bar D Brandy, Colorados Brandy Bow, contributed a number of great dogs to the Kintyre Cattle Co. operation. These include Kintyre Holly Smoke, Kintyre Hot Chocolate of Chianti, and Kintyre Blue Suede Shoes. These dogs carried on the working dog legacy for Kintyre and others through the southeast United States.

Telles Carmel Candy Girl, a Brandy daughter who was sent to the west coast influenced a number of bloodlines there as well as in the Midwest and Colorado. Her daughter, Nelson's Peanut, produced a number of dogs for Kaitaia kennel and others. Her son, Telles' Whiskey High, produced Red Bear, sire of many foundation dogs for Czers kennel in Oregon. Her son Walters Cricket produced the well known sire L7 Goodtime Charlie, who became an important foundation for L7 kennel and produced a number of good dogs for Las Rocosa kennel.

Barrs' Colorado Kelly, another Brandy daughter, made her way to Texas. There, she brought the Colorado bloodlines to Kenneth Barrs' K-Barr kennel, Joel Finch's cattle operation and Tex-Star kennel, and a number

Jerry Rowe and Jim Porter working cattle with Commanche in Gypsum, Colorado. Courtesy of Rowe

of other working ranches. Barr was also influential in bringing more Colorado dogs to the Oklahoma and Texas area and creating a foundation of good working dogs in those areas.

Brandy left her mark on the Sorensen's breeding program with one other terrific producer, a daughter named Weimers' Colorado Bonnie. Her story is also told in depth in the Sorensen's chapter but bears a quick review here. Bonnie produced a number of fine dams for several kennels in the Midwest, Colorado, and California. McCormick's Colorado Ona produced a male named McCormick's Smokey Seminole, who was an early foundation sire for Pincie Creek in Alabama. Daughter Hiawatha Copper Stub was the granddam of the good cattle working dog Loida's Just Plain Charlie, and the dam Tina Horejsi who was bred to a number of good Wood dogs to produce top dairy dogs in Minnesota. Another daughter named Jon-Jo produced dogs which are behind the Dakota Blue and Zephyr kennels. Daughter Adam's Colorado Babe produced another of Betty Nelson's foundation females, Mary's Rocky Mountain Babe. Bonnie's most important daughter was Colorado's Ravishing Ruby who was the foundation female of Powder River kennel in Colorado. When bred to another Sorensen male, Powder River City Slicker, the two produced dogs which influenced the

Rowe's Commanche Warrior. Courtesy of Rowe

Commanche demonstrating his finesse and eye on ducks. Courtesy of Rowe

Balance, control, and authority. Commanche was the total Aussie. Courtesy of Rowe

J Bar D Red Chief, red merle male of outstanding type, temperament and conformation. Courtesy of Rowe

J Bar D Misty Chenee. Courtesy of Rowe

Rowe's Hondo Warrior (McGills Slagle Blue x J Bar D Choctaw Chilli) as a mature dog with one of his many show ring wins. Courtesy of Rowe

breeding programs of working kennels all over the country.

Stanna Rhynard of the Dakota Blue kennel obtained a Chilli daughter by the male J Bar D Briggs' Chisum—a Rowe's Warrior Bow grandson by J Bar D Red Chief. She was named J Bar D Misty Chenee and became a grand dam of the breed through her puppies sold from coast to coast.

J Bar D Choctaw Chilli would have secured her place in history if she had only produced J Bar D Brandy but her legacy extends far beyond this one great bitch. Brandy's offspring and further descendants were bred back to other Chilli descendants for generations, which brought back together the strong genetic heritage of Rowe's Warrior Bow over and over. Chilli was the dam of several great sires whose influence runs deep in the Australian Shepherd. Chief among these dogs is her son, Piz's Little Joe. An intensely linebred Wood dog, Little Joe became the sire of one of the most famous red males in the entire breed, Ch Jones' Reddy Teddy, CDX, CWD. Through Little Joe and Reddy Teddy, Chilli's influence was spread far and wide and when Reddy Teddy was bred back to daughters, granddaughters, great granddaughters, etc. the genetic strength was preserved and strengthened with the Wood line. Little Joe and Reddy Teddy will be discussed in greater depth in the chapter devoted to Cathy Jones and her Ruff N Reddy kennel, but the contribution of the J Bar D breeding in the success of these kennels can never be forgotten.

Chilli produced another top sire in the Rowes own male, Rowe's Hondo Warrior. Hondo was not only the son of this top bitch but also a grandson of Sisler's Shorty through the Shorty son, McGill's Slagle Blue. Hondo was a dark blue merle with minimal white, much like Wood's Jay, and he was another cross to Shorty, to add to the Wood's Jay blood everyone was using. Hondo stood about 21-1/2 inches tall and was a very dark merle with

J Bar D Crow Chief, solid black male (J Bar D Red Chief x J Bar D Choctaw Chilli). Courtesy of Rowe

J Bar D Choctaw Princess, solid red female, full sister to J Bar D Crow Chief and Cortez. Courtesy of Rowe

little white. He had a beautiful head typical of the Sisler/Wood dogs and was shown in conformation successfully by his co-owners, Stuart and Gerri Dykstra. The Dykstras used Hondo on a daily basis at a local sale barn where he became an indispensable hand working all kinds of livestock. He had a wonderful temperament and loved children. Hondo's daughters were bred back to Commanche and Warrior Bow and other dogs of this type to produce many good dogs both for the J Bar D families and other kennels around the country.

J Bar D Cortez, red merle and white female (J Bar D Red Chief x J Bar D Choctaw Chilli). Full sister to J Bar D Crow Chief. Courtesy of Rowe

Chili produced the foundation dam of the Dakota Blue kennel, J Bar D Misty Chenee, a red merle female of high quality. Misty was by the Warrior Bow grandson, J Bar D Briggs Chisum (a Ch J Bar D Red Chief son), concentrating the Warrior Bow quality which in turn made her a success in the whelping box. The half brother to half sister cross of Chilli bred to J Bar D Red Chief produced 3 offspring, including the stunning black male J Bar D Crow Chief and the beautiful red female J Bar D Choctaw Princess. Chilli and Red Chief also produced a number of great dogs for Rowes and Dick and Sue Dennis of Roadrunner kennel. Red Chief was the sire of Roadrunner's R-Lucky Chipawa and Roadrunner's Cinnamon Bear. Chilli produced J Bar D Roadrunner's Hopi Squaw and these dogs were crossed back with other J Bar D dogs to produce another legacy for J Bar D. A number of these dogs are illustrated in the book, *Pups 'N Stuff*, co-written by Sue Dennis and Sharon Rowe in 1971.

PEDIGREE OF Rowe's Hondo Warrior

Born: ______ Sex: ______
Owned by: ______

Parents	Grandparents	Great-grandparents	Great-great-grandparents
Sire: McGills Slagle Blue	Sire: Sisler's Shorty	Sire: Keno	Sire: unknown
			Dam: unknown
		Dam: Blue Star	Sire: unknown
			Dam: unknown
	Dam: Weedmeyer's Weedie	Sire: unknown	Sire: unknown
			Dam: unknown
		Dam: unknown	Sire: unknown
			Dam: unknown
Dam: J Bar D Choctaw Chilli	Sire: Rowe's Warrior Bow	Sire: Ronsleys Charcoal Smoke	Sire: Ritters Streak
			Dam: Ritters Lass
		Dam: Murphy's Cochina	Sire: Ronsleys Charcoal Smoke
			Dam: Rorsleys Mitzi Buck
	Dam: Sorensens Queenie	Sire: Sorensens Gunsmoke	Sire: Williams Laddie
			Dam: Jeffreys Sheila
		Dam: Nightingales Tammy	Sire: Ely's Chico
			Dam: Hollands Nikki

Bred by: ______ Address: ______ Phone: ______

The details given above are correct to the best of my knowledge. Signed: ______(Breeder)

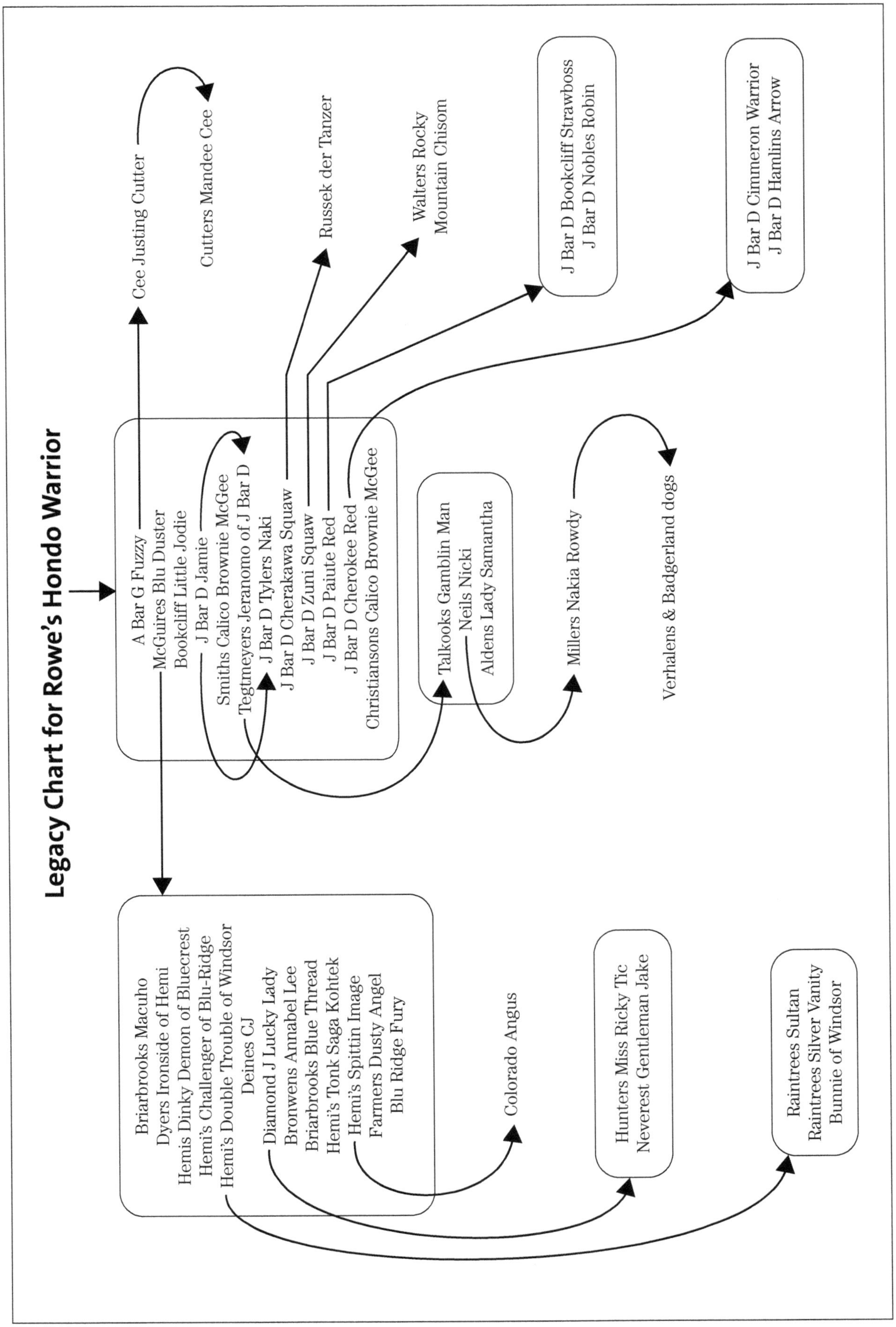
Legacy Chart for Rowe's Hondo Warrior
A Bar G Fuzzy
McGuires Blu Duster
Bookcliff Little Jodie
J Bar D Jamie
Smiths Calico Brownie McGee
Tegtmeyers Jeranomo of J Bar D
J Bar D Tylers Naki
J Bar D Cherakawa Squaw
J Bar D Zuni Squaw
J Bar D Paiute Red
J Bar D Cherokee Red
Christiansons Calico Brownie McGee
Cee Justing Cutter
Cutters Mandee Cee
Russek der Tanzer
Walters Rocky
Mountain Chisom
J Bar D Bookcliff Strawboss
J Bar D Nobles Robin
J Bar D Cimmeron Warrior
J Bar D Hamlins Arrow
Talkooks Gamblin Man
Neils Nicki
Aldens Lady Samantha
Millers Nakia Rowdy
Verhalens & Badgerland dogs
Briarbrooks Macuho
Dyers Ironside of Hemi
Hemis Dinky Demon of Bluecrest
Hemi's Challenger of Blu-Ridge
Hemi's Double Trouble of Windsor
Deines CJ
Diamond J Lucky Lady
Bronwens Annabel Lee
Briarbrooks Blue Thread
Hemi's Tonk Saga Kohtek
Hemi's Spittin Image
Farmers Dusty Angel
Blu Ridge Fury
Colorado Angus
Hunters Miss Ricky Tic
Neverest Gentleman Jake
Raintrees Sultan
Raintrees Silver Vanity
Bunnie of Windsor

The unusual J Bar D Red Chief (Rowe's Warrior Bow x Porter's [Rowe] Yakima Blue Squaw) with Sharon Rowe won Novice Dog (Corona, California, 1970). Courtesy of Rowe

J Bar D Crow Chief 10 months Winners Dog 5 pt major, 1976. Courtesy of Rowe

This book came about because Sharon and Sue were asked by their veterinarian, Dr John Pallaro, to create a book about breeding that he could give to his clients since there was very little information at the time about breeding and whelping dogs.

J Bar D Red Chief was the most influential sire produced by Rowe's Warrior Bow. A stunning red dog who stood about 23 inches tall, Red Chief was the first breed champion for the Rowes and the first red breed champion in the Australian Shepherd breed. Red Chief was not only a beautiful champion but produced his good movement, balance and type in his puppies. He had the correct hard coat which never matted, not even behind his ears. Red Chief either had an unusual genetic makeup or produced a genetic mutation in his son J Bar D Crow Chief. Red Chief was bred to his red merle sister J Bar D Choctaw Chilli to produce Crow Chief. Crow Chief was a solid black dog with only a few white hairs on his chest—normally an impossibility!

Crow Chief was about 21 inches tall and was also taken into the show ring by Sharon. She had a hard time winning with him due to his color but she kept trying for awhile. When her patience was nearly at an end she met Pat Clark (later Goodwin) who offered to take her dog and show him. She had no better luck than Sharon. Crow Chief never finished his Championship but garnered countless reserves. Solid black dogs were not often seen and judges were reluctant to put him up due to his color! Crow Chief was later traded to a person on the east coast for the influential dam Mary's Rocky Mountain Babe. Pat Goodwin continued breeding dogs descended from the original J Bar D dogs.

Rowe's Warrior Bow had only a short time to make his mark on the Australian Shepherd and his sound genetics and individual quality assured his place in history. Like too many dogs of this era his time as a sire was cut short leaving us to wonder how much different the breed might be today if he had been used for much longer than he was. Thankfully he was rescued from a life of obscurity by Sharon Rowe and wisely used as were his descendants!

Another quality sire who was only lightly used at stud was Rowe's Hondo Warrior. As discussed previously he was the son of J Bar D Choctaw Chilli by a son of Sisler's Shorty, McGill's Slagle Blue. Blue was a Wyoming ranch dog owned by Owen McGill. McGill worked with Sharon Rowe for many years breeding ranch dogs. He was a horse trainer who first met Sharon at the National Western Stock Show when he brought some of his horses and cattle to exhibit. McGill raised good cattle working Quarter Horses from his great Palomino cutting horse stallion, Hollywood Gold. As a stock breeder he had a good working knowledge of genetics and, with Sharon's growing knowledge of Australian Shepherds, the two bred a number of top dogs.

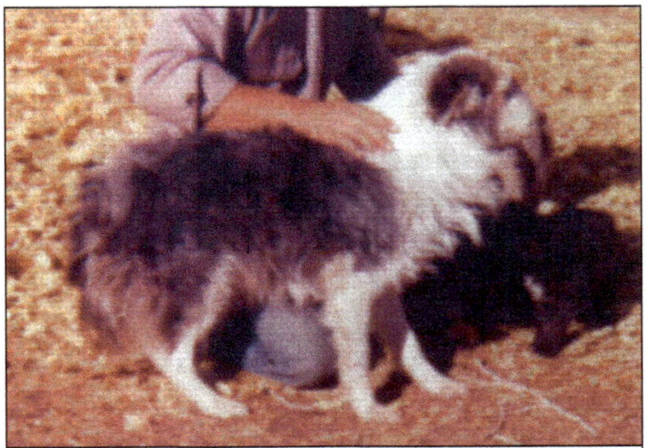

Rare photo of McGill's Slagle Blue—a Sisler's Shorty son and sire of Rowe's Hondo Warrior. Courtesy of Rowe

McGill's Slagle Blue. Courtesy of Rowe

Hondo's quality and ability were no surprise and he proved his worth as a working dog at the sale barn with Stuart Dykstra, who later became his owner. He was used at stud lightly and left a few good dogs which added to the expanding J Bar D legacy. One of his most beautiful daughters, typifying the Wood or mountain-type dog was Ch McGuire's Blu-Duster. A blue merle, copper and white female standing about 191/2 inches tall, she easily finished her breed Championship by the time she was two years old. Owned by Ruth McGuire and later by Loyd & Linda (Wilson) Sloan, she produced foundation dogs for a number of kennels in the Midwest and Southwest United States. Although Briarbrook kennel is known today for Flintridge descended Aussies, they originally owned many top working dogs from mostly Colorado based bloodlines. Before going to Briarbrook, Blu-Duster was used to produce some top dogs with Ruth McGuire's founding sire, Ch McGuire's SeMeGo Hemi UDT, OTDc, STDsd. Many of these puppies went to Arizona to found a number of kennels. Hemi's Double Trouble of Windsor, a black and white female, produced

Rowe's Hondo Warrior, (McGills Slagle Blue x J Bar D Choctaw Chilli), dark blue merle and white male as a young dog. A very typical Wood type dog. Courtesy of Rowe

J Bar D Cherokee Red, dark red merle male. (Rowe's Hondo Warrior x Rector's Samantha Ann). Courtesy of Rowe

foundation dogs for Raintree and Windsor kennels. Diamond J Lucky Lady produced a number of dogs for breeders in Michigan and the Upper Midwest area. Hemi's Spittin Image produced the wonderful Colorado Angus for the Sorensens. Deine's CJ stayed in Colorado with the Deine family and many other sons and daughters were good ranch and family dogs around the country.

Hondo left some important progeny for the Rowes, too. J Bar D Cherokee Red produced a dog named J Bar D Cimmeron Warrior who was responsible for finally opening Jerry Rowe's eyes to the potential of a strong heading dog on livestock. Cimmeron was a top working dog who was originally sold to a local dairy. The dairyman, like most cattlemen of the day, wanted a dog that would heel exclusively. He used, or tried to use, Cimmeron until he was about a year of age. He finally returned him to Jerry as useless, complaining the dog wanted only to run to the heads of cattle. Because Cimmeron didn't work the way he wanted, the dog had spent most of his time tied to a chain on a spring. There he became frustrated, bouncing off the chain continuously all day long. The dairyman had not only failed to teach Cimmeron how to heel cattle but he had failed in teaching him anything else, either. He was completely disobedient, refusing to come or stop when told. Sharon tried tying the dog to herself with a rope and was rewarded with a rough ride across snow covered ground when she tried to take him to sheep! Her drastic attempt at control was prompted by Cimmeron's rough treatment of sheep. He had apparently been so encouraged to bite by the dairyman that he knew nothing else and would constantly unkindly pluck wool from sheep. The frustration of both dog and trainers escalated daily.

During this time, the Rowes and Sorensens had explored the world of border collie trials and traveled cross country to view these events which were still mostly held in the eastern United States. During their travels they developed a friendship with Lewis Pence, the only border collie trainer who would even speak to them about training this new breed called Australian Shepherd. Pence taught Jerry most of his lasting lessons about stock dog training and was invited often to Colorado to share his insights with another fledgling herding dog club the Rowes had helped form—Stockdog Fanciers of Colorado—the first all breed herding club in the United States. It was during one of Pence's visits that he asked the Rowes to bring him the worst dog they had. Cimmeron fit the bill completely! This poor misunderstood dog was finally deciphered by Mr. Pence. When he entered the sheep filled arena, the Rowes prepared themselves to witness the worst. But Pence immediately recognized Cimmeron's potential and maneuvered him into the comfortable position of fetching the sheep. Cimmeron's change of attitude was almost immediate. He settled into the style he had been fighting for all his life and once Pence allowed him to use his strong heading instincts, he became an obedient and willing partner in working sheep.

Cimmeron was the first dog to teach Jerry how a dog would read not only livestock but also his human partner. This talented dog could do things Jerry wanted him to do just by thinking them through. He would perform many jobs from repetition, but from time to time he would do something new without any special training. He just understood what was wanted and figured it out on his own. It was this experience which illustrated perfectly to the Rowes how to work dogs using their natural instinct and started them on the path which would eventually become a livelihood. This story is typical of the many tales told by early breeders about the strong working instinct, intelligence, and natural ability of the founding dogs. Once they understood how to work with dogs such as Cimmeron, the Rowe's knowledge of training blossomed and they were able to teach others and develop the breed into what it is today.

J Bar D Cimmeron Warrior—once he was allowed to use his natural instincts he became a top working dog. Courtesy of Rowe

PEDIGREE OF J Bar D Choctaw Chilli

Born: ______ Sex: ______
Owned by: ______

Parents	Grandparents	Great-grandparents	Great-great-grandparents
Sire: Rowe's Warrior Bow	Sire: Ronsleys Charcoal Smoke	Sire: Ritters Streak	Sire: Ritters Duke
			Dam: Ritters Babe
		Dam: Ritters Lass	Sire: Ritters Duke
			Dam: Ritters Nickle
	Dam: Murphys Cochina	Sire: Ronsleys Charcoal Smoke	Sire: Ritters Streak
			Dam: Ritters Lass
		Dam: Ronsleys Mitzi Buck	Sire: Byes Rusty II
			Dam: Byes Penny
Dam: Sorensen's Queenie	Sire: Sorensen's Gunsmoke	Sire: Williams Laddie	Sire: Uhlendorfs Tippy
			Dam: Thorntons Chrissie
		Dam: Jeffreys Sheila	Sire: Ely's Major
			Dam: Williams Georgina
	Dam: Nightingale's Tammy	Sire: Elys Chico	Sire: Byes Rusty II
			Dam: Byes Grypsie
		Dam: Hollands Nikki	Sire: Woods Jay
			Dam: Knolls Kiki

Bred by: ______ Address: ______ Phone: ______

The details given above are correct to the best of my knowledge. Signed: ______(Breeder)

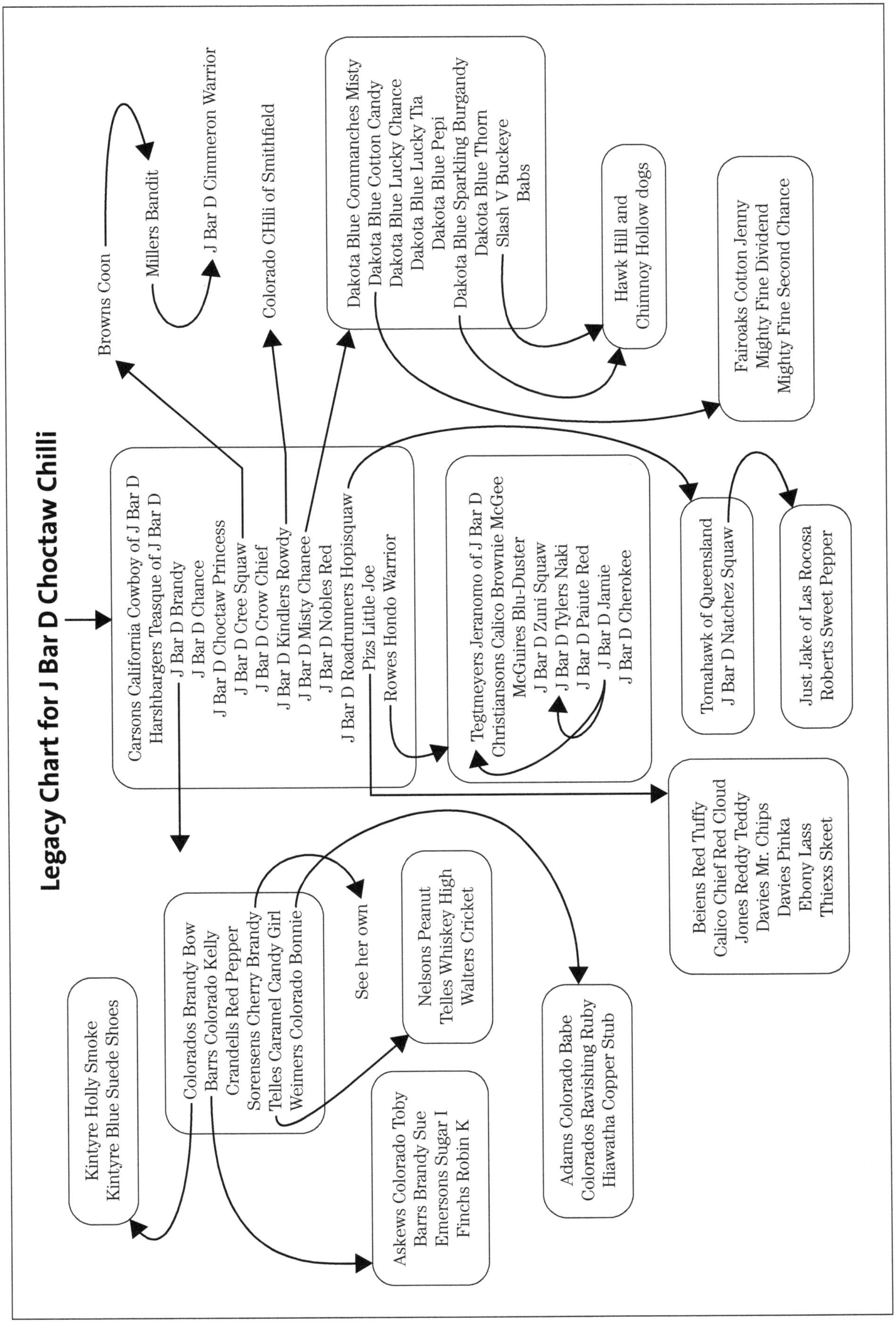
Legacy Chart for J Bar D Choctaw Chilli
Carsons California Cowboy of J Bar D
Harshbargers Teasque of J Bar D
J Bar D Brandy
J Bar D Chance
J Bar D Choctaw Princess
J Bar D Cree Squaw
J Bar D Crow Chief
J Bar D Kindlers Rowdy
J Bar D Misty Chanee
J Bar D Nobles Red
J Bar D Roadrunners Hopisquaw
Pizs Little Joe
Rowes Hondo Warrior
Browns Coon
Millers Bandit
J Bar D Cimmeron Warrior
Colorado CHili of Smithfield
Dakota Blue Commanches Misty
Dakota Blue Cotton Candy
Dakota Blue Lucky Chance
Dakota Blue Lucky Tia
Dakota Blue Pepi
Dakota Blue Sparkling Burgandy
Dakota Blue Thorn
Slash V Buckeye
Babs
Hawk Hill and
Chimnoy Hollow dogs
Fairoaks Cotton Jenny
Mighty Fine Dividend
Mighty Fine Second Chance
Tegtmeyers Jeranomo of J Bar D
Christiansons Calico Brownie McGee
McGuires Blu-Duster
J Bar D Zuni Squaw
J Bar D Tylers Naki
J Bar D Paiute Red
J Bar D Jamie
J Bar D Cherokee
Tomahawk of Queensland
J Bar D Natchez Squaw
Just Jake of Las Rocosa
Roberts Sweet Pepper
Beiens Red Tuffy
Calico Chief Red Cloud
Jones Reddy Teddy
Davies Mr. Chips
Davies Pinka
Ebony Lass
Thiexs Skeet
Kintyre Holly Smoke
Kintyre Blue Suede Shoes
Colorados Brandy Bow
Barrs Colorado Kelly
Crandells Red Pepper
Sorensens Cherry Brandy
Telles Caramel Candy Girl
Weimers Colorado Bonnie
See her own
Nelsons Peanut
Telles Whiskey High
Walters Cricket
Askews Colorado Toby
Barrs Brandy Sue
Emersons Sugar I
Finchs Robin K
Adams Colorado Babe
Colorados Ravishing Ruby
Hiawatha Copper Stub

Hondo produced a few other dogs which were influential in breeding programs around the country but he was primarily a wonderful companion and working partner for his owners. In a sense, the greatness of most dogs should be measured in this way and their breeding prowess a secondary consideration. Hondo was a superior individual and a credit to the breed. While his contribution may be less tangible than that of his kennel mates, Commanche and Warrior Bow, he was an important dog nonetheless.

While it is common to give a lot of credit to the male lines in any breed, top females of a breed likewise deserve credit. A contemporary of Commanche was another of the Rowe's original dogs, Porter's Yakima Blue Squaw. Yak was actually the second Australian Shepherd the Rowes obtained after Dingo. She was purchased from Jim Healey in 1964 for $35. A granddaughter of Wood's Jay, through her dam Healey's Pepsi Cola, and by Wood's Teddy Blue Boy, she was the only puppy in her litter, a blue eyed blue merle with lots of white trim. At the time Fletcher Wood had determined there were two types of Australian Shepherds—the Thoroughbred type vs. the Quarter Horse type. The Thoroughbreds were tall and leggy, typical of many of his own dogs and the Sisler dogs and the Quarter Horse type which were short legged and stubby. Yak was the short Quarter Horse type and in Fletcher Wood's eyes she exhibited all the faults he was trying to avoid in his own dogs!

Porter's Yakima Blue Squaw at about 4 months of age. Courtesy of Rowe

Porter's Yakima Blue Squaw, blue and white female. Courtesy of Rowe

Yak had a multifaceted personality. The Rowes describe her as snotty, believing she was better than most and deserving of royal treatment at all times! When they first brought her home as a puppy she refused to be caught, having her own agenda and being perfectly content to follow her own whims. It took the lure of human food to convince her to come when she was called. It was perhaps this early experience which started her lifelong habit of being willing to eat anything a human would eat!

As Yak matured she continued to develop her own way of doing things. While Fletcher Wood may have considered her undesirable in many physical qualities, she was typical in mental abilities! The Wood dogs were all tough, intelligent and never gave up. The Rowes once had some neighbors with children who would hit the Rowe children with a stick over the fence. Yakima was the playmate and caretaker of the Rowe kids and decided she didn't like what she saw the neighbor children were doing to her charges. When the brats continued their one sided game, Yakima decided to end it by taking the stick away from them. Although she did not harm the neighbor children they threatened to sue the Rowes because of her actions. So they decided to sell Yakima through an ad in the paper. A rancher came to look at her and asked whether she had a black roof in her mouth. At the time there was a widespread belief this trait meant a dog was more intelligent. The Rowes had never looked at the roof of her mouth so they invited the man to have a look for himself. When he tried, Yak was insulted and bit him. In spite of this the rancher decided she was worth the $35 asking price and chose to take her home. Jerry put her in his car and told her to stay. On the way home Yak bit him several more times when he reached over the seat to try to pet her. When they arrived home, Yak refused to get out of the car—Jerry had told her to stay and stay she would.

Yakima. Courtesy of Rowe

At a loss for a solution, the rancher decided to try leaving his car door open overnight to see if Yakima would come out on her own. By morning she had exited the car and was seen running along a hilltop behind his house and exercising her disobedient streak by refusing to come back. With a day's work ahead of him, he saddled his horse and set out. Yakima decided this looked interesting enough to her to tag along. After following him around the ranch all day, the two became best of friends. It was this ranch where Jerry and Sharon spent many days helping to gather, sort, brand and vaccinate cattle, working their young dogs and absorbing many lessons about training them.

Yakima was a terrific chute dog in those days. When crowding cattle through loading chutes most ranchers didn't have hotshots to get them moving again when they would jam up in tight quarters. This is where the dogs came in. Yakima was a pro at unsticking a crowded chute. Jerry remembers a time when a group of Herefords were in the chutes being loaded in a truck. Somebody was trying to close a gate and Yak dashed in to heel a cow from between one man's legs. The cow kicked, the rancher

Yakima and Commanche teaming up to work the chutes. Courtesy of Rowe

went flying, knocking the gate down, but Yakima held the cattle until the gate was shut.

At the end of one day the cattle were getting tired and more stubborn than usual. One particular cow was hanging back and threatening to undo all the hard day's work. The rancher sic'ed Yakima on this cow who was not tired enough for it to affect her aim. Her kick was on target and knocked Yakima completely out. Not wanting to leave her behind the rancher put her on the floorboard of his truck. His wife was driving the truck and not paying attention to the comatose dog on the floorboard. Suddenly Yak woke up, bit his wife and was promptly evicted from the truck. Yak caught up with the herd and hit the heels of every bovine in the herd, apparently in revenge for being kicked!

While Yakima was a never-say-die cattle working dog, she, like a lot of the Wood dogs, hated sheep. When the Rowes young herding group was just beginning, they were trying to develop different ways of working their dogs without losing control or harming the few sheep they had. One of their techniques involved working the dogs on a long rope, especially the tougher dogs that liked to bite stock. Yak was one of these dogs. She was so tough and so quick she still managed to kill at least one sheep even while on this leash. The group didn't give up but struggled for many years until Lewis Pence befriended Jerry and began helping the group learn the ropes of working with their dog's natural instincts.

Yakima working on her obedience around livestock. Courtesy of Rowe

Most Wood dogs had very tight cat feet with black pads. Yak, of course, had white long toed feet and was the squatty Quarter Horse type of dog Fletcher Wood avoided. His reasoning was sound, as Yakima would tire easily and get sore footed when she was worked. Her owners would lock her up in the house to recover but her heart and mind were not going to settle for that. She would break a window in order to escape and come to work or just be with her folks. The other thing Fletcher faulted in Yakima, of course, was her extensive white trim and blues eyes. He believed dogs with brown eyes had better vision, especially at a distance. All existing records show Yakima's sire, Wood's Teddy Blue Boy, was only bred the one time because of the faults that Fletcher believed appeared in Yak. Sharon confirmed this in a conversation with Fletcher, who didn't have much to say about Teddy Blue Boy and likely never used him again.

Commanche and Yakima doing a full day's work at Jim Porter's ranch. Porter is seen here waiting for the branding iron with the two dogs looking on. Courtesy of Sorensen

Yak was the dam of the unusual dog Ch J Bar D Red Chief. He was a very dark color when he was born and to all appearances was a black dog. As he got older he turned a dark mahogany and was to all appearances a very dark red dog. He is the dog who, when bred to a red merle bitch, J Bar D Choctaw Chilli, produced the black dog J Bar D Crow Chief! Nobody was ever able to determine why this happened but it was a genuine oddity in a time when Aussie people believed they understood the genetics of coat color inheritance in the breed.

J Bar D Red Chief. Courtesy of Rowe

Red Chief sired a dog named Zip or Zipper who showed the same talent at clearing fences as Commanche. He actually climbed the fence until he could hook his feet on the top to push himself over. Some of the fences he cleared were eight feet high—a reason jumping ability is something the original breed standard writers wanted to include in the standard!

Another important female obtained in trade from the east coast was Mary's Rocky Mountain Babe. A daughter of Piney Wood's Colorado Valor and Adams' Colorado Babe, she was nevertheless all J Bar D/Sorensen in pedigree. Valor's sire was a son of Sorensen's Red Man and Sorensen's

J Bar D Brandy, red merle female and producer of a dynasty of working dogs and foundation breeding dogs. Courtesy of Rowe

Cherry Brandy and her dam was a daughter of Weimer's Colorado Bonnie who was by Sorensen's Red Man out of J Bar D Brandy. Babe was a 19 inch solid red female with very little white and no copper. She was a very fast dog who loved to work and would do anything asked of her. Her speed had to do with her low to the ground running ability. She wasted no motion and could catch anything on four legs. She was a very low heeler but also had a lot of eye when working livestock. Unfortunately she did not have well set eyes and a horse accidentally kicked one out early in her life. Because of this she wouldn't take direction on her blind side when working. Bred to both Commanche and Cimmeron, Babe's largest contribution to the breed was made after she went to live with Betty Nelson. There she produced a dynasty of good dogs for Betty and from her dogs a number of good dogs for Fourwinds and other west coast kennels. Her daughter by Cimmeron, Moore's Cinnamon Stick, was also the dam of Anita Laycock's foundation dam, Laycock's Cheyenne Maurine. Babe became one of Betty's most beloved dogs and went everywhere with her and would do anything for her. This well traveled dog was a real credit to her J Bar D background.

From these beginnings the J Bar D breeding efforts grew and spread through the country, literally from coast to coast. The inbred Commanche son and grandson, J Bar D Apache Warrior was the founding sire for the Swaner family in Texas. The blue merle dog with good conformation has descendants throughout the south. Another well bred female, J Bar D Arapahoe Lady accompanied Apache to the Swaner's. The Rowe's bought her after breeding to her. After seeing the quality she was able

Mary's Rocky Mountain Babe, red and white female who took the J Bar D genetics to the West Coast and left a wide legacy there. Courtesy of Rowe

to produce from Commanche, they decided she would be a good match for Apache to help the Swaner's establish their own breeding program. She was by Olson's Clare Blue Chips out of Healey's Sissy II. Chips was, of course, Wood breeding while Sissy was from the Halls breeding which is behind some of the Sorensen dogs. Lady produced offspring by both Commanche and Apache during her lifetime.

Bar D Arapahoe Lady (Olson's Clare Blue Chip x Healey's Sissy II), blue merle female of beautiful type and top Wood breeding. Courtesy of Rowe

The red merle Red Chief son, J Bar D Briggs' Chisum was sold to Don and June Briggs and worked their ranch for many years. He was bred back to Choctaw Chilli once to concentrate his Warrior Bow heritage. He also produced a number of good dogs before he was stolen from the Briggs' ranch.

J Bar D Cherokee Red, owned by local rancher Glen Noble, was a pick of the litter choice from the breeding of Rowe's Hondo Warrior to Rector's Samantha Ann. He eventually sired J Bar D Cimmeron Warrior out of Miller's Bandit, a linebred Commanche Warrior female sired by Brown's Coon. Coon was the good dog bred by Rowe's and originally owned by Marvin and Paula Brown, later owned by Dick Sorensen for a time and subject of many amusing stories in the Sorensen chapter of this book. Coon's dam was a litter sister of Hondo Warrior, J Bar D Cree Squaw. The dam of Miller's Bandit, Miller's Sheba, was also a daughter of Green's Mr Cobar, another Colorado dog of great character and out of Rowe's Cheyenne Squaw, another Comanche daughter out of Kelly's Blue Rockette.

Green's Mr Cobar, (Green's Mr Ezekiel x Green's Copper Fluff), a dog of good type that crossed well with the other mountain-type dogs of his era. Courtesy of Rowe

Green's Mr Cobar was of largely unknown breeding but typical of the dogs found in Colorado at the time. He was the constant companion of Gloria Green, a friend of the Rowes and local character in the Aussie world. Gloria worked at the local sale barn and Mr Cobar was her right hand. She was working the pens in the back one day when a Holstein bull hit a heavy gate, knocked it down and trapped the unconscious

Green's Mr Cobar moving the cattle toward the chute to load them into a truck at one of the first All-Breed Stockdog Trials (1973) at the Rowe Farm, Broomfield, Colorado. Courtesy of Rowe

Gloria underneath. The bull was not finished with Gloria and began to work her over from atop the gate. Mr Cobar was busy at the time taking a group of cows down another alley but heard the commotion and ran back to Gloria. Analyzing the situation in an instant, he latched onto the nose of that bull and took him all the way back down the alley, saving Gloria's life. Gloria's husband, Bob, was a horseshoer who bought horses on the side. Gloria was a small woman so she would get on anything he brought home, including young colts. Bob once brought home a rank horse that Gloria couldn't stay on. She was bucked off and knocked out cold. Like the Holstein at the sale barn, this horse wasn't satisfied with just unloading Gloria and began to trample her. Mr Cobar would have none of that and this horse was taken as far away from Gloria as he could get him. Mr Cobar was a black tri dog with intense brown eyes who was never far from Gloria's side—which was apparently very fortunate for her!

When he wasn't rescuing Gloria, Mr Cobar spent his time at other pursuits. He was a great kids dog and played games with them like his contemporary Comanche and others. Gloria's son disappeared one day and after looking for him for quite awhile, Mr Cobar discovered him sleeping in a hiding place. He uncharacteristically disappeared from Gloria's side on another day and she began searching for him around her house. He was discovered under the bed, happily sleeping with some kids. He did have one other trait he shared with Commanche which was a mutual hatred. The two males never got along but were well behaved enough not to cause trouble when they had to be in the same vicinity. Sharon remembers one amusing incident when she had driven over to the Green's with Commanche in the back of her truck. Sharon got out and left Commanche in the truck and Mr Cobar rushed over to inspect the vehicle. He didn't realize Commanche was in the bed and sprang up to look inside. Commanche, hearing footsteps outside, prepared to guard the truck from the intruder. The two dogs comically screeched to a halt as their faces met at the tailgate and, obeying their good training, both turned away from each other in mid-air! Sharon and Gloria were in tears laughing at the sight.

A foundation dog owned by Betty Nelson's neighbor, John Brown, was named Brown's Rocky who had been taught a wide repertoire of tricks. Rocky was invited to perform the "halftime show" at the first Invitational trial put on by Bob Carrillo in Santa Rosa, California. In the spirit of Jay Sisler, John created an act of difficult and amusing tricks which amazed and delighted audiences. In addition to the standard cute dog tricks of sitting up to beg, rolling over, and shaking hands, Rocky would climb a ladder and even achieved the strength and balance to leap to the top of a fence post and pose there! Rocky's sire was of vague Oregon background being by a blue merle ranch dog and out of a black ranch bitch. However, his dam was of proven strong Aussie background. Her sire, Lee's Savage Sam was ¾ old Ritter breeding, closely related to Rowe's Warrior Bow.

Brown's Rocky (Gillworth's Stub x Nelsons Samantha Ann CDX) exhibiting one of his many talents and tricks. Courtesy of Rowe

The Australian Shepherd community in the 1960s and '70s was a close knit one and everyone got along, bred dogs together and tried to learn together about their breed. One of these families was the Dennis family who had moved to Colorado from California and brought a new strain of dogs to the area. Their kennel name was Roadrunner and over the years they produced a number of dogs with Sharon and Jerry Rowe, mixing the lines with great success. The Dennis family's primary sire was a dark mahogany sable dog named Roadrunner's Picacho. He was from the California McConkey breeding but was also a grandson through his sire of Jeane Farned's dog Farned's Dead Eye. Dead Eye was a Colorado dog who sired one of Tom and Charlene Jeffreys' foundation females! This sable male, Picacho, was crossed with a red merle J Bar D female and produced the blue merle female J Bar D Hopi Squaw. Hopi was both a well made dog and a good working dog, and Sharon competed successfully with her for quite some time. These experiments preserved a good line of dogs while teaching the breeders more and more about genetics. A number of good dogs were produced by the Rowes and the Dennis family, most of which have lived on through a number of descendants.

Farned's Brownie, Jefferys' Dead Eye Jr. (Ted), Jefferys' Dolly (Dee), Farned's Dead Eye, Jefferys' Sheila, 1965. Courtesy of Rowe

J Bar D Hopi Squaw (Roadrunners Picacho x J Bar D Choctaw Chilli), Best Puppy. Courtesy of Rowe

A daughter of Hopi Squaw and by Commanche was J Bar D Natchez Squaw. She was bred a number of times and produced many of Jeanne Roberts' foundation dogs for her Mountain High kennel. Natchez Squaw was also the dam of Just Jake of Las Rocosa along with a number of other well known dogs. She eventually ended up working a ranch in southern Colorado.

One of Sharon's favorite dogs was the blue merle female, J Bar D Chinook Squaw. This daugher of Roadrunner's R-Lucky Chipawa and Rowe's Cheyenne Squaw was one of the successful combinations of the Roadrunners and J Bar D dogs. Her dam was inbred on Commanche from a father to daughter cross while her sire was a Red Chief son from a Roadrunner's bred female. Sharon took a special interest in this very bright puppy from the beginning and taught her a wide range of tricks by the time Chinook was only 3 months old. She became a trial dog for Sharon. One memorable day another dog crawled under the arena fence and started chasing Chinook's sheep. Sharon stopped her dog while the stray was caught, then continued on as if nothing happened. In those days, competitors did not get "reruns" and you were expected to handle anything that happened during your run and just accept it as part of trialing! Chinook's one phobia was thunderstorms and she once escaped while traveling with Sharon's parents and was lost. Fortunately she was later found hiding in the showers on the campground—and it was the exact same shower her mother had used earlier in the day! Eventually Chinook became the much loved companion of Sharon's parents. She was a trustworthy dog who made friends wherever she went and never met a person she could not convince to love her.

Another memorable dog produced by the Rowes was J Bar D Skippy. Skippy's sire was a linebred Commanche dog with a dash of Roadrunners breeding and his dam was Floyd and Marie Leavitt's fine female, Leavitt's Patches. The Leavitt's were friends of the Rowe's and are another important part of Aussie history during the early years. Skippy was a nicely conformed dog who was used lightly for breeding. He was another of the good J Bar D dogs who helped found Jeanne Robert's Mountain High kennel in addition to producing dogs for the Rowes and others. Although he was a good looking dog, a broken leg early in life prevented a show career for him. He was eventually sold to Ron Harrison, a local horse trainer where he was used around the riding stable for working stock.

In the late 1970s, Sharon was convinced by Barrie Abbott of Queensland kennel to take a top bitch named J Bar D Kickapoo Squaw. Kickapoo was a truly top quality blue merle shown successfully by Sharon. She quickly became a favorite of Sharon's and was trained to do tricks, work ducks, and perform in obedience. Like all breeding programs, the J Bar D line benefited from this infusion of outside blood and it was a smart addition which had already been proven to work well with their existing dogs. Kickapoo was by the well known California sire Queensland's Coondog Jimmy Blue and out of a kennelmate named Queensland's

J Bar D Natchez Squaw (Rowe's Commanche Warrior x J Bar D Hopi Squaw). Courtesy of Rowe

J Bar D Chinook Squaw (Roadrunners R-Lucky Chipawa x Rowe's Cheyanne Squaw). Courtesy of Rowe

The blue merle dog, J Bar D Skippy (Kenner's Blue of J Bar D x Leavitt's Patches). Courtesy of Rowe

PEDIGREE OF Roadrunners R-Lucky Chipawa

Born: ______ Sex: ______
Owned by: ______

Parents	Grandparents	Great-grandparents	Great-great-grandparents
Sire: J Bar D Red Chief	Sire: Rowe's Warrior Bow	Sire: Ronsleys Carcoal Smoke	Sire: Ritters Streak
			Dam: Ritters Lass
		Dam: Murphys Cochina	Sire: Ronsleys Charcoal Smoke
			Dam: Ronsleys Mitzi Buck
	Dam: Porters Yakima Blue Squaw	Sire: Woods Teddy Blue Boy	Sire: unknown
			Dam: unknown
		Dam: Healeys Pepsi Cola	Sire: Woods Jay
			Dam: Woods Little Kola
Dam: Roadrunners Secoya	Sire: Roadrunners Picacho	Sire: Bradleys Calico Shep	Sire: Farneds Dead Eye
			Dam: Hays Sugar
		Dam: Bradleys Beauty Queen	Sire: McConkeys Sir Blue Silver
			Dam: McConkeys Sweet Sue
	Dam: Howes Little Peedee	Sire: Elys King of Kings	Sire: Elys Chico
			Dam: Williams Georgina
		Dam: Howes Little Popo Agie	Sire: Sniffs Dandy Danny
			Dam: Sniffs Tina

Bred by: ______ Address: ______ Phone: ______

The details given above are correct to the best of my knowledge. Signed: ______ (Breeder)

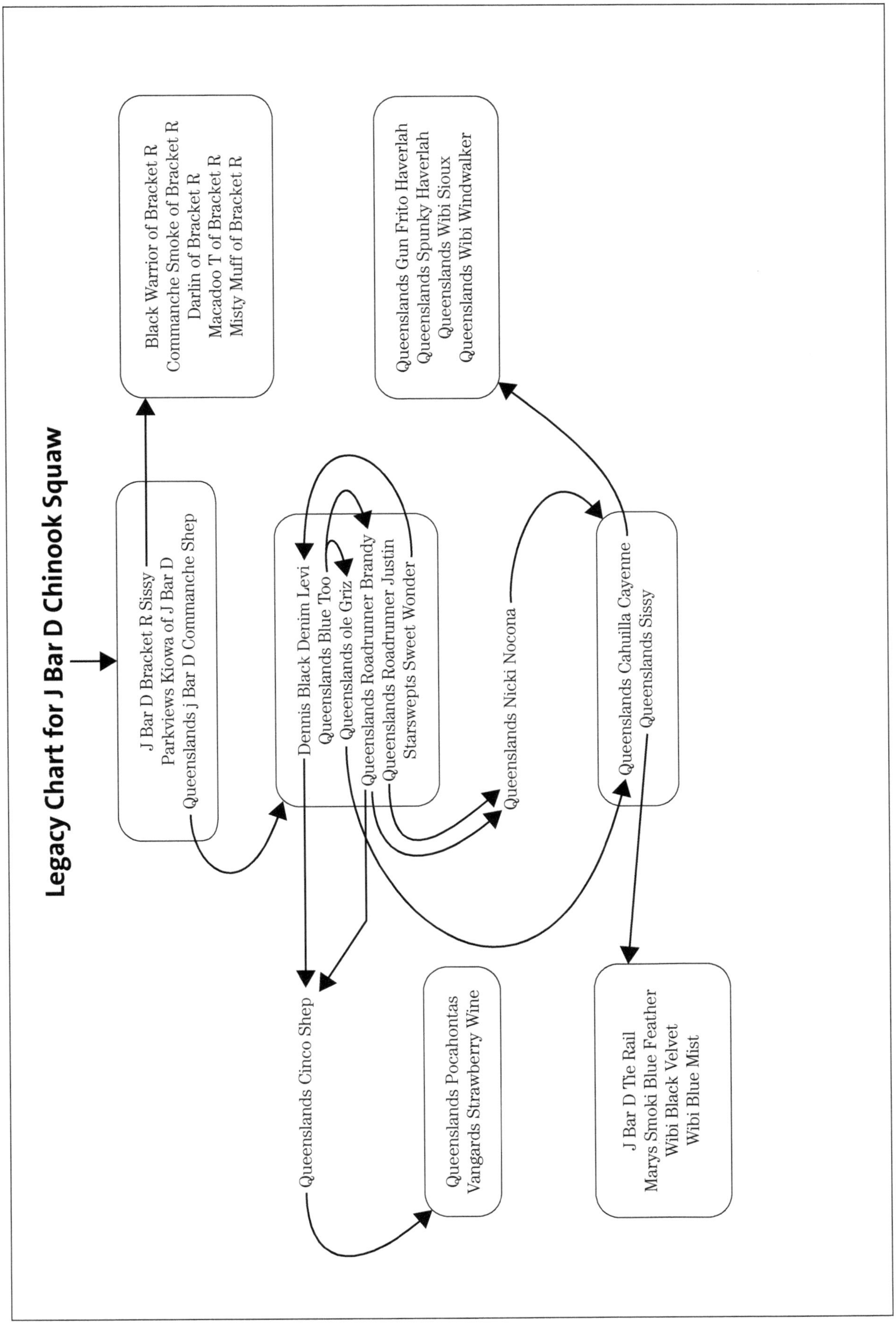
Legacy Chart for J Bar D Chinook Squaw
J Bar D Bracket R Sissy
Parkviews Kiowa of J Bar D
Queenslands j Bar D Commanche Shep
Black Warrior of Bracket R
Commanche Smoke of Bracket R
Darlin of Bracket R
Macadoo T of Bracket R
Misty Muff of Bracket R
Dennis Black Denim Levi
Queenslands Blue Too
Queenslands ole Griz
Queenslands Roadrunner Brandy
Queenslands Roadrunner Justin
Starswepts Sweet Wonder
Queenslands Gun Frito Haverlah
Queenslands Spunky Haverlah
Queenslands Wibi Sioux
Queenslands Wibi Windwalker
Queenslands Nicki Nocona
Queenslands Cahuilla Cayenne
Queenslands Sissy
Queenslands Cinco Shep
Queenslands Pocahontas
Vangards Strawberry Wine
J Bar D Tie Rail
Marys Smoki Blue Feather
Wibi Black Velvet
Wibi Blue Mist

J Bar D Kickapoo Squaw (Queensland's Coondog Jimmy Blue x Queensland's Tequila). Courtesy of Rowe

Tequila. Jimmy Blue was by another widely used California male named King's Boo.

A number of good dogs and bitches from the Rowes were also owned and used by the Sorensens and are discussed at length in Sorensen's chapter. These two groups and many others were all breeding dogs in the same period of time and much of this history happened in parallel. After establishing such a rich foundation for themselves and many other kennels in America, the Rowes experienced some difficulties in life and were forced to get out of dogs for a number of years. During this time the kennels they helped found continued breeding dogs. When the Rowe's life settled down again they were able to find dogs descended from their own foundations to pick up and continue on.

By relying on old friendships, the Rowes were able to start again with dogs of their breeding. The first was J Bar D Mystic Warrior. Sired by Queensland's Wibi Windwalker, who descends multiple times from Commanche, Jones' Reddy Teddy and many other Colorado foundation dogs and out of Mt Blue Christians Praise, a Newman's Blue Sprog daughter out of a Commanche granddaughter. Sprog was also descended from a long line of Wood and foundation Colorado dogs and will receive more attention on the chapter on Casey Masterson Newman and her HS & W kennel. Mikey, as Mystic Warrior was known, was Jerry's trial and farm dog for many years. He was an extremely social dog which sometimes worked to his disadvantage when trialing. If Jerry became frustrated

Queensland's Coondog Jimmy Blue, (King's Boo x McConkey's La Tigra) dark blue with bright copper and white male. Sire of J Bar D Kickapoo Squaw. Courtesy of Rowe

with him, he would often decide to walk the fence looking for friends in the crowd rather than continuing to work in the arena! He especially loved children, in a manner typical of his great ancestor Commanche. Mikey was a dark blue self merle like many of the old mountain-type dogs and stood about 20-21 inches tall. He was a strong heading dog who would only fetch for Jerry in the beginning of his training. Jerry describes an unusual technique Mikey could use on sheep which were more timid or flighty than normal. He had a way of humbling himself and luring the sheep to Jerry instead of pushing them. Jerry describes this as sucking them to him like a pied piper. While he might have been a frustrating dog to trial he proved that you never stop learning when it comes to training dogs and Jerry is quick to grant that he learned a number of valuable lessons from Mikey. Used only sparingly at stud, Mikey nonetheless left a number of valuable progeny which are producing dogs today for HS & W and other breeders around the world.

The Queensland kennel of Barrie Abbott of California kept a number of the old bloodlines going for many years. A major sire for them who can be found behind dogs in other kennels was Queensland's Wibi Windwalker—owned by Pat Goodwin. A large dark self blue merle Windwalker was a typical dog of the mountains. An amazing story is told about his sire, the solid black Queensland's Bear Paws (Nelson's Ben Reddy x Nelson's Sweet Sue). He was given to Barrie's boyfriend at that time who chained him up in his yard. Bear Paws broke the chain and ran away, disappearing and surviving for 4 months. Eventually thirst drove him to seek water at a local Thoroughbred ranch and the owner was able to catch him. At the time of his capture he had managed to lose the chain but somehow had gotten his front leg caught in his collar. He was reunited with the boyfriend who

Mt Blue Christian's Praise, blue, copper and white female (Newman's Blue Sprog x Misty Mountain Blue). Dam of J Bar D Mystic Warrior. Courtesy of Rowe

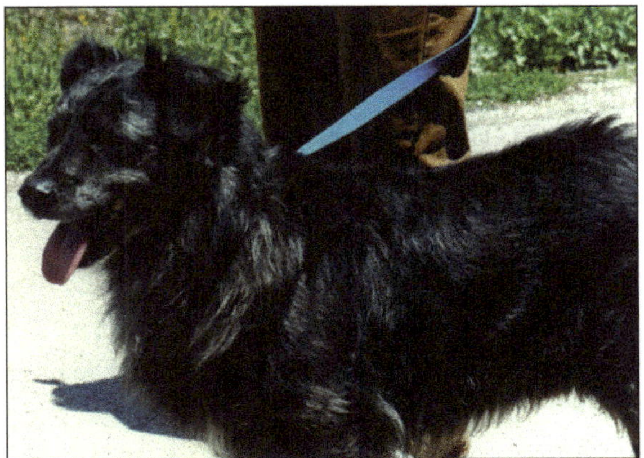

J Bar D Mystic Warrior as a young dog (Queensland's Wibi Windwalker x Mt Blue Christian's Praise). Courtesy of Rowe

Queensland's Wibi Windwalker, blue merle male (Queensland's Bear Paws x Queensland's Cahuilla Cayenne). Courtesy of Rowe

promptly chained Bear Paws again. Barrie was in the process of moving and left the dog with the boyfriend. Bear Paws again broke the chain and 4 days later appeared at Barrie's new home 5 miles away—a place Bear Paws had never been—convincing her the dog would never leave her side again.

The Rowes moved to Kansas to establish Twin Creek Farm where they run week-long sheep camps to teach others how to train stock dogs for trialing or every day ranch work. They continue to breed on a small scale. From a letter written by Jean Taylor to Betty Nelson, "Then there's the Rowes, bless them. I hope they manage to gather a group of dogs before long that they can get excited about. Having had what they had, anything less is disappointing." These dogs are forming the nucleus of a new wave of J Bar D dogs which combines the old with the new. The Rowes have added some dogs of Bob Vest's breeding to these original founding dogs they got from Betty. This addition was to produce working dogs for Jerry to trial and use on their Kansas sheep farm. Jerry has refined his training techniques learned from Commanche, Cimmeron, Lewis Pence, and hundreds of other dogs and people he has met over the years. He and Sharon now run a successful training facility where they freely share their knowledge with new generations of people and dogs. Giving them something they wish they had when they were starting out with Australian Shepherds. They continue to judge working trials and conformation shows, as they have done for decades, as well as competing in trials. Jerry has qualified for the Finals a number of times in recent years, as have many of his students. They, like the Sorensens, are now revered mentors and have earned the respect of Aussie fanciers throughout the world. Their place in the history of the Australian Shepherd is immeasurable and their contributions to the breed are legend. They have served the breed well and hold a place matched by few others in history. We are grateful for what they left us and for what they continue to do for Australian Shepherds and all working dogs.

J Bar D Chinook Squaw turning cattle during a trial. Courtesy of Rowe

PEDIGREE OF J Bar D Mystic Warrior

Born: ____________ Sex: ________
Owned by: ______________________

<table>
<tr><td rowspan="8">Sire
Queenslands Wibi Windwalker</td><td rowspan="4">Sire
Queenslands Bearpaws</td><td rowspan="2">Sire
Nelsons Ben Ready</td><td>Sire Jones Reddy Teddy</td></tr>
<tr><td>Dam Nelsons Gin Fizz</td></tr>
<tr><td rowspan="2">Dam
Nelsons Sweet Sioux</td><td>Sire Rowe's Commanche Warrior</td></tr>
<tr><td>Dam Nelsons Sweet Tam</td></tr>
<tr><td rowspan="4">Dam
Queenslands Cahnilla Cayenne</td><td rowspan="2">Sire
Queenslands Ole Griz</td><td>Sire Queenslands J Bar D Commanche Shep</td></tr>
<tr><td>Dam Queenslands Blue Too</td></tr>
<tr><td rowspan="2">Dam
Queenslands Nicki Nocona</td><td>Sire Queenslands Roadrunner Justin</td></tr>
<tr><td>Dam Queenslands Roadrunner Brandy</td></tr>
<tr><td rowspan="8">Dam
Mt Blue Christians Praise</td><td rowspan="4">Sire
Newmans Blue Sprog</td><td rowspan="2">Sire
Adairs Black Bear</td><td>Sire McGuires SeMeGo Hemi</td></tr>
<tr><td>Dam Mastersons Dotclijee Shunka</td></tr>
<tr><td rowspan="2">Dam
Mastersons Dotclijee Chunka</td><td>Sire Jeffreys Little Pooh Bear</td></tr>
<tr><td>Dam Poudre Condesa</td></tr>
<tr><td rowspan="4">Dam
Misty Mountain Blue</td><td rowspan="2">Sire
Las Rocosa Sydney</td><td>Sire Taylors Whiskey</td></tr>
<tr><td>Dam Las Rocosa Leslie</td></tr>
<tr><td rowspan="2">Dam
Stonehenge Phantasy</td><td>Sire Rowes Commanche Warrior</td></tr>
<tr><td>Dam Shanahans Phantom</td></tr>
</table>

Bred by: ____________ Address: ______________________ Phone: ____________

The details given above are correct to the best of my knowledge. Signed: ______________________(Breeder)

PEDIGREE OF J Bar D Cimmeron Warrior

Born: ______ Sex: ______
Owned by: ______

Parents	Grandparents	Great-grandparents	Great-great-grandparents
Sire J Bar D Cherokee Red	Sire Rowes Hondo Warrior	Sire McGills Slagle Blue	Sire Sislers Shorty
			Dam Weedmeyers Weedie
		Dam J Bar D Choctaw Chili	Sire Rowe's Warrior Bow
			Dam Sorensens Queenie
	Dam Rectors Samantha Ann	Sire Hartnagles Hud	Sire Taylors Whiskey
			Dam Taylors Buena
		Dam Colberts Black Pepper	Sire Clingers Ely Sage
			Dam Savers Canberra
Dam Millers Bandit	Sire Browns Coon	Sire Rowes Commanche Warrior	Sire Sorensens Gunsmoke
			Dam Van Gordons Blue Asta
		Dam J Bar D Cree Squaw	Sire McGills Slagle Blue
			Dam J Bar D Choctaw Chili
	Dam Millers Sheba	Sire Greens Mr Cobar	Sire Greens Mr Ezekiel
			Dam Greens Copper Flyer
		Dam Rowes Cheyanne Squaw	Sire Rowe's Commanche Warrior
			Dam Kelly's Blue Rockettz

Bred by: ______ Address: ______ Phone: ______

The details given above are correct to the best of my knowledge. Signed: ______(Breeder)

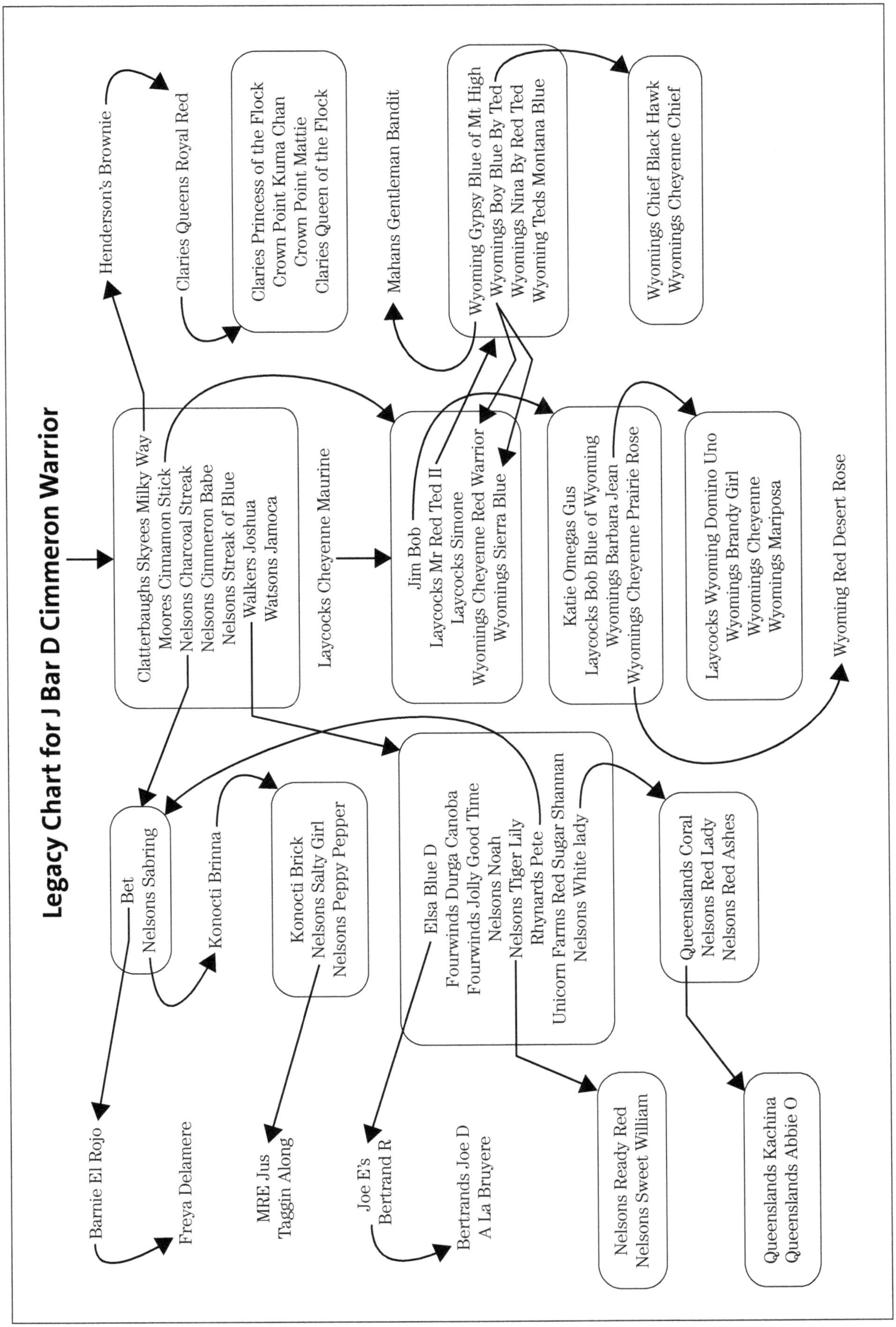
Legacy Chart for J Bar D Cimmeron Warrior
Clatterbaughs Skyees Milky Way
Moores Cinnamon Stick
Nelsons Charcoal Streak
Nelsons Cimmeron Babe
Nelsons Streak of Blue
Walkers Joshua
Watsons Jamoca
Laycocks Cheyenne Maurine
Henderson's Brownie
Claries Queens Royal Red
Claries Princess of the Flock
Crown Point Kuma Chan
Crown Point Mattie
Claries Queen of the Flock
Mahans Gentleman Bandit
Wyoming Gypsy Blue of Mt High
Wyomings Boy Blue By Ted
Wyomings Nina By Red Ted
Wyoming Teds Montana Blue
Wyomings Chief Black Hawk
Wyomings Cheyenne Chief
Jim Bob
Laycocks Mr Red Ted II
Laycocks Simone
Wyomings Cheyenne Red Warrior
Wyomings Sierra Blue
Katie Omegas Gus
Laycocks Bob Blue of Wyoming
Wyomings Barbara Jean
Wyomings Cheyenne Prairie Rose
Laycocks Wyoming Domino Uno
Wyomings Brandy Girl
Wyomings Cheyenne
Wyomings Mariposa
Wyoming Red Desert Rose
Bet
Nelsons Sabring
Konocti Brinna
Konocti Brick
Nelsons Salty Girl
Nelsons Peppy Pepper
Elsa Blue D
Fourwinds Durga Canoba
Fourwinds Jolly Good Time
Nelsons Noah
Nelsons Tiger Lily
Rhynards Pete
Unicorn Farms Red Sugar Shannan
Nelsons White lady
Queenslands Coral
Nelsons Red Lady
Nelsons Red Ashes
Barnie El Rojo
Freya Delamere
MRE Jus
Taggin Along
Joe E's
Bertrand R
Bertrands Joe D
A La Bruyere
Nelsons Ready Red
Nelsons Sweet William
Queenslands Kachina
Queenslands Abbie O

Born: ______ Sex: ______
Owned by: ______

PEDIGREE OF J Bar D Arapahoe Lady

Parents	Grandparents	Great-grandparents	Great-great-grandparents
Sire: Clare Olsens Blue Chips	Sire: Woods Adobe	Sire: Woods Dandy	Sire: Finleys Texas Bob
			Dam: Merinos Blue Bitch
		Dam: Woods Little Blue	Sire: Blue Boy
			Dam: Blue Shadow
	Dam: Woods Little Kola	Sire: Woods Jay	Sire: Sislers Shorty
			Dam: Sislers Trixie
		Dam: Woods Little Blue	Sire: Blue Boy
			Dam: Blue Shadow
Dam: Healeys Sissy II	Sire: Halls Smokey Blue	Sire: Farneds Dead Eye	Sire: Gregg
			Dam: Whiskey Bell
		Dam: Hoopers Blue Keno	Sire: Lilly's Mr Blue Chips
			Dam: Lilly's Pecos Jane
	Dam: Halls Amy	Sire: Lighters Rusty	Sire: Elys Major
			Dam: Lighters Asta
		Dam: Lighters Asta	Sire: Woods Jay
			Dam: Woods Little Blue

Bred by: ______ Address: ______ Phone: ______

The details given above are correct to the best of my knowledge. Signed: ______(Breeder)

Porter's Yakima Blue Squaw and Rowe's Dingo. Courtesy of Rowe

The first two Rowe females, Rowe's Dingo and Porter's Yakima Blue Squaw. Courtesy of Rowe

Rowe's Dingo and Porter's Yakima Blue Squaw. Courtesy of Rowe

Exhibiting the good training which was the pride of dog owners of the day! Left to right: Jeffreys' Sheila, Porter's Yakima Blue Squaw, Rowe's Dingo, and Jeffreys' "Dee." Courtesy of Rowe

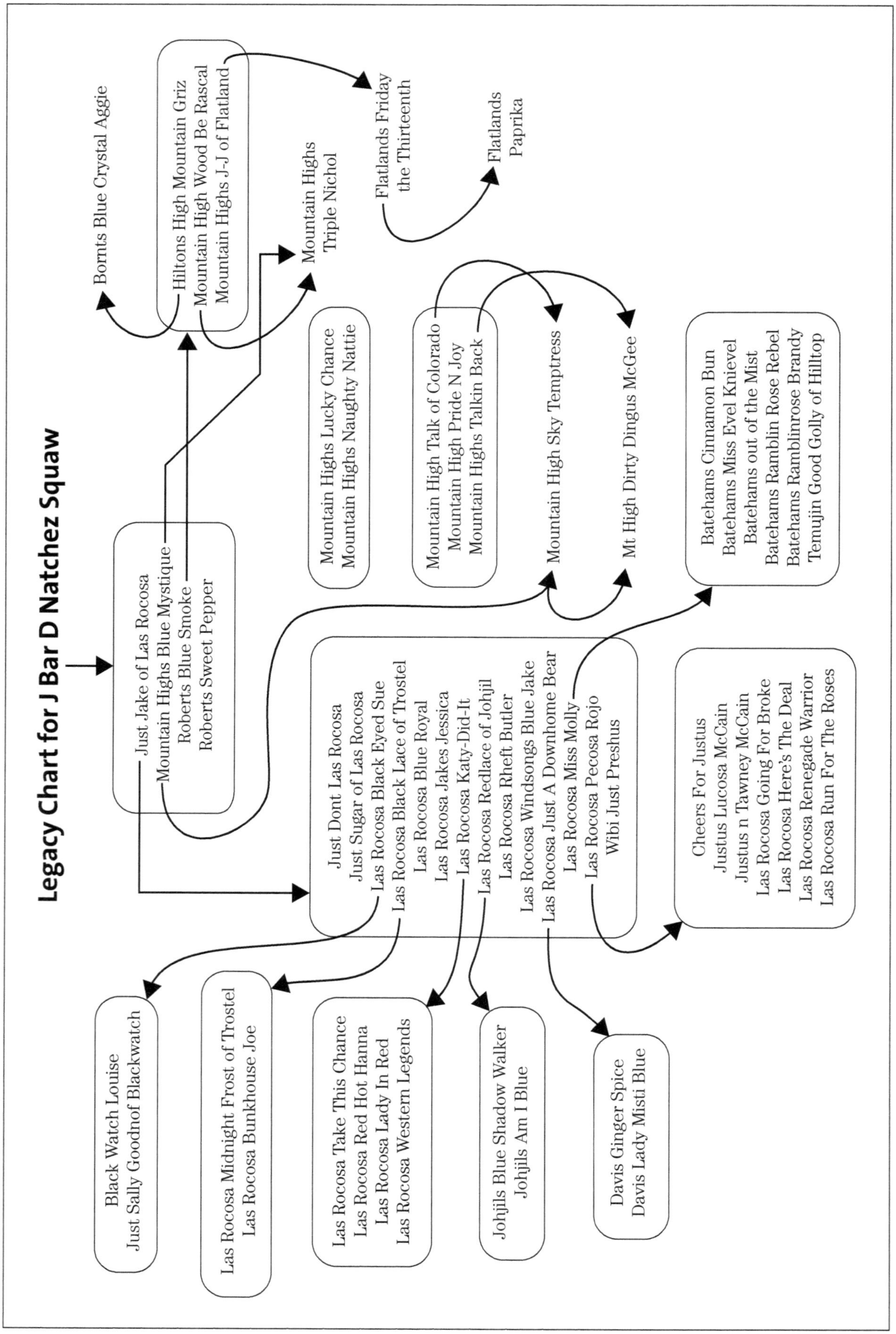
Legacy Chart for J Bar D Natchez Squaw
Just Jake of Las Rocosa
Mountain Highs Blue Mystique
Roberts Blue Smoke
Roberts Sweet Pepper
Bornts Blue Crystal Aggie
Hiltons High Mountain Griz
Mountain High Wood Be Rascal
Mountain Highs J-J of Flatland
Mountain Highs Triple Nichol
Flatlands Friday the Thirteenth
Flatlands Paprika
Mountain Highs Lucky Chance
Mountain Highs Naughty Nattie
Mountain High Talk of Colorado
Mountain High Pride N Joy
Mountain Highs Talkin Back
Mountain High Sky Temptress
Mt High Dirty Dingus McGee
Batehams Cinnamon Bun
Batehams Miss Evel Knievel
Batehams out of the Mist
Batehams Ramblin Rose Rebel
Batehams Ramblinrose Brandy
Temujin Good Golly of Hilltop
Just Dont Las Rocosa
Just Sugar of Las Rocosa
Las Rocosa Black Eyed Sue
Las Rocosa Black Lace of Trostel
Las Rocosa Blue Royal
Las Rocosa Jakes Jessica
Las Rocosa Katy-Did-It
Las Rocosa Redlace of Johjil
Las Rocosa Rheft Butler
Las Rocosa Windsongs Blue Jake
Las Rocosa Just A Downhome Bear
Las Rocosa Miss Molly
Las Rocosa Pecosa Rojo
Wibi Just Preshus
Cheers For Justus
Justus Lucosa McCain
Justus n Tawney McCain
Las Rocosa Going For Broke
Las Rocosa Here's The Deal
Las Rocosa Renegade Warrior
Las Rocosa Run For The Roses
Black Watch Louise
Just Sally Goodnof Blackwatch
Las Rocosa Midnight Frost of Trostel
Las Rocosa Bunkhouse Joe
Las Rocosa Take This Chance
Las Rocosa Red Hot Hanna
Las Rocosa Lady In Red
Las Rocosa Western Legends
Johjils Blue Shadow Walker
Johjils Am I Blue
Davis Ginger Spice
Davis Lady Misti Blue

Commanche as a youngster. Courtesy of Rowe

J Bar D Choctaw Chilli, legendary dam of many breed greats. Courtesy of Rowe

Rowe's Commanche Warrior, J Bar D Shoshone and Sharon Rowe on an Aussie club outing in the mountains. Courtesy of Rowe

Kindler's Rowdy of J Bar D (J Bar D Briggs' Chisum x J Bar D Choctaw Chilli). Courtesy of Rowe

Mahana's Barney, a Commanche son out of Merz's Genevieve, a female of unknown pedigree. Courtesy of Rowe

J Bar D Shoshone Squaw, (J Bar D Briggs' Chisum x J Bar D Choctaw Chilli) red merle female. Courtesy of Rowe

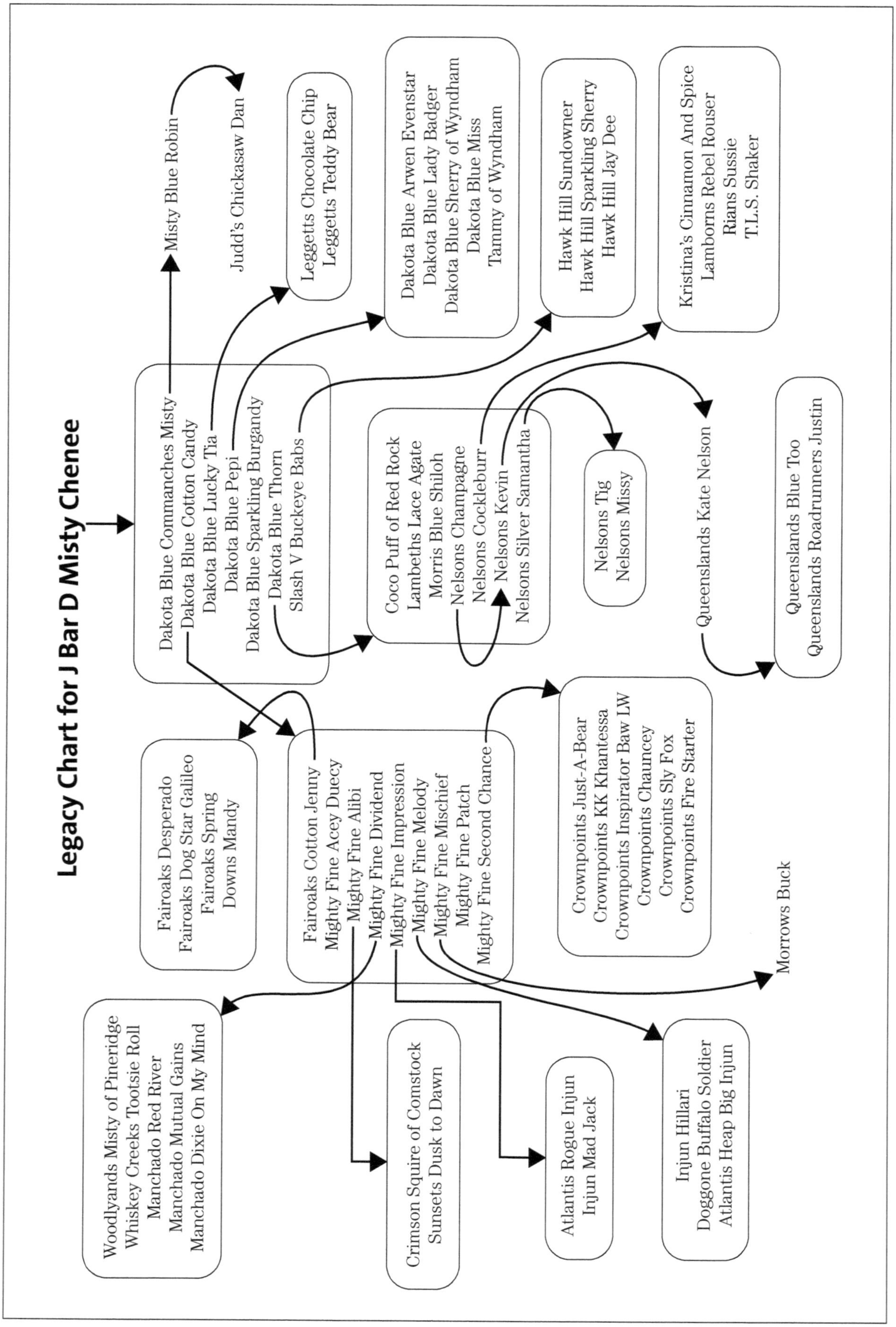
Legacy Chart for J Bar D Misty Chenee
Dakota Blue Commanches Misty
Dakota Blue Cotton Candy
Dakota Blue Lucky Tia
Dakota Blue Pepi
Dakota Blue Sparkling Burgandy
Dakota Blue Thorn
Slash V Buckeye Babs
Misty Blue Robin
Judd's Chickasaw Dan
Leggetts Chocolate Chip
Leggetts Teddy Bear
Dakota Blue Arwen Evenstar
Dakota Blue Lady Badger
Dakota Blue Sherry of Wyndham
Dakota Blue Miss
Tammy of Wyndham
Hawk Hill Sundowner
Hawk Hill Sparkling Sherry
Hawk Hill Jay Dee
Kristina's Cinnamon And Spice
Lamborns Rebel Rouser
Rians Sussie
T.L.S. Shaker
Coco Puff of Red Rock
Lambeths Lace Agate
Morris Blue Shiloh
Nelsons Champagne
Nelsons Cockleburr
Nelsons Kevin
Nelsons Silver Samantha
Nelsons Tig
Nelsons Missy
Queenslands Kate Nelson
Queenslands Blue Too
Queenslands Roadrunners Justin
Fairoaks Desperado
Fairoaks Dog Star Galileo
Fairoaks Spring
Downs Mandy
Fairoaks Cotton Jenny
Mighty Fine Acey Duecy
Mighty Fine Alibi
Mighty Fine Dividend
Mighty Fine Impression
Mighty Fine Melody
Mighty Fine Mischief
Mighty Fine Patch
Mighty Fine Second Chance
Crownpoints Just-A-Bear
Crownpoints KK Khantessa
Crownpoints Inspirator Baw LW
Crownpoints Chauncey
Crownpoints Sly Fox
Crownpoints Fire Starter
Morrows Buck
Woodlyands Misty of Pineridge
Whiskey Creeks Tootsie Roll
Manchado Red River
Manchado Mutual Gains
Manchado Dixie On My Mind
Crimson Squire of Comstock
Sunsets Dusk to Dawn
Atlantis Rogue Injun
Injun Mad Jack
Injun Hillari
Doggone Buffalo Soldier
Atlantis Heap Big Injun

Rhynard's Tammys Patience, blue and white female (Rhynard's Blue Smokey Sky x Rhynard's Patchie). Rhynard's Blue Smokey Sky was a Clinger's Royal Blue son out of Clinger's Ely Blue Queenie (Clinger's Royal Blue x Shockley's Blue Samantha—a Olson's Clare Blue Chip's daughter). Very typical mountain bred bitch. Courtesy of Rowe

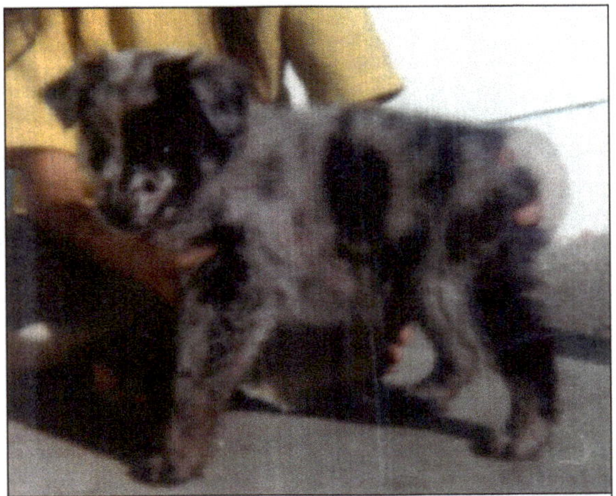

J Bar D Shoshone Squaw, red merle female Courtesy of Rowe

Nordstrom's Tammy, an unregistered ranch dog of unknown pedigree but excellent Aussie type, dam of the first litter by Rowe's Commanche Warrior. Courtesy of Rowe

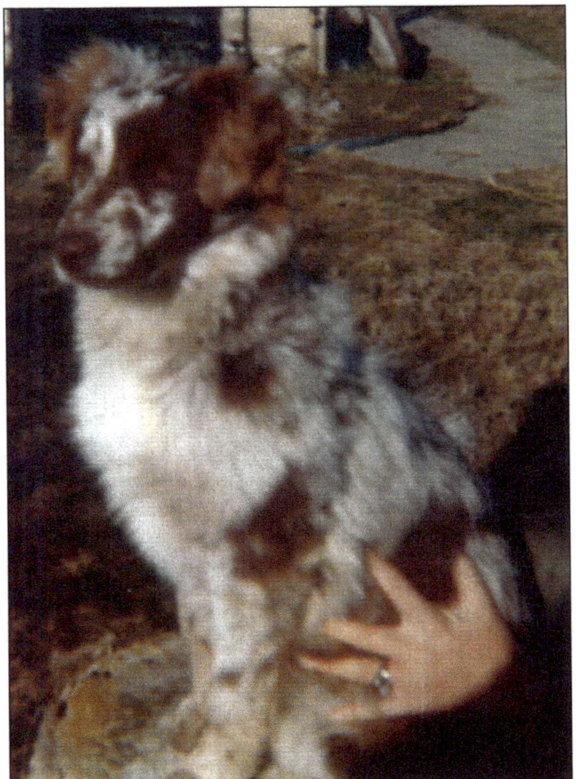

J Bar D Shoshone Squaw, red merle. Courtesy of Rowe

J Bar D Cree Squaw, blue and white female full sister to Rowe's Hondo Warrior (McGill's Slagle Blue x J Bar D Choctaw Chilli). Courtesy of Rowe

Rowe's Apache, blue and white male—pick of Bum's 1st litter, July 1968, (Rowe's Commanche Warrior x Nordstrom's Tammy). Courtesy of Rowe

Mahana's Barney as an adult. Courtesy of Rowe

Roadrunner's R-Lucky Chipawa. Courtesy of Rowe

J Bar D Duke (J Bar D Red Chief x Rowe's Cheyanne Squaw). Courtesy of Rowe

Rowe's Cherokee Squaw. Out of Dingo's first litter. Sharon teaching her to jump thru her arms. Courtesy of Rowe

J Bar D Noble's Red (J Bar D Briggs' Chisum x J Bar D Choctaw Chilli). Briggs' Chisum was a J Bar D Red Chief son. Courtesy of Rowe

J Bar D Zipper son of J Bar D Red Chief and Rowe's Cheyanne Squaw. Courtesy of Rowe

J Bar D Shawnee Squaw (Rowe's Commanche Warrior x Merz's Genevieve). Courtesy of Rowe

J Bar D Apache Warrior. A top Commanche son. Courtesy of Rowe

J Bar D Red Chief after one of his many wins. Courtesy of Rowe

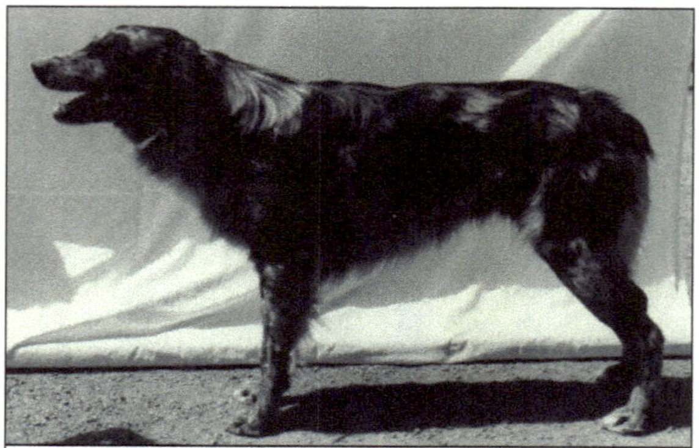

Rowe's Apache Warrior. Pick of litter from Nordstrom's Tammy. Courtesy of Rowe

J Bar D Bookcliff's Strawboss, red merle, copper and white male son of J Bar D Red Chief x J Bar D Paiute Red (Rowe's Hondo Warrior x Rector's Samantha Ann). Courtesy of Rowe

Roadrunner's R-Lucky Chipawa (J Bar D Red Chief x Roadrunner's Secoya) Blue and white male. Courtesy of Rowe

J Bar D Crow Chief, the stunning solid black male who was something of an oddity in his day. Although solid black dogs occurred in litters, blue merles were more popular and it was rare to see the solid dogs in the show ring. Courtesy of Rowe

Rowe's Commanche Warrior working sheep in the quiet, authoritative style he was known for. Courtesy of Rowe

J Bar D Chipawa. Courtesy of Rowe

J Bar D Chipawa. Courtesy of Rowe

Ronsley's Johnny Ringo, one of the few remaining Ritter bred dogs owned by the Ronsleys. Courtesy of Rowe

J Bar D Echo, a nice working dog by Commanche x Anderson's Blu LaKita. Courtesy of Rowe

Commanche was a winner in all venues, including conformation. Aussies were expected to be good at everything they tried in the beginning of the breed. Courtesy of Rowe

The award winning Commanche with some of his many trophies. Courtesy of Rowe

Commanche was a tough cow dog when needed. Courtesy of Rowe

Jerry and Commanche working ducks during an early demo. Courtesy of Rowe

Silvercast Rumor of J Bar D, blue, copper and white male (Rowe's Commanche Warrior x Silvercast Apache Blue). His dam was a daughter of Ch Moeller's Blue Chip CD, a dog of unknown background but very typical in looks of the mountain dogs. Courtesy of Rowe

Tegtmeyer's Jeranomo of J Bar D (Rowe's Hondo Warrior x J Bar D Jamie) Jamie was also a Hondo daughter. Courtesy of Rowe

Commanche was a top trial dog in any age. Courtesy of Rowe

Tegtmeyer's Jeranomo of J Bar D. Courtesy of Rowe

Aussies were used for any job that presented itself. Working chutes and alleys were all in a days work. Commanche was good at everything. Courtesy of Rowe

Commanche showing off a room full of his winnings. Courtesy of Rowe

Commanche with some of his many awards. Courtesy of Rowe

A mature and well-oiled livestock working team—Commanche and Jerry Rowe. Courtesy of Rowe

The iconic Commanche looking remarkably like his sire, Sorensen's Gunsmoke. Courtesy of Rowe

Harshbarger's Tesque of J Bar D, blue and white female (Rowe's Commanche Warrior x J Bar D Choctaw Chilli). Courtesy of Rowe

Kelly's Blue Rockette, (Harper's Corky x Brown's Clarida Violet) blue and white female and winner of many obedience awards. Courtesy of Rowe

The beautiful and remarkably intelligent J Bar D Chinook Squaw with Terry Rowe and Judge Nan Boerup. Courtesy of Rowe

The Commanche and Jerry team with another win. Courtesy of Rowe

Three top dogs. Rowe's Cheyanne Squaw (center) her dam—Kelly's Blue Rockette (left) and her daughter J Bar D Chinook Squaw (right). Courtesy of Rowe

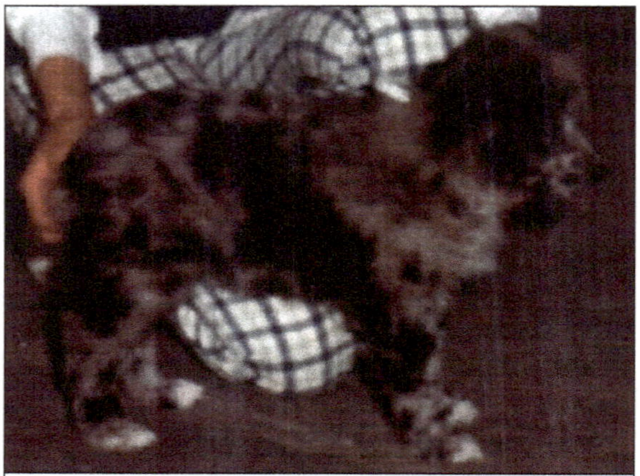

Tyler Nyki of J Bar D (Rowe's Hondo Warrior x J Bar D Jamie). Courtesy of Rowe

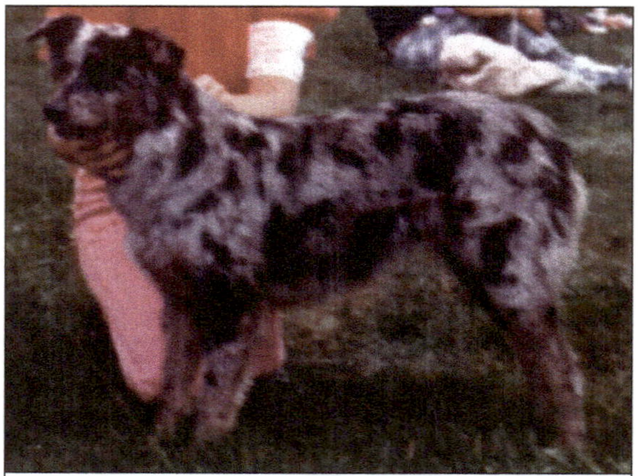

J Bar D Chinook Squaw (Roadrunner's R Lucky Chipawa x Rowe's Cheyenne Squaw). Courtesy of Rowe

J Bar D Cimmeron Warrior (J Bar D Cherokee Red x Millers Bandit). Courtesy of Rowe

Rowe's Cheyanne Squaw. Courtesy of Rowe

Kelly's Blue Rockette. Courtesy of Rowe

Left to Right, Rowe's Dingo, Tina, and Yakima. Courtesy of Rowe

The inbred Hondo son/grandson Tegtmeyer's Jeranomo of J Bar D working sheep in the same quiet manner the Rowe's strove for in their breeding. Courtesy of Rowe

J Bar D Zuni Squaw, red merle female, (Rowe's Hondo Warrior x Rector's Samantha Ann). Courtesy of Rowe

J Bar D Apache Warrior. (Rowe's Commanche Warrior x J Bar D Shawnee Squaw—by Commanche x Merz's Genevieve) Courtesy of Rowe

Rowe's Commanche Warrior. Courtesy of Rowe

Shannahan's Phantom, a nice all around dog typical of dogs of that era. Courtesy of Rowe

Shannahan's Phantom working calmly and quietly on sheep. Courtesy of Rowe

Shannahan's Phantom, a good working bitch who produced well when bred to Commanche. Courtesy of Rowe

Rowe's Commanche Warrior taking charge of a flock. Courtesy of Rowe

Shannahan's Phantom, (Deine's Spark x Deine's Lucky Lady) blue merle female. Bred to Commanche she produced Stonehenge Phantasy, an important breeding female. Courtesy of Rowe

Chipawa, blue and white daughter of Yakima and Sorensen's Charlie. Born in 1966, Sharon bred her once and and every puppy she produced had multiple dewclaws on every leg—which they called cluster dewclaws. This is an example of just one of many oddities which could appear in the early Aussies due to their varied and unknown background. Courtesy of Rowe

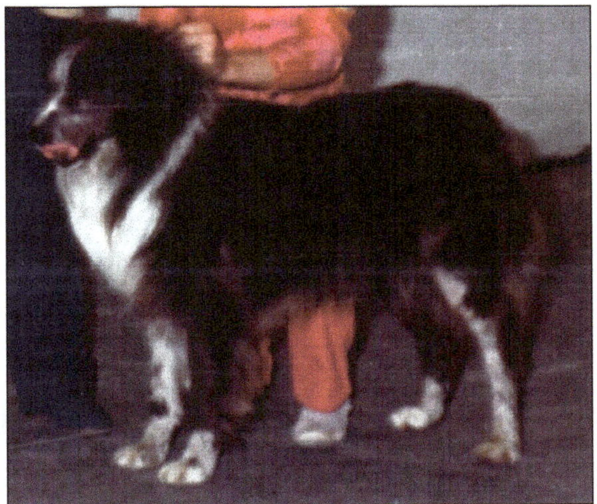

J Bar D Red Chief. Courtesy of Rowe

1973 puppy by Roadrunner's R-Lucky Chipawa x J Bar D Arapahoe Lady. Courtesy of Rowe

Seminole Warrior of J Bar D, blue merle male son of J Bar D Red Chief. Courtesy of Rowe

J Bar D Arapahoe Lady. Courtesy of Rowe

A painting of the great Rowe's Commanche Warrior with plaques honoring the titles he won. Painted in August of 1975 by Stiefel, this work hangs in the Rowe's clubhouse at their Kansas ranch and was a welcome gift from Frank and Becky Anderson (Powers and Lor A kennels). Courtesy of Rowe

J Bar D Arapahoe Lady. Courtesy of Rowe

J Bar D Keener's Blue, dark blue merle male by Mahana's Barney, out of Peterson's Missy of J Bar D (Roadrunner's R-Lucky Chipawa x Rowe's Cheyanne Squaw). Courtesy of Rowe

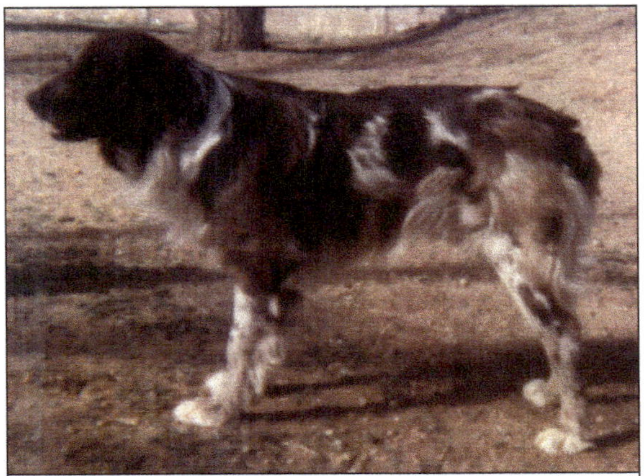

Rowe's Warrior Bow. Courtesy of Rowe

Rowe's Warrior Bow. Courtesy of Rowe

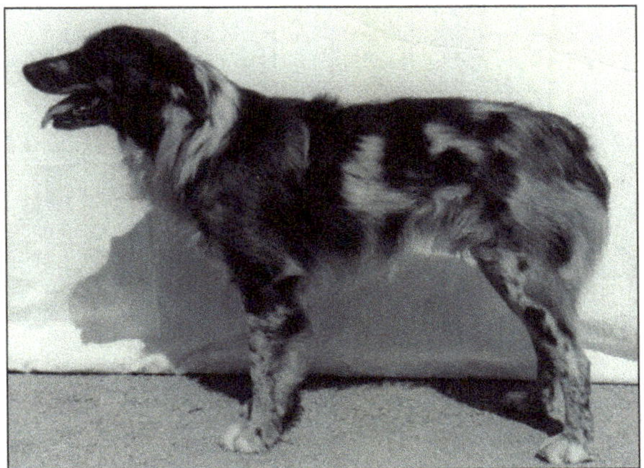

Rowe's Warrior Bow. Courtesy of Rowe

Rowe's Warrior Bow. Courtesy of Rowe

Shannahan's Phantom. Courtesy of Rowe

J Bar D Brandy. Courtesy of Rowe

Mary's Rocky Mountain Babe, red and white female and founder of a number of great dogs on the west coast. Courtesy of Rowe

Rowe's Warrior Bow, stunning even as a young dog. Courtesy of Rowe

Rowe's Warrior Bow. Courtesy of Rowe

Green's Mr Cobar working cattle at home. Courtesy of Rowe

Green's Mr Cobar, called "Charlie" by those who knew him. Courtesy of Sorensen

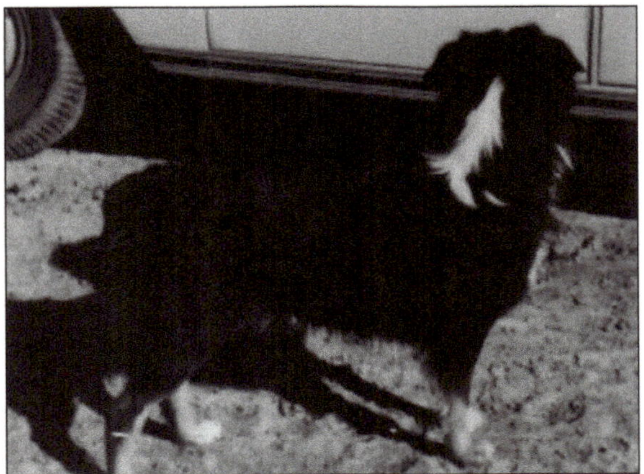

Green's Mr Cobar, a black tri dog of good type and structure. Courtesy of Sorensen

J Bar D Brandy. Courtesy of Rowe

Mr Cobar daughter Harshbarger's Tesque of J Bar D. Courtesy of Rowe

Roadrunner's Cisne Rojo, (Roadrunner's Picacho x Howe's Little Peedie). Courtesy of Rowe

Harshbarger's Tesque of J Bar D. Courtesy of Rowe

Green's Mr Cobar. Courtesy of Rowe

Green's Mr Cobar. Courtesy of Rowe

Harshbarger's Tesque of J Bar D. Courtesy of Rowe

A blue merle daughter of Green's Mr Cobar and Harshbarger's Tesque of J Bar D. Courtesy of Rowe

Roadrunner's Picacho, sable grandson of Farned's Dead Eye and McConkey's California breeding. Very influential dog in the J Bar D breeding program. Courtesy of Rowe

J Bar D Hopi Squaw (Roadrunners Picacho x J Bar D Choctaw Chilli). Courtesy of Rowe

J Bar D Hopi Squaw, blue merle and white female. Courtesy of Rowe

J Bar D Hopi Squaw and Roadrunner's "Oso." Courtesy of Rowe

Hopi and Oso. Models who were used to illustrate *Pups 'N Stuff*—the book authored by Sharon Rowe and Sue Dennis. Courtesy of Rowe

J Bar D Natchez Squaw, (Rowe's Commanche Warrior x J Bar D Hopi Squaw), blue merle female as a youngster (February 1973). A very influential dog in a number of kennels. Courtesy of Rowe

J Bar D Hopi Squaw as an award winning puppy. Courtesy of Rowe

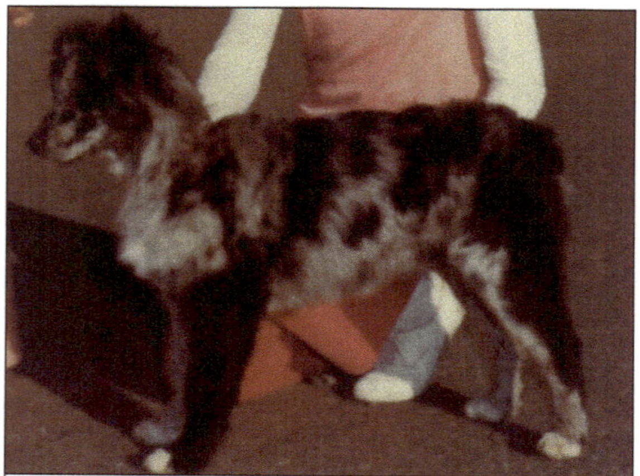

J Bar D Natchez Squaw. Courtesy of Rowe

J Bar D Natchez Squaw. Courtesy of Rowe

Walker's Joshua, blue merle male who was influential in many west coast breeding programs and a top working dog in his own right. Courtesy of Sorensen

J Bar D Natchez Squaw. Handled by Dena Harshbarger at a local show with Nan Boerup. Courtesy of Rowe

Sanger's Smokem of J Bar D (J Bar D Skippy x J Bar D Natchez Squaw) as a 6 month old male. Courtesy of Rowe

Rowe's Cherokee Squaw (female from Rowe's Dingo and Sorensen's Gunsmoke first litter). Courtesy of Rowe

J Bar D Aztec "TEX" (Rowe's Commanche Warrior x J Bar D Natchez Squaw). Courtesy of Rowe

J Bar D Arrow Warrior, blue and white male (Rowe's Commanche Warrior x J Bar D Natchez Squaw). Courtesy of Rowe

J Bar D Sundance grandson J Bar D Red Chief. Courtesy of Rowe

J Bar D Acoma Squaw (Rowe's Commanche Warrior x J Bar D Natchez Squaw). Courtesy of Rowe

J Bar D Chinook Squaw. Courtesy of Rowe

J Bar D Chinook Squaw. Courtesy of Rowe

J Bar D Chinook Squaw heading cattle during a trial. Courtesy of Rowe

J Bar D Chinook Squaw working sheep in the same quiet, authoritative manner as her grandsire Commanche. Courtesy of Rowe

J Bar D Chinook Squaw with some of her winnings. Courtesy of Rowe

J Bar D Kickapoo Squaw. Courtesy of Rowe

J Bar D Skippy (Kenner's Blue of J Bar D x Leavitt's Patches). Courtesy of Rowe

J Bar D Chinook Squaw and Pat Gordon prepare for one of their many successful dog shows. Courtesy of Rowe

J Bar D Mystic Warrior as an older dog in June 2000. Photo by Author

J Bar D Kickapoo Squaw (Queensland's Coondog Jimmy Blue x Queensland's Tequila) Blue merle female of McConkey's and old California breeding. Courtesy of Rowe

J Bar D Mystic Warrior "Mikey" with the typical old fashioned mountain dog head type. Photo by Author

The kind and happy expression that reflected J Bar D Mystic Warrior's friendly temperament. Photo by Author

J Bar D Kickapoo Squaw as a puppy. Courtesy of Rowe

Queensland's Tequila, dam of J Bar D Kickapoo Squaw. (Bradleys Joker x Queensland's Querida de Picacho—Querida was by Roadrunner's Picacho x J Bar D Fancy and Joker was by Ubrun's Johnny x Ubrun's Blackie). Courtesy of Rowe

King's Boo in 1972. Courtesy of Rowe

J Bar D Kickapoo Squaw. Courtesy of Rowe

J Bar D Smoke Signal showing his working talent at a young age. (Queensland's Wibi Windwalker x Wibi Black Velvet) He went on to have a successful trial career with owner Terri Daniel. Courtesy of Rowe

1972 photo of Kings Boo, a very influential West Coast sire with strong ties to the old Colorado dogs. His daughter and granddaughter, Faxon's Smokey Blue Buttons, was the dam of the Sorensen's Colorado Hotline Paige. He was also the sire of Queensland's Coondog Jimmy Blue, who sired the Rowe's J Bar D Kickapoo Squaw. Courtesy of Rowe

J Bar D Kickapoo Squaw. Courtesy of Rowe

J Bar D Dakota (Anderson's [Lor A] Captain B Teddy Blue x J Bar D Kickapoo Squaw). Courtesy of Rowe

J Bar D Kickapoo Squaw. Courtesy of Rowe

Mikey was a good working dog, especially on sheep. A lesson Jerry learned from this dog was his ability to draw sheep toward him as well as push them away. Mikey was a master at this technique. Courtesy of Michelle Murvai

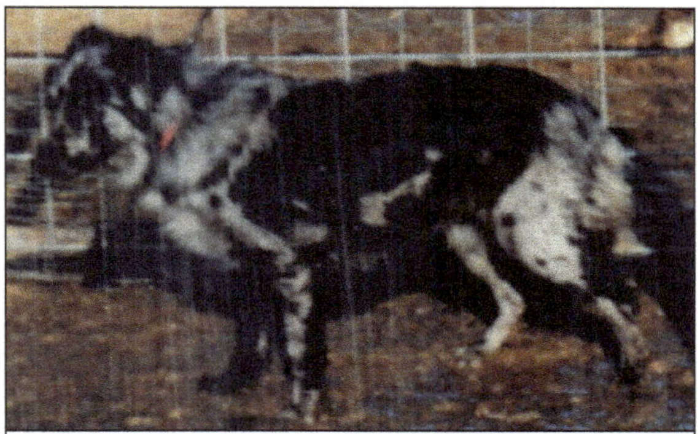

Queensland's J Bar D Little Bit, (Queensland's Bear Paws x Ruff N Reddy Too Small Jones) blue merle female. Courtesy of Rowe

J Bar D Tomahawk, (Wibi Crystal Blue Smoke x Wyoming Gypsy of Mt High), blue merle male. Courtesy of Rowe

J Bar D Mystic Warrior ("Mikey") in June 2000 at a working trial. Photo by Author

Queensland's Wibi Windwalker as a young dog. (Queensland's Be ar Paws x Queensland's Cahuilla Cayenne). Courtesy of Rowe

Queensland's Wibi Windwalker. Courtesy of Rowe

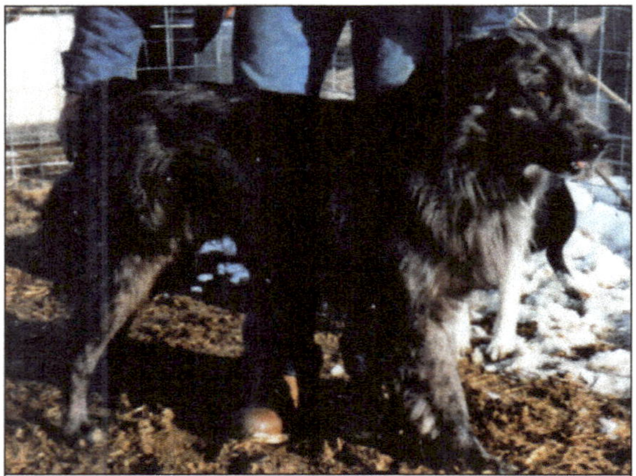

Queensland's Wibi Windwalker. Courtesy of Rowe

Choc son of Rowe's Commanche Warrior and Harshbarger's Tesque of J Bar D. Courtesy of Rowe

J Bar D Blue Smoke. Courtesy of Rowe

J Bar D Smoke Signal. Courtesy of Rowe

Sharon and J Bar D Chinook Squaw, 1st place Sheep Class, CASA Spring Specialty, May 1974. Courtesy of Rowe

J Bar D Blue Smoke at about 1 year of age. Courtesy of Rowe

J Bar D Smoke Signal, blue merle male (Queensland's Wibi Windwalker x Wibi Black Velvet). Courtesy of Rowe

Chevreherd Smokin Flintlock in the prime of life—2009. Photo by Author

Mystiques Smokin Tomahawk, blue and white male (J Bar D Smoke Signal x Rebel's Mystique Bear Paws) a classic example of the old Queensland's dogs descending from Commanche and Reddy Teddy. Photo by Author

Mystiques Smokin Tomahawk had the perfect Aussie head profile. Photo by Author

Mystiques Smokin Tomahawk showing his beautiful blocky, square head. The type that marked the old foundation mountain-type dog. Photo by Author

Mystiques Smokin Tomahawk as an older dog. Photo by Author

Sharon with J Bar D Mystic Warrior HS STDcd OTDs

Chevreherd Smokin Flintlock, blue and white male (J Bar D Smoke Signal x Chevreherd Cowhock Nibbler—a Jones Reddy Teddy/Hemi/Vest cross female). Photo by Author

Jerry Rowe and Commanche, Kathy Jones and Reddy Teddy, Colorado, 1976. Courtesy of Rowe

Jerry and Sharon Rowe. Photo by Danita Jones

chapter 16

Tom and Jean Taylor
SILVERLEDGE

Unlike the majority of "Woods" breeders, Tom and Jean Taylor of Silverledge fame were headquartered on the east coast of the United States in Pennsylvania. In the early 1970s, Tom was a western horse trainer, saddlemaker, and silversmith who traveled the country showing and Jean managed a 50 horse stable while he was away. In those days everybody had a "horse show dog" and the Taylors were no different. Their first Aussie was a red merle puppy bitch Tom had picked up in Iowa on a horse buying trip. It was uncertain whether she was purebred but she was enough Aussie to convince the Taylor's they never wanted to be without one. This "little doll" as Jean called her was named Skunk. She was killed when she was about six months old so was never bred and left a hole in the hearts of the Taylors.

In search of something to fill that hole, Tom set out on his next horse trading trip and returned with another female Aussie. Buffie came from an Iowa horse ranch and was from the local type of "farm dogs" found all over the country. Like their first, she was unregistered but unlike their first they knew she was purebred. She was soon registered as Taylor's Silver Buffie through the ASCA Hardship registry program. Her "tryout" for a registration number consisted of a home movie submitted to the ASCA viewing committee. Working ability was the primary criteria at that time so the Taylors staked a neighbor's calf to a post and captured Buffie heeling him for the camera!

After acquiring Taylor's Luke the Drifter, the red merle Buffie was bred to him and produced an outstanding red merle male named Walter's Colorado Cash. He was later bred by Leslie Sorensen to a Sorensen's Red Man daughter to produce the top foundation matron for the Powder River kennel, Ch Colorado's Ravishing Ruby CDX STDc ATDds. This same cross produced McCormick's Colorado Ona, who was the dam of Pincie Creek's original founding sire, McCormick's Smokey Seminole.

Buffie was eventually traded to Leslie Sorensen who was looking for some good quality red dogs, which were not abundant at the time. Buffie had just come out of heat when she left for Colorado and they didn't think she was bred but she surprised Leslie two months later with a litter out of Luke. Tom informed Leslie she owed him a stud fee so she sent as payment a nice puppy which soon became one of Jean's favorite dogs, Taylor's Calamity Jane CD. The Luke and Buffie litter produced a beautiful dog named Ch. Colorado Sunset.

Buffie was used judiciously by Leslie with Sorensen's Red Man, a Red Man son, a Gunsmoke son and Slash V Little Rock to produce a lasting legacy of top dogs. One daughter, Mini Acre Wooden Nickel, was an important founding female for the Mini Acre kennel of Ken Claussen. Other offspring became the foundation for kennels in the Midwest and California—her granddaughter Mary's Rocky Mountain Babe was a tremendous producer for Betty Nelson when crossed to J Bar D Cimmeron Warrior. A magnificent male from this cross, Walker's Joshua, was an influential working sire on the west coast for many years as well as a top working dog in his own right.

The Taylor's next Aussie, another red merle, launched their serious involvement with the breed and became an influential sire for them and eventually for the Sorensens in Colorado.

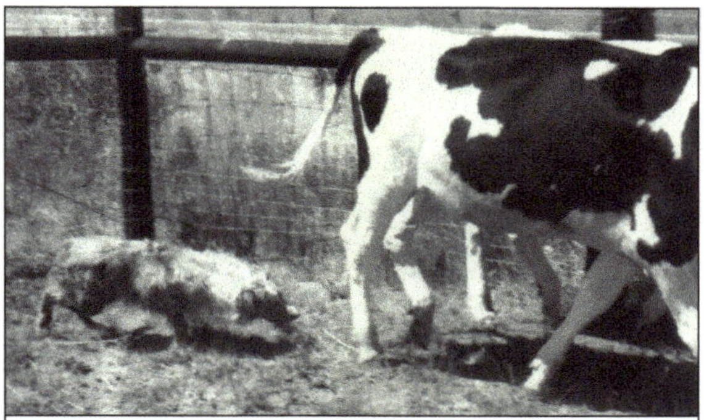

Taylor's Silver Buffie showing her natural heeling style. Courtesy of Sorensen

Taylor's Red Wilbur, aka Sorensen's Red Man on the left and a young Taylor's Silver Buffie on the right. Courtesy of Jean Taylor

Taylor's (Colorado's) Silver Buffie, one of the first red merle females at Colorado kennel and producer of legions of top working dogs. Courtesy of Sorensen

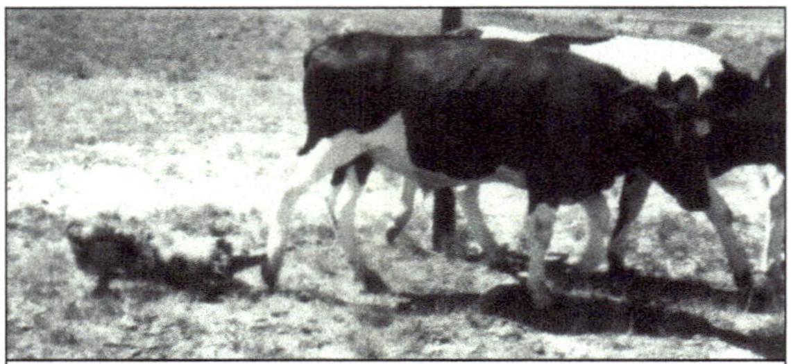

Taylor's Silver Buffie became a useful working dog after she moved to Colorado kennel. Courtesy of Sorensen

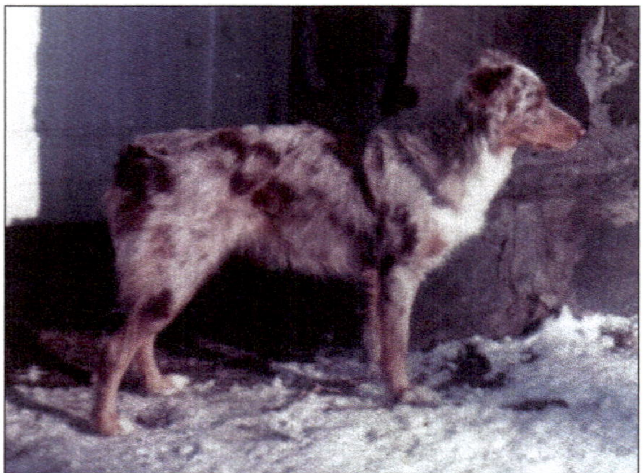

Taylor's Silver Buffie in Pennsylvania. Courtesy of Jean Taylor

Mini Acre Wooden Nickel (Colorados Joshua x Taylor's Silver Buffie), blue merle female who later became the great granddam of the popular sire, Ch Zeller's Winterhaven Rusty. Courtesy of Sorensen

He was first known as Taylor's Red Wilbur but became more famous after he found his way to Colorado and became Sorensen's Red Man.

"Willie," as he was known, came from a barrel racer by the name of Ardith Bruce. She was headquartered in Utah but traveled the country competing at horse shows and rodeos. Her family ranched in Utah and raised Aussies from local dogs. Willie was about a year old when Tom found him and he was not registered, like many of the dogs in those early days. Jean describes him as a "cute little devil" and doesn't know why Ardith let him go but Tom was a skilled trader who "could talk the stars out of heaven if he wanted something." Apparently he took a liking to Willie and brought him home in about 1971.

Jean got Willie registered with the Bruce's dogs listed as parents. In later years, after seeing Taylor's Whiskey she became convinced it was likely Willie was related to the early dogs of Joe Taylor. He had an uncanny resemblance to Whiskey and even shared some behavioral traits with him. The physical traits he had in common with Whiskey were size—about 35-40 pounds, type—a very tight topline, good angulation, tight feet, and a pretty head. It was the head most of all which led Jean to believe there must be some relationship between the two. They had the same squint of the eyes, wide topskull, and expression. Jean swears they could have been twins based on this alone. They shared a dark liver coloration that was the icing on the cake. It all added up to too much similarity to be a coincidence. The Aussie wasn't very numerous at that time and not many of those dogs of that day shared this type. Add to that the proximity of the Bruce's to the Taylor ranch and it becomes even more likely.

Most of Willie's story is told in the chapter on the Sorensens Colorado kennel because he was used most extensively there and it was there he achieved his greatest breeding success as Sorensen's Red Man. Between the get he left behind on the east coast and the dogs he produced in Colorado, Willie's influence was felt in kennels all over the United States.

About a year after bringing home Willie, Tom made a stop at the Sorensens to look at their dogs. Jean had been corresponding with them because she and Tom had admired the photos of their dogs which were appearing in the *Aussie Times*. Leslie had admired photos of Willie and he rode along with Tom when he went to Phoenix to deliver a horse. On the way home he detoured to the Sorensens to see what they had. Sorensens lived up in the mountains near Toponas, Colorado, at that time and Tom just intended to go for a look at the dogs. Leslie happened to have a big dark blue rangy male who eventually was registered as Taylor's Luke the Drifter and Tom fell in love with him at first sight. The stunning 9 month old blue bitch who became Taylor's Blue Prissy was also in residence and Tom must have seen something he liked in her, too, because he traded something for the pair and they were soon on their way to Pennsylvania. Luke's dam and Prissy's sire were full brother and sister owned by Polly Bjork who was also an early breeder of the mountain-type Aussie.

This foundation could not have been a better start in Aussies. Jean wasn't looking for a particular type of dog at that time and when Tom showed up with Luke and Prissy they weren't like anything Jean had ever seen before. At about a year of age, Luke was already 21-1/2 inches tall, rangy and immature. Bred by Polly Bjork, Luke was born in March of 1971 and was by the great Wood's Stubbie and out of Bjork's Chula. Chula was by Bjork's Baron, a classically bred Colorado dog by Olson's Clare Blue Chips. Blue Chips was out of Wood's Little Kola by Wood's Adobie and his dam was a dog

The red merle male, Taylor's Red Wilbur "Willie," later made famous under the name of Sorensen's Red Man. Courtesy of Jean Taylor

June 1980 photo of a mature Taylor's Luke the Drifter CDX. Courtesy of Jean Taylor

Taylor's Luke the Drifter CDX in June of 1980. Courtesy of Jean Taylor

PEDIGREE OF Taylor's Luke the Drifter

Born: ________ Sex: ________
Owned by: ________

Parents	Grandparents	Great-grandparents	Great-great-grandparents
Sire Woods Stubbie	Sire Woods Jay	Sire Sislers Shorty	Sire Keno
			Dam Blue Star
		Dam Sislers Trixie	Sire Blue Khan
			Dam Emmetts Blue
	Dam Woods Slate	Sire Woods Jay	Sire Sislers Shorty
			Dam Sislers Trixie
		Dam Woods Specks	Sire Woods Jay
			Dam Trumps Stubbie
Dam Bjorks Chula	Sire Bjorks Baron	Sire Olsons Blue Chips	Sire Woods Adobie
			Dam Woods Little Kola
		Dam Rohenbach's Banjo	Sire Thornton's Smokey Blue
			Dam Thornton's Blue Blue
	Dam Roadrunners Cholla	Sire McConkey's Blue Pecos	Sire McConkeys Sir Blue Silver
			Dam McConkeys Sweet Sue
		Dam McConkey's Calico Cameo	Sire Little's Mr. Robert
			Dam Little's Pimenta Chica

Bred by: ________ Address: ________ Phone: ________

The details given above are correct to the best of my knowledge. Signed: ________(Breeder)

of Thornton's breeding, another early Colorado breeder of the mountain-type dog whose dogs also traced back to Wood's Jay. Chula's dam was Roadrunner's Cholla, by a McConkey dog out of a Mr. Roberts dog. This made Cholla a California bred dog of lineage that was a proven niche with the mountain-type dogs.

A young Taylor's Luke the Drifter when he still lived with the Sorensen's. Photo from either late 1971 or early 1972. Courtesy of Sorensen

The dark blue merle female, Taylor's Blue Prissy (DeWitt's Tuffy x J Bar D Fancy) Courtesy of Jean Taylor

Polly Bjork gave Chula's history in a personal letter to Betty Nelson. Polly was active in the early Colorado clubs and showed her dogs like everyone else. Chula had a CD and was trained through Open but was never shown for her CDX. She was shown in conformation as well and Polly proudly claimed her show record as "…2 thirds in Open bitch, 1 second and one first with a Reserve Winners," then she stopped showing her. Shows were not very plentiful and most people had better things to do than show their Aussies! Polly did not live on a ranch so she had to find opportunities for Chula to work. The few times she was able to try her on stock and do a little training she states she was a very good worker. Bred to several good working dogs, she produced puppies who, in Polly's opinion, were good workers "at everything." She did sell a number of puppies to ranchers and says they were "working fools" and "wonderful watch dogs." The title of working dog included much more in those days than just livestock work. She kept a blue male from Chula which she stood at stud for awhile and considered him a good dog because "nobody comes close to our yard."

In Chula's lifetime she had 3 litters with 10 puppies in each litter. She was bred to a blue merle male each time and only had 5 white puppies out of 30. She never had a red puppy but did have a few with copper trim—fairly rare for her lineage. Polly owned a dog named Sage she bred to Chula once. He may have been a Wood dog or sold to Fletcher Wood at some point because she tells that he worked cattle and horses and was a good watch dog for Fletcher.

Luke was one of the last dogs produced by Stubbie and so much like him in appearance he once attracted the attention of Fletcher Wood when Jean was walking him around a horse show grounds in Colorado.

Taylor's Luke the Drifter CDX , easily recognizable in any age as a dog of Wood breeding. By L.M. Gray, Courtesy of Jean Taylor

Blue merle female, Taylor's Calamity Jane CD as a young dog. Courtesy of Jean Taylor

Wood proclaimed him the spittin' image of Stubbie and admired him greatly. Jean remembers there were a number of different types in the Wood line and Luke was of the large, rangy, extremely angulated type with strong, classic head. He was slightly under 22 inches in height at maturity but his deep chest and length held a hefty 70-75 pounds of weight. In typical Wood's blue fashion he was born with a well-defined merle pattern but darkened to almost solid black by the time he was around five years of age. He had very little white in his intensely blue coat and two brown eyes.

Working trials were in their infancy at this time and Jean was not able to travel to trials with Luke so she found a performance outlet for him in the obedience ring. He was Tom's dog until the time Jean decided to get into obedience with him. Luke excelled in obedience and was the type of dog who thrilled judges and audiences alike. His attention was intense and his movements so exceedingly precise he never scored under 196. He never achieved a perfect 200 but came within 1/2 point of that goal on one occasion. His biggest win was an Open A class of 29 entries and he missed out on High In Trial at this show by losing to his son Rocky (Ch Colorado's Shamrock of Silverledge CDX)!

The timeless stories about Luke are best told by Jean. His prowess in obedience would have taken him to the top in another era. As it was, he dominated any competition Jean could find for him at that time. She spent six weeks training Luke for obedience and then went to their first show where Luke was High In Trial! She has this to say about his drive and ability. "Got into obedience after reality forced us to give up the horses. Had a person nearby that had done a lot of obedience, and I had the dogs, and it was a pretty good substitute for showing horses. That's when I took Luke over, and he focused on me. He was an incredibly powerful dog, and an electric presence at my side. We didn't train with treats, we just drilled until he got it right, and his goal in life was to get it right. Part of it was his appearance...big, really dark blue dog...Is he black, or what? What kind of dog IS that? A lot of it was the way he moved, absolutely glued to my side, my slightest move he responded to (bad news when I clutzed). I wasn't aware at the time, but we went to a big "B" Match (AKC, you had to find a show listed and get it approved for ASCA), where there were 36 dogs in Open A, with 3 rings going. I had been warming the dog up outside the rings a little, and they told me later that the other 2 rings stopped to watch his go. He was spectacular. We won the Open A class."

Jean also had Taylor's Calamity Jane entered in Novice B and a 5 month old Luke and Jane pup—Shamrock of Silverledge—entered in Novice Y. This may sound unusual but in those days Jean had been showing horses and thought nothing of just loading up the dogs and heading for a show. No conditioning, no proofing, she just expected her dogs to handle things and, amazingly, they did! She says it this way, "I guess I took superdog for granted. Aren't they all like that?" As if that wasn't enough, not only did Luke take High In Trial that day, but Jane was second in Novice B and the puppy, Rocky, won the Novice Y class! Jean laughs, "There were people going up to the show steward and saying, 'What are these dogs doing here?' They did not appreciate us, because we took all their hardware that day. That was fun!"

Taylor's Luke the Drifter CDX and Jean after one of their many obedience wins. Photo by L.M. Gray. Courtesy of Jean Taylor

The linebred Luke male, Silverledge Slide Me To Da Bar CD STDs as an older dog. Photo by Author

Jean knew Luke best and describes his personality like this "Luke was just a lotta lotta dog. Having been around horses a lot, you sort of judge them the same way. Luke had the kind of personal power that Tigger (his grandson Silverledge Slide Me To Da Bar CD, STDs) had, he could judge a person on sight. Everything about him was quality. Intensity of color, clean feet and ears, good angulation, very deep chest, powerful hindquarters. And that intense focus on you. Always knew exactly where you were—not bothersome, but he knew."

Working trials were still in their infancy and there was virtually no opportunity on the east coast for titling. Nevertheless, Jean did her best to give her dogs a chance to show what they could do. She tells this story about Luke's limited exposure to stock. "At one point in Pennsylvania, I built a duck pen—not any

Silverledge Slide Me To Da Bar CD STDs. Photo by Author

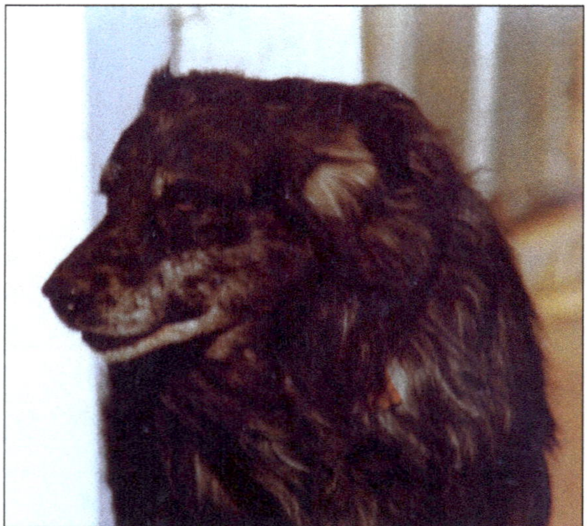

Taylor's Luke the Drifter CDX. Courtesy of Jean Taylor

bigger than a good sized round pen, and got some ducks. He was actually very good with the ducks, just wanted to gather them up, no duck-for-dinner. That was the only time in his lifetime, at least till he was really old, that I had access to stock of any kind. Fletcher Wood mentioned, that time I got to meet him, that one of the things he valued about Stubbie was that he would be gentle with baby calves, even just shoving them with his nose to move them…no force needed."

Luke's intuitive senses were not limited to livestock or obedience training. He had the legendary guardian instincts the Aussie was once known for and Jean tells this story of his judgment in an encounter Tom had. "Tom at one point was teaching students how to make saddles. Had an old factory building rented, there were 6 or 7 students, I think, each with his own work table. Luke had a personal vendetta going with one of them. He would steal his lunch almost every day, and although he was polite enough to him, he made it a point to nip him pretty much every day. His judgment of the guy's character was proven correct a year or so later, when the guy stole a lot of our stuff. Should have listened to the dog. Luke really didn't like people who smoked pot. Must have altered their brain waves, and that made them not okay." Luke always "rode shotgun" on the many trips made to major horse shows with the couple's silver business. A would-be thief once followed their truck and trailer away from the Quarter Horse Congress with the obvious intention of stealing their profits. He never counted on meeting a big blue dog who took his guardian duties very seriously!

Bred to only nine bitches in his lifetime, Luke produced a dynasty of great dogs in all venues of competition. And, true to his bloodline, these dogs proved to be just as great as producers as their sire. An early breeding in 1975 was to the Wood's Stubbie granddaughter Taylor's Calamity Jane CD. This cross produced Ch Colorado's Shamrock of Silverledge CDX and Silverledge Cattle Kate CDX, ATDdsc. Shamrock (Rocky) was a 2-1/2 year old trained obedience dog when he was sold to Leslie Sorensen. Leslie picked him up one weekend and the following weekend she took him to a specialty in Utah. He was entered in Open obedience and won the class and High in Trial—his first time in the open ring! He spent time showing with Leslie's daughter Becky and won several top Junior Showmanship awards with her. With Leslie he proved his worth as a sire, producing a great number of outstanding show and working dogs with her.

Taylor's Shamrock of Silverledge, red merle male later known as Ch. Colorado's Shamrock of Silverledge CDX, at an early show, shown by Jean Taylor. Photo by LM Gray. Courtesy of Jean Taylor

At Jean's, Rocky had been a kennel dog but when he moved into the Sorensen home he became a family dog. Leslie had a chain link fenced yard when Rocky arrived. According to Jean, he had never shown a strong protective instinct while he was with her, but after a few months that all changed. Rocky's close relationship with the Sorensen family brought out his latent protectiveness and he became a serious guard dog when people approached his yard! He still loved people and was a perfect gentleman at all times but he had decided his new job was protecting his new home. He had been part of a group at Jean's and didn't get as much individual attention as he did when he became the special house dog at Sorensens. He was also reaching the normal age of maturity of these dogs—3-4 years old—and reached his potential in life as the special family dog of Dick and Leslie Sorensen.

Rocky had his CD when he came to Colorado but Leslie finished his Championship and CDX. He was a big winner in the 1980s and became a favorite teammate for the Sorensen's daughter, Becky, during her Junior Showmanship career. One year at the ASCA Nationals, the pair won Top Junior Handler! He had the intelligence and steadiness necessary to be a good companion and show dog for a young girl and his name said it all. He was a "rock" when it came to his temperament. Although he showed a lot of talent on livestock, he was never shown in that arena. Jean confesses, "(I had)…obedience trained him so much that I brainwashed him so that he wouldn't take initiative and go away from me. He thought he always had to be at my command, which is detrimental to a stock dog."

Although Rocky was the type of dog Jean was seeking to breed she let him go to Leslie when she and Tom decided to move west. Her efforts

PEDIGREE OF Colorado's (Taylor's) Shamrock of Silverledge

Born: ______ Sex: ______
Owned by: ______

Parents	Grandparents	Great-grandparents	Great-great-grandparents
Sire: Taylor's Luke the Drifter	Sire: Wood's Stubbie	Sire: Woods Jay	Sire: Sislers Shorty
			Dam: Sislers Trixie
		Dam: Woods Slate	Sire: Woods Jay
			Dam: Woods Specks
	Dam: Bjorks Chula	Sire: Bjorks Baron	Sire: Olsons Blue Chips
			Dam: Rohenbach's Banjo
		Dam: Roadrunners Cholla	Sire: McConkey's Blue Pecos
			Dam: McConkey's Calico Cameo
Dam: Taylor's Calamity Jane	Sire: Slash V (George's) Buckeye Bobby	Sire: Tartaglia's Chocolate	Sire: Scaltritti's Dado
			Dam: Scaltritti's Stormy
		Dam: Caligari's Lady	Sire: Tartaglia's Chocolate
			Dam: Duart's Silver
	Dam: Hall's (Sorensens) Apache	Sire: Wood's Stubbie	Sire: Wood's Jay
			Dam: Wood's Slate
		Dam: J Bar D Flirt	Sire: Sorensen's Gunsmoke
			Dam: Jeffrey's Missy Blue

Bred by: ______ Address: ______ Phone: ______

The details given above are correct to the best of my knowledge. Signed: ______(Breeder)

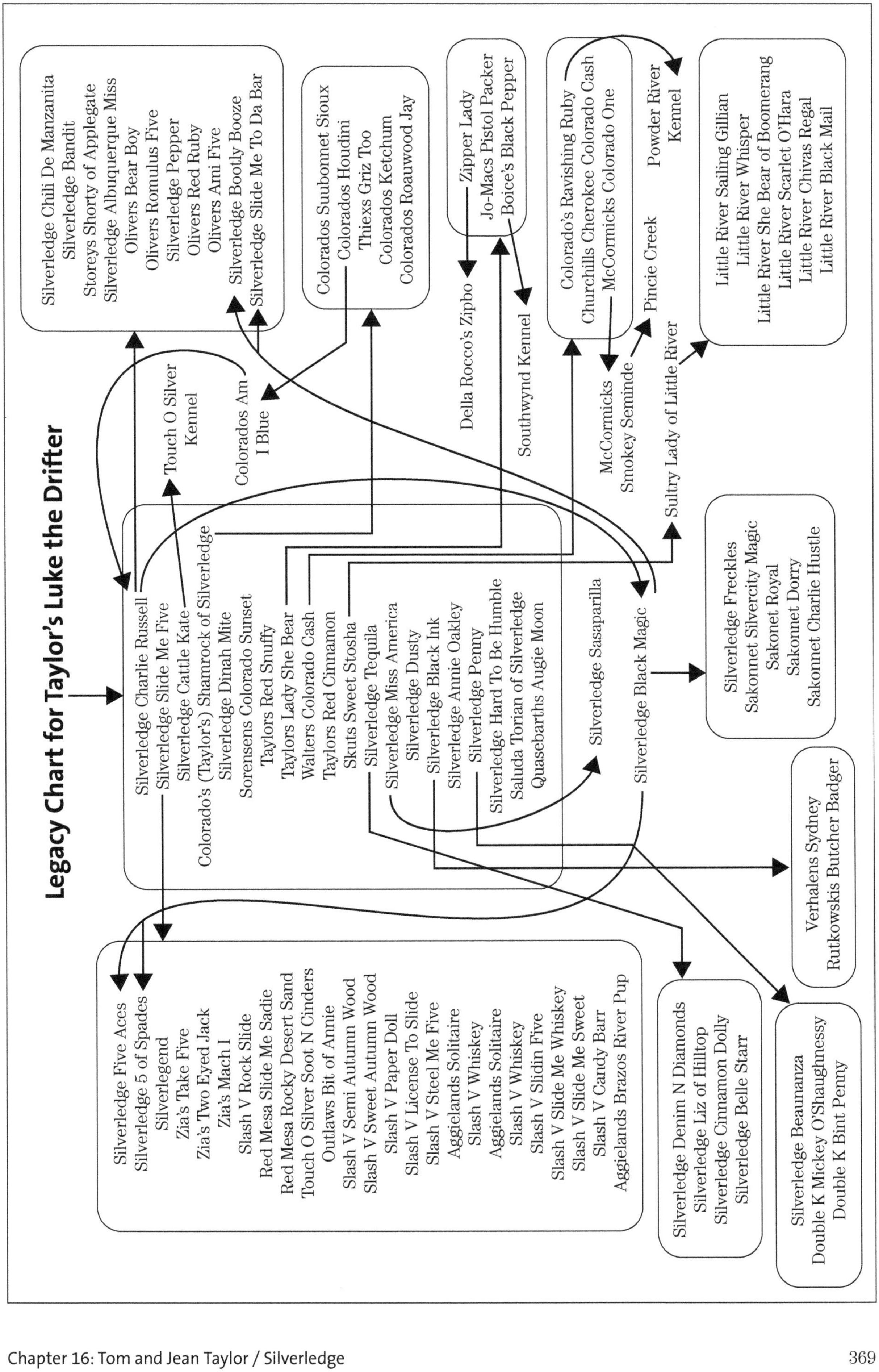
Legacy Chart for Taylor's Luke the Drifter
Silverledge Chili De Manzanita
Silverledge Bandit
Storeys Shorty of Applegate
Silverledge Albuquerque Miss
Olivers Bear Boy
Olivers Romulus Five
Silverledge Pepper
Olivers Red Ruby
Olivers Ami Five
Silverledge Bootly Booze
Silverledge Slide Me To Da Bar
Colorados Suubonnet Sioux
Colorados Houdini
Thiexs Griz Too
Colorados Ketchum
Colorados Roauwood Jay
Zipper Lady
Jo-Macs Pistol Packer
Boice's Black Pepper
Colorado's Ravishing Ruby
Churchills Cherokee Colorado Cash
McCormicks Colorado One
Powder River Kennel
Little River Sailing Gillian
Little River Whisper
Little River She Bear of Boomerang
Little River Scarlet O'Hara
Little River Chivas Regal
Little River Black Mail
Touch O Silver Kennel
Colorados Am I Blue
Della Rocco's Zipbo
Southwynd Kennel
McCormicks Smokey Seminde
Pincie Creek
Sultry Lady of Little River
Silverledge Charlie Russell
Silverledge Slide Me Five
Silverledge Cattle Kate
Colorado's (Taylor's) Shamrock of Silverledge
Silverledge Dinah Mite
Sorensens Colorado Sunset
Taylors Red Snuffy
Taylors Lady She Bear
Walters Colorado Cash
Taylors Red Cinnamon
Skuts Sweet Stosha
Silverledge Tequila
Silverledge Miss America
Silverledge Dusty
Silverledge Black Ink
Silverledge Annie Oakley
Silverledge Penny
Silverledge Hard To Be Humble
Saluda Torian of Silverledge
Quasebarths Augie Moon
Silverledge Sasaparilla
Silverledge Black Magic
Silverledge Freckles
Sakonnet Silvercity Magic
Sakonnet Royal
Sakonnet Dorry
Sakonnet Charlie Hustle
Verhalens Sydney
Rutkowskis Butcher Badger
Silverledge Five Aces
Silverledge 5 of Spades
Silverlegend
Zia's Take Five
Zia's Two Eyed Jack
Zia's Mach I
Slash V Rock Slide
Red Mesa Slide Me Sadie
Red Mesa Rocky Desert Sand
Touch O Silver Soot N Cinders
Outlaws Bit of Annie
Slash V Semi Autumn Wood
Slash V Sweet Autumn Wood
Slash V Paper Doll
Slash V License To Slide
Slash V Steel Me Five
Aggielands Solitaire
Slash V Whiskey
Aggielands Solitaire
Slash V Whiskey
Slash V Slidin Five
Slash V Slide Me Whiskey
Slash V Slide Me Sweet
Slash V Candy Barr
Aggielands Brazos River Pup
Silverledge Denim N Diamonds
Silverledge Liz of Hilltop
Silverledge Cinnamon Dolly
Silverledge Belle Starr
Silverledge Beaunanza
Double K Mickey O'Shaughnessy
Double K Bint Penny

to produce this type of dog were thwarted by the lack of a good gene pool on the east coast at that time. There were no other dogs like hers and she did not care for the show type dogs which were beginning to show up in the area. Other than the Sorensens, she didn't know where to find the type of dogs she wanted and needed to establish her bloodlines. In hindsight she thinks her breeding program would have been much different if she had lived in Colorado where she could see many more of these dogs firsthand. This was the home of the Wood breeding and mountain-type of Aussie—the type she wanted. She says, "There is a certain look and the way they think that is totally different from today's average Aussie."

The other Luke x Calamity Jane puppy who was a standout at everything she tried was Silverledge Cattle Kate CDX ATDdsc. Kate was a superior performer with 47 HIT awards in obedience and working trials before her untimely death in 1984. Luke's most famous son, Ch. Silverledge Slide Me Five STDc was linebred on Wood's Stubbie and became one of the most famous and successful sires of working trial dogs in the history of the breed. Luke's daughter, Silverledge Miss America, was the granddam of one of Jean's most prolific and top producers Silverledge Black Magic. And his last son, Silverledge Charlie Russell was Jean's link to the future.

When Jean began breeding seriously she had a good idea of what she wanted to produce. Most of all she wanted to retain a "certain quality of mind" she already had in her foundation dogs. Likening it to telepathy, she describes the feeling of just looking at the dog and knowing it was trying to read your mind. It has never been a common trait but one that could be found in the breed at that time. Luke had this quality and was the main reason she valued him so highly. Amusingly, Jean admits this could be a detriment in obedience if she could not keep her wits about her. If she lost concentration, Luke lost his. An embarrassing event happened at the 1980 ASCA Nationals when Jean was showing Luke in Open B obedience. She was having an excellent go but lost her focus when it came to the retrieve over the high jump exercise. She sent Luke to retrieve his dumbbell and he went around the jump before retrieving the dumbbell, taking the jump on his way back. Even the judge was distressed over this mistake and told Jean about his disappointment!

Luke's kennel mate, Taylor's Blue Prissy (DeWitt's Tuffy—full brother to Bjork's Chula x J Bar D Fancy), became another important breeding dog left on the east coast when the Taylors moved west. Describing Prissy in a letter to Betty Nelson "The bitch we got with Luke was a dandy dog. Wish I had her now." Prissy was Jean's barn dog for several years while she managed the 50 horse riding and lesson stable in Pennsylvania. Jean says she had the perfect temperament for the job—discriminating but friendly. Prissy was her right hand around the barn, helping to bring in horses, lunge them, chase cats up into the hay and whatever fun stuff she could stir up around the stable. She had a sweetness toward those she knew and was trustworthy around people and horses "unless they misbehaved," says Jean. She also relates this about Prissy's appearance, "She would have been a show dog, too, lovely mover, the best coat you could ever want. Used to be in the barn all day, nasty and wet. She'd go plop in the pond, shake, lay down to dry, and she was clean. All the dirt fell off where she lay. Never even smelled bad."

Prissy was bred to Luke once and although she was happy with the pups she got, they were, in her words "too much, too rangy, too spooky," not the ideal for her area. She describes them as "really super" but not the type of dog she could find the right homes for on the east coast. These were dogs bred for the isolated open ranges of the American west. Dog owners in the east wanted obedience dogs and pets for largely urban environments. Jean eventually sold both Prissy and a half sister to Luke, but later regretted the move. In later years she ran into a former client, Bill Haggis, at a National Specialty. Even after all those years he still raved about the Luke and Prissy pup he had from this litter. "But I wasn't smart enough to know what I had" she says. She still regrets not repeating the cross but the timing and her limited knowledge about the breed stopped her. Her search for outcrosses to temper what she liked would have been as effective with these fine bitches as it was with Luke.

Prissy became a famous producer in the northeast when bred to Willie (Sorensen's Red Man). Jean was still learning about the breed and not really aware of the high quality of her dogs but, like Fletcher Wood, seemed to instinctually make excellent crosses which

Taylor's Blue Prissy, blue merle female pictured at about 1 year of age in 1972. Courtesy of Jean Taylor

Taylor's Blue Prissy made her mark on the east coast of the United States. Her descendants can still be found in that area today. Courtesy of Jean Taylor

Taylor's Blue Prissy. Courtesy of Jean Taylor

Ch Sakonnets American Revolution CD, red and white male (Sorensen's Red Man x Taylors Blue Prissy) Foundation sire of Sakonnet kennel. Photo taken February 1982 by LM Gray. Courtesy of Sorensen

Ch Sakonnets American Revolution CD finishing his ASCA Championship on May 27, 1985. Photo by LM Gray. Courtesy of Sorensen

are still legend in that part of the country. The puppies mostly went to ranchers, horseshoers, horse trainers and the like. Many were never registered and it isn't really known if they were bred or what the results were. It is likely some were used for breeding as it is still possible to find dogs of this type cropping up in rural areas of the northeast and it is not likely this is an accident. Jean was visiting the area many years later and went to see a local farrier who once purchased dogs from her. The farrier's daughter pulled out a pedigree with all Jean's old dogs on it and wanted to know if she had any more like them.

Jean felt she had nowhere to go with Prissy in her area. Dogs which might have made good mates for her were too far away and Jean barely knew where her type of dogs were coming from anyway. Having no idea at the time where she and Tom might end up with their move she decided to sell Prissy to Linda Gray in 1976. Linda was in love with Prissy and bought her bred to Sorensen's Red Man. The most well known dog from this cross was Ch Sakonnet American Revolution, a solid red male. This was from the already proven cross of Prissy to Willie and was Prissy's second litter by Willie. Prissy's red male became the foundation sire for the Sakonnet kennel.

When Jean decided to become a dog breeder she also decided she needed to establish an identity for her kennel. In 1975, she and Tom were still in the horse business but were also creating fine silver trophy belt buckles and horse show gear for the western horse market. Tom suggested "Put Silver in the name. Some of the dogs are kind of silver colored and we are dealing in silver." Jean thought the name Silverledge had a good ring to it and it met all the requirements laid down by Tom. Two short years later, the Taylors decided to abandon Pennsylvania's cold winters in the Pocono

Mountains to look for a warmer climate. Tom was originally from Oklahoma and longed for more sunshine than he was seeing in the east. The horse business was not what it once had been and the big western horse shows and best Australian Shepherds were out west. So Tom and Jean loaded up their household with a few of their best Aussies and headed to an unknown destination. While passing through Albuquerque, New Mexico their car broke down and required some lengthy repairs. While waiting for the car to be fixed, they looked around their temporary home and found they were falling in love with it. Once the car was running again, the decision had been made to stay right where they were. Although Tom continued to train horses on the side and travel the show circuit with his silver gear, leather belts and custom cowboy boots, the couple also put down roots in Santa Fe. Opening a store for their goods, they were able to make a good living taking orders for their fine work as well as selling on the road.

Jean stayed in touch with a few people from the east coast after the move. One acquaintance had a Willie/Prissy male "big, nice Pooh-Bear type" in Jean's words. His name was Circle K's Big Blue. They also had a red female "very much red Wood's type" from Luke and Jane named Silverledge Penny. Jean got a puppy from the cross of Big Blue and Penny after moving to New Mexico and named him Silverledge Beaunanza. "Beau" became one of Calamity Jane's mates and produced Ch. Silverledge Butch Cassidy, a handsome red merle who reminded Jean of Willie. Butch later sired a WTCH, CD, RDX dog out of a purely show line bitch, showing the prepotence of his genetics!

Although Jean appreciated and sought to retain the qualities in her original dogs, she also had determined she wanted to alter a few things about them in her breeding program.

A big dark blue dog recognizable as "Woods" anywhere was Circle K's Big Blue. Courtesy of Jean Taylor

Silverledge Beaunanza (Circle K's Big Blue x Silverledge Penny), blue merle male. Courtesy of Jean Taylor

Ch Silverledge Butch Cassidy, red merle and white male, as a young puppy in New Mexico. Courtesy of Jean Taylor

Ch Silverledge Butch Cassidy as a mature dog. His coat is darkening as all merles do with age, making him appear to be a solid color instead of a merle. Courtesy of Jean Taylor

She says in a 1993 letter to Betty Nelson, .".we were exposed to some of the best of the Colorado dogs of their day, but living on the East Coast at the time, never did see any of their relatives. We only knew we liked what we had, but didn't know how good they really were." The *Aussie Times* was Jean's only method of seeing other dogs and the breeding accomplishments of other breeders at that time. She was especially interested in the crosses Terry Martin was making with Slash V (George's) Buckeye Bobby. By now the bitch she had gotten as a stud fee puppy from Leslie Sorensen, Taylor's Calamity Jane CD, had proven her worth and Jean saw her future in crossing her two foundation lines.

Ch Slash V Buckeye Bobby, red merle male who was Jean's choice of an outcross for her Wood foundation dogs. Courtesy of Sorensen

Jane, nicknamed "Miss Audacity," had become the epitome of an Aussie bitch for Jean. Her sire was Buckeye Bobby and her dam was a Woods/Gunsmoke cross female named Hall's (Sorensen's) Apache. Apache's story has been told in greater detail in the chapter on the Sorensen dogs but to summarize she was a top quality working dog who produced a number of great dogs for Colorado kennel. Jean describes Calamity Jane, "She had personality plus. She was an itty-bitty thing, about eighteen inches on tiptoe, but she was all heart and a good little dog." Steady, willing to tackle anything, she was Jean's choice to breed to Luke and the cross was everything she hoped for. She got the "Total Aussie—brains, soundness, conformation." Two outstanding performers from that cross were Silverledge Annie Oakley who earned a UD on the east coast and Silverledge Cattle Kate CDX ATDdsc who earned 47 High In Trials with Sharon Woolery. Cattle Kate was a truly outstanding working dog who never quit under any circumstance. Jane lacked only one leg to earn her CDX when Jean stopped showing her. Like most of Jean's dogs at that time, Jane had little opportunity to work stock, except ducks. But when she had the chance, she proved to be a keen worker and dedicated fetcher. As Jean says, "she would have been a dandy little dog to train."

Jean was considering her long range plans for breeding and Jane was to figure largely in those plans. Still looking to add a few more traits she felt she lacked, she continued to research what other breeders were doing and how it was working for them. Terry Martin was using Jane's sire to cross with Taylor's Whiskey daughters and getting puppies which Jean liked very much. She first saw Bobby while visiting Terry in 1977 and found him to be just as he looked in all the photos she had seen. Bobby was a "Caligari" type dog, referring to a type of dog being bred in California by Lois George and epitomized by their top winning Ch Copper Canyon Caligari. Buckeye Bobby added moderate size (20-21 inches), a blocky head, heavy bone, flat toplines, a mild but bold and confident temperament and large ears—all things Jean was trying to introduce in her dogs. She relates the liver red color would disappear with the Bobby line and the reds become more orange but the addition of the physical qualities was a priority and color could be fixed later. Whiskey's puppies had a reputation for silent low heeling, nice heads, nice ears and good toplines. So Jean decided to give this cross a try with her own Bobby daughter bred to Taylor's Whiskey. The Taylors had just recently moved to New Mexico in the spring of 1977 but there was no such thing as easy shipping to Moab, Utah where Taylor's Whiskey resided. They were planning a business trip to the area and contacted Whiskey's owner, Joe Taylor (no relation to Tom and Jean Taylor), asking "She is not quite in heat, but since we are going to be up there, can we drop her off on our way and you go ahead and keep her until she comes in heat?" Joe okayed the arrangement and Jane was left in Moab in early April. By mid May Jane had not come into heat and Joe called to find out what Jean wanted to do with her. Joe was planning to move cattle to the mountains and couldn't leave her at home as Whiskey was going with him. Jean asked him to just take her along on the drive and breed her when she came in heat. So Calamity Jane spent her summer in the mountains of Colorado and came home at the end of the season bred to Taylor's Whiskey.

Typical Silverledge puppies. Mature beyond their age showing their excellent type and structure. Courtesy of Rowe

This was to be Jane's most important litter in spite of the superior dogs she had already produced. She produced another litter or two for Jean but mostly spent the rest of her life as a beloved companion and remained sound and healthy to the age of 14. She died suddenly of a stroke or heart attack at the Taylor ranch in Edgewood, New Mexico and is buried there.

Born: ____________ Sex: ________
Owned by: ____________

PEDIGREE OF Silverledge Beaunanza

Parents	Grandparents	Great-grandparents	Great-great-grandparents
Sire: Circle K's Big Blue	Sire: Sorensens Red Man	Sire: Bruce's Blue	Sire: unknown
			Dam: unknown
		Dam: Bruce's Blue Susie	Sire: unknown
			Dam: unknown
	Dam: Taylor's Blue Prissy	Sire: DeWitts Tuffy	Sire: Bjorks Baron
			Dam: Roadrunners Cholla
		Dam: J Bar D Fancy	Sire: J Bar D Red Chief
			Dam: J Bar D Flirt
Dam: Silverledge Penny	Sire: Taylor's Luke the Drifter	Sire: Woods Stubbie	Sire: Woods Jay
			Dam: Woods Slate
		Dam: Bjorks Chula	Sire: Bjorks Baron
			Dam: Roadrunners Cholla
	Dam: Taylor's Calamity Jane	Sire: Slash V (George's) Buckeye Bobby	Sire: Tartaglia's Chocolate
			Dam: Caligari's Lady
		Dam: Hall's (Sorensens) Apache	Sire: Wood's Stubbie
			Dam: J Bar D Flirt

Bred by: ____________ Address: ____________ Phone: ____________

The details given above are correct to the best of my knowledge. Signed: ____________(Breeder)

PEDIGREE OF Taylor's Calamity Jane

Born: ______ Sex: ______
Owned by: ______

Parents	Grandparents	Great-grandparents	Great-great-grandparents
Sire: Slash V (George's) Buckeye Bobby	Sire: Tartaglia's Chocolate	Sire: Scaltritti's Dado	Sire: Oregon Ranch dog
			Dam: Oregon Ranch bitch
		Dam: Scaltritti's Stormy	Sire: Oregon Ranch dog
			Dam: Oregon Ranch bitch
	Dam: Caligari's Lady	Sire: Tartaglia's Chocolate	Sire: Scaltritti's Dado
			Dam: Scaltritti's Stormy
		Dam: Duart's Silver	Sire: Smees Ranch dog
			Dam: Smees Ranch bitch
Dam: Hall's (Sorensens) Apache	Sire: Wood's Stubbie	Sire: Wood's Jay	Sire: Sisler's Shorty
			Dam: Sisler's Trixie
		Dam: Wood's Slate	Sire: Wood's Jay
			Dam: Wood's Specks
	Dam: J Bar D Flirt	Sire: Sorensens Gunsmoke	Sire: William's Laddie
			Dam: Jeffrey's Sheila
		Dam: Jeffreys Missy Blue	Sire: Williams Laddie
			Dam: Picraux's Juliet

Bred by: ______ Address: ______ Phone: ______

The details given above are correct to the best of my knowledge. Signed: ______(Breeder)

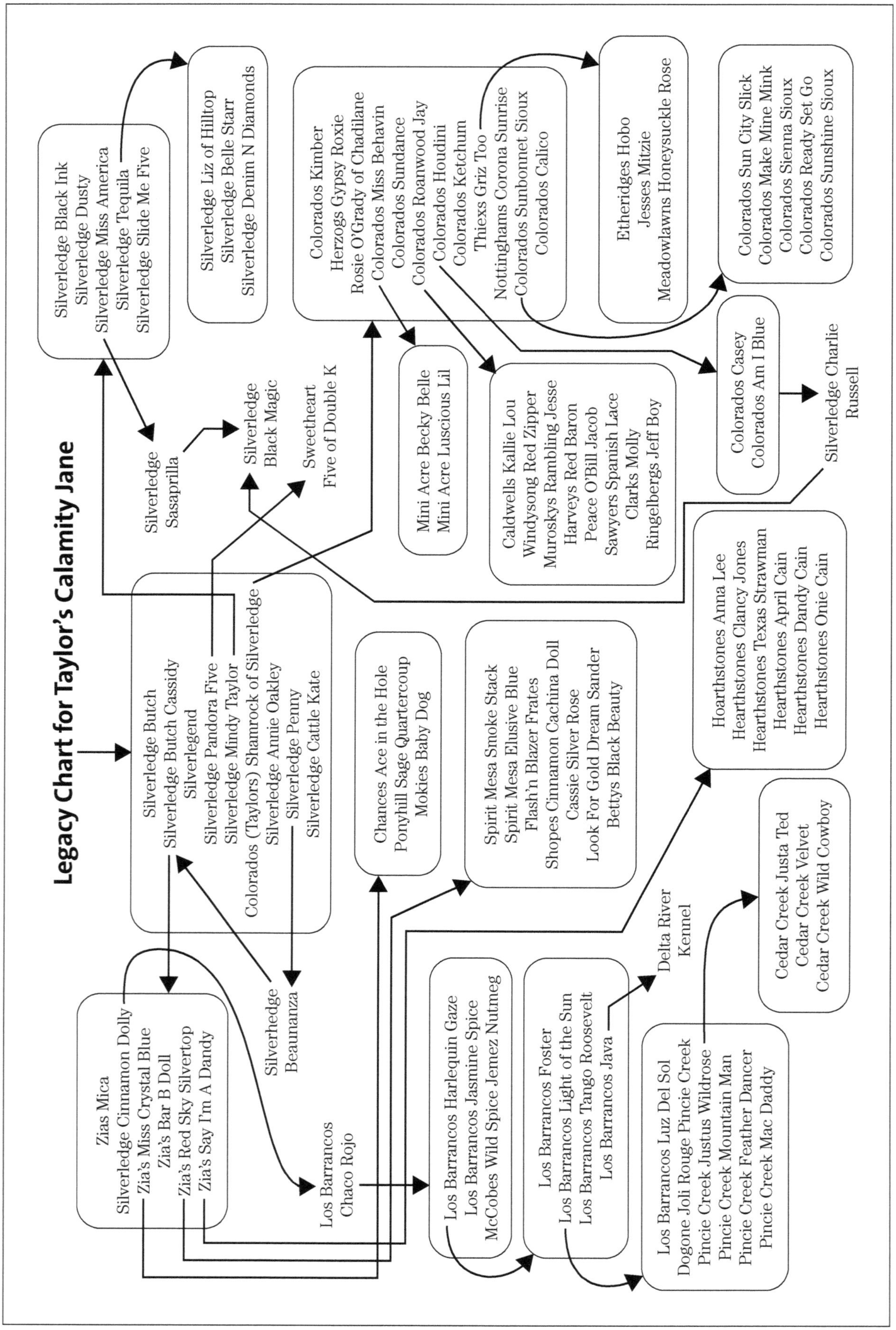
Legacy Chart for Taylor's Calamity Jane
Silverledge Black Ink
Silverledge Dusty
Silverledge Miss America
Silverledge Tequila
Silverledge Slide Me Five
Silverledge Liz of Hilltop
Silverledge Belle Starr
Silverledge Denim N Diamonds
Colorados Kimber
Herzogs Gypsy Roxie
Rosie O'Grady of Chadilane
Colorados Miss Behavin
Colorados Sundance
Colorados Roanwood Jay
Colorados Houdini
Colorados Ketchum
Thiexs Griz Too
Nottinghams Corona Sunrise
Colorados Sunbonnet Sioux
Colorados Calico
Etheridges Hobo
Jesses Mitzie
Meadowlawns Honeysuckle Rose
Colorados Sun City Slick
Colorados Make Mine Mink
Colorados Sienna Sioux
Colorados Ready Set Go
Colorados Sunshine Sioux
Silverledge Sasaprilla
Silverledge Black Magic
Sweetheart Five of Double K
Mini Acre Becky Belle
Mini Acre Luscious Lil
Caldwells Kallie Lou
Windysong Red Zipper
Muroskys Rambling Jesse
Harveys Red Baron
Peace O'Bill Jacob
Sawyers Spanish Lace
Clarks Molly
Ringelbergs Jeff Boy
Colorados Casey
Colorados Am I Blue
Silverledge Charlie Russell
Silverledge Butch
Silverledge Butch Cassidy
Silverlegend
Silverledge Pandora Five
Silverledge Mindy Taylor
Colorados (Taylors) Shamrock of Silverledge
Silverledge Annie Oakley
Silverledge Penny
Silverledge Cattle Kate
Chances Ace in the Hole
Ponyhill Sage Quartercoup
Mokies Baby Dog
Spirit Mesa Smoke Stack
Spirit Mesa Elusive Blue
Flash'n Blazer Frates
Shopes Cinnamon Cachina Doll
Cassie Silver Rose
Look For Gold Dream Sander
Bettys Black Beauty
Hoarthstones Anna Lee
Hearthstones Clancy Jones
Hearthstones Texas Strawman
Hearthstones April Cain
Hearthstones Dandy Cain
Hearthstones Onie Cain
Zias Mica
Silverledge Cinnamon Dolly
Zia's Miss Crystal Blue
Zia's Bar B Doll
Zia's Red Sky Silvertop
Zia's Say I'm A Dandy
Silverhedge Beaunanza
Los Barrancos Chaco Rojo
Los Barrancos Harlequin Gaze
Los Barrancos Jasmine Spice
McCobes Wild Spice Jemez Nutmeg
Los Barrancos Foster
Los Barrancos Light of the Sun
Los Barrancos Tango Roosevelt
Los Barrancos Java
Delta River Kennel
Cedar Creek Justa Ted
Cedar Creek Velvet
Cedar Creek Wild Cowboy
Los Barrancos Luz Del Sol
Dogone Joli Rouge Pincie Creek
Pincie Creek Justus Wildrose
Pincie Creek Mountain Man
Pincie Creek Feather Dancer
Pincie Creek Mac Daddy

From a letter to Betty Nelson, Jean says, "I bred Jane to Taylor's Whiskey in 1977.......they were typey, and low heelers. One of them produced Slide Me Five, by Luke. He was a really nice dog, and a good sire." Of the four puppies in this litter—two males and two females—the one from this litter who became famous as Slide's dam was Silverledge Mindy Taylor. Mindy was a wonderfully correct dog with quite a lot of angulation and fantastic movement. She captured many of the qualities Jean was trying to introduce into her Wood foundations and she was soon bred to Luke the Drifter to produce one of the most famous dogs of the Silverledge line, Ch. Silverledge Slide Me Five STDc.

Black and white female, Silverledge Mindy Taylor, the dam of a number of top dogs. Courtesy of Jean Taylor

Slide stood out from the start. Jean remembers, "Slide was the 'special' one in the litter. He had Luke's powerful personal attachment, even from a baby puppy. I would go in to feed the puppies, slip out the gate, and guess who would be on my side of the gate? I would think 'How did you do that?' So of course he got to come in the house with me. So either he was special because I treated him special, or I treated him special because he just WAS special, who knows? His personality was a lot like Luke, in that he was just 'there'—quiet when nothing was happening, ready to go when you were. Liked people a lot, but not pushy, not crabby at all. I would let Holly (Jean's daughter) lead him around the show grounds some when she was only five or six." Terry Martin of Slash V kennel later leased Slide and was largely responsible for making him famous. She also recognized his unique quality, saying "He was a beautiful, correct dog and had kind of a regal bearing—one of those dogs who quietly gives the impression that he is 'something special'."

Silverledge Mindy Taylor. Courtesy of Jean Taylor

An interesting story surrounds Slide's name. Jean had sold Shamrock of Silverledge "Rocky," a red merle, about 8 months prior to Slide's birth. Red dogs were not common, especially when breeding a blue merle (Luke) to a black and white (Mindy). When the red merle Slide appeared in the litter, Jean thought it was fate and gave him the gambler's name "Slide Me

The red merle male, Ch Silverledge Slide Me Five STDc, around 1980 before he became famous. At home in New Mexico. Courtesy of Jean Taylor

Ch Silverledge Slide Me Five STDc as a youngster. Courtesy of Jean Taylor

Five." The name is a celebration of sorts, like a verbal "high five" or congratulations!

Jean recognized Slide's quality early on and she was showing her dogs locally so he joined the "show string" for awhile. In those days Slide was unbeatable in the New Mexico area. She attended the 1980 CASA (Colorado) Specialty when Slide was a little over one year old. He won his class but was uneasy in the crowded ring used for the Winners class and didn't show well enough to take the big prize that day. He was a well built, well angulated, beautiful moving dog and Jean managed to get to enough shows to finish his breed Championship. But a long term commitment to this dog was not in the cards. Jean now had a small child and she and Tom were in the process of changing over their business from a traveling booth to a permanent, full-time store in Santa Fe. Life was hectic to say the least, there was little show activity and no stock dog trials in her area and dogs dropped to the bottom of her priority scale. Jean didn't forget this special dog however. Her focus on the breeding program of Terry Martin and her mimicry of it made Slide a good fit for the Slash V kennel. While manning her booth at the Fort Worth Cutting Horse Futurity, Jean ran into Terry and offered to lease Slide to her for awhile. It didn't take long for Terry to accept the offer and in the fall of 1982 Slide moved to Texas where he would be used as he should be for the next two years.

Terry wasted no time using Slide at stud. His first breeding was to her top bitch, Slash V Semi Sweet. If he had never been used again, this cross would have assured his place in Aussie history. From this litter came the legendary WTCH Slash V Slide Me Sweet CD "Slip," owned, trained and trialed by Red Oliver. Four puppies from this litter were entered and competed in the 1985 Washington State Stockdog Futurity, placing first, second, fourth and sixth. Slip's brother, Slash V Sliden Five ATDd OTDcs was one of these puppies who were trialed for the very first time in this big venue. Answering to "Rowdy," this talented dog also competed in the all three classes of livestock in the regular National Specialty trial

Born: ______ Sex: ______
Owned by: ______

PEDIGREE OF Silverledge Slide Me Five

Parents	Grandparents	Great-grandparents	Great-great-grandparents
Sire: Taylor's Luke the Drifter	Sire: Woods Stubbie	Sire: Woods Jay	Sire: Sislers Shorty
			Dam: Sislers Trixie
		Dam: Woods Slate	Sire: Woods Jay
			Dam: Woods Specks
	Dam: Bjorks Chula	Sire: Bjorks Baron	Sire: Olsons Blue Chips
			Dam: Rohenbach's Banjo
		Dam: Roadrunners Cholla	Sire: McConkey's Blue Pecos
			Dam: McConkey's Calico Cameo
Dam: Silverledge Mindy Taylor	Sire: Taylor's Whiskey	Sire: Mansker's Turk	Sire: Mansker's Smokey
			Dam: Mansker's Duchess
		Dam: Mansker's Ana Lee	Sire: Green's Kim
			Dam: Manser's Freckles
	Dam: Taylor's Calamity Jane	Sire: Slash V (George's) Buckeye Bobby	Sire: Tartaglia's Chocolate
			Dam: Caligari's Lady
		Dam: Hall's (Sorensens) Apache	Sire: Wood's Stubbie
			Dam: J Bar D Flirt

Bred by: ______ Address: ______ Phone: ______

The details given above are correct to the best of my knowledge. Signed: ______(Breeder)

Ch Silverledge Slide Me Five STDc around 1980 as a young dog in New Mexico. Courtesy of Jean Taylor

Slide as a "teenager" maturing nicely and showing his beauty and good Wood type. Courtesy of Jean Taylor

Ch Silverledge Slide Me Five STDc working ducks at the 1983 ASCA National Specialty with Terry Martin. Photo by Author

following the futurity. Rowdy made his mark by winning every class he entered and tied for High in Trial at this National!

Slip went on to become a legend in the breed in her own lifetime, becoming the only bitch and the very first Supreme Champion Stockdog Finals dog in ASCA history. Bred to a number of different males of different bloodlines, Slip produced a dynasty of Working Trial Champions herself. In the 1988 Centex Stockdog Futurity, 7 of the 32 entries were Slide puppies or grand puppies, many from Slip. Her brother Rowdy had a very short career at stud, producing a number of good dogs before he was tragically lost to an accident at a young age.

Jean's decision to lease Slide to Terry was one of the best choices she could have made for him. This superior dog got his chance to prove his greatness and in Terry's capable hands, he achieved it. After the success of the Slide and Ticki cross, Terry decided to repeat it, intending to keep a puppy for herself. Instead, all the puppies were sold before they were born and Terry was left with a mismarked female named Slash V Paper Doll. Although not trialed, Paper Doll became a top producer for Slash V and is behind many of the best dogs being produced there today. She never achieved Hall of Fame status herself but produced three daughters who have earned that honor. Slide was bred to a variety of other Slash V bitches and a few outside bitches during his time in Texas, including Slash V Steel Rockette STDc, Slash V Steel Curtain STDsc, Black Powder's Blue Sussy, Slash V Kattle Kate, Manchado Silver Doll (owned by Sharon Woolery), Powder River's Outlaw Annie and Russ Ford's foundation bitch, Ford's Blue Bonnet. No matter the breeding of his mate, he produced highly trainable, strong working puppies with his good looks and soundness.

Ch Silverledge Slide Me Five STDc working ducks for the first time at about one year of age. Courtesy of Jean Taylor

While in Texas, Terry hauled Slide to a few shows along with several other dogs she and Steve Winn were showing at the time. On one memorable day she was handling Slide and Steve was showing his own champion female, Ch. Brushwood Buckeye Charm CD ATDds, OTDc, in the Best of Breed class. In a wonderful testament to the soundness and correctness of these two top working dogs they won Best of Breed and Best of Opposite Sex on that day! In addition to earning attention in the breed ring while in Texas, it was his stay with Terry and his breedings with "Ticki" that eventually earned him Hall of Fame honors as a sire.

WTCH Slash V Slide Me Sweet CD "Slip," red merle female who became a legend in her own time. Courtesy of Jean Taylor

Slide's two years with Terry soon ended and Terry's friend Barb Nielsen (Hager), who had admired him for some time, was interested in using him herself. She contacted Jean to see if she, too, could lease Slide to use at her Blue Isle kennel in Washington State. Jean was still overloaded with business and family obligations and readily agreed to allow Barb to continue promoting him for her. He continued producing top quality working dogs in the northwest United States just as he had done in Texas.

Ch Silverledge Slide Me Five STDc showing the typical strong Wood head type found in all the Silverledge dogs. Courtesy of Jean Taylor

He had been with Barb for only about 6 months when she took him to a show with a

Slash V Slidin Five ATDd OTDcs "Rowdy," red merle male, littermate to Slip. A superior male who was lost at a young age to an accident. Courtesy of Jean Taylor

Ch Silverledge Slide Me Five STDc as a mature dog while living with Terry Martin. Courtesy of Terry Martin

number of her other dogs. One evening she was away from her campsite on the showgrounds when someone shot off a huge burst of fireworks. Her dogs were in the camper, including Slide. Although a confirmed escape artist to begin with, he and one of Barb's females were terrified of this barrage and escaped in panic. A few hours later the female returned but there was not a trace of Slide. Barb was devastated and looked for him everywhere but he was gone. She was about 150 miles from home and Slide did not know the area so she was sure she would never see him again. She returned home wondering how she was going to tell Jean the dog of a lifetime was gone. Incredibly, Slide appeared at Barb's house 2 weeks later, thinner than he started out but safe after crossing two major freeways on his journey! And Jean didn't hear the story of his adventure until many years later!

Slide eventually returned home to New Mexico where Jean kept him for the rest of his life as a beloved companion. He settled in, keeping her life interesting with his "sense of what he had to do" and his sociability with other people. She calls him an "opinionated" dog and remembers him as very similar to Luke in personality—always wanting to be with her and please her. The cross which produced Slide was made twice and Jean remembers a consistent quality in all the puppies. She believes "The best way to tell what a dog is genetically is to look at all its littermates."

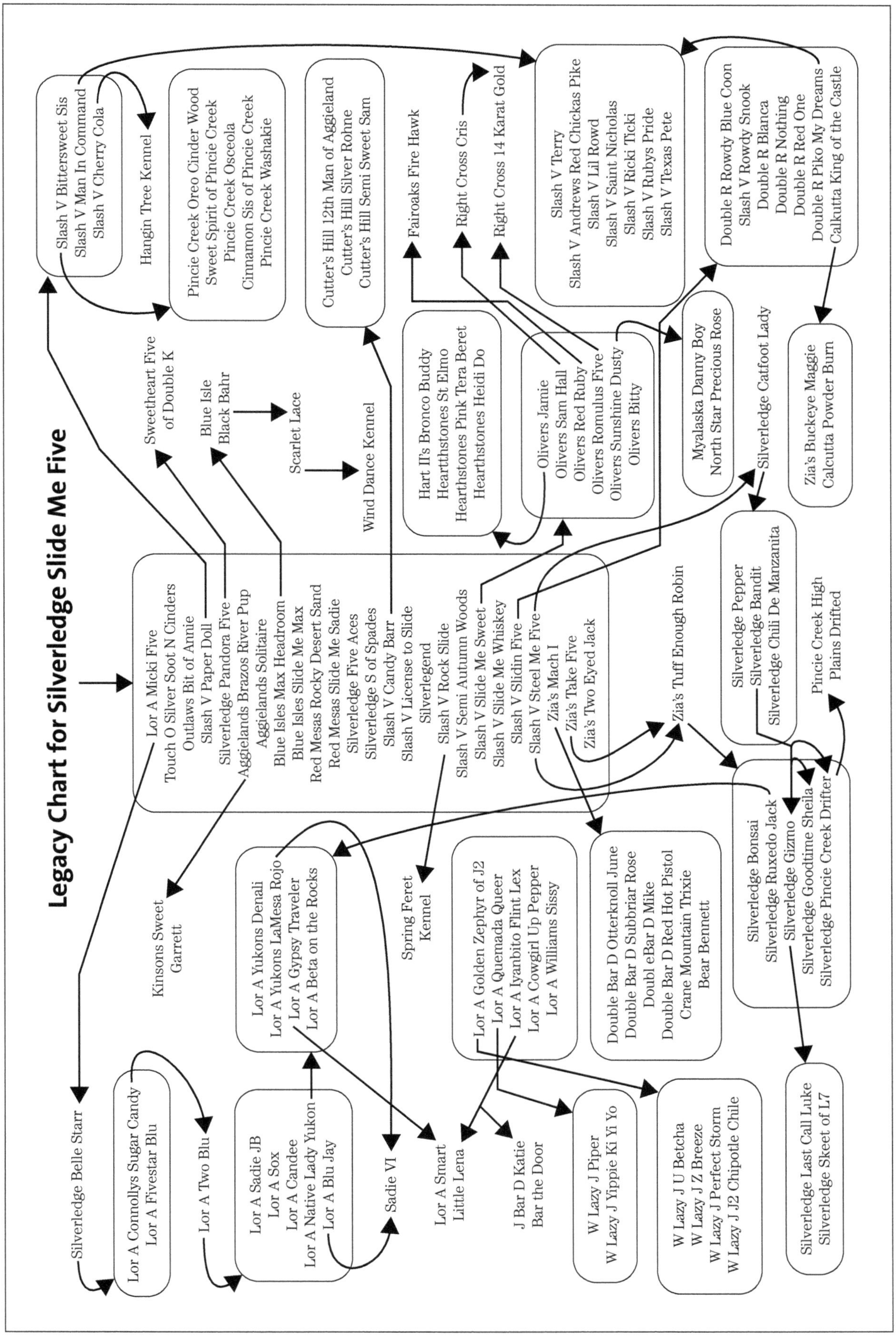
Legacy Chart for Silverledge Slide Me Five
Lor A Micki Five
Touch O Silver Soot N Cinders
Outlaws Bit of Annie
Slash V Paper Doll
Silverledge Pandora Five
Aggielands Brazos River Pup
Aggielands Solitaire
Blue Isles Max Headroom
Blue Isles Slide Me Max
Red Mesas Rocky Desert Sand
Red Mesas Slide Me Sadie
Silverledge Five Aces
Silverledge S of Spades
Slash V Candy Barr
Slash V License to Slide
Silverlegend
Slash V Rock Slide
Slash V Semi Autumn Woods
Slash V Slide Me Sweet
Slash V Slide Me Whiskey
Slash V Slidin Five
Slash V Steel Me Five
Zia's Mach I
Zia's Take Five
Zia's Two Eyed Jack
Slash V Bittersweet Sis
Slash V Man In Command
Slash V Cherry Cola
Hangin Tree Kennel
Pincie Creek Oreo Cinder Wood
Sweet Spirit of Pincie Creek
Pincie Creek Osceola
Cinnamon Sis of Pincie Creek
Pincie Creek Washakie
Cutter's Hill 12th Man of Aggieland
Cutter's Hill Silver Rohne
Cutter's Hill Semi Sweet Sam
Fairoaks Fire Hawk
Right Cross Cris
Right Cross 14 Karat Gold
Slash V Terry
Slash V Andrews Red Chickas Pike
Slash V Lil Rowd
Slash V Saint Nicholas
Slash V Ricki Ticki
Slash V Rubys Pride
Slash V Texas Pete
Double R Rowdy Blue Coon
Slash V Rowdy Snook
Double R Blanca
Double R Nothing
Double R Red One
Double R Piko My Dreams
Calkutta King of the Castle
Sweetheart Five of Double K
Blue Isle Black Bahr
Scarlet Lace
Wind Dance Kennel
Hart II's Bronco Buddy
Hearthstones St Elmo
Hearthstones Pink Tera Beret
Hearthstones Heidi Do
Olivers Jamie
Olivers Sam Hall
Olivers Red Ruby
Olivers Romulus Five
Olivers Sunshine Dusty
Olivers Bitty
Myalaska Danny Boy
North Star Precious Rose
Silverledge Catfoot Lady
Zia's Buckeye Maggie
Calcutta Powder Burn
Silverledge Pepper
Silverledge Bandit
Silverledge Chili De Manzanita
Zia's Tuff Enough Robin
Pincie Creek High Plains Drifted
Kinsons Sweet Garrett
Lor A Yukons Denali
Lor A Yukons LaMesa Rojo
Lor A Gypsy Traveler
Lor A Beta on the Rocks
Spring Feret Kennel
Lor A Golden Zephyr of J2
Lor A Quemada Queer
Lor A Iyanbito Flint Lex
Lor A Cowgirl Up Pepper
Lor A Williams Sissy
Double Bar D Otterknoll June
Double Bar D Subbriar Rose
Doubl eBar D Mike
Double Bar D Red Hot Pistol
Crane Mountain Trixie
Bear Bennett
Silverledge Bonsai
Silverledge Ruxedo Jack
Silverledge Gizmo
Silverledge Goodtime Sheila
Silverledge Pincie Creek Drifter
Silverledge Belle Starr
Lor A Connollys Sugar Candy
Lor A Fivestar Blu
Lor A Two Blu
Lor A Sadie JB
Lor A Sox
Lor A Candee
Lor A Native Lady Yukon
Lor A Blu Jay
Sadie VI
Lor A Smart Little Lena
J Bar D Katie Bar the Door
W Lazy J Piper
W Lazy J Yippie Ki Yi Yo
W Lazy J U Betcha
W Lazy J Z Breeze
W Lazy J Perfect Storm
W Lazy J J2 Chipotle Chile
Silverledge Last Call Luke
Silverledge Skeet of L7

Silverledge Pandora Five, blue merle, copper and white female by Slide x Taylor's Calamity Jane. Courtesy of Jean Taylor

Blue merle, copper and white male Lor A Mickey Five (Ch Silverledge Slide Me Five STDc x Akins Lor A Ellie Blu) Courtesy of Jean Taylor

The consistent high quality among Slide's littermates attested to his greatness as a sire. He was always known best for producing an extreme willingness to work with the handler, the low heeling ability Jean was aiming for and when crossed with good working bitches the puppies were almost always that perfect combination of work and willingness. His physical correctness was also an asset which passed to his puppies consistently and his prepotency was not an accident.

Life still didn't allow much "dog time" and Jean never kept a puppy from Slide, although she used him a few times after his return. A few outside bitches came to Slide during this time including Red Mesa's Kachina Doll and Akins Lor A Ellie Blu. Jean also co-owned a Slide female out of a Slash V bitch with her good friend, Liz Winn. The female was Slash V Steel Me Five "Steeler" by Slide out of Slash V Steel Curtain (a Slash V Rocky Top daughter, Rocky Top being another "Caligari" type dog like Buckeye Bobby). Steeler was a nice female standing about 19 inches tall, well built and pretty. She had a nice head, no white markings and was just an overall good bitch. She never had the opportunity to work livestock because Liz's son, Steve, was working Buckeye Charm and his Slide son, Rowdy and had no time for training any more dogs.

Liz had also bred her own Buckeye Bobby bitch, Ch. Brushwood Buckeye Charm CD ATDds OTDc [Slash V (George's) Buckeye Bobby x Slash V Fancy Face)], to Slide and kept a male named Zia's Take Five. Buckeye Charm, called "Kate," was another bitch Jean admired greatly and got to know well. She describes Kate as a well built and pretty moving bitch, although not as pretty headed as she would have liked. She is the bitch Liz's son, Steve, was showing the day she and Slide took top wins at the Texas show. She says Kate was "quite a dog," tough, trainable and came very close to being the highest titled bitch in the breed at the time. She lacked only one cattle score to become a triple ATD due to her young owner, Liz's son Steve, getting busy with college and other obligations and never having time to finish her. Kate was Steve's first trial dog and he was working with Red Oliver at the time while the two were both trying to learn how to train stock dogs. Jean says they trained and retrained their dogs "dozens of times." They trained one way for awhile, then decided it was wrong and started again. All that time, Kate never quit.

The red merle female, Ch. Brushwood Buckeye Charm CD ATDds OTDc as a young dog. Courtesy of Jean Taylor

In 1989 the decision was made to cross the two Slide puppies, Zia's Take Five and Slash V Steel Me Five. This cross was to produce Jean's next generation, a dog named Zia's Tuff Enough Robin CD STDd OTDs RTDs, known as "Tuffy." Tuffy fit the goal Jean had been seeking for many years. He had the boldness, confidence and ambition of the Slash V and Caligari lines and the sweetness and "tune-in-ability" from Slide and Luke. He was a stocky, well-built dog, very athletic with lots of working ability. He was an outcross even though he was 50% Slide Me Five blood and showed the characteristics of the Buckeye Bobby side strongly.

Ch. Brushwood Buckeye Charm CD ATDds OTDc as a mature dog. Courtesy of Jean Taylor

At the time Jean got him he was the closest thing to a Slide puppy a person could get. The two sides of the family brought soundness and freedom from major faults, good type and bone, athleticism and solid, excellent toplines to the mix. Standing about 20 inches tall and weighing about 52 pounds he was a well balanced moderate dog. He was correct and a sound mover, very showy but not quite well angulated enough to do well in the show ring. He almost always won his class when Jean showed in the New Mexico area and even got a Reserve or two. The Buckeye Bobby line tempered the shyness which could appear in the Wood dogs and added to the natural tough, forward and ambitious personality found in the Wood dogs. They were very trainable like the Wood dogs but able to take more "abuse" and mistakes in training without becoming upset.

Tuffy shone in obedience with all the drive and ambition needed for high level work and the ability to concentrate and take a lot of repetition. Daughter Holly won the State Fair 4-H obedience with him when she was

Ch. Brushwood Buckeye Charm CD ATDds OTDc owned by Liz Winn. Courtesy of Jean Taylor

Zia's Take Five, red merle male (Ch Silverledge Slide Me Five STDc x Ch. Brushwood Buckeye Charm CD ATDds OTDc), sire of Zia's Tuff Enough Robin. Courtesy of Jean Taylor

10 years old with a score of 196. Jean took him over for stock work and found a determined fetcher with excellent balance, good wear and even a little eye. He worked close naturally but could take some discipline to modify that tendency. He was gutsy and would take on anything which challenged him without being overly free with his teeth. He would use a body block to turn ornery sheep instead of biting in many cases. Tuffy used his brain to figure out the solutions to problems and turned into a great dog for Jean to begin learning stock dog training. No matter how many mistakes she made he never quit or turned off!

Tuffy produced another good generation of dogs for Jean even though her breeding activities had slowed down considerably. His son, Silverledge Pincie Creek Drifter—commonly known as "Hawkeye," out of the Charlie Russell daughter Silverledge Bandit, was sold as a puppy to Roger Stevens of Pincie Creek. He was used lightly there before being sold. One of Hawkeye's sons has recently produced a litter for Jean out of descendant of another Tuffy son, Silverledge Tuxedo Jack (who was out of the Sprog daughter, Silverledge Freckles).

Silverledge Tuxedo Jack was a striking black bi dog who was a beautiful show quality individual. Once again Jean's life became complicated and although she adored Jack, worked him on stock and in obedience, and thought very highly of him, she was not able to compete with or promote him. Jean found him a loving home for him and he was used for breeding by Becky Anderson of Lor A kennel. Becky kept several puppies by Jack and the line continues in her kennel.

A Tuffy daughter, littermate to Hawkeye, was kept by Jean and bred lightly. Her name was Silverledge Gizmo and she was crossed once to the inbred Charlie Russell male, Silverledge Slide Me To Da Bar CD STDs, to produce two exceptional red merle males sold to Wood fanciers in the east. Michelle Murvai obtained Silverledge Last Call Luke and Buster and Donna Berryhill of L7 kennel obtained Silverledge Skeet of L7. Neither male was used extensively but both were good examples, once again, of the Luke and Buckeye Bobby cross which had been so successful for Jean for so many years.

Although Jean's Buckeye Bobby cross dogs brought in things she felt the Wood dogs lacked she was not seeking to radically change her line from the type she began with. Those early dogs, especially Luke, were still her ideal. Having looked for some time for the "perfect outcross" she had not found any other lines she liked. After moving to New Mexico, the Taylors were exposed to more of the bloodlines of the day, but none of those dogs had what they wanted for their line. Jean states if she had lived in Colorado "it might have been different, but I didn't see many of the Colorado dogs and the ones I did see, I wasn't that wild about."

Zia's Tuff Enough Robin CD STDd OTDs RTDs as a puppy. Courtesy of Jean Taylor

Zia's Tuff Enough Robin CD STDd OTDs RTDs "Tuffy," red merle male, as a youngster. Courtesy of Jean Taylor

But some Colorado dogs never lost their allure for Jean. The Wood dogs she had in the beginning and those she saw later still drew her back to her foundations. One of the earliest dogs she had in the mid-seventies was a Jeffreys' Pooh Bear daughter out of Luke's dam named Taylor's Nifty Babe. Babe came from Leslie Sorensen at about 9 months of age and she remembers her as a very nice bitch. "Strong, sensible and gorgeous" is how she describes Babe and she liked Pooh Bear because she liked this female so much. She never forgot this impressive bitch and decided she wanted another like her, saying "There was something special in her temperament and attitude." She was bred to Luke at least once but all the puppies were sold locally to ranchers and horse people and she never knew what became of them.

Jean also experimented with Babe and breeding to outside males. Dogs like hers or dogs she wanted to cross with hers were not to be found on the east coast. She had no choice but try shipping and for the first time she shipped her prize female to Oregon in order to cross her with Ch. Sharp's Ragnar. This was breeding on faith as she had only seen and admired photos of this dog and read some things about him. She was looking for that right outcross and Ragnar seemed to have the good, solid, responsible temperament, well balanced, moderate type, good topline and bone, and a nice earset she was looking for. It was a big risk

Zia's Tuff Enough Robin CD STDd OTDs RTDs as a young dog. Courtesy of Jean Taylor

Zia's Tuff Enough Robin CD STDd OTDs RTDs "Tuffy," like all of the Silverledge dogs, was an excellent obedience dog. Courtesy of Jean Taylor

Silverledge Pincie Creek Drifter, red merle male, as a puppy. Courtesy of Jean Taylor

and expense for those days but she was looking at some of the Aussies which were starting to show up on the east coast and none of them appealed to her. She had no choice but to give Babe and Ragnar a shot.

She was pleased with the results but still had little space for keeping many dogs and let these puppies go to the local horse and pet people. She was still new to the breed, Luke was young and she thought, "This is good, but I can probably do better." This inexperience cost her what she realizes now were puppies that were "better than good, they were VERY good, but I sold them for pet dogs or ranch dogs." She remembers one puppy sold to a cutting horse trainer as a turn back dog as being a very nice dog, with good type, that had everything she would later be hoping for! This bitch was later bred back to Luke and she didn't even keep a puppy from that litter! Now that she has a sense of history of the breed and the experience of seeing many more dogs, she realizes what she has lost and that they are nearly gone for good. Today she is trying to recapture the quality she once had and get back to "where they were when they weren't smart enough to know it."

When she later had more time and space she tried to reintroduce this blood into her program by breeding a sister to Ch Silverledge Slide Me Five STDc, Silverledge Tequila, to the beautiful Ragnar son, Ch Hilltops L. R. Apollo. This time she kept a bitch from the cross for a short time but eventually let her go, too. Once again the too-many-dogs-too-little-time gremlin appeared. Later regretting that decision, she speculates on how this outcross might have added to her bloodline. The female was named Silverledge Denim 'N Diamonds and Jean comments this was a good outcross on the Wood lines.

Babe was eventually sold to a saddle maker in West Virginia along with a Willie/Prissy female. Jean later wished she had another like her once she gained experience in the breed and was trying to find more like her original dogs! When this realization occurred Jean decided to became acquainted with Charlene Jeffreys hoping to reintroduce this quality into her dogs through a breeding to Jeffreys' Pooh Bear.

Silverledge Tuxedo Jack. Courtesy of Jean Taylor

In 1980, Jean decided to take one of Slide's littermates, Silverledge Miss America, back to Pooh Bear. He was 10 years old by then and died a few months after the breeding. When she saw him he was everything she hoped for. A gorgeous dog, he had the heavy coat, the strong, noble head she had in Luke, and he produced a noble, responsible temperament. Jean states, "I have tried to stay in the old lines, when I could find a dog I could get crazy about."

Silverledge Tuxedo Jack, stunning black bi male who combined the Wood and Caligari lines Jean favored. Courtesy of Jean Taylor

Silverledge Miss America was a lovely self blue merle, well built and good looking. She was on the leggy side as a youngster but certain to mature nicely. She was kept specifically to breed but Jean hadn't done much with her. She was nice tempered and easy to get along with so it was no problem to keep her around just for breeding. The Taylors were headed for the Denver Merchandise Mart show and Miss America was in heat. She was young, not quite two years old, but the opportunity to breed was too important. It was January and the weather was poor so there was only one opportunity to get her to Pooh Bear. They got one good breeding and that's all it took. The puppies were spectacular and she kept a couple of them but let the rest go to pet homes since she was now pregnant herself and didn't know anyone at the time who was interested in her type of dog. As she says, "I knew they were good—but not how good—but they had to go."

In keeping with her habit of that time, Jean kept a good female from this cross, Silverledge Sasparilla, and let her dam go. She was pleased with all the puppies in this litter, citing their good quality of mind and nice dark Wood's blue color with no white (a dark merle that darkens

Blue merle and white female, Taylor's Nifty Babe (Jeffreys' Pooh Bear x Bjorks Chula). Early foundation female for Silverledge. Courtesy of Jean Taylor

Taylor's Nifty Babe. Courtesy of Jean Taylor

more with age) as something she was glad to add back into her line. It was an elegance that had been disappearing from her dogs the further removed they became from Luke.

Jean says this about the female she kept from this cross, Silverledge Sasaparilla, "I kept one of the Pooh Bear pups and bred her to Charlie (Silverledge Charlie Russell). Too bad I wasn't in the mood to raise puppies to sell in those days. Those were good pups! Sensible, mellow, typey, strong. The ones that went on to work were highly spoken of. As it was, I kept one bitch, and sold the mother."

The female Jean kept, Silverledge Black Magic, became almost a second foundation for both her and others, producing a number of excellent puppies by her sire Silverledge Charlie Russell, Ch Silverledge Slide Me Five STDc, Newman's Blue Sprog CD STDc, and Sakonnet Times.

Magic was leased to Linda Gray in 1987 and produced two pups by her male Sakonnet Times. A black tri female, Sakonnet Royal competed in a stockdog futurity in Texas shown by Red Oliver. A black tri son, Sakonnet Charlie Hustle (Pete) earned a breed Championship and a CDX. From her litter to her sire, Charlie Russell, she produced the magnificent red merle male, Silverledge Slide Me To Da Bar, CD, STDs. Her cross to Newman's Blue Sprog CD STDc produced the beautiful blue female, Silverledge Freckles.

Magic was a black tri with minimal white. She had the strong Wood type head—broad skull, dry, clean, nice earset—with a good, square body and solid, level topline. A strong "alpha" personality made her a good cross with a variety of males and she always produced top quality puppies no matter who she was bred to.

Another great sire who was largely unknown outside of the few fanciers of the Wood type dogs was Silverledge Charlie Russell. A son of Luke and a female named Colorados Am I Blue, he was produced when Luke was aging in order to preserve his genes. At the time Jean didn't have the right female to cross with Luke so she began looking for something to use in order to keep something for her next generation. Her good friend, Liz Winn, had acquired a bitch with the perfect pedigree, Colorado's Am I Blue. From the Whiskey/Caligari mix Jean favored on one side and a granddaughter of her own Ch Colorado's Shamrock of Silverledge CDX on the other, she was a good choice to go back to Luke.

Blue merle puppy from the cross of Ch. Sharp's Ragnar and Taylor's Nifty Babe. Courtesy of Jean Taylor

Bred by Leslie Sorensen, this blue merle appeared to be a black tri, having only a small spot of merling on her face. Liz was trying to preserve the old bloodlines herself and didn't want to lose this valuable investment. When Jean learned of Liz's purchase she arranged to lease Am I Blue for her litter to Luke in 1982. The result was two puppies, one of which was Luke's last son, the very handsome blue merle Silverledge Charlie Russell.

Blue merle, copper and white female, Silverledge Tequila, sister of Slide and another example of the quality produced by the cross of Luke with other working bloodlines. Courtesy of Jean Taylor

It was fortunate Jean had the foresight to make such a cross before it was too late. Less than 3 years later, in March of 1985, Luke died. Unfortunately for Charlie (and for the breed, in a sense), he was born into a hectic time in Jean's life. Her daughter, Holly, was only a year old and the family was living in a cramped motor home while in the process of building their new house. Charlie was born in September and the weather was cold, wet, muddy and nasty. As a puppy, he had a habit of jumping on the baby and with the demands of their business and the needs of a youngster, Charlie got shoved outside and ignored. He became just a backyard dog and never received the socialization he needed to become a show dog. By the time Jean was able to find time for him, he was two years old and she was never able to undo the early neglect. If only a benefactor had been available, like Terry Martin became for Slide Me Five, this story might have a different ending.

In spite of Charlie's obscurity, Jean had the top quality dog she needed to continue her breeding program. Although she used and appreciated him, she felt he was wasted as a breeding dog. If only he had been more well known he would have been crossed with many more high quality bitches who would have carried on his legacy. In spite of this, he did

Silverledge Tequila. Courtesy of Jean Taylor

Silverledge Tequila as a mature dog. Courtesy of Jean Taylor

Silverledge Belle Starr, blue merle, copper and white female, (Lor A Mickey Five x Silverledge Tequila) as a puppy. Courtesy of Jean Taylor

have one topnotch outside mate, WTCH Slash V Slide Me Sweet CD, winner of the Washington Futurity and at that time the only female to have ever won all three classes of livestock at the ASCA National Finals—becoming the first Supreme Champion Finals Stockdog in the history of the breed. Charlie was her first mate and this line breeding produced an excellent litter, one of which, WTCH Oliver's Romulus Five, became a WTCH and foundation sire for Nebraska's Bruce Nelson. The other puppies went to ranchers and all were sturdy and good tempered—just as their pedigree dictated.

Thankfully Jean used him judiciously and his contribution, though small, was topnotch. As described previously, his daughter, Silverledge Black Magic, became a top matron for Jean, producing a number of excellent offspring both for her and others. His last son, Silverledge Slide Me To Da Bar, CD STDs, became an important addition to both the HS & W and Kristone kennels.

As a backyard dog, Charlie was still part of the Taylor family and exhibited the sweet and noble character the line has always been known for. Like his sire, he was an intense and attentive companion. He ignored strangers unless they would approach him making eye contact, when his protective instincts would surface and he would warn them off. He was neither people aggressive nor dog aggressive and was always an easy dog to live with. He traveled well and always had quiet, perfect manners.

In Jean's promotional material she describes Charlie's conformation this way "Medium bone, very deep chest, powerful rear. Very well angulated, a little tall for his length, but he doesn't crab. Moves like a dream. He has an elegance about him. Color—very clear and intense, no cast or mud to him. Dark 'Wood's blue', born with lots of contrast, darkens with age. Very little white or copper trim in this line." This description is one seen time and time again of dogs of this line.

Charlie didn't languish in his backyard full time. From time to time Jean would take him out on stock to see what he would do and she even did some elementary training with him at Steve Winn's. His first exposure was at about a year of age and it took him a few times to learn what he was to do. Once he figured out his job he was equally as good on the head or the heel. Like his sire, he was very sensitive to direction and had an almost telepathic ability to understand what Jean wanted. Also like Luke, he was devastated if she became angry or upset, so she always took care not to ruin him by creating a master/slave relationship. An equal partnership was her aim and when she achieved it she says it was "pretty incredible."

Silverledge Belle Starr as a mature dog. Courtesy of Jean Taylor

Even with little opportunity, Charlie showed his ability and genetic background with his strong instinct and style. He had a good gather or fetch and ran fairly wide for an Aussie. He was sensitive and easy to direct but still had a lot of power. Although he would bite, he also would listen and his power emanated from confidence and will, not aggression and punishment. Jean tells the story of one memorable day when she was working a group of about 15 mama cows with calves. She was pleased with the lesson, seeing some nice fetlock heeling and good wear from Charlie. At one point a cow challenged him and he took her in the face. Charlie had mama cow with the entire group heading down the road at a high trot and Jean was proudly thinking she had quite a dog. Her smile turned to a frown when she realized "…if they hit the highway at that speed it would be hamburger patties, so I called him off. Lo and behold, he got off and came back to me."

Two of the finest females ever produced in the breed. (standing) Silverledge Miss America (Taylor's Luke the Drifter CDX x Silverledge Mindy Taylor) and her daughter (lying down) by Jeffreys' Pooh Bear, Silverledge Sasaparilla. Both dark blue merles who kept the old Wood type alive through Sasaparilla's daughter Silverledge Black Magic. Courtesy of Jean Taylor

Charlie's hit or miss opportunities on stock continued through his life. Although he had the talent, nobody ever had the time to develop it so he remained one of the many talented ranch dogs who never achieve

Silverledge Miss America. Courtesy of Jean Taylor

The beautiful, correct head type of Silverledge Sasaparilla. Courtesy of Jean Taylor

Silverledge Sasaparilla. Courtesy of Jean Taylor

fame except among their own family of admirers. Jean took him out in her arena once when he was 11 years old and he was still the same old worker she had always seen. He still tried as hard as he ever had but at that age he was not fit and could not make the effort without harming himself. She finally called him off and put him up to protect him from himself. As she says, "They are athletes and you have to keep their welfare in mind."

The last dogs of Jean's original line were almost all Charlie pups or grandpups. In addition to Silverledge Black Magic, Jean had an outcross bitch named Silverledge Bandit. A dark blue self merle, this Charlie Russell daughter out of Silverledge Catfoot Lady was used to cross back to the Buckeye Bobby type dog, Tuffy—Jean's favorite cross. Catfoot Lady was also a half sister to Tuffy out of Tuffy's dam Slash V Steel Me Five.

From the cross of Tuffy to Bandit came a mismarked blue merle female named Silverledge Gizmo and a stunning red merle male named Silverledge Pincie Creek Drifter, known as Hawkeye. Gizmo remained at Silverledge while Hawkeye, as his name implies was sold to Roger Stevens of Pincie Creek. He was used lightly there, then sold to Bill Wyatt where he was used extensively at stud, producing well with several outcross females. Gizmo was bred to the inbred Charlie Russell son and grandson, Silverledge Slide Me To Da Bar, to produce two beautiful red merle males, Silverledge Last Call Luke, owned by Michelle Murvai and Silverledge Skeet of L7, owned by Buster and Donna Berryhill.

Silverledge Bandit was also leased by Linda Gray and produced a nice litter when bred to her Silverledge Black Magic son. Bandit was, in appearance, typical of her Wood breeding, about 19 inches tall and 38 pounds, well made, very muscular and extremely quick and agile.

She had the typical wariness of strangers but traveled well and was unconcerned in public situations. On stock she was very intense and had a precise balance. She would shove her way between sheep and a fence to get where she thought she should be. She was only used on cattle a few times and proved to be quick, determined and more than able to turn and move them where she wanted. She had a strong fetching instinct with the biddability to make training her a pleasure. Jean later bred her to Silverledge Tuxedo Jack resulting in another superb obedience dog named OTCH Silverledge Lady of Spain. Shown by her owner she earned every agility and obedience title available in AKC, ASCA and CKC!

Another valuable cross was made with the great Silverledge Black Magic to the old Wood bred dog Newman's Blue Sprog CD STDc. From this cross, Jean kept a striking self blue merle bitch named Silverledge Freckles. She says this about Freckles, "The old Colorado lines tend to be deep-thinkers, steady, intense, and tuned in. They can take a lot of discipline, but tend to be more sensitive to criticism than some. Not quitters, they just care what you think. Wider workers, harder and sharper with more bite than most. They have a lot of stock sense and want-to, and are willing to think with you." Jean describes Freckles as having plenty of power on stock with both bite and some eye. Well balanced and having good self control she was trainable and biddable. Jean liked the intelligence of the whole litter, stating, "I have raised some smart dogs, but these were something special, when it came to thinking independently."

The cross to Sprog gave Jean back the close old Wood breeding that was slipping away. She describes Freckles as "mellow, loving personality, fairly protective, VERY personable, sound, solid…a keeper." She was about 20 inches in

Silverledge Sasaparilla as a young dog. Courtesy of Jean Taylor

Silverledge Sasaparilla showing the beautiful head type that was her birthright. Courtesy of Jean Taylor

Silverledge Black Magic, as beautiful in motion as standing still. Courtesy of Jean Taylor

Silverledge Sasaparilla as a mature dog. Courtesy of Jean Taylor

Silverledge Sasaparilla as a mature dog. Her coat was turning dark and she looked nearly black at this time—typical of merles of this line. Courtesy of Jean Taylor

Silverledge Black Magic, black tri female of excellent type. Courtesy of Jean Taylor

height and 45 pounds, nicely balanced, well angulated and very strong. Jean praised this lineage in a letter to Betty Nelson in the early 90s, "Then I have gone back to Sprog to try to hold the old-time Colorado blood. I like that look best of all, and the good ones were fabulous. Only recently have I had an opportunity to train a whole lot on stock. Have been trying to improve temperment and structure all this time. I might have lost something along the way, but I doubt it. I'm pretty pleased at what I am seeing in my young dogs."

Freckles produced two litters in her lifetime. First bred to Zia's Tuff Enough Robin CD STDd OTDs RTDs, her son, Silverledge Tuxedo Jack became an important addition to Jean's kennel. He was eventually used by Becky Anderson of Lor A kennel and his lines continue today through this link. Freckles was also bred once to a beautiful solid black male obtained from Joe Gahl of Minnesota. Gahl's Silverledge Smokin' Jose blended the lines of Ch. Jones Reddy Teddy CDX CWD with Newman's Blue Sprog CD STDc. His sire was Pincie Creek Rembrandt (Bearfoot's Cinder and Ash x Slash V Bittersweet Sis) and his dam was Gahl's Casey, a daughter of Sprog and Gahl's Terra (Ruff N Reddy Shade—Ruff N Reddy Shadey Joe x Ruff N Reddy SeMe Sashay out of a largely Sorensen bred female named Tina Horejsi)

A last attempt to preserve the wonderful lines of Luke and Charlie was the cross Jean made of Magic back to her sire, Charlie Russell. This litter resulted in two puppies, a male and a female. The female became a companion at a horse barn and Jean kept perhaps the greatest dog she ever bred, Silverledge Slide Me To Da Bar, called "Tigger."

When Tigger was still a youngster, Jean had this to say about him in her promotional materials, "Not a year old yet. He must be 22" and

65 lbs already. He's going to be a big dog. Very impressive, too. Classic old Colorado style, powerful and angulated.

Temperament—confident and assertive. Likes most people, adores little kids. I'll be showing him...will take some work to get him to accept the exam, but not a big problem.

He is a very late bloomer. Has a massive head, and even though he's barely adolescent, there's a sense of power about him. He's very obedient and eager to please, fortunately. A very personal dog...loves to be with you, patient and sensible, nice to be with.

Stock—very assertive, very powerful. Pretty wild. Had him in with the sheep 3 or 4 times and decided to wait and get a really good down on him, so I have a little control. Going to take it slowly...As I said, he is still very young."

Silverledge Sasaparilla, a beautiful, correct female. Courtesy of Jean Taylor

Jean bred this magnificent dog once with Silverledge Gizmo before her husband, Tom, died. Shortly after Tom died, Tigger was sent to Kris Toft to breed to a Sprog granddaughter, Kristone Mity Black Medicine CD (Ch Fairoaks Medicine Wolf CD x Masterson's Gitel). Once this breeding was complete, Tigger remained with Kris for the remainder of his life. This litter produced 6 puppies. One female, Kristone Medicine Woman, became a new foundation for Kristone, HS&W and Remuda kennels. In addition, Tigger was bred to another Gitel daughter, HS& W Blazing Star, and this litter formed the nucleus of the Wood bloodlines now found in England and Europe.

Tigger, like his sire/grandsire Charlie, was a great dog who was an incredibly intuitive dog, he was as honest as Jay Sisler and spent his life looking for ways to please. Trained in obedience prior to coming to Missouri, it took very little to prepare him for his CD title. He earned his first two scores in fine fashion but his third leg was one for the record books. It was a rainy midsummer day and Kris just wanted to finish that CD and go home. A large Novice class at a popular show gave the pair time to warm up but the handler was not concentrating fully on the task ahead. By the time their number was called, Kris was wet, cold and tired and just wanted to be done with the job. It's a shame nobody was paying much attention except the judge because when the twenty-some qualifiers returned to the ring for their scores, Tigger's number was called first.

Silverledge Black Magic, a superior individual with a superior background. Courtesy of Jean Taylor

Phantom blue merle, copper and white female, dam of Silverledge Charlie Russell, Colorado's Am I Blue. Courtesy of Jean Taylor

Colorado's Am I Blue. Courtesy of Jean Taylor

Amazingly he had overcompensated for the lackluster attitude of his handler and pulled off a first place win in this huge class!

His talent on stock knew no limits except those of his handler. A powerful dog who was 100% a team player, he constantly tried to understand what was wanted of him even when his handler was not always sure. He worked close but was easy to push out and easily learned to rate his stock. Cattle were his forte and he could handle anything which faced him. He only achieved an STD on sheep but this was due to lack of opportunity, not lack of talent.

Tigger's temperament was second to none. Afraid of nothing he was friendly to people and dogs alike. Aloof but cordial he did not push his affections on anyone but was happy to receive attention when it was offered. But it was his regal bearing which was unique in the dog world. He controlled any room he walked into and was that real rarity—a genuine "alpha" dog. He never found a need to fight other dogs—they instinctively knew he was in charge. He once was walking through a large covered show arena filled with dogs being prepared for showing. As he passed by, every head—human and canine—turned to look at him. He once nearly stopped in mid-air two surly country dogs racing through a hedge to attack him. He merely stood still by the opening in the hedge with his lips pulled back from his substantial teeth daring them to come near him. Not a sound came from him and the two attackers minced slowly away, not wishing to find out what fate might await them should they foolishly try to take him down.

He once stopped a fight in which four bitches were attempting to kill a male dog. Kris was panicking because there was no way to pull off the attackers without leaving the poor male to the rest. Tigger came to the rescue. His body slammed the entire group, which disappeared

Two prodigal sons of Taylor's Luke the Drifter CDX. (Left) Ch Silverledge Slide Me Five STDc and (right) Silverledge Charlie Russell. Courtesy of Jean Taylor

instantly, and came to sit in front of Kris. It took a moment to decipher what had happened but Tigger waited patiently until he was told what a good boy he was! He terrified his ill-mannered daughter once when she dared to race by him and nip him on the back end. She had been raised in a household where she was never disciplined and never learned respect for her elders—or anyone else for that matter. She found it amusing to pick on Tigger until the day he decided to explain the proper way to treat a better. He literally pretended to ignore her while sniffing around the ground. As she made her pass he rounded on her, knocking her to the ground and rolling her physically down a hill yelping hysterically every step. Tigger had not harmed her in any way but his physical dominance of her along with his menacing voice convinced her to abandon her improper game.

Beautiful Wood type is seen in this head shot of Silverledge Charlie Russell as a young dog. Courtesy of Jean Taylor

Tigger taught many more lessons than he was taught by any person. A true alpha dog is at the top because of his innate confidence and understanding of how polite canine society can work. Physical dominance is not necessary unless used as a last resort. Alpha dogs are not required to submit. They understand hierarchy and respect it as long as those around them also understand and respect it. Tigger never shrank

Silverledge Charlie Russell as a young dog. Courtesy of Jean Taylor

Silverledge Charlie Russell, the epitome of good type for the breed and his line. Courtesy of Rowe

Silverledge Charlie Russell as a young dog. Courtesy of Jean Taylor

Silverledge Slide Me To Da Bar CD STDs. Photo by Alex Kapsemalis

from his responsibilities as an alpha dog, wore the burden with nobility and good judgment and accepted the perks of being top dog as well. He expected the same respect in return and, although forgiving of fools, he did not tolerate injustice even though he realized we all make occasional mistakes. Although he was almost always serious, he was a happy dog who came by his name, Tigger, honestly. He was a once in a lifetime dog who left an unforgettable mark on anyone who ever knew him.

Today Jean has retired from business and finally has the time to devote to learning about stock dog training and she is still breeding on a small scale. Her lifetime of experience in the breed is now guiding her breeding choices and she has dogs which fill her criteria well. In a 1995 interview in the Australian Shepherd Annual she relates many of her views on breeding, where the breed is, where it is going, and what she has found to be the best type of dog. The following are some of her observations and conclusions. Most are excellent guidelines for anyone wishing to breed and preserve the best of the Australian Shepherd.

Silverledge Charlie Russell. His timeless beauty and Wood type are evident. Courtesy of Jean Taylor

The Aussie has been bred for show ring "perfection" for many decades now. The emphasis on outward appearance is based solely on what appeals to judges and has bypassed things that make a dog durable or useful in working situations. "Pretty" is in the eye of the beholder and Jean has always found the elegance of the Wood dogs a striking and beautiful contrast to the dogs being bred for show. Though appearance is what one notices first, she demands a dog who will look you in the eye and relate to you. A dog that can go all day if needed. Only a dog with good, sturdy conformation can do that. Traits selected for by the show ring detract from that in many ways.

Silverledge Bandit, blue merle female, copper and minimal white. Courtesy of Jean Taylor

Over the years Jean never veered away from her goal of preserving the real honesty of the breed along with versatility. She selected for good structure, good enough to meet the show ring breed standard, but did not neglect versatility in her dogs. Although she didn't always have the opportunity to work stock, she always took advantage of what opportunities she DID have to test her dogs and train them to the best of her ability. Her dogs didn't let her down and the good crosses she made preserved the instincts and trainability she was fortunate to start with.

The Aussie was largely founded as a cattle working dog. Jean recognizes that in order to preserve the real Aussie, there is a need for real "dyed-in-the-wool cattle dogs and ranch dogs that will go out and work every day." Breeding for sports and trial work is helpful and may be one of the few methods left to really evaluate and preserve this type of dog. Until there is more understanding of the need and use of these dogs, it will be difficult to preserve what is being quickly lost.

Silverledge Catfoot Lady, blue merle, copper and white female (Zia's Heartbreaker x Slash V Steel Me Five), dam of Silverledge Bandit. Courtesy of Jean Taylor

Silverledge Catfoot Lady. Courtesy of Jean Taylor

As the world changes, the variety of uses for Aussies continues to expand. From working trial sports to agility to companionship to watch dog or ranch dog, a good Aussie should be able to do a fairly wide range of duties. A trait Jean has found to be valuable and a good indicator of intelligence and ability is the desire to carry things. Sprog was a clever dog who had a natural retrieving ability and Casey Newman capitalized on this ability for a wide range of uses. Jean's Sprog daughter, Freckles, inherited this trait and so did many of her puppies. This is the kind of unseen ability Jean is hoping to preserve in her dogs into the future.

Ultimately Jean still aims for a quality mind in a sound body—exactly the type of dog she had when she started in the breed. She insists on keeping the old Colorado type of conformation, color, head, and overall look. Even though her dogs were seldom famous she was able to keep the type she wanted through blending three lines which proved to cross well. The Caligari/Wood crosses where her ideal but she always selected from those litters the dogs which retained the qualities she already had and didn't want to sacrifice. She has found the new bloodlines aren't proving superior to the old lines and has chosen to stick with her foundations to keep good, determined low heeling ability for cattle work in her dogs.

Conformationally, Jean has always preferred a moderate sized and built dog. Bone should relate to that size and should not be heavy, nor should a dog overall be heavy. Beefy dogs do not hold up to all day hard work and get muscle fatigue sooner than a moderate and balanced dog. Good angulation has always been important to her, especially in the shoulder. Angulation and muscling have always been intertwined and she has always recognized the need for plenty of good quality muscle to balance and control the bone structure in dogs. She wants a fairly level topline on a somewhat long bodied dog. A hard coat is a must. Her ideal was one of her foundation bitches who could be wet, muddy and disgusting, yet lie down on the floor for a nap and upon awakening, rise up, leaving all traces of dirt in a pile on the floor. Her coat not only shed dirt and water but odor, too. This is the functional type of coat a working dog needs in a real world situation.

Head type is important to the character and definition of a breed and Jean's early experience with many of the founding dogs stamped her preference. She likes a blocky head but not a heavy head with a fair back skull. She breeds for a heavier ear than the usual higher Wood ear but prefers something between the high Wood ear and the (too heavy) Caligari type ears. Most of all she wants a clean head with good expression and there must be intelligence in the eyes that says, "Yes, I am here. I am myself." The personality of the dog should show through in the entire expression and demeanor of the dog. She does not see the Aussie as a dog for everybody. They require effort and shouldn't just be a house decoration. Her ideal dog is a demanding one who wants to serve it's human and needs a job. If Aussies are expected to tolerate a modern lifestyle, temperament is of the utmost importance. While preserving the true type Aussie, owners must be prepared to direct these personality traits to proper outlets and not allow the dog to create inappropriate activities from boredom.

Silverledge Catfoot Lady. Courtesy of Jean Taylor

Black tri litter sister to Silverledge Bandit. The consistency of this line is evident in this female's similarity to her half sister Silverledge Black Magic. Courtesy of Jean Taylor

Jean sums up her goals in an excerpt from a letter she wrote to Betty Nelson: "Someday maybe I will be a connoisseur of working style, but at the moment I am just learning fast. As long as the instinct is there, I'm not willing to sacrifice attitude or structure for a certain style. My aim is to raise a middle-of-the-road dog that a novice can work, or obedience, or do therapy or agility or whatever, and still be tough enough to be at least adequate on cattle." Sounds like a good definition of an Aussie for anyone!

From this same letter Jean comments on what she sees at shows and her satisfaction with the standards she has set for herself. "I do alright myself at shows. I have to have a good one before I will take the trouble, but I love the sensation one of these dark merles with no white causes... especially when they are built like a brick outhouse. I have to tell you,

Silverledge Gizmo in spite of her markings a very correct and high quality female. Courtesy of Jean Taylor

Silverledge Gizmo at about 6 months of age. Courtesy of Jean Taylor

Silverledge Last Call Luke as a young puppy. Courtesy of Michelle Murvai

Silverledge Last Call Luke (Silverledge Slide Me To Da Bar CD STDs x Silverledge Gizmo. Photo by Author

Silverledge Last Call Luke, red merle male. Courtesy of Michelle Murvai

whenever I go to a show, I come home and kiss ALL my dogs! Have never regretted NOT breeding for show dogs."

Jean Taylor is one of those breeders who never sought fame and notoriety but who always had a firm idea of her own likes and goals in breeding dogs. She never fell for the many fads that came and went but observed and learned from the top breeders whose skill had been proven over time. She chose what she liked from those programs and developed a preservation breeding program on a small scale that produced many top dogs. Although she bred and owned one of the most famous and influential sires in the history of the breed she did not base her breeding decisions on fleeting show or trial success. Her personal standards were timeless and proven and she was dedicated to breeding first of all to that standard. This dedication and integrity served both her and the breed exceedingly well.

Silverledge Bandit, blue merle, copper and white female. Courtesy of Jean Taylor

Silverledge Bandit as a youngster. Courtesy of Jean Taylor

Silverledge Bandit. Courtesy of Jean Taylor

Silverledge Freckles, blue merle female, daughter of Newman's Blue Sprog CD STDc who she strongly resembled. Courtesy of Jean Taylor

Silverledge Freckles, correct structure, good type and a regal bearing were abundant in this female. Courtesy of Jean Taylor

Silverledge Bandit. Photo by Author

Solid black male, Silverledge Ace, by Newman's Blue Sprog CD STDc x Silverledge Black Magic, littermate to Silverledge Freckles. Courtesy of Jean Taylor

Silverledge Ace combined the Silverledge and HS & W lines in a beautiful, correct package. Courtesy of Jean Taylor

Silverledge Tuxedo Jack, black bi male, as a young puppy. Jack was a blend of the Wood and Caligari type dogs Jean preferred. Photo by Author

Silverledge Freckles in the foreground with her son, Silverledge Tuxedo Jack in the background. Both dogs are watching intently while another dog works sheep. Photo by Author

Silverledge Tuxedo Jack as a stunning 9 month old. Courtesy of Jean Taylor

Gahl's Silverledge Smokin Jose. Courtesy of Jean Taylor

Gahl's Silverledge Smokin Jose. Courtesy of Jean Taylor

Gahl's Sam, Red merle littermate to Jose, owned by Joe Gahl. The Gahl's raise market hogs and use the dogs every day on their Minnesota farm. Courtesy of Jean Taylor

Gahl's Sam (Pincie Creek Rembrandt x Gahl's Casey

Silverledge Heartbreaker STDdsc , blue merle male by Gahl's Silverledge Smokin Jose x Silverledge Freckles. Courtesy of Jean Taylor

Blue merle female from Gahl's Silverledge Smokin Jose x Silverledge Freckles. Courtesy of Jean Taylor

Jose x Freckles puppies. Courtesy of Jean Taylor

Silverledge Heartbreaker STDdsc, blue merle male by Gahl's Silverledge Smokin Jose x Silverledge Freckles. Courtesy of Jean Taylor

PEDIGREE OF Gahl's Silverledge Smokin' Jose

Born: ____________ Sex: ________
Owned by: ____________________

Parents	Grandparents	Great-grandparents	Great-great-grandparents
Sire: Pincie Creek Rembrandt	Sire: Bearfoot's Cinder N' Ash	Sire: Ruff N Reddy Joe Willy Wang	Sire: Jones Reddy Teddy
			Dam: Ruff N Reddy Mity Happy Hocker
		Dam: Ruff N Reddy Shadey Wood Ash	Sire: Jones Little Redwood
			Dam: Shadey Mt Mity Miss
	Dam: Slash V Bittersweet Sis	Sire: Slash V Spirit of Ag-gieland	Sire: Slash V Little Rock
			Dam: Slash V Semi Sweet
		Dam: Slash V Paper Doll	Sire: Silverledge Slide Me Five
			Dam: Slash V Semi Sweet
Dam: Gahl's Casey	Sire: Newman's Blue Sprog	Sire: Adair's Black Bear	Sire: McGuire's Semego Hemi
			Dam: Mastersons Dotclijee Shunka
		Dam: Masterson's Dotelijee Shunka	Sire: Jeffrey's Pooh Bear
			Dam: Poudre Condesa
	Dam: Gahl's Terra	Sire: Ruff N Reddy Shade	Sire: Ruff N Reddy Shady Joe
			Dam: Ruff N Reddy Se Me Sashay
		Dam: Tina Horejsi	Sire: Hiawatha Copper Stub
			Dam: Volza Blue Lady Booch

Bred by: ____________ Address: ____________ Phone: ____________

The details given above are correct to the best of my knowledge. Signed: ____________(Breeder)

Silverledge Slide Me To Da Bar CD STDs as a puppy. His direct confident gaze is an omen of the noble self assured dog he was to become. Courtesy of Jean Taylor

Silverledge Slide Me To Da Bar CD STDs as a puppy. Courtesy of Jean Taylor

Silverledge Slide Me To Da Bar CD STDs. Courtesy of Jean Taylor

Silverledge Slide Me To Da Bar CD STDs as a "teenager." Courtesy of Jean Taylor

Silverledge Slide Me To Da Bar CD STDs at about 2 years of age. Courtesy of Jean Taylor

PEDIGREE OF Silverledge Slide Me To Da Bar

Born: ______ Sex: ______
Owned by: ______

Parents	Grandparents	Great-grandparents	Great-great-grandparents
Sire: Silverledge Charlie Russell	Sire: Taylor's Luke the Drifter	Sire: Woods Stubbie	Sire: Woods Jay
			Dam: Woods Slate
		Dam: Bjorks Chula	Sire: Bjorks Baron
			Dam: Roadrunners Cholla
	Dam: Colorado's Am I Blue	Sire: Slash V Wild Timber	Sire: Martin's Tim Tim
			Dam: Hoyt's Wildwood Flower
		Dam: Colorado's Houdini	Sire: Colorado's Shamrock of Silverledge
			Dam: Las Rocosa Colorado Genie
Dam: Silverledge Black Magic	Sire: Silverledge Charlie Russell	Sire: Taylor's Luke the Drifter	Sire: Woods Stubbie
			Dam: Bjorks Chula
		Dam: Colorado's Am I Blue	Sire: Slash V Wild Timber
			Dam: Colorado's Houdini
	Dam: Silverledge Sasparilla	Sire: Jeffrey's Pooh Bear	Sire: Jeffreys Shade
			Dam: Sorensens Blue Mist
		Dam: Silverledge Miss America	Sire: Taylor's Luke the Drifter
			Dam: Silverledge Mindy Taylor

Bred by: ______ Address: ______ Phone: ______

The details given above are correct to the best of my knowledge. Signed: ______(Breeder)

Silverledge Slide Me To Da Bar CD STDs "Tigger," so named for the way he bounced with joy in anticipation of any job! Courtesy of Jean Taylor

Tigger winning a very large and competitive AKC Novice B class to complete his CD title

Silverledge Last Call Luke, 2 years old. Photo by Author

Silverledge Skeet of L7, red merle dog by Tigger x Silverledge Gizmo. Photo by Author

Silverledge Last Call Luke, approximately 10 years of age. Courtesy of Michelle Murvai]

Silverledge Last Call Luke, red merle dog by Tigger x Silverledge Gizmo at about 6 months of age. Courtesy of Michelle Murvai

Kristone CR Mity Blue Magic "Charlie," blue merle Tigger son out of Kristone Mity Black Medicine CD "Shadow." About 6 months of age. Photo by Author

Charlie at age 13 years. Photo by Author

Charlie at about 9 months of age. His resemblance to his grandfather/great-grandfather, Silverledge Charlie Russell is unmistakable. Photo by Author

Charlie (Tigger x Shadow) at about 9 months of age. Photo by Author

Charlie at age 13 years—very dark now and still a noble, magnificent dog. Photo by Author

Tigger at home in New Mexico in 1996. Courtesy of Michelle Murvai]

Charlie showing the strong head type Tigger produced consistently. Photo by Author

Silverledge Slide Me To Da Bar CD STDs at about 2 years of age. Courtesy of Jean Taylor

Mimi at age 11 years. Sound with good type typical of her line. Photo by Author

Kristone Medicine Woman "Quinn," Tigger daughter out of Kristone Mity Black Medicine CD—a Sprog granddaughter. Approx 4 months old

Black female, Kristone Mity Screamin Mimi, daughter of Tigger x Kristone Mity Black Medicine CD. Producing consistent good conformation, solid minds and excellent type were Tigger's hallmarks. Photo by Author

Quinn as a young dog

Tigger epitomized the nobility and dignity of the Wood line. Courtesy of Jean Taylor

Lena showing her superb head type. Courtesy of Jean Taylor

PEDIGREE OF Lor A Smart Little Lena

Born: ________ Sex: ________
Owned by: ________

Parents	Grandparents	Great-grandparents	Great-great-grandparents
Sire: Lor A Iyanbito Flint Lex	Sire: Lor A Yukons Denali	Sire: Silverledge Tuxedo Jack	Sire: Zia's Tuff Enough Robin
			Dam: Silverledge Freckles
		Dam: Lor A Native Lady Yukon	Sire: Lor A Two Blu
			Dam: Lor A Native Lady Raven
	Dam: Lor A Star Blu Lady Quinn	Sire: Lor A Five Star Blu	Sire: Connollys Sweet Smiley
			Dam: Silverledge Belle Starr
		Dam: Lor A Native Lady Jane	Sire: Lor A Native King Zeb
			Dam: Lor A Shamrock Queen
Dam: Lor A Gypsy Traveler	Sire: Silverledge Tuxedo Jack	Sire: Zia's Tuff Enough Robin	Sire: Zia's Take Five
			Dam: Slash V Steel Me Five
		Dam: Silverledge Freckles	Sire: Newmans Blue Sprog
			Dam: Silverledge Black Magic
	Dam: Lor A Native Lady Yukon	Sire: Lor A Two Blu	Sire: Haggands Dusty Rose Canyon
			Dam: Lor A Connbllys Sugar Candy
		Dam: Lor A Native Lady Raven	Sire: Lor A Five Star Blu
			Dam: Lor A Native Lady Jane

Bred by: ________ Address: ________ Phone: ________

The details given above are correct to the best of my knowledge. Signed: ________(Breeder)

Lor A Smart Little Lena (Lor A Iyanbito Flint Lex x Lor A Gypsy Traveler), blue and white female exhibiting the blend of Wood and Buckeye Bobby blood which has been her lifelong goal. Photo by LM Gray, Courtesy of Jean Taylor

Lor A Smart Little Lena—Preserving the future requires a dedication to understanding the past! Courtesy of Jean Taylor

Lena has retained all the natural working ability of her bloodline. She is furthering Jean's education about training stockdogs today. Courtesy of Jean Taylor

chapter 17

Cathy Jean Jones
RUFF-N-REDDY

Lifelong horsewoman and Fort Collins, Colorado native, Cathy Jean Jones, was most responsible for popularizing the red Wood dogs through her beloved male, Ch Jones Reddy Teddy CDX OTDdsc SCH HA. Cathy grew up in a Colorado ranching family learning to ride and train horses, eventually bringing this love of animals and her talents as a trainer to the Australian Shepherd. She met her husband, Dick, through their mutual interests in horses, showing and rodeo and for many years they showed and bred Arabians and Quarter Horses all over the United States.

The two had always had dogs around their home ranches growing up but they had been "just dogs." Dick was likely more familiar with the Aussie from his experience working for a variety of ranchers and being around stock dogs but neither he nor Cathy had any real experience with the Aussie before they found their first dog through a newspaper ad. They were living near Fort Collins, Colorado in the late sixties and just wanted a pet for their home.

From the litter they chose a blue merle and copper trimmed female they named Smokey. She was just chosen as a pet and filled this role perfectly. Cathy states she was neither purebred nor registered but she did have all the traits of any good Aussie and turned out to be a good introduction to the breed.

Ted as a young puppy at one of his first shows—starting off a winner! Courtesy of Raymie Jones

The red merle full sister of Ch Jones Reddy Teddy CDX OTDdsc, SchHA, Davies' Pinka. Photo by Sue Noll, courtesy of Raymie Jones

When they moved to their Wellington place in 1973, Smokey was about 5 years old. With more room in their home and their hearts, they decided to search for a second dog to be a mate to Smokey and companion for Dick. Some friends in Fort Collins knew some people who raised purebred Aussies and when they called Dick and Cathy in 1973 to tell them about a litter of puppies, they were ready to go shopping. There was a puppy leftover from the latest litter and the breeder, Limon Davies, was having trouble finding a home for him. Cathy went to look at him, liked what she saw and brought him home. In hindsight she wondered about the wisdom of adding to their household so quickly after their move and the start of her husband's new business. But fate has its own timetable and the red male who was to change Cathy's life was hers.

"That one dog started everything," says Cathy. Good fortune brought her not only a perfect companion and teammate but a stud dog with the pedigree to become one of the finest producers of top working dogs in breed history. This acquisition also happened during a time when Aussie activities were taking off in this part of Colorado and Cathy quickly joined the fun. Her friendly personality made her friends wherever she went and her growing attachment to her red male blossomed into a competitive partnership second to none.

As an accomplished horsewoman, Cathy knew the value of pedigree and she soon learned her new buddy had a bloodline well known in the area. Fletcher Wood was an important influence and everyone knew the value of his breeding. Cathy states, "this particular dog was not bred for show—he was bred for protection. They had all the white and copper bred out of them. For many years they were bred strictly for working; if they didn't work, they didn't stick around. They were well-known for being very rough cattle dogs and very protective of their place. It had come to the point where some of them had a reputation for being real rotten, but they were bred that way. The people who had them wanted them to be that way. They were not sociable with strange people. When I got the little dog, I didn't get him for that purpose. I purchased him only to bring him home, raise him here, use him with the other dog and have him as a family pet."

It's likely Cathy learned this early on from Joe Piz, who owned Ted's sire, Piz's Little Joe. Little Joe was the type of "one man dog" the line was known for, the dog having picked him when he went to claim his stud fee puppy from his litter. Little Joe was very devoted to Piz, who used him as a guard dog at the Piz bakery, as a turn back dog to train his cutting horses, and to guard the family race cars on the weekends. Although Joe called him "rotten" he states he was not vicious as long as you knew your place and made no wrong moves. His intelligence was such that he could merely watch other dogs do something and learn how to do it himself. Ted seemed to have inherited much of his character from this side of the family.

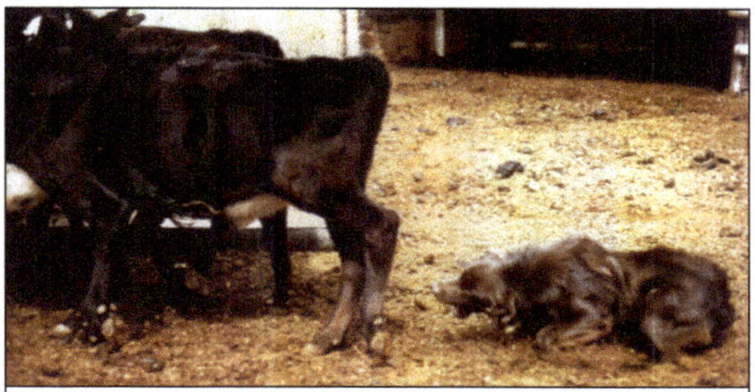

Ch Jones Reddy Teddy CDX, OTDdsc, SchHA early on in his training on cattle. Courtesy of Rowe

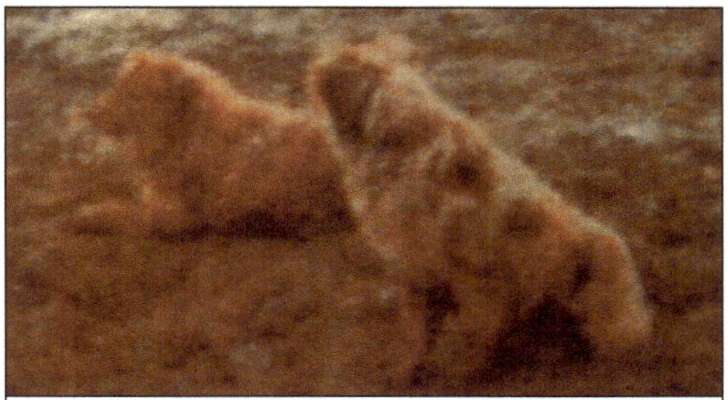

Ted's sire, Piz's Little Joe on the left. Unknown dog on the right. Courtesy of Joe Piz

Ted and his full sister Davies' Pinka. Courtesy of Raymie Jones

Piz's Little Joe. Courtesy of Joe Piz

PEDIGREE OF Ch Jones Reddy Teddy, CDX, OTDc,s,d, SchHA

Born: ____________ **Sex:** ____________
Owned by: ____________

<table>
<tr><td rowspan="8">Sire
Piz Little Joe</td><td rowspan="4">Sire
Piz Joseph</td><td rowspan="2">Sire
Woods Jay</td><td>Sire Sislers Shorty</td></tr>
<tr><td>Dam Sislers Trixie</td></tr>
<tr><td rowspan="2">Dam
Woods Chili</td><td>Sire Woods Jay</td></tr>
<tr><td>Dam Greens Dingo</td></tr>
<tr><td rowspan="4">Dam
J Bar D Choctaw Chilli</td><td rowspan="2">Sire
Rowes Warrior Bow</td><td>Sire Ronsleys Charcoal Smoke</td></tr>
<tr><td>Dam Murphys Cochina</td></tr>
<tr><td rowspan="2">Dam
Sorensens Queenie</td><td>Sire Sorensens Gunsmoke</td></tr>
<tr><td>Dam Nightingales Tammy</td></tr>
<tr><td rowspan="8">Dam
Davies' Ginger Red</td><td rowspan="4">Sire
Montoyas Rowdy</td><td rowspan="2">Sire
Clingers Royal Blue Merle</td><td>Sire Sorensens Gunsmoke</td></tr>
<tr><td>Dam Van Gordons Blue Asta</td></tr>
<tr><td rowspan="2">Dam
Kalbs Princess</td><td>Sire Clingers Royal Blue Merle</td></tr>
<tr><td>Dam Shockleys Blue Samantha</td></tr>
<tr><td rowspan="4">Dam
Montoyas Snipper</td><td rowspan="2">Sire
Piz Joseph</td><td>Sire Woods Jay</td></tr>
<tr><td>Dam Woods Chili</td></tr>
<tr><td rowspan="2">Dam
Pizs Lady II</td><td>Sire Piz Joseph</td></tr>
<tr><td>Dam Piz Lady I</td></tr>
</table>

Bred by: ____________ **Address:** ____________ **Phone:** ____________

The details given above are correct to the best of my knowledge. Signed: ____________ (Breeder)

In later years, after Cathy became educated about genetics and bloodlines she considered the Wood dogs as the outstanding dogs in her area. Specifically she admired the Sisler dogs, Wood's Jay, Wood's Stubbie, Piz's Joseph and Piz's Little Joe as the top foundation working dogs of her area. These dogs were well known by this time in the ranching community and had been strictly bred for ranch work and not showing. The attitude of the day was ruthless when it came to culling litters and any dog that didn't work just didn't stay around for very long. They were certainly not used for breeding! Cathy's comment about this practice, "I don't know that that is all wrong."

In 1990, Mark Lane interviewed Joe Piz who made the following statement about Cathy and Ted. "The biggest thing that ever happened to me to make my dogs successful was Cathy Jones, God bless her hide. She was more determined than I will ever be when she got hold of one of these dogs. I sold this litter and she got this dog off another man. That was her first dog. Well, he took her. I went back to this guy and said, 'You rotten son-of-a-bitch, you didn't do right by that lady.' He replied, 'All she wanted was a dog.' I said, 'And you took her, but you have a certain amount of respect you have to give people." Cathy said, 'I'll make something of that dog.' She did. That gal took that dog and took it as far as you can take a dog. She did it, nobody else. She deserves all of the credit, not part of the credit, all of the

Ted's full sister, Davies' Pinka. Courtesy Joe Piz

Davies' Pinka was a tough working cattle dog just like her brother. Courtesy of Joe Piz

Cathy bred for dogs which turned onto stock at a young age. This dark red puppy has an intense focus on the ducks and is even showing a little eye. Courtesy of Mark Lane

Ch Jones Reddy Teddy CDX, OTDdsc, SchHA showing his hard heeling style. Courtesy of Rowe

Ch Ruff N Reddy Mity Muhammad CD STDds, solid black with white male (Ted x Jones Mity Bite) a dog "made" by Cathy and Ruth McGuire in all venues—a multiple versatility winner. Courtesy of Melissa Couch

Ted working a lot full of Holsteins. Courtesy of Raymie Jones

credit. Nobody else stuck their nose in there. Oh, she'd come down and visit with me. We got to be real great friends and I respected her so much because she was a hell of a lady. I think she could have taken anybody's dog and done well with it. It's fortunate that she got a dog that fit her. The disposition fit her. She was able to take that particular dog and do things with it. Once she got Reddy Teddy and she got the rest of the pups, she had it made. They all had the same type of disposition. She learned, she knew it, she went on with it, Reddy Teddy. She's the one that made that dog. She's the one that made that bloodline, nobody else."

After researching Ted's background, it was obvious he was meant for more than just siring a few puppies from their Smokey. He was a working dog through and through and the decision was easy to follow this route in Ted's training. Dick and Cathy were both from ranch communities where working cattle, calf roping and rodeo was part of daily life. They both enjoyed it and were naturally drawn to weaving Ted into this fabric. As time went on, Cathy began to see the value of preserving Ted's talents and breeding was added to the couple's agenda.

The little red "leftover" wasted no time in showing what he was made of. He had attached himself to Cathy in spite of her intent to make him Dick's companion. When the Jones' headed off to their first rodeo after Ted's purchase, Cathy was not going to leave her new friend behind. An incident at this event was her first glimpse at the dynamo she had. She tells the story like this, "When he was about four months old, something happened that was really cute. He was just getting into his gangly stage and was kind of an ugly-looking thing. He looked like a little liver red bear. We were involved in rodeos quite a little bit at the time, and we went to one that happened to be up by Laramie. So, of course, since this dog had latched onto me, I was bound and determined that he was going to the rodeo with us. Dick was out roping cows. When it was his time to come up, I was watching from the side of the arena. The little dog was sitting there, and I had, for some reason, taken him off the leash. The stock was running, and the horse was after the calf. When I looked down, this little four-month-old puppy was there trying to get into that rodeo arena. Thank God, it had mesh fence. He was running up and down that fence trying to get those cows and that horse. If he had gotten in there, we would have had a terrible wreck. I had my eyes off him for about a second. Four months old is not very old to have that kind of drive and instinct. Actually he had more guts than he had brains at that age. I finally got my hands on him, and I sat there while he jerked and tried to run. Every time a calf and horse would go through there, he would try to follow. He was that way all of his life, which really attracted people to him as a stock dog. He was bred that way."

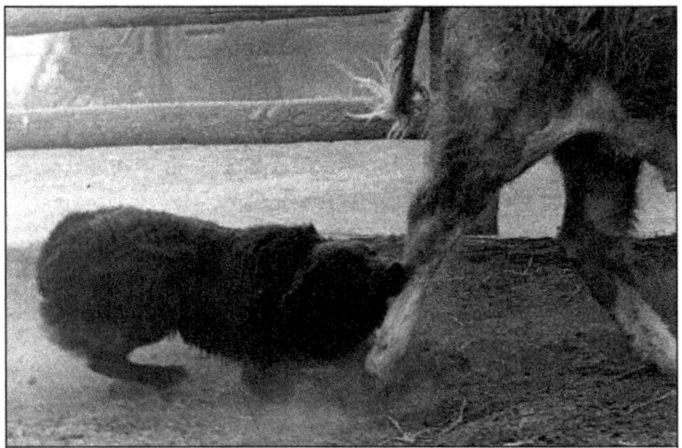

A young Ted learning about the correct place to hit on cattle—young dogs make mistakes, hopefully escape injury, and learn valuable lessons from those mistakes. Courtesy of Raymie Jones

Ch Jones Reddy Teddy CDX OTDdsc ScHA showing the eye and control he had when working sheep. Courtesy of Raymie Jones

Perhaps Cathy forgot to tell husband Dick about this incident or perhaps he didn't quite believe her because she tells another story about his experience with Ted a short month later. "One time, Dick was out fixing fence. Our neighbors had some cattle. He used to take the two dogs out with him when Reddy Teddy was just a five-month-old pup. When Dick took him out there, he was gone. Dick went out there to try to find this dog, and he found him lying there in the weeds. Like a Border Collie, he was sneaking up on these cattle out there. It was just a natural thing for him."

As he grew, Ted showed no signs of slowing down. Indeed he became stronger and stronger when it came to livestock and was beginning to show a good deal of independence when it came to working with people. He had learned his new owners, in spite of their skill and experience with horses and livestock, knew very little about dealing with a strong willed, independent thinking, one-of-a-kind working dog. He was becoming more and more uncontrollable because he had no real direction or job. His instinct at this age was stronger than most dogs of his age, making him "impossible" to be around if any livestock were in sight. Cathy decided she had to do something and do it quick or he might not live to see his first birthday. As Cathy says, "He would take those animals out and bring them back constantly. He was just a terror."

Fortunately she found obedience classes in Fort Collins and began driving there once a week to learn how to channel all the energy,

Ch Jones Reddy Teddy CDX, OTDdsc, SchHA. Often criticized for "aggression," Cathy Jones recognized that good cattle working dogs had to have the grit, stamina and never-say-die attitude to do the job. She never apologized for favoring this type dog and delighted in showing Ted hitting cattle low and hard. Courtesy of Sorensen

The little red heathen eventually became an accomplished and successful obedience dog. Here Ted performs the retrieve in an Open obedience class at a dog show trial. Photo by Sue Noll. Courtesy of Raymie Jones

intelligence and instinct in this little prodigy. Ted was around 7 or 8 months old at this time and took to the training with enthusiasm. Cathy was a natural trainer from her lifelong experience training horses and the two soon blended into a close knit team.

As Cathy progressed with Ted's training, her competitive side began to emerge. She was learning more about Aussies and was going to some shows to learn about the competitions. In her travels she met Ruth McGuire, who was showing her legendary Ch McGuire's SeMeGo Hemi UDT OTDc STDsd in all the venues he would nearly own in his career. The versatility attracted Cathy and she knew her solid red male could compete with the best of them. In the beginning she found herself competing with the great Hemi but over time she and Ruth eventually became good friends. Ruth taught Cathy to value a dog built for function and the two shared many viewpoints when it came to what made a good dog. Ruth helped Cathy understand things other people never considered—mainly understanding what motivated a dog to do what it did. She also started Cathy on the right path when she decided to begin breeding.

The Ted and Cathy partnership was immediately successful in the show rings and arenas. Cathy's meticulous attention to detail in training translated into many wins in performance from the beginning. It happened Ted was doing very well with his stock training when the very first ASCA sanctioned stockdog trial was to be held in Phoenix, Arizona. Cathy set her sights on attending this trial with her young dog and competing in every venue she could. This was the first out of state show she had ever attended and Ted was only about 18 months old. She tells Betty Nelson in a 1975 letter she was "apprehensive showing against the BEST" as she put it and only hoped she and Ted would do their best against them. Much to her surprise and delight her young dog was the star of the trial and show. *The Phoenix Gazette* was covering the trial and featured Ted and his successes in a half page article. In an article she wrote for the *Aussie Times*, Terry Martin remembers Ted attracted attention for another reason—red dogs were rare at that time and most people had never seen a solid red with no trim. Between his looks and ability, Martin says Cathy and Ted, "made a real impact with his unusual appearance, silent low heeling style on cattle and controlled fetch on sheep and ducks."

A young Ted with an early win under judge Ernie Hartnagle. Courtesy of Raymie Jones

Terry Martin begins her 1975 *Aussie Times* article on Ted by noting he was a controversial dog throughout the 1970s. Cathy never shied away from calling him aggressive on stock and in personality. It was a trait she valued and sought to preserve. His success as the first Schutzhund titled dog in the breed only served to enhance this reputation. In Cathy's hands this aggression—which, in the politically correct world of today, now goes by such terms as "drive" or "enthusiasm"—was an asset in the many venues in which she competed with Ted. But this temperament also earned it's share of critics who denigrated Ted for the very things bred into his line for generations.

Ted was almost always in the ribbons in obedience. Here he is tied for first place with a sheltie at an obedience trial! Courtesy of Raymie Jones

This former asset was turning into a liability in an increasingly urbanized world. Ted was no different than Jeffreys' Shade, Wood's Jay or any number of his relatives from a more rural age. His only real fault was the timing of his birth. Martin states he was "ferocious" when anyone approached Cathy's Datsun pickup or if someone came near his crate when he was in it. This was nothing new to those who deliberately chose an Aussie as a "horse show" dog to be left with truck, trailer, or tack to prevent it's theft. Cathy's interest in showing and competing in performance with Ted brought this typical Wood dog into the spotlight and introduced his line to many in the breed who knew little about it. This ignorance of the breed foundation lines was not so surprising. Outside of Colorado, the Wood dogs were little known to the rest of the fancy. Aussie shows and trials had only been organized a short time and travel by

Ch Jones Reddy Teddy CDX, OTDdsc, SchHA
Despite Ted's reputation for aggressive cattle work, his trainability and team player mentality made him a truly versatile working dog no matter what Cathy asked of him. Courtesy of Sorensen

Ted working ducks. Courtesy of Raymie Jones

Ted doing what he did best—low heeling cattle! Courtesy of Ruth McGuire

competitors with the resultant intermingling of lines was just beginning.

Ted was often the first introduction to the Wood dogs many people had. Their naivete about his bloodlines and personality naturally came as a surprise. As a result, grossly overstated generalities sprang up regarding him and the Wood dogs as a whole. Derogatory terms such as vicious, an alligator and other criticisms were applied to him and many of these misconceptions persist to this day. But Ted was nothing but a typical Wood dog. Endowed with good judgment, ample brains and good trainability, so long as his owner was, indeed, in charge, he was a well behaved and safe dog. As with all Wood dogs, in the hands of a weak human he could have become what the rumors implied. In the hands of Cathy Jones he became one of the most important and influential working dogs of his day.

This amusing story told by Terry Martin illustrates Ted's core personality perfectly, "I remember Scott and Randy (Terry's sons) feeding him cinnamon rolls at breakfast and then running around the kitchen with Ted 'heeling' them wildly but never biting. Ted would do all his obedience commands for them; sit, down, etc." She remembers Ted as a ruthless beggar who would root your hand incessantly to be petted. If you stopped he would nose your hand until you began petting again. But if you talked to him he would curl his lips back and growl. Although it was a startling message he never escalated this threat into action. In the 1970s Terry and Cathy, along with Terry's son Scott and all of their dogs, attended the ASCA National Specialty in Kansas City. They were sharing a room and young Scott was lying on the bed with his dog, Martini, petting Ted who was on the floor by the bed. Scott said something to Ted which elicited the trademark growl, but Martini was having none of that directed at "her" boy. She bailed off the bed into Ted—who was easily twice her size—to explain to him her idea of proper behavior toward Scott! Ted recognized his mistake and did not retaliate but took his correction from this little fireball because it was justified.

Terry describes another easily misunderstood game Cathy used to play with Ted. Again, his Schutzhund experience came to the fore in an attack game the two greatly enjoyed. The "play fight" had all the earmarks of a genuine attack in the eyes of anyone who did not understand the relationship they shared. Ted's lips would be curled back, showing every one of his sharp, jagged teeth with fierce growling backing up the display. To the uninitiated it appeared to be a dangerous situation. But the jig was up as Cathy erupted into laughter, clearly enjoying the episode as much as Ted. She had only to say "Quit" and the whole thing ended instantly. Terry's comment was an understated "Quite an interesting dog."

Ch Jones Reddy Teddy CDX OTDdsc SchHA striking a very "western" pose. Courtesy of Melissa Couch

For all her personal contact with Ted, Terry admits she never quite understood him. She acknowledges his "unique" personality left him open for a lot of criticism from those who disliked him. But his reputation as a silent low heeler with extremely strong instinct to group and fetch (the strongest she had ever seen) and ability to produce it drew her to breed a number of females to him over the years, adding these traits to her own Slash V bloodline. She closes her article with this accolade, "He is without a doubt one of the exciting early working sires promoted during the infancy of our stock dog program."

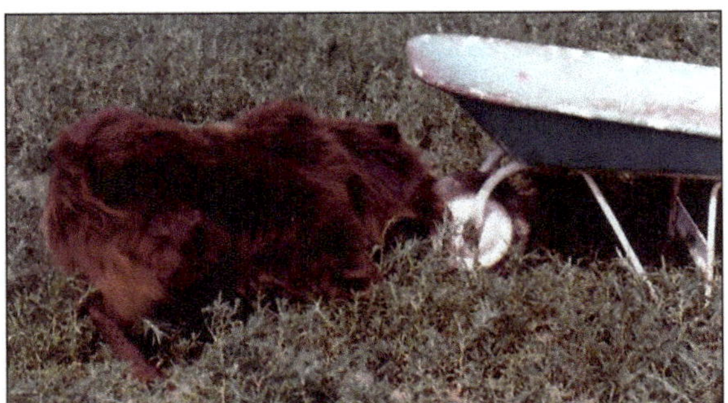

Is Ted low heeling that wheelbarrow or practicing his Schutzhund training? Courtesy of Raymie Jones

As Cathy pursued more titles for her talented male she began to encounter the prejudice against his solid color found in the conformation show ring. With the optimistic attitude typical of Cathy, she wore this prejudice proudly and took pride in taking her quality dog with the unpopular color into the ring. His personality could add to his handicap with some growling at judges in the beginning of his competitive career. In both conformation and obedience this trait was not appreciated and Cathy worked hard to convince Ted to allow the required examinations without commenting.

Ted at Schutzhund training working the sleeve. Well controlled and highly obedient, this type of game suited Ted well. Courtesy of Raymie Jones

Cathy with Ted working cattle at a trial. Ted wasn't purely a "sic'em" dog but once trained, worked cattle effectively and precisely. Courtesy of Raymie Jones

She must have succeeded because Ted eventually became a conformation Champion and progressed through the Open obedience classes to earn his CDX.

Jean Taylor describes a little of what Cathy was up against. She defines the Wood's Stubbie type of dog as what people of the area were more familiar with when they thought of a Wood bred dog. This type was longer and more angulated with a noble personality, extremely broad blocky head and generally a very dark blue color. The Piz Joseph type of dogs tended to have the look typified by Ted—less blockiness of the head, higher earset and less angulation. The personality of the Piz/Ted type dogs was harder compared to the Stubbie type dogs, too. Even though they were closely related by pedigree, the selection process of each breeder had emphasized different traits and Joe Piz quickly created a branch of this family to suit a different purpose. Ted eventually was the unofficial standard bearer for the "Red Wood's type" which defined them as unique from the dogs Fletcher Wood was breeding. Cathy must have recognized these differences, too, and her admiration for the Stubbie type of dog led her to choose a number of females to breed to Ted which favored the more classic Wood type of dog.

Cathy never gave up being a proud promoter of solid colored dogs. Ted not only lacked copper and white trim himself, he never produced a dog with copper trim even when bred to females which had it. She quickly learned why Ted had been a "leftover" and realized she was lucky in more ways than one to have him. The breed was almost exclusively seen as a blue merle colored breed. So much so that Fletcher Wood had tried to produce a line which would "breed true" for merle color. Of course this could not happen and Mr. Wood eventually realized the value of the solid colored dogs but the rest of the fancy was much slower in coming to this conclusion.

In Ted's day he would often be the only solid red dog at an entire show. In fact, he was one of only a handful of solid reds even being shown in the entire breed! He stuck out like the proverbial sore thumb. Judges, most of whom had little real knowledge of the breed, would see him and think, in Cathy's words, "Hmmm, you know, I can't do anything with this dog. This must be a mistake." This attitude extended to solid blacks as well and was not limited to judges in the show ring. Breeders put down solid colored dogs because they could not sell them and it took Cathy a long time and a lot of effort to educate both judges and breeders the solid colored dogs were as good or better than any others.

As hard as Cathy worked at promoting solid colored dogs, the prejudice never completely faded. In later years she would state she did not breed for color, which might have hampered her puppy sales. She realized people wanted "frills and chrome" and although they may have needed a good working dog they wanted it to have a fancy look. She says, "I always had that shoved down my throat ever since I had this dog. He was a plain red dog, and so I kind of went off the other end." But her persistence paid off to some degree.

Ted's signature style. Courtesy of Raymie Jones

Ch Jones Reddy Teddy CDX, OTDdsc SchHA. Cathy Jones believed working dogs should be photographed doing their job but occasionally allowed good shots to be taken of her correct male. Courtesy of Sorensen

Ch Ruff N Reddy Mity Muhammad CD STDds, black and white male who was a top winner (Ted x Jones Mity Bite) in spite of his solid black color. Owned by Patty Ennen and showed and trialed by her, Ruth McGuire and Cathy Jones. Courtesy of Melissa Couch

The distinct head type associated with Ted and something he passed on consistently to many of his progeny. Ted is pictured here as a mature dog. Courtesy of Raymie Jones

Judges at least became familiar with the non-frilly colored dogs and she was not only able to finish Ted but other solid colored dogs she bred including Ch Ruff-N-Reddy Mity Muhammad CD STDds and others. Her work also made reds and red merles more acceptable because they, too, suffered a lack of popularity. Her success in breeding strong colored dogs eventually convinced many the non-merle colored dogs were necessary in the breed to improve fading color in merle focused programs.

So the course was firmly set early on for the Ruff-N-Reddy kennel. Cathy recognized the obstacles she faced, embraced them and turned them into assets. The rapport she had with Ted was an unstoppable force. He was the first dog she had ever let into her heart and he led her into competitive dog events. She soon turned around her young hooligan, using his overwhelming desire to please to train him in any sport which attracted her interest. For stock trial work he became the ultimate team player she could control across a large arena at any time. He taught her she only thought she knew what a good stock dog was before he came along. His natural tendency on stock was as a natural low heeler but she was able to teach him to head for their trial work. He took to this eagerly,

An historic union of classic Wood dogs. L to R: Charlene Jeffreys with Jeffreys' Shade, Cathy Jones with Davies' Blue Kola, Patty Ennen with Piz's Sheba, Sue Foster with Clare Olson's Blue Joe. Photo by Sue Noll. Courtesy of Sorensen

showing the style which took him to the top, described by Cathy, "When he works his stock, he crawls like a border collie, especially on ducks and sheep."

Cathy was learning about the finer points of what makes a good working dog at a rapid pace and Ted's continued success in every ring pushed her even harder. Shows and trials were still few and far between and the pair were training every day to be ready for the next big opportunity no matter how far they had to travel. After their debut triumphs in Phoenix, they traveled to a large show in Utah. As usual Ted was entered in every class available. Not only were fanciers forced to take advantage of the few shows/trials by entering every dog in every class they could but they believed in the versatility of the breed and put their dogs to the test at every turn. Ted didn't let her down, winning the Open Red dog conformation class, the Open B Obedience class, High Scoring Dog in Show, Open Cattle and Open Sheep!

For Cathy, working ability always remained her highest priority in breeding. She bred dogs for cattle work and did her best to make sure they found homes with people who worked them. Seldom would she allow one of her puppies to go into a pet home although under special circumstances, to the right owner, she

Cathy showing one of her many red bitches. Courtesy of Raymie Jones

The versatile, beautiful black and white Reddy Teddy son, Ch Ruff N Reddy Mity Muhammad CD STDds. Courtesy of Sorensen

A gathering of some of the best of the best. L to R: Jeffreys' Pooh Bear, Jones' Mity Bite, Ch Ruff N Reddy Mity Muhammad CD STDds, Piz's Sheba, Ch Jones' Reddy Teddy CDX, OTDdsc SchHA, Davies' Blue Kola, Clare Olson's Blue Joe, Ch Ruff N Reddy Blue Tinker Toy. Courtesy of Sorensen

would do it. The owner still had to be prepared to work the puppy in some fashion, usually obedience, and to understand the sort of dog they were getting. She kept track of the progress of her puppies and was always available to help out with problems or take back dogs which were not being handled correctly.

This emphasis on cattle work was in line with the development of the Wood dogs going all the way back to Sisler's Shorty. This was cattle country and the demand was for dogs who could do the tough jobs. There was no compromise on this requirement from either breeders or buyers. Ted was typical of the cattle dogs of that day who were used almost exclusively as driving dogs. Ted did not naturally go to the head and would not grab a nose. Through training he could be placed where Cathy needed him in order to work a trial course or work stock on the ranch but he was bred as a driving dog and it was in this capacity he excelled.

Interestingly enough, Ruth McGuire remembers Ted as having a lot of eye in certain situations. In some cases he would lock up and Cathy would have to make some kind of physical contact with him to break his trance. In hindsight the possibility exists that the state of knowledge of training encouraged driving over fetching and through constant use as a drive

Ted facing down a rogue sheep! Courtesy of Raymie Jones

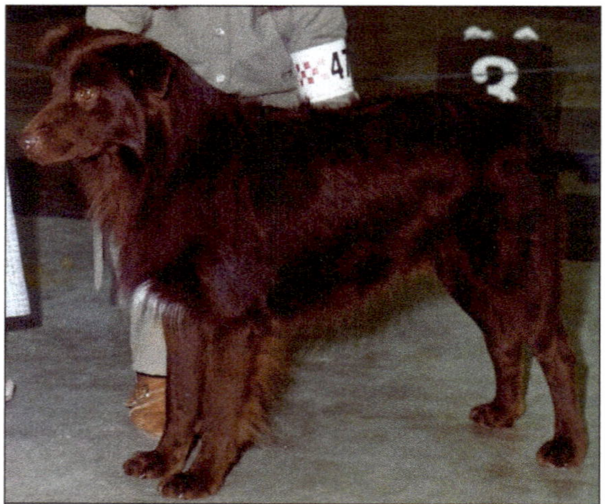

Ted winning yet another Best of Breed. Courtesy of Raymie Jones

A famous and often used photo of Ted low heeling on cattle. Courtesy of Raymie Jones

Ted working the blind at a Schutzhund training session. Courtesy of Raymie Jones

dog, the heading ability of the dogs was never developed as it might have been.

As Ted's fame and list of accomplishments was growing, Cathy was already beginning to think about breeding. The Jones' first Aussie, Smokey was what Cathy described as 1/2 bred Aussie and, although she and Dick considered her a good dog, she was not in Ted's league. If they were to maintain his quality they needed quality bitches to breed to him. Ruth McGuire had provided guidance in training and competition and now stepped up to help Cathy learn about breeding.

In 1974, the 19 month Ted won the Colorado Australian Shepherd Association's first All-Around Dog title. In order to achieve this he had to earn the highest combined scores in conformation, obedience and stock trialing. He had earned his CD title by then in three consecutive shows with all of his scores 195-1/2 or higher, earning three High In Trial awards along the way. To top it all off he earned the High Scoring Stock Dog award from CASA for the highest combined scores in stock work for 1974.

All of this early success was attracting nationwide attention and Ted became a father for the first time in 1975 when Beth Alden of Almostheaven kennel brought Red Lady of Cattle Call to him for an all red litter. The cross was a mix of California's Tartaglia/Caligari breeding with Ted's Wood background and proved to be a successful one for Almostheaven, producing dogs which became the foundation for this bloodline.

Another breeding in 1975 was to Leslie Sorensen's female Colorados Buckeye Barbie (George's Buckeye Bobby x Sorensen's Cherry Brandy). Cathy tells Betty Nelson in a letter that she kept a solid colored bitch from the litter, who was trying to work as a young puppy just like her sire. She described her as "full of

Ted's talents on livestock were not limited to hard, low heeling on cattle. When working sheep and ducks he was focused, deliberate and well under control. Here he turns a group of sheep just using his presence and probably a little eye. Courtesy of Mark Lane

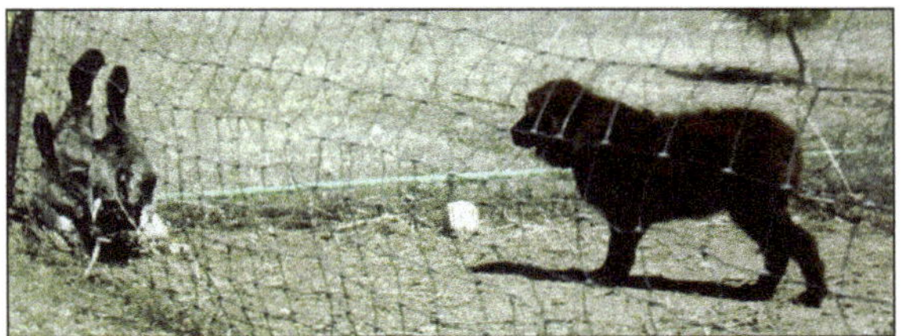

A young solid red Ruff N Reddy puppy getting an early exposure to ducks and showing a lot of self control. Notice he is not under any type of physical restraint. Courtesy of Mark Lane

Another look at the famous low heeling style of Ch Jones Reddy Teddy CDX OTDdsc SchHA. Courtesy of Raymie Jones

Almostheaven's Sweet Honesty CD STDc (11 pts) at age 2, red merle and white female (Ted x Almostheavens Red Lady of Cattle Call) Photo by Olan Mills. Courtesy of Mark Lane

Colorado's Buckeye Barbie, red merle and white female (George's Buckeye Bobby x Sorensen's Cherry Brandy) one of Ted's first mates and producer of a number of good working dogs. Courtesy of Melissa Couch

hell" with good conformation and the temperament needed for working stock. Also in this litter was a solid colored red male who went to Wyoming and was said to have the best conformation, aggressive temperament and was darker than Ted!

The puppy Cathy kept was eventually sold to Sue Noll and became Ruff N Reddy Teddy's Tuffit. Sue, like everybody else, put her talented puppy to work on livestock where she proved to be a little more mellow than Ted and did not work quite as aggressively. Tuffit found her real calling when she became certified in Search and Rescue work and tracking. Tuffit had a good temperament and was well suited for this work. She was called to track a criminal at least once and her ability led to a conviction in court. In the early 80s, a blue merle son of Tuffit's by Jeffreys' Pooh Bear, named Boomer, was called to track a band of thieves who had stolen a truck and broken into a television store. He and 4 other dogs were used to track the gang and the search resulted in the arrest of all 3 criminals and recovery of all the merchandise.

Two other good females were bred to Ted in 1975—Parmenters Dixie of Bar Lazy K (Bar Lazy K's Red Baron x Kolzs Blue Favor), a Deines and Las Rocosa cross, who produced a male by the name of Parmenters Latigo. Latigo was used by Pat Lambeth in California and was behind many Smithfield dogs—a kennel which later acquired a full brother of Ted (Ch Calico Chief Red Cloud) to use as the main sire in their breeding program.

The legendary J Bar D Brandy was also bred to Ted in 1975, producing another top working and producing female named Colorado's Brandy Bow. Brandy Bow became a top producer for the Kintyre Cattle Co. and was the beginning of their breeding program which eventually centered around the Ted son Ch Kintyre Jeb Stuart STDd OTDc.

At about the same time Cathy acquired what she described as her first "Piz bred" bitch—a silver blue merle with large black patches, four white feet, a small white snip down her face, no copper and "quite a little lady." This was Jones' Mity Bite, outstanding matron of Cathy's line and founder of a dynasty of lines through her progeny. Mity Bite went to the big Utah show to make her debut and was entered in the puppy class. Cathy says she was the smallest of 15 entries in the class and was overlooked by the judges. Cathy's friend Patty owned a litter mate male to Mity Bite and they entered the pair in the brace class even though they were only about 2-1/2 months old. Amazingly they won this class and made a big hit for an incident during the class. The two puppies had not worked together as a brace very much and between their puppy antics and resistance to being show dogs for the day became irritated with each other. Their patience reached an end and a raucous puppy fight broke out in the middle of the ring. Cathy was helpless with laughter watching Patty trying to break up this fight while maintaining some semblance of decorum and trying to present the puppies to the judge!

Ruff N Reddy Shake N Bake, blue merle and white female (lying down) and Tuffernhel's Shadow black and white male standing on right (Jeffreys' Pooh Bear x Ruff N Reddy Teddy's Tuffit). Courtesy of Melissa Couch

Ch Calico Chief Red Cloud, red merle male, full brother to Ch Jones' Reddy Teddy CDX, OTDdsc SchHA. Foundation sire for Smithfield kennel. Courtesy of Rowe

Ch Calico Chief Red Cloud, red merle male full brother to Ted as a youngster. Courtesy of Raymie Jones

Kintyre Blue Suede Shoes "Tweedles" blue merle female (Jeffreys' Pooh Bear x Colorado's Brandy Bow) Photo by Sue Noll. Courtesy of Raymie Jones

Ch Kintyre Jeb Stuart STDd OTDc, dark self blue merle male (Ch Jones' Reddy Teddy CDX OTDdsc SchHA x Davies' Blue Kola). Classically bred straight Wood male who established a dynasty of top cattle working dogs for the Kintyre Cattle Co. Courtesy of Sorensen

Jones' Mity Bite, likely a fairly young dog in this photo. Her silvery merling still stands out as it only does in young dogs of this line. Courtesy of Melissa Couch

Mity Bite, even in those days, was not the type of dog which would win at dog shows. She was an "old time doggy bitch" in the words of Ruth McGuire with the typical Wood deep gun-metal blue color. Her head type was not the impressive blocky head of Sisler's Shorty or Wood's Stubbie but a more moderate type. With no copper and just the tiniest bits of white on the chest and tips of toes, it was easy for judges to dismiss her in favor of the "flash" that was becoming popular in the show ring. In spite of this she did take some points as a young bitch, going Winners Bitch at a show when she was 9 months old! But when it came to producing top quality working dogs with excellent conformation this show ring reject was worth her weight in gold. With a sweet temperament and the right mate in Ted, Mity Bite left a legacy second to none. This issue of temperament was nearly as important to Cathy as working ability. Melissa Couch says this about the Ted x Mity Bite female she had, "She was a good watch dog but not people aggressive. And she didn't miss anything that went on around our place." Even when crossed with outside males, such as Slash V Frosted Skotch, her quality shone through. Cathy's pick from this cross, Ruff N

Jones' Mity Bite, dark blue merle and white foundation female for Ruff N Reddy kennel (Jeffreys' Cinnamon Man x Piz's Sheba). Courtesy of Sorensen

Reddy Mity Happy Hocker CD STDd, became a second foundation and another favorite in the generations that followed.

Ruff N Reddy Mity Happy Hocker CD STDd, a blue merle and white female, became known as "Skidder." As a youngster she earned this name after catching her foot in a fence and breaking it. The break was bad enough it required casting and she learned to get around by skidding the cast across the floor! Skidder was a happy dog with a very sweet temperament. Although Cathy loved Ted's toughness she always tried to select females with solid, sound but sweeter temperaments to balance him out. Skidder filled this bill in every way. She handled fair discipline, never sulked or held a grudge. Melissa Couch, who kept her for 3 years, emphasizes she was just an overall very stable tempered dog who "never met a stranger." After her stay in Texas with Melissa, Skidder went back to Cathy, who placed her in a retirement home as was her belief with all her dogs. She didn't believe in relegating them to kennel life and always made a point of bringing each one in the house to have individual time so they could enjoy just being dogs.

Jones' Mity Bite became the standard for all the females who followed at Ruff N Reddy. Color, type, breeding—the consistent high quality of the females in Cathy's kennel was truly amazing. Courtesy of Raymie Jones

Jones' Mity Bite had not only the look Cathy sought but the tough cattle working ability that was a must for her breeding plans. Courtesy of Joe Piz

Piz's Sheba, dam of Jones' Mity Bite and a legion of other great dogs. Shown here in old age and very pregnant so not at her best. Photos of Sheba are rare so it is included here to show her overall good conformation and type in spite of the wear and tear of age. Courtesy of Melissa Couch

PEDIGREE OF Jones Mity Bite

Born: ______ Sex: ______
Owned by: ______

Parents	Grandparents	Great-grandparents	Great-great-grandparents
Sire: Jeffreys Cinnamon Man	Sire: Beiens Red Tuffy	Sire: Pizs Little Joe	Sire: Piz Joseph
			Dam: J Bar D Choctaw Chilli
		Dam: Pizs Lady II	Sire: Piz Joseph
			Dam: Piz Lady I
	Dam: Jeffreys Cinnamon Bear	Sire: Montoyas Rowdy	Sire: Clingers Royal Blue Merle
			Dam: Kalbs Princess
		Dam: Montoyas Snipper	Sire: Piz Joseph
			Dam: Piz Lady II
Dam: Pizs Sheba	Sire: Piz Joseph	Sire: Woods Jay	Sire: Sislers Shorty
			Dam: Sislers Trixie
		Dam: Woods Chili	Sire: Woods Jay
			Dam: Greens Dingo
	Dam: Pizs Lady II	Sire: Piz Joseph	Sire: Woods Jay
			Dam: Woods Chili
		Dam: Piz Lady I	Sire: Woods Jay
			Dam: Sorensens Tammy

Bred by: ______ Address: ______ Phone: ______

The details given above are correct to the best of my knowledge. Signed: ______(Breeder)

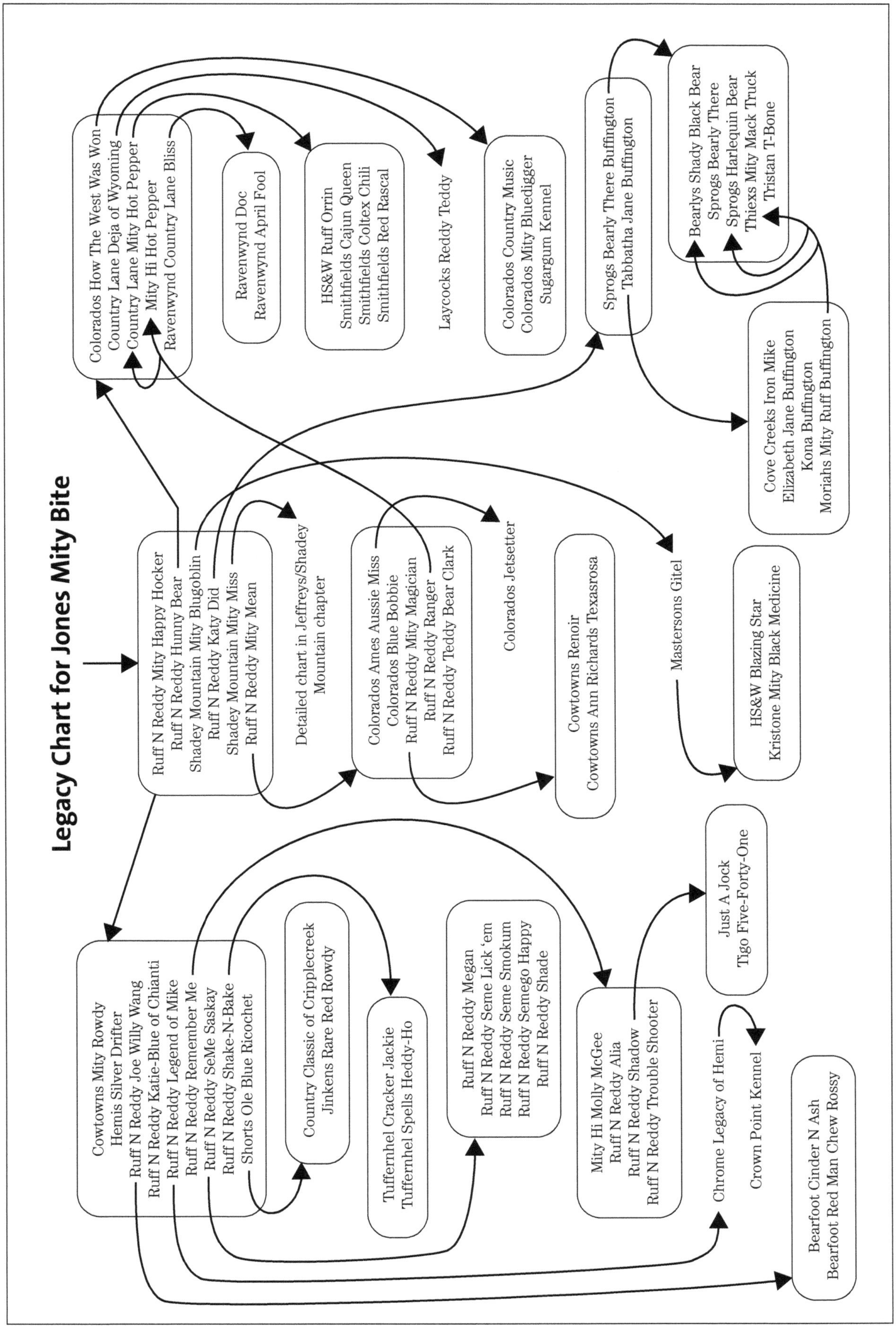
Legacy Chart for Jones Mity Bite
Ruff N Reddy Mity Happy Hocker
Ruff N Reddy Hunny Bear
Shadey Mountain Mity Blugoblin
Ruff N Reddy Katy Did
Shadey Mountain Mity Miss
Ruff N Reddy Mity Mean
Colorados How The West Was Won
Country Lane Deja of Wyoming
Country Lane Mity Hot Pepper
Mity Hi Hot Pepper
Ravenwynd Country Lane Bliss
Ravenwynd Doc
Ravenwynd April Fool
HS&W Ruff Orrin
Smithfields Cajun Queen
Smithfields Coltex Chili
Smithfields Red Rascal
Laycocks Reddy Teddy
Colorados Country Music
Colorados Mity Bluedigger
Sugargum Kennel
Sprogs Bearly There Buffington
Tabbatha Jane Buffington
Bearlys Shady Black Bear
Sprogs Bearly There
Sprogs Harlequin Bear
Thiexs Mity Mack Truck
Tristan T-Bone
Cove Creeks Iron Mike
Elizabeth Jane Buffington
Kona Buffington
Moriahs Mity Ruff Buffington
Detailed chart in Jeffreys/Shadey Mountain chapter
Colorados Ames Aussie Miss
Colorados Blue Bobbie
Ruff N Reddy Mity Magician
Ruff N Reddy Ranger
Ruff N Reddy Teddy Bear Clark
Colorados Jetsetter
Cowtowns Renoir
Cowtowns Ann Richards Texasrosa
Mastersons Gitel
HS&W Blazing Star
Kristone Mity Black Medicine
Cowtowns Mity Rowdy
Hemis Silver Drifter
Ruff N Reddy Joe Willy Wang
Ruff N Reddy Katie-Blue of Chianti
Ruff N Reddy Legend of Mike
Ruff N Reddy Remember Me
Ruff N Reddy SeMe Saskay
Ruff N Reddy Shake-N-Bake
Shorts Ole Blue Ricochet
Country Classic of Cripplecreek
Jinkens Rare Red Rowdy
Tuffernhel Cracker Jackie
Tuffernhel Spells Heddy-Ho
Ruff N Reddy Megan
Ruff N Reddy Seme Lick 'em
Ruff N Reddy Seme Smokum
Ruff N Reddy Semego Happy
Ruff N Reddy Shade
Mity Hi Molly McGee
Ruff N Reddy Alia
Ruff N Reddy Shadow
Ruff N Reddy Trouble Shooter
Just A Jock
Tigo Five-Forty-One
Chrome Legacy of Hemi
Crown Point Kennel
Bearfoot Cinder N Ash
Bearfoot Red Man Chew Rossy

Raymie Jones with Ruff N Reddy Mity Happy Hocker CD STDd continuing the winning ways of the Ruff N Reddy dogs. Courtesy of Raymie Jones

Ruff N Reddy Shake N Bake, blue merle and white female (Ted x Ruff N Reddy Mity Happy Hocker CD STDd) Courtesy of Melissa Couch

Skidder was a nice clear blue merle with very dark eyes who was about 20 to 20-1/2 inches in height. She was not a low heeler but had a very strong gathering instinct with a very strong natural wear. Melissa describes her as practicing her wearing ability even when just following her people around. She would gather anything she was sent after and was an excellent all around working dog. In her spare time she loved to run rabbits and excelled as a Frisbee dog. When Cathy's son, Wade, would show her in conformation, her favorite "bait" was always a small, concealed Frisbee.

Bred to both Ted and Hemi, Skidder produced important breeding offspring from every litter. Her son by Ted, Ruff N Reddy Joe Willy Wang made an important mark after he was sold to Al Combee in Florida to work his dairy farm. He was bred to another top Ruff N Reddy female, Ruff N Reddy Shadey Wood Ash, to produce two top sires. Bearfoots Cinder N Ash was leased for several years by Pincie Creek and bred to their good Slash V and Hangin Tree bred females to establish a new bloodline in the southeast United States, combing the best of two good working bloodlines. After trying to purchase Cinder N Ash but being denied, in later years Pincie Creek was able to acquire his

Ruff N Reddy Mity Happy Hocker CD STDd. Courtesy of Melissa Couch

litter brother, Bearfoots Red Man Chew Rossy. Chewy gave a new infusion of close up old Wood breeding to the Pincie Creek program, fixing the Wood type firmly in their line and bringing back the good cattle working ability Cathy's dogs were always known for.

Skidder was also bred to Hemi producing three notable progeny. Ruff N Reddy Legend of Mike produced Chrome Legacy of Hemi, who sired two influential dogs for Crown Point—Crown Point Keeper of the Flock and Crown Point Banner O Braemar. Ruff N Reddy

Ruff N Reddy Mity Happy Hocker CD STDd. Courtesy of Melissa Couch

Ruff N Reddy Joe Willy Wang, solid black male (Ted x Ruff N Reddy Mity Happy Hocker CD STDd) Courtesy of Raymie Jones

The blue merle Ted daughter Ruff N Reddy Shake and Bake (left) with her dam Ruff N Reddy Mity Happy Hocker CD STDd. (right). Shake and Bake was eventually owned by Sue Noll. She got her name when she born and got chilled. Cathy put her in the oven to warm her up—creating visions of the popular pre-made dinner mixes of the day. Courtesy of Raymie Jones

Ruff N Reddy Mity Happy Hocker CD STDd, a very correct clear blue merle. Courtesy of Melissa Couch

Ruff N Reddy Mity Happy Hocker CD STDd showing a good controlled gather on ducks. She has lowered her head and appears to be showing a little eye, too. Courtesy of Raymie Jones

PEDIGREE OF Ruff N Reddy Mity Happy Hocker, CD STDd

Born: ______ Sex: ______
Owned by: ______

Parents	Grandparents	Great-grandparents	Great-great-grandparents
Sire: Slash V Frosted Skotch	Sire: Poudre Silver Sage of Coppertone	Sire: Salt of Flintridge	Sire: Harpers Old Smokey
			Dam: Blue Spice of Flintridge
		Dam: Shanks Ginks	Sire: Proud Buck of Flintridge
			Dam: Iacovettas Shasta
	Dam: Martins Buster Brown Shoe	Sire: Martins Tim Tim	Sire: Taylors Whiskey
			Dam: Taylors Buena
		Dam: Martins Josie	Sire: Taylors Whiskey
			Dam: Taylors Buena
Dam: Jones Mity Bite	Sire: Jeffreys Cinnamon Man	Sire: Beiens Red Tuffy	Sire: Piz Little Joe
			Dam: Piz Lady II
		Dam: Jeffreys Cinnamon Bear	Sire: Montoyas Rowdy
			Dam: Montoyas Snipper
	Dam: Pizs Sheba	Sire: Piz Joseph	Sire: Woods Jay
			Dam: Woods Chili
		Dam: Piz Lady II	Sire: Piz Joseph
			Dam: Piz Lady I

Bred by: ______ Address: ______ Phone: ______

The details given above are correct to the best of my knowledge. Signed: ______(Breeder)

Remember Me produced Mity Hi Molly McGee, a foundation female for Anita Laycocks kennel and dam of Wyomings Gypsy Blue of Mt High, Wyomings Boy Blue By Ted and Wyomings Nina By Red Ted.

Another Hemi x Skidder daughter was named Ruff N Reddy Remember Me. She was originally owned by Patty Ennen but traded back to Cathy for a female Patty had always admired named "Tweedles"—Kintyre Blue Suede Shoes. Remember Me was called "Trouble" and although a well bred female, she wasn't what Patty was looking for and when the opportunity came to get Tweedles—a dog she had admired for a long time—she took it.

When Cathy got Remember Me back she was bred to Patty's Mity Muhammad. She sold one of the male puppies from the litter to a local rancher. The man drove a Jeep and when he was not in the vehicle, the dog would wait for him by crawling onto the floor under the steering wheel where he could not be seen. One day the owner had purchased a new Jeep and the dog was in his usual place. It was a warm day so the owner had left the windows open for the dog while he went into a store. He returned to find quite a scene around his new Jeep.

Ruff N Reddy Joe Willy Wang inherited his sire's trademark head type. Courtesy of Raymie Jones

Ruff N Reddy Joe Willy Wang, solid black male with just a small white patch on the chest. (Ted x Ruff N Reddy Mity Happy Hocker CD STDd) Courtesy of Melissa Couch

Ruff N Reddy Mity Happy Hocker CD STDd learning the ropes on ducks. Courtesy of Raymie Jones

Kintyre Blue Suede Shoes, blue merle female (Jeffreys' Pooh Bear x Colorado's Brandy Bow—Ted daughter) Courtesy of Raymie Jones

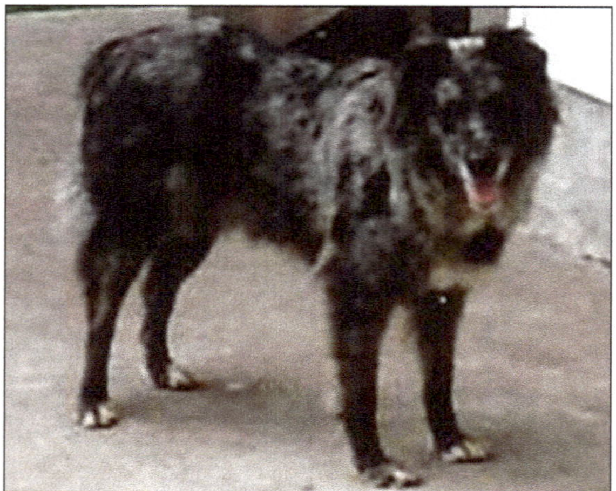

Gahl's Terra, dark blue merle and white female (Ruff N Reddy Shade x Tina Horejsi—an old J Bar D/Sorensen bred female), owned by Joe Gahl of Minnesota. Courtesy of Jean Taylor

A stranger was standing with his arm in the window of the Jeep screaming and telling the gathering crowd that someone had put a dog into HIS Jeep and the dog had hold of his hand and would not let go! The real owner called the police while the dog continued to hold the thief firmly by the hand. When the police arrived, the lie was told again but immediately revealed when the real owner gave a release command and the dog instantly let go of the crook.

Skidders most important daughter was Ruff N Reddy SeMe Sashay, a blue merle female. Known as "Bunny" she produced several important offspring which spread the Ruff N Reddy influence as far away as Alaska. Bunny's daughter, Ruff N Reddy Seme Lick'em was owned by Ruth McGuire when she moved to Alaska with her male, Jones Stockyard Hustler. Both of these dogs were used extensively to help found the North Star breeding program of Carla Schroeder. North Star stuck closely to the Wood type and breeding in their selection and breeding and carried the Ruff N Reddy legacy into the future. Another Bunny offspring, Ruff N Reddy Seme Smokum was also sent to Alaska to add to the North Star gene pool via the Myalaska kennel. Lastly a male named Ruff N Reddy Shade sired the indispensable working female Gahl's Terra for hog producer Joe Gahl of Minnesota. Terra eventually produced a female by Newman's Blue Sprog CD STDsc via a primitive method of chilled semen named Gahl's Casey. Casey was eventually crossed to a Pincie Creek dog of close up Wood breeding to produce Jean Taylor's black male, Gahl's Silverledge Smokin Jose.

One of Ted x Mity Bite's most beautiful daughters was a very dark blue female with very little merling named Ruff N Reddy Hunny Bear. She was sold to Mark Lane of Country Lane kennel and was his first Australian Shepherd and the only direct Ted puppy he ever owned. He describes her as guardy but well controlled. What he remembers most is her utter fearlessness in the face of any threat. A pack of three local strays once attacked two other dogs owned by Mark and were chasing them as they were heading for home. When Hunny Bear heard the commotion she passed Mark's two dogs going in the opposite direction to take on the whole pack of Chow mixes all alone. Although she likely had the strength to actually whip the pack, it was her attitude and heart that made them change their collective mind about finishing the fight they had started. Her guardiness with humans never went past the growling stage if she decided she did not trust you…unless you decided not to heed the warning.

Hunny Bear was bred 4 times and although Mark kept one nice female named Country Lane's Mity Hot Pepper, a read merle female, most of the other puppies went to ranchers who knew the Ruff N Reddy line and wanted good working dogs. Bred once to a show line male, Hunny Bear's other 3 breedings were all to Ruff N Reddy dogs—twice to Ruff N Reddy Ranger and once to Jones Little Redwood. She produced the magnificent Colorados How the West Was Won who sired Colorados Mity Bludigger.

Ruff N Reddy Hunny Bear as a puppy. Quite the pretty silver blue merle as a baby she would turn nearly black as an adult, like most dogs of this line. Her superior conformation is evident even at this age. Courtesy of Mark Lane

Ruff N Reddy Hunny Bear who closely resembled her wonderful dam, Jones' Mity Bite. Courtesy of Casey Newman

Ruff N Reddy Mity Happy Hocker CD STDd showing her good gathering style on ducks. Courtesy of Raymie Jones

Legacy Chart for Ruff N Reddy Mity Happy Hocker

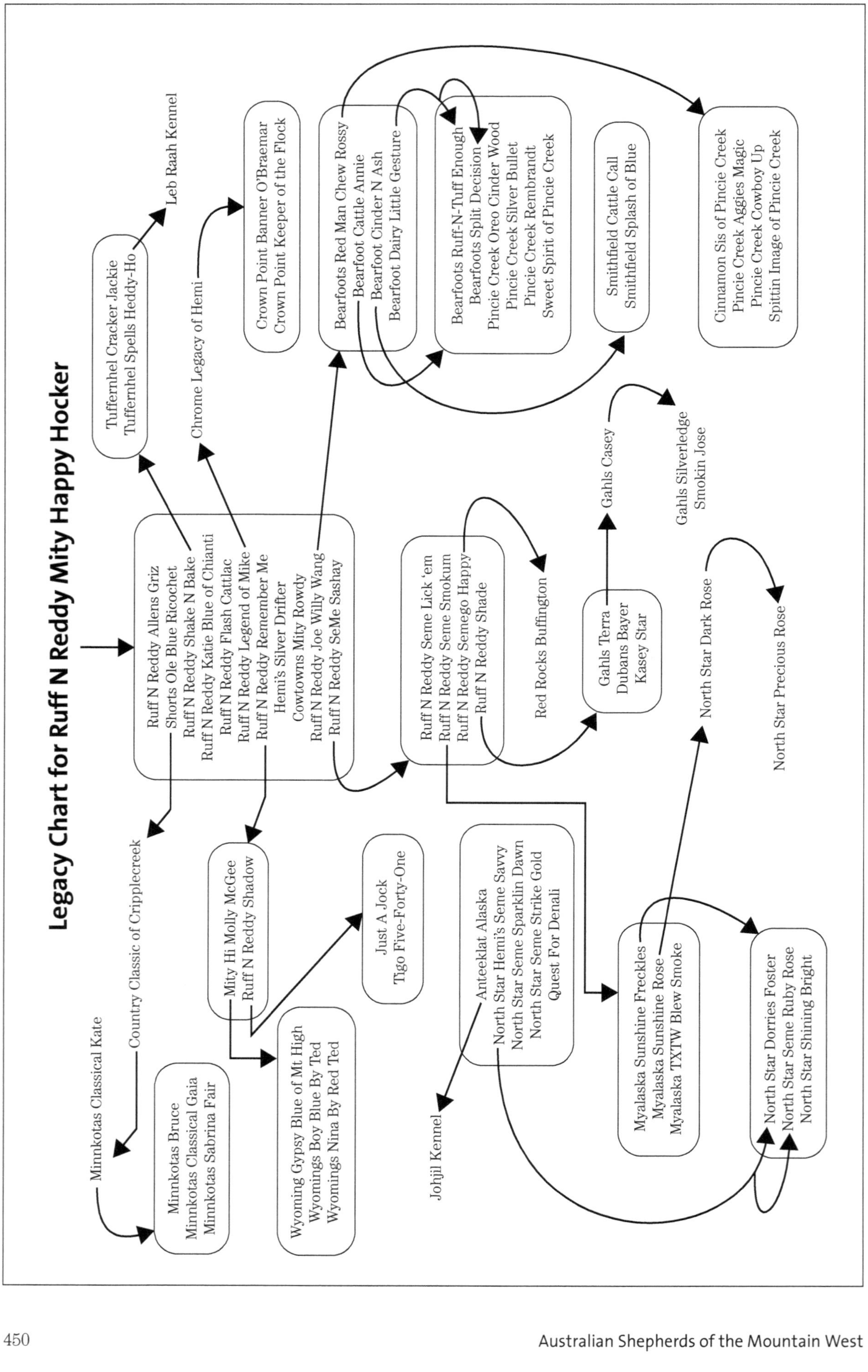

When Jeffreys' Shade was very old, Mity Bite was bred to him right before he died. Once again, Cathy's intuition was perfect and this cross produced Shadey Mountain Mity Miss. Her story is told in more detail in the Tom and Charlene Jeffreys' chapter but suffice it to say that the Mity Miss Legacy was second to none for not only Cathy Jones but also the Shadey Mountain kennel and the HS&W breeding program of Casey Masterson Newman.

Ruff N Reddy Hunny Bear, blue merle and white female (Ted x Mity Bite). Mostly black with small spots of merling through her body and greater amounts on her head and feet. Her son, Colorado's How the West Was Won, was even more sparsely merled. Courtesy of Melissa Couch

Country Lane's Mity Hot Pepper; red merle and white female who produced a number of good offspring for Mark Lane, Casey Newman and Vicki Smith of Smithfield kennel. (Ruff N Reddy Ranger x Ruff N Reddy Hunny Bear) Courtesy of Casey Newman

Ruff N Reddy Hunny Bear as an older dog, nearly black but she was a dark blue merle (Ted x Mity Bite). Courtesy of Casey Newman

Ruff N Reddy Hunny Bear didn't have many opportunities to work livestock but when she did she made the most of it! Courtesy of Mark Lane

Ruff N Reddy Hunny Bear. Courtesy of Melissa Couch

Colorado's How the West Was Won as a puppy when you could still see some of his blue merle markings (Jones' Little Redwood X Ruff N Reddy Hunny Bear). Although obtained as a puppy from breeder Mark Lane, he was all Cathy Jones breeding. Courtesy of Sorensen

Colorado's How the West Was Won about 8 months of age. This dog was such a dark blue merle he was nearly black as an adult. In this puppy picture it is possible to see he did have some merling at an early age. Courtesy of Sorensen

Cathy's first breeding choices proved to be wise ones. Jones' Mity Bite is now behind a multitude of top dogs who continue to influence the breed today. The philosophy of that day among many breeders was to not overproduce dogs. Culling was a normal facet of breeding and although some breeders had frivolous criteria for culling (such as color), the breed population was kept in check and the quality overall remained high due to this practice.

Pups out of Jeffreys' Shade x Jones' Mity Bite. The blue puppy on the left is Shadey Mountain Mity Miss. Courtesy of Raymie Jones

Colorado's How the West Was Won, very dark blue self merle male. Bred by Mark Lane of Country Lane Kennel, this dog was all old Wood breeding from dogs bred by Cathy Jones. (Jones' Little Redwood x Ruff N Reddy Hunny Bear) Of the big, strong, broad-headed type established by Jay Sisler and Fletcher Wood as their ideal. Courtesy of Sorensen

Country Lane's Mity Hot Pepper. Although never trialed, Pepper was a competent and sensible working dog. Courtesy of Melissa Couch

Ted's popularity and reputation was continuing to grow and 1976 saw a full book for his stud services. It was becoming quickly apparent his tough aggressive style on cattle was highly heritable. His puppies turned on early and intensely. This was a real plus not only to Cathy but to breeders who needed this trait blended into their lines. Several outside females were brought to Ted in 1976. The inbred George's Red Rustler daughter/granddaughter, Moor's Rustlin Ruby, produced a litter by Ted for the Gundersons of Missouri. A repeat of the successful cross with Colorado's Buckeye Barbie was also done, producing another fine litter of great working dogs, including Ruff N Reddy Magic Sparkle, a good female that went to the Kintyre Cattle Co. Cathy also leased Mity Bite's dam, Piz's Sheba, from Joe Piz for a cross to Ted producing the great Ch Ruff N Reddy Mity Muhammad CD STDds, owned by Patty Ennen. This solid black dog was the first of many excellent dogs from this cross, which was repeated several times by Cathy. She was quickly learning the value of good females in her breeding program. Sheba was another "old time doggy bitch" in Ruth McGuire's words but this type had a knack for being a good mother and her pedigree was second to none. She was a sweet tempered, heavy boned female who was never shown but her worth did not come from titles, rather from the superior progeny she produced no matter the sire of her litters. When she was about 9 years old Ruth McGuire leased her to breed to her great sire Hemi and once again Sheba out produced herself.

Ruff N Reddy Mity Blu Goblin, dark blue merle and white female (Ted x Mity Bite) an accidental acquisition by Casey Newman who became an important foundation for the HS&W breeding program. Courtesy of Casey Newman

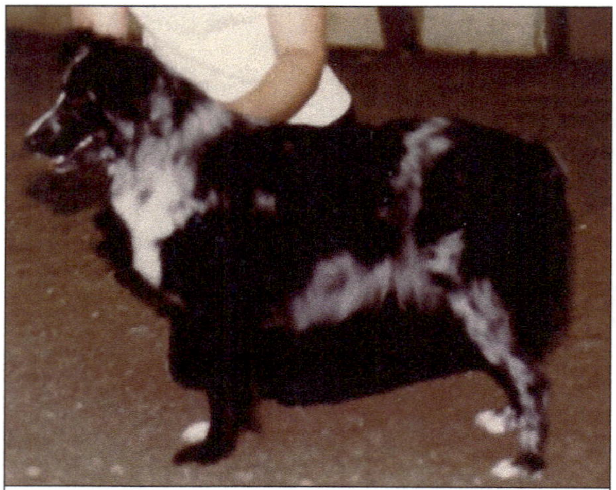

Ch Ruff N Reddy Blue Tinker Toy as a younger dog. Her merling is still light and contrasts sharply with the black spots. Like most dogs of this line she darkened to nearly black as an adult and aged dog. Photo by Sue Noll. Courtesy of Raymie Jones

Briarbrooks Diamond Jim a red merle male born in 1976. Full brother of Briarbrooks Melbourne who sired Couch's Lacy Mae and Couch's Belle Starr. (Ted x Bar Lazy K's Diamond Lil of Las Rocosa) Courtesy of Raymie Jones

Ch Ruff N Reddy Blue Tinker Toy winning Best of Breed. Her merling is already showing signs of getting darker even though she is in the prime of life in this photo. Courtesy of Raymie Jones

Piz's Sheba—the great matron who produced some of Cathy and Ruth's best dogs. Blue merle and white (Piz Joseph x Piz Lady II—a Piz Joseph daughter). Sheba had 9 crosses to Wood's Jay in 5 generations. Courtesy of Raymie Jones

Before 1976 was through, Ted was bred to several other important females producing a number of dogs which were to make a name both for themselves and their famous sire. Another solid black dog was born this year out of Nielsens Holly II (Del Ray Diamond Bob x Weigants Holly), a very high quality female of old Northwest Diamond A breeding. His name was Ch Slash V Easy Jet CD STDds and his success in the show ring and as a working dog for his owners Doug and Linda Bailey just increased Ted's reputation as an all around dog and top producer. In the late 1970s when Cathy was serving as Show Secretary for the Annual CASA Show she wanted to show Ted in the Get of Sire class with Easy Jet and Muhammad. She enlisted Ruth to step in and handle Ted, a request Ruth was hesitant to accept. She needn't have worried. Ted's impeccable training and Cathy's careful instructions got them through the class without incident, winning first in the process. Unfortunately, during the class, Ruth stepped in a hole, injuring her back and Cathy had to return the favor and show Ruth's female, Magic, in the stock trial the next day!

Ruth was often enlisted to assist Cathy with her dogs and since she was one of Cathy's early mentors and a good friend, this arrangement worked out well. One job Ruth never minded was helping with Mity Muhammad. Although this stunning black dog was owned by Patty Ennen, Patty had moved to Canada and was unable to get to many shows. Cathy arranged for Mo to live with her for awhile to be shown and Ruth was often the handler. She relished the job because she dearly loved this big, loving dog. Traveling all over the western plains states to shows, Ruth eventually finished Mo in Santa Fe, New Mexico, earning a breeding to Mo for her own Hemi daughter, Magic, for her efforts. The success of this cross encouraged

Ruff N Reddy Magic Sparkle, solid red female at age 6 months (Ch Jones' Reddy Teddy, CDX, OTDdsc SchHA x Colorado's Buckeye Barbie). One of many top dogs owned by Colville Jackson. Courtesy of Sorensen

Colorado's Reddy Raider and Ruff N Reddy Magic Sparkle at 9 months and 6 months of age respectively (October 1976). Both were by Ch Jones' Reddy Teddy CDX OTDdsc, SchHA and owned by Colville Jackson. Reddy Raider is at the head with Magic Sparkle bringing up the rear. Courtesy of Sorensen

Ch Slash V Easy Jet CD STDds, black male (Ted x Nielsens Holly II)—moderate, well balanced, what an Aussie should be. Courtesy of Raymie Jones

Ch Slash V Easy Jet CD STDds as a young dog showing the same eye on ducks as his famous sire, Ted. Courtesy of Raymie Jones

Ch Ruff N Reddy Mity Muhammad CD STDsd, black male (Ted x Piz's Sheba) winning Best Working Dog at the 1980 CASA Specialty with Ruth handling. Courtesy of Ruth McGuire

Ruth to breed Magic back to his sire, Ted, 3 times in later years.

The remainder of 1976 saw several more excellent litters by Ted. Mini Acre Freckles (J Bar D Millers Rex x Las Rocosa Penne of Mini Acre), owned by cattleman Ken Claussen produced one of the many Ted litters born at Mini Acre. Another outcross from Parmenters Dixie of Bar Lazy K (Red Rustler/Las Rocosa cross female) produced Lambeths Krispy Kritters, an important foundation for Pat Lambeth's Little Spot kennel in California. And to top off the year, Terry Martin made the first of several crosses to Ted with her female Martin's Red Fudge (Martin's Tim Tim x Martin's Josie). This cross produced Slash V Fudge Cookie and Terry's Slash V Semi Sweet.

Cathy states her viewpoint about breeding in this succinct way, "The breeding end of things has always been separate from showing. I have always had a slightly different attitude about breeding than most people; I've probably been a little bit of an outlaw in that area. I am known around here as having a very large hammer. In other words, I don't raise that many puppies, and I am very particular about the ones I do raise. If there is a puppy that I don't especially care for, it is not around long. I am considered to be kind of hard-headed or hard-hearted, but I have never raised a litter of ten puppies. Breeding, to me, is a whole different situation than showing. Of course, they both intermingle, but breeding is a much more serious undertaking than the shows ever were. You are laying a foundation for the bloodline of a breed." She showed more than a little intuition about where the breed was headed when she made this statement, "I think right now is the time when we formulate what this breed is going to be all about in the future. There just needs to be some soul searching by both breeders and the national club." Her words should be repeated every year by every breeder and every breed club on earth.

This was not merely lip service for Cathy. She would only breed 2-3 litters per year and always culled about half the puppies at birth. Although merle whites are not as common in the intensely pigmented lines such as the Wood lines, they would still occur and would be the first to go. This was seldom an issue as Cathy tried to make merle to solid breedings in order to avoid it. After evaluating the puppies for color she would determine if there were any other genetic or breathing problems with any puppies. She would not try to save puppies with problems as she believed in only keeping the best. Normally she would have from 8-10 puppies in her litters so they would be culled down to 4 or 5 based on size (she preferred keeping the middle sized puppies), head and overall type. The dogs were always first and foremost working dogs to Cathy. She thought there were too many pets in the world and she didn't want her breeding to contribute to that overpopulation. Her dogs were not appropriate for most pet homes anyway and any that ended up there always seemed to come back to haunt her.

Ch Ruff N Reddy Mity Muhammad CD STDsd winning Best of Breed at the 1980 CASA Specialty. Courtesy of Ruth McGuire

On very rare occasions she would allow a puppy to go to a pet home. She once met a lady who wanted a dog specifically from her bloodline for an obedience competition dog and pet for her children. She knew the woman and she knew the dog would have the work it needed. He would be neutered and sold without registration so he would not be used for breeding. Even these conditions were not always enough and Cathy stuck to her plan in placing puppies most of the time.

"Mo" working cattle with same grit and toughness as his sire. Courtesy of Melissa Couch

By this time in the breed's history the Wood's line was already beginning to become scarcer. Cathy never had any difficulty in selling puppies because there was still a demand among cattle ranchers for good, strong dogs to work their stock. She knew of only two other people breeding to keep the Wood line going and since they were all located in her area the breeding was not spread all over the country. She was always able to get good prices for her puppies because the people who came to her for a dog knew the dog would be what they needed. As Cathy puts it, "When you breed for a special type you can dictate the price.." Her emphasis on a low heeling dog and her ability to consistently

Legacy Chart for Slash V Semi Sweet

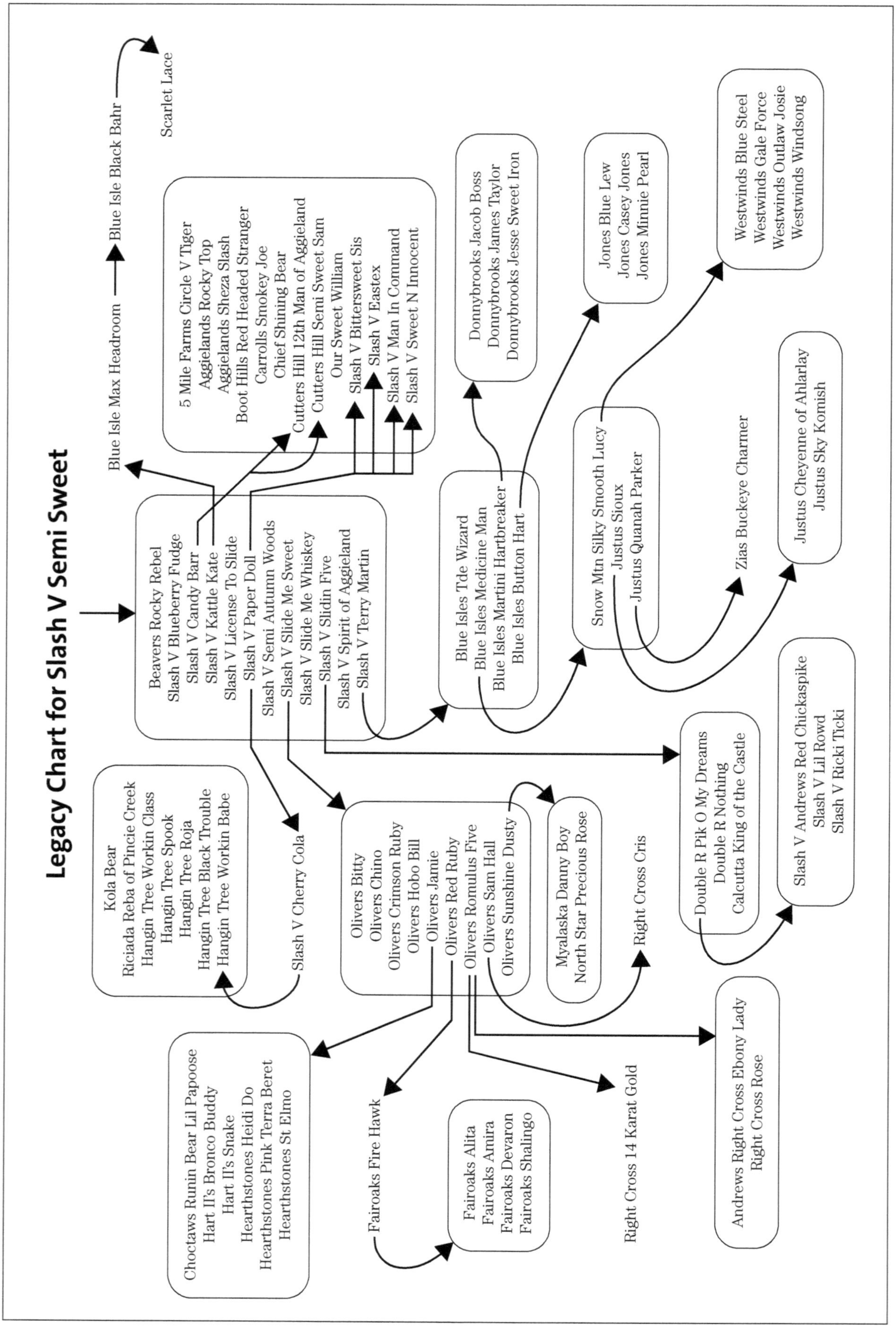

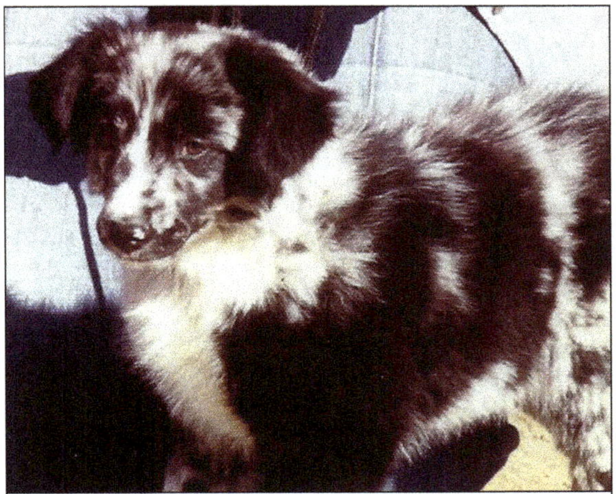

Ch Ruff N Reddy Blue Tinker Toy as a high quality puppy. Using only the best dogs for breeding was the "secret" to the success of this line. Courtesy of Raymie Jones

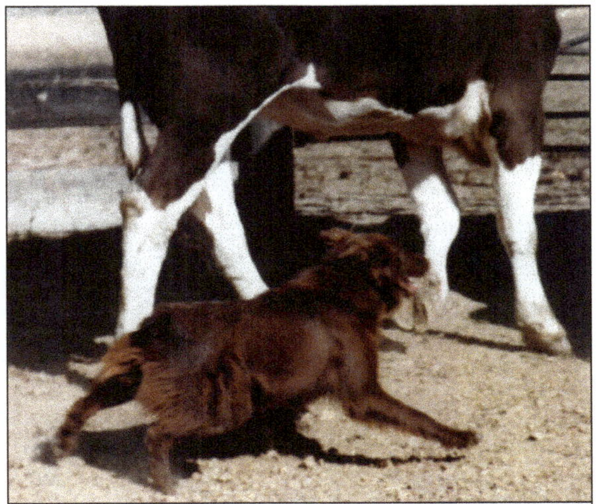

Solid red Ted daughter, Slash V Fudge Cookie CRD, STD, out of Terry Martin's female, Martin's Red Fudge. Courtesy of Sorensen

produce it made her dogs highly valued and worth their price. So much so that most of her puppies were sold before she made a breeding. If she ever had a puppy left she knew she could go ahead and begin their training and eventually sell the dog to a good home as a started or trained dog.

The kind expression that reflects the wonderful temperament of Ch Ruff N Reddy Mity Muhammad CD STDds, black male (Ted x Piz's Sheba). Courtesy of Raymie Jones

By 1977, Cathy appeared to be getting more and more selective about her breeding choices. The previous few years had shown her what Ted was capable of producing with a number of good females but Cathy's standards were always much higher than just "good." She always went for the best and she and Ruth had studied and learned enough about pedigrees and what lines crossed well to produce higher and higher percentages of very very good dogs. In addition to a repeat of the Martin's Red Fudge

Ch Ruff N Reddy Mity Muhammad CD STDds working sheep. Courtesy of Raymie Jones

A young solid red puppy showing the natural eye that often appeared in puppies of Cathy's breeding. Courtesy of Mark Lane

cross, Ted was bred again to Piz's Sheba and then to the incomparable Davies' Blue Kola. Betty Nelson also sent her prized foundation female Nelson's Gin Fizz (Flying X Half Hitch x Nelson's Daiquiri) to Ted, producing a sire who would make a lasting mark on her breeding program—Nelson's Ben Reddy. Last but certainly not least, Jim and Sue Foster brought their top female, Ch Colorado's Ravishing Ruby to Ted, producing dogs which left their mark all over the working world from coast to coast.

Always active in local clubs and activities, during this time Cathy was volunteering for a number of jobs in the CASA club. One of these opportunities was working a promotional booth for the club and the breed at the National Western Stock Show held every year in January in Denver. Club members would work the booth in shifts of several hours at a time and the traffic through this venue made for tremendous exposure for the breed. It was during one of Cathy's shifts she met the great Jay Sisler. During a break in his performances, he was taking in the sights and saw Reddy Teddy sitting with Cathy in the booth. He was drawn to Ted and

Ch Ruff N Reddy Blue Tinker Toy hamming it up for the camera at 12 years of age. Courtesy of Melissa Couch

Ch Kintyre Jeb Stuart STDd OTDc. Courtesy of Raymie Jones

Ch Kintyre Jeb Stuart STDd OTDc, dark self blue merle (Ted x Davies' Blue Kola) a winner as a show dog and working dog. Shown here at age 2 years in November 1979. Owned and handled by Colville Jackson. 22" and 56 pounds of pure working dog. Courtesy of Sorensen

"Monty"—good, moderate type in a strong working dog. (Ch Kintyre Jeb Stuart STDd OTDc x Colorado's Lace Leggins) Courtesy of Sorensen

Ted taking a nice square flank to bring in a straggler. Courtesy of Mark Lane

"Monty," blue merle Jeb Stuart son out of Colorado's Lace Leggins owned by Jerry Morrow and described as an aggressive low heeler. Courtesy of Sorensen

Ch Kintyre Jeb Stuart STDd OTDc as a precocious 6 month old puppy showing his strong working instincts. Courtesy of Sorensen

Nelson's Ben Reddy, an important sire for Betty Nelson by Ted x Nelson's Gin Fizz

PEDIGREE OF Davies Blue Kola

Born: ____________ Sex: ________
Owned by: ____________________

Sire	Sire	Sire	Sire
Woods Stubbie	Woods Jay	Sislers Shorty	Keno
			Dam Blue Star
		Dam Sislers Trixie	Sire Blue Khan
			Dam Emmetts Blue
	Dam Woods Slate	Sire Woods Jay	Sire Sislers Shorty
			Dam Sislers Trixie
		Dam Woods Specks	Sire Woods Jay
			Dam Trumps Stubbie
Dam Davies Checkers	Sire Jeffreys Shade	Sire Woods Stubbie	Sire Woods Jay
			Dam Woods Slate
		Dam Jeffreys Missy Blue	Sire Williams Laddie
			Dam Picrauxs Juliette
	Dam Woods Little Kola	Sire Woods Jay	Sire Sislers Shorty
			Dam Sislers Trixie
		Dam Woods Little Blue	Sire (Sislers) Blue Boy
			Dam (Sislers) Blue Shadow

Bred by: ____________ Address: ____________ Phone: ____________

The details given above are correct to the best of my knowledge. Signed: ____________(Breeder)

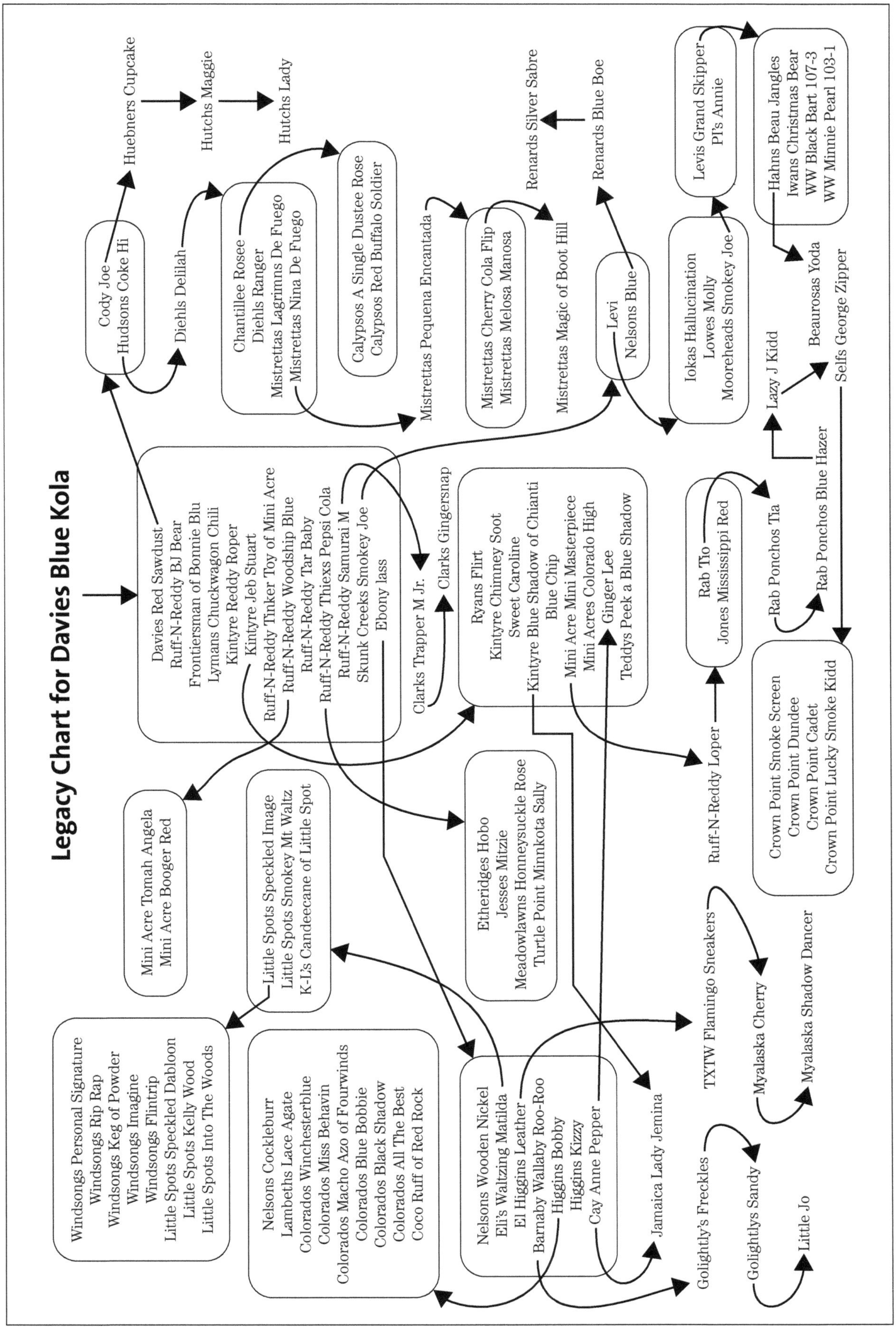
Legacy Chart for Davies Blue Kola
Davies Red Sawdust
Ruff-N-Reddy BJ Bear
Frontiersman of Bonnie Blu
Lymans Chuckwagon Chili
Kintyre Reddy Roper
Kintyre Jeb Stuart
Ruff-N-Reddy Tinker Toy of Mini Acre
Ruff-N-Reddy Woodship Blue
Ruff-N-Reddy Tar Baby
Ruff-N-Reddy Thiexs Pepsi Cola
Ruff-N-Reddy Samural M
Skunk Creeks Smokey Joe
Ebony lass
Cody Joe
Hudsons Coke Hi
Huebners Cupcake
Hutchs Maggie
Hutchs Lady
Diehls Delilah
Chantillee Rosee
Diehls Ranger
Mistrettas Lagrimns De Fuego
Mistrettas Nina De Fuego
Calypsos A Single Dustee Rose
Calypsos Red Buffalo Soldier
Mistrettas Pequena Encantada
Mistrettas Cherry Cola Flip
Mistrettas Melosa Manosa
Mistrettas Magic of Boot Hill
Renards Silver Sabre
Renards Blue Boe
Levi
Nelsons Blue
Iokas Hallucination
Lowes Molly
Mooreheads Smokey Joe
Levis Grand Skipper
PI's Annie
Hahns Beau Jangles
Iwans Christmas Bear
WW Black Bart 107-3
WW Minnie Pearl 103-1
Beaurosas Yoda
Lazy J Kidd
Selfs George Zipper
Clarks Trapper M Jr.
Clarks Gingersnap
Ryans Flirt
Kintyre Chimney Soot
Sweet Caroline
Kintyre Blue Shadow of Chianti
Blue Chip
Mini Acre Mini Masterpiece
Mini Acres Colorado High
Ginger Lee
Teddys Peek a Blue Shadow
Rab Tio
Jones Mississippi Red
Rab Ponchos Tia
Rab Ponchos Blue Hazer
Ruff-N-Reddy Loper
Crown Point Smoke Screen
Crown Point Dundee
Crown Point Cadet
Crown Point Lucky Smoke Kidd
Mini Acre Tomah Angela
Mini Acre Booger Red
Little Spots Speckled Image
Little Spots Smokey Mt Waltz
K-L's Candeecane of Little Spot
Etheridges Hobo
Jesses Mitzie
Meadowlawns Honneysuckle Rose
Turtle Point Minnkota Sally
TXTW Flamingo Sneakers
Myalaska Cherry
Myalaska Shadow Dancer
Windsongs Personal Signature
Windsongs Rip Rap
Windsongs Keg of Powder
Windsongs Imagine
Windsongs Flintrip
Little Spots Speckled Dabloon
Little Spots Kelly Wood
Little Spots Into The Woods
Nelsons Cockleburr
Lambeths Lace Agate
Colorados Winchesterblue
Colorados Miss Behavin
Colorados Macho Azo of Fourwinds
Colorados Blue Bobbie
Colorados Black Shadow
Colorados All The Best
Coco Ruff of Red Rock
Nelsons Wooden Nickel
Eli's Waltzing Matilda
El Higgins Leather
Barnaby Wallaby Roo-Roo
Higgins Bobby
Higgins Kizzy
Cay Anne Pepper
Jamaica Lady Jemina
Golightly's Freckles
Golightlys Sandy
Little Jo

Cathy with Blue Tinker Toy manning a booth at one of the many promotional events held by her club. Courtesy of Raymie Jones

Two youngsters making a promising start. Wade Jones with Ch Ruff N Reddy Blue Tinker Toy winning a first place. Courtesy of Raymie Jones

ASCA National Specialty 1978—Benton, Michigan. Blue Tinker Toy winning BOB Puppy over 80+ entries. Courtesy of Raymie Jones

struck up a conversation with Cathy. He was so fascinated with Ted and Cathy's stories of his abilities, the two went outside for a quick demo of Ted's talents! Surely they each recognized a kindred spirit in the other as both Jay and Cathy were trainers who never settled for less than the best from their dogs and horses. Leslie Sorensen describes Cathy's training methods as "meticulous" whether she was training her horses or her dogs. She did everything carefully, step by step and always did things right the first time. Her success in competition resulted from this careful attention to detail, just as Jay Sisler's success as a performer resulted from his own reach for perfection.

It was also during one of these shifts Cathy met a rancher from Wyoming who had a good female of Sisler/Woods breeding he was looking to breed. He saw Ted, picked up one of Cathy's business cards and following the show called her up to book the breeding. The female was Keefers Miss Nitzi (Cards Ruff x Iacovettas Shasta) and Cathy's pick from the litter became one of her most beloved dogs—after Ted, of course. This was Ch Ruff N Reddy Blue Tinker Toy, a dark clear silver blue female. Known as "Boo Boo" to her friends, she was a very petite dog who might have matured at 18 inches, although Ruth estimates she may have been closer to 17-1/2 inches in reality. This only added to her natural "cuteness" along with an excellent, type head and good bone. Her size was never a handicap in the show ring. Cathy attended the 1978 ASCA Nationals in Berrien Springs, Michigan where Boo Boo promptly defeated every Non-Regular puppy at the show, winning Best Puppy! This show turned out to be a phenomenal success for Cathy with Boo Boo's kennelmate, Ted, taking second in Utility, and winning both Open Ducks and Advanced Cattle.

Cathy and Ch Ruff N Reddy Blue Tinker Toy winning Best Puppy at the 1978 CASA Specialty over 30 entries. Courtesy of Raymie Jones

Ch Ruff N Reddy Blue Tinker Toy at the 1978 CASA Specialty. Tiny but attracting attention from the very start. Courtesy of Raymie Jones

A red letter day at the 1978 CASA Specialty—Cathy Jones with Ted winning Best of Breed, Patty Ennen with Mity Muhammad winning WD and Blue Tinker Toy taking Best Puppy. Courtesy of Raymie Jones

Ted (center) with two of his daughters "Bebe Babe" (left) and Slash V Fudge Cookie (right) at the 1978 National Specialty. Courtesy of Raymie Jones

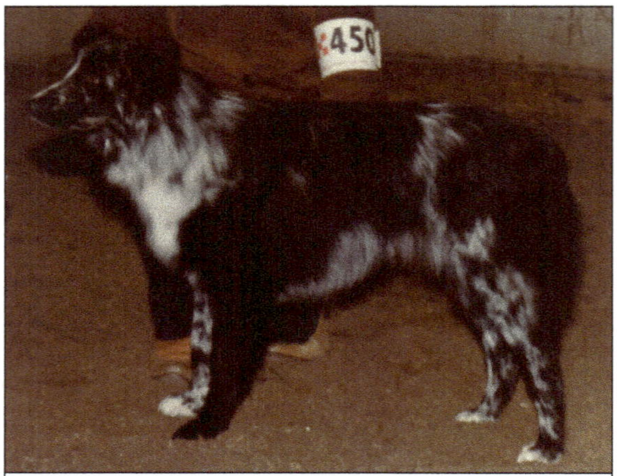

Ch Ruff N Reddy Blue Tinker Toy, blue and white female by Ted out of Keefer's Miss Mitzi (Card's Ruff x Iacovetta's Shasta—Card's Ruff was a son of Wood's Little Blue, Shasta was largely Sisler bred) One of Cathy's favorite girls and dam of Jones Little Redwood and Ruff N Reddy Too Small Jones. Courtesy of Sorensen

Tinker Toy became a competent working dog under Cathy's excellent training and handling, Courtesy of Raymie Jones

Ch Ruff N Reddy Blue Tinker Toy still loved to play the show dog at 12 years of age. Courtesy of Melissa Couch

It turned out Boo Boo was a natural in the show ring, not only finishing her Championship—the first of Ted's offspring to do so—but also stepping in to help Ruth's daughter win Best All Around Jr Handler. Tinker Toy was her stock and obedience dog, while Ruth's own Magic was her conformation and showmanship dog. Competing at the same time was the Sorensen's daughter Leigh Ann, who was always a consistent winner, so this achievement was especially meaningful to the Jones/McGuire team.

Boo Boo had a charming personality and became Cathy's husband's favorite dog. She became his constant companion and learned to love a favorite blue ball which was never far away. When it was time for her to eat she would drop the ball in the bowl and eat around it. It went in the water dish when she needed a drink and was always in her mouth for safekeeping at all other times. On livestock she proved to be a good low heeler like her famous sire.

In 1978 Cathy acquired a very dark blue merle female named Shadey Mt Su-Mac from Charlene Jeffreys. She was apparently a stud fee puppy from Charlene's male Olson's Blue Joe and out of Leslie's Sorensen's superb female Colorado's Perky Miss. Su-Mac wasted

Ch Jones Reddy Teddy CDX, OTDdsc, SchHA and Ch Ruff N Reddy Blue Tinker Toy. Courtesy of Sorensen

Born: ____________ Sex: ________
Owned by: ____________________

PEDIGREE OF Ch Ruff N Reddy Blue Tinker Toy

Parents	Grandparents	Great-grandparents	Great-great-grandparents
Sire: Jones Reddy Teddy	Sire: Piz Little Joe	Sire: Piz Joseph	Sire: Woods Jay
			Dam: Woods Chili
		Dam: J Bar D Choctaw Chilli	Sire: Rowes Warrior Bow
			Dam: Sorensens Queenie
	Dam: Ginger Red	Sire: Montoyas Rowdy	Sire: Clingers Royal Blue Merle
			Dam: Kalbs Princess
		Dam: Montoyas Snipper	Sire: Piz Joseph
			Dam: Piz Lady II
Dam: Keefers Miss Nitzi	Sire: Cards Ruff	Sire: Woody Dandy	Sire: Finleys Texas Bob
			Dam: Merinos Blue Bitch
		Dam: Woods Little Blue	Sire: Blue Boy
			Dam: Blue Shadow
	Dam: Iacovettas Shasta	Sire: Sislers Dan	Sire: Teton
			Dam: Sal
		Dam: Wilcoxs Cindy	Sire: Hubbards Shep
			Dam: Taylors Buena

Bred by: ____________ Address: ____________________ Phone: ____________

The details given above are correct to the best of my knowledge. Signed: ____________________(Breeder)

Ch Ruff N Reddy Blue Tinker Toy, blue merle female (Ted x Keefer's Miss Nitzi) This little dynamo inherited her sire's ability as a low heeler on cattle. Courtesy of Raymie Jones

Shadey Mountain Sumac with Cathy at the 1978 Colorado Specialty winning Best of Breed Puppy over 48 entries. Blue merle female (Olson's Blue Joe x Colorado's Perky Miss) Photo by Sue Noll. Courtesy of Raymie Jones

no time taking the show ring by storm, winning Best Puppy at the 1978 Colorado Specialty over 48 entries! She finished her Championship easily at a young age and was soon a favorite of Cathy's. Ruth McGuire remembers her as heavy coated with a typical good hard Wood coat. Her color was a deep gunmetal blue like her grandsire Gunsmoke, although she had a muddy undercoat which Cathy faulted. Of course she had a nice sweet temperament like most of the females Cathy obtained. And, typical of the very dark Wood dog, she had no copper or white trim. She had the blocky head and square muzzle typical of the Wood dogs all down the line. Her bite was tight but correct.

Not much is known about how Cathy used her in the Ruff N Reddy breeding program. She was bred to Ted at least once producing at least one female which was bred from later named Sun Shadows Silhouette. Perhaps the muddy cast to her color, which appeared in the

Cathy Jones with Ted at the 1978 CASA Specialty shows. Patty Ennen with Ch Ruff N Reddy Mity Muhammad CD, STDds (Ted x Piz's Sheba) winning All Around Dog. Courtesy of Raymie Jones

litter she had with Ted, eliminated her from the breeding program. Cathy culled the litter heavily and kept only the solids and the clearer merles. She always preferred good clear color and apparently did not want to risk adding the muddy cast to the good color she already had. A dog named Flatlands Friday the Thirteenth is also shown to be out of Su-Mac by a dog named Mountain Highs J-J of Flatland (a J Bar D and Las Rocosa cross male) but little else is known about her produce or her fate.

Success was now a constant companion to Cathy Jones and demand for Ted's puppies was high. She repeated the breedings to the two great foundation females Piz's Sheba and Davies' Blue Kola, which did not fail to produce another generation of top working and breeding dogs. A solid black female, Ruff N Reddy Voodoo Queen, (Ted x Piz's Sheba) was sold to Rob Brown of Throckmorton, Texas. The Browns raised horses and cattle and needed a tough dog to help work their stock. According to Melissa Couch this little powerhouse was full of character and worked the ranch every day like a little machine. She says she resembled her full brother Mity Muhammad and was a top

Shadey Mountain Sumac as a young dog. A winner in the show ring like most of Cathy's dog. Courtesy of Raymie Jones

Shadey Mountain Sumac as a mature female showing the typical darkening of her coat as she aged. She looks very much like her sire in this photo. Courtesy of Raymie Jones

Red merle Ted x Sumac puppy Courtesy of Raymie Jones

Red merle Sumac pup working ducks Courtesy of Raymie Jones

Red merle Sumac pup. Courtesy of Raymie Jones

quality dog. Unfortunately she was mostly bred to the ranch border collies to produce dogs for working their cattle and contributed little to the Aussie gene pool. She was crossed at least once to a dog named Rab Tio (by Ruff N Reddy Loper, a Ted x Mini Acre Mini Masterpiece son out of Hewes Cricket, a Jeffreys' Pooh Bear daughter out of a Sorensen bred female). This cross produced a female named Rab Ponchos Tia who is behind a number of Crown Point dogs.

Nineteen-seventy-eight was also the year of the first litter for another of Cathy's treasured females, Ruff N Reddy Mity Happy Hocker CD STDd. Skidder rewarded Cathy's intuition about her and produced a nice solid black male named Ruff N Reddy Joe Willy Wang. Cathy kept this good dog for some time, showing him around the country when she traveled. According to Ruth he was a long legged and sound dog but did not have as good type as she might have liked. Nevertheless, he proved his genetic worth later as did most of Cathy's dogs. His amusing name was taken from a top winning Greyhound who was competing in Colorado at the time. When Cathy listened to the race results, the Greyhound Joe Willy Wang was winning consistently and she could not help being taken in by the alliteration of "Joe Willy Wang won again."

Cathy kept Joe Willy for several years but eventually sold the good natured dog to a dairyman in Florida named Al Combee, along with a beautiful red merle female named Ruff N Reddy Shadey Wood Ash to form the foundation of the Bearfoot working dogs. This pair eventually produced the two excellent sires Bearfoots Cinder N Ash and Bearfoots Red Man Chew Rossy, who still figure prominently today in the breeding program of Pincie Creek. Although Cinder N Ash was used just for

Born: ____________ Sex: ________
Owned by: ____________

PEDIGREE OF Shadey Mountain Sumac

Parents	Grandparents	Great-grandparents	Great-great-grandparents
Sire: Olsons Blue Joe	Sire: Piz Joseph	Sire: Woods Jay	Sire: Sislers Shorty
			Dam: Sislers Trixie
		Dam: Woods Chili	Sire: Woods Jay
			Dam: Greens Dingo
	Dam: Piz Lady II	Sire: Piz Joseph	Sire: Woods Jay
			Dam: Woods Chili
		Dam: Piz Lady I	Sire: Woods Jay
			Dam: Woods Chili
Dam: Colorados Perky Miss	Sire: Sorensens Gunsmoke	Sire: Williams Laddie	Sire: Uhlendorfs Tippy
			Dam: Thorntons Chrissie
		Dam: Jeffreys Sheila	Sire: Elys Major
			Dam: Williams Georgina
	Dam: Nichols Little Bonnie	Sire: Arington's Stub	Sire: Woods Jay
			Dam: Harpers Jinx
		Dam: Mchean's Lady	Sire: unknown
			Dam: unknown

Bred by: ____________ Address: ____________ Phone: ____________

The details given above are correct to the best of my knowledge. Signed: ____________(Breeder)

Ruff N Reddy Joe Willy Wang, black male (Ted x Ruff N Reddy Mity Happy Hocker CD STDd) and Ch Ruff N Reddy Blue Tinker Toy (Ted x Keefer's Miss Nitzi). Courtesy of Raymie Jones

breeding at Pincie Creek, Chewy became a loved member of the Stevens family. When Roger and Elaine would take him to trials, people were naturally curious about his breeding. When they were told he was a Ted grandson, their disbelief was overwhelming! As Chewy crawled into their laps begging for petting they emphatically declared he could not possibly be related to that "nasty tempered Reddy Teddy dog"! In fact, the vast majority of Ted's offspring had very sound, stable, friendly temperaments.

Bearfoots Cinder N Ash, black and white male (Ruff N Reddy Joe Willy Wang x Ruff N Reddy Shadey Wood Ash) A dog who became a great cross for Pincie Creek kennel with their Hangin Tree foundation dogs. This bloodline has preserved a strong Wood type dog. Courtesy of Pincie Creek

Bearfoots Cinder N Ash. A dog of nice moderate head type. Courtesy of Pincie Creek

Bearfoots Cinder N Ash. Courtesy of Pincie Creek

It appears Cathy was becoming more selective about breeding outside females to Ted, but a few still met her approval, including a female named Slash V Wild Firewood, owned by Barb Nielsen. It is likely the success of the crosses Terry Martin had made influenced her decision to allow this cross and it, too, rewarded

Pincie Creek Silver Bullet, dark blue self merle male (Bearfoots Cinder N Ash x Slash V Bittersweet Sis) His Wood background is unmistakable with his broad square head, deep blue clear color and strong build. Courtesy of Roger and Kathy Stevens

Bearfoots Red Man Chew Rossy, red and white male (Ruff N Reddy Joe Willy Wang x Ruff N Reddy Shadey Wood Ash) the second male of this cross to leave his mark at Pincie Creek. A fairly tall dog like his sire but well proportioned and with a temperament second to none. Courtesy of Roger and Kathy Stevens

Pincie Creek Silver Bullet, a ¾ bred Wood dog. His sire was all Wood and his dam was by Slash V Spirit of Aggieland x Slash V Paper Doll, half brother and sister out of the legendary Slash V Semi Sweet, a Ted daughter. Courtesy of Roger and Kathy Stevens

Bearfoots Licorice Stick Rossy. Courtesy of Melissa Couch

Bearfoots Licorice Stick Rossy, solid black female (Ruff N Reddy Joe Willy Wang x Ruff N Reddy Shadey Wood Ash) Courtesy of Melissa Couch

Bearfoots Red Man Chew Rossy showing his good head type and correct front. Courtesy of Roger and Kathy Stevens

the breeders with a nice litter of good working dogs. 1979 was consistent with this philosophy, too, with another cross to Davies' Blue Kola and a repeat breeding to the Foster's top producer Ch Colorado's Ravishing Ruby.

The classic cross of Ted and Mity Bite produced a top sire in 1980, Ruff N Reddy Mity Mean. Charlene Jeffreys and Cathy had been collaborating on breeding their dogs for several years and after the loss of Shade and Olson's Blue Joe, Charlene was looking for a blue male to replace them. Mity Mean fit the bill although the name she chose became something of a self fulfilling prophecy. Charlene's expectations were very high for Mity Mean—she wanted another Shade—but he was his own dog. Since she did not bond with him like she did Shade he ended up with a lack of discipline growing up.

Because Charlene wasn't using him and he was 7 or 8 years old, Cathy decided to take him back to use for breeding. Cathy quickly learned what Ted could have been like without all the careful training and firm discipline she had given him. Mity Mean had never worked on livestock or at anything else, and by the time

Bearfoots Red Man Chew Rossy as a fairly young dog when he first came to Pincie Creek. A dark liver red, this trait stamps his progeny even after several generations. Courtesy of Roger and Kathy Stevens

Bearfoots Red Man Chew Rossy as an older dog. The color of red dogs tends to fade slightly especially when they are outdoors a lot. Chewy developed an impressive coat as an older dog. Courtesy of Roger and Kathy Stevens

she got him he was set in his ways. Nonetheless she liked him but had to learn to live with his peculiarities. One of these traits was a territoriality about his own space such as a dog house or crate. After a fight with another dog in which he got his head through a fence, got it stuck and couldn't get out, Cathy rescued him from the trap. He crawled off to his dog house and Cathy just had to leave him alone and hope he would be alright. She didn't dare put her hand in the dog house to pull him out! This guardiness of his own space was a quirk he would pass along to his male offspring Mity Magician. Eventually Mity Mean went to live with Leslie Sorensen where he was possibly used a time or two for breeding and then lived out his life with her.

The friendship and breeding philosophies of Cathy Jones and Ruth McGuire had resulted in several decisions to experiment in the whelping box. Ruth, of course, had been competing with her great dog, Ch McGuire's SeMeGo Hemi UDT OTDc STDds for many years. He was bred sparingly and although Ruth had a great knowledge of dogs and bloodlines she

Pincie Creek Punch Cigar, black and white male (Bearfoots Red Man Chew Rossy x Peters Ranch Teal Sydney Bear) a beautiful example of the successful crosses Pincie Creek has made with Wood dogs and Hangin Tree dogs. Photo by and courtesy of Alex Kapsemalis

Pincie Creek Punch Cigar, an excellent working dog. Strong but biddable. Photo by and courtesy of Alex Kapsemalis

Ruff N Reddy Mity Mean, blue merle and white male. Courtesy of Sorensen

Ch Kintyre Jeb Stuart STDd OTDc, dark blue merle male anchor of the Kintyre Cattle Co. breeding program. Shown here at 6 months of age. By Ch Jones Reddy Teddy CDX, OTDdsc, SchHA x Davies' Blue Kola. Courtesy of Sorensen

Multi-talented father and son, Ted and Mity Mo. "Mo" was the All Around Dog at both the 1978 and 1979 CASA Specialty. Courtesy of Ruth McGuire

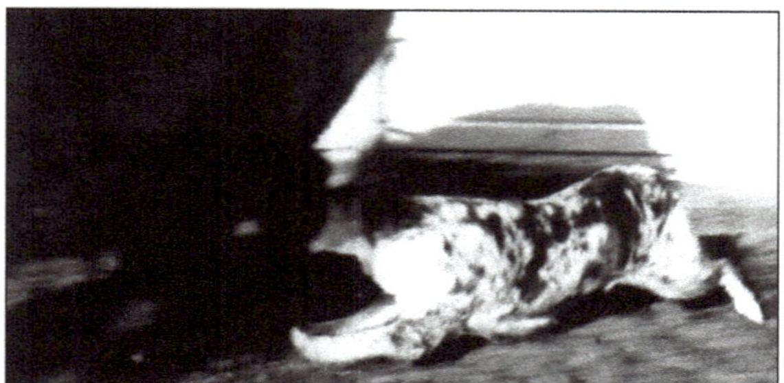

An enthusiastic young Magic learning how to heel those cows! Courtesy of Ruth McGuire

was not breeding a lot of dogs herself. She had been mentoring Cathy and many discussions of breeding and lines and team evaluation of litters had occurred. Ruth's Hemi daughter, Ch Hemi's Mountain Magic of J & J CDX STDcsd, became her next competition dog when Hemi retired. Ruth was generous with her knowledge of dogs and her love of the dog she showed for Cathy and Patty Ennen, Mity Muhammad. She decided to try a cross of her female with this sweet lapdog male. The breeding was a success and Ruth decided to go back to Mo's sire, Ted, with Magic 3 times after that!

In 1981 that first cross produced a pick puppy for Cathy named Ch Ruff N Reddy Pee Dee Que, a solid red female who was nearly a clone of her sire! Ruth remembers her as a small dog with very tiny ears which Cathy was always concerned might turn into prick ears. She did try working with the ears to some extent but although they broke very high there was never an issue with them turning into prick ears. To be on the safe side, when Cathy

Ruff N Reddy Mity Mean. Dark blue with the typical broad Wood head desired in the line. Courtesy of Sorensen

or Ruth showed her they were always careful to use the bait at eye level or lower just to be sure she didn't tip her head back and cause the ears to flip up. Another smaller bitch, Ruth remembers her as somewhat taller than Tinker Toy but only slightly. She likely topped out at under 19 inches.

Of course, all breeders try to recreate their best and favorite dogs and Cathy was no different. She nearly achieved it with Pee Dee Que after years of trying. Pee Dee Que was not only nearly identical to Ted but had several qualities which were an improvement over Ted. Ted was a well balanced dog but did not have a lot of angulation. Pee Dee Que was improved in this trait and had an excellent front and better

Ruff N Reddy Mity Mean, blue and white male (Ted x Jones Mity Bite) as a younger dog showing his good clear color. A very dark merle with little merling he became nearly solid black as an older dog. Courtesy of Melissa Couch

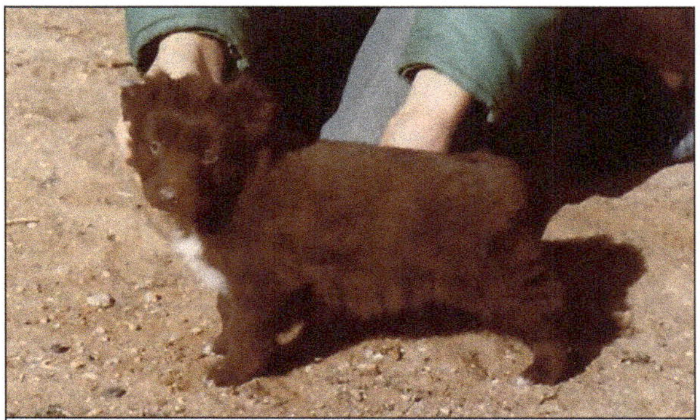

Ch Ruff N Reddy Pee Dee Que as a tiny puppy. Courtesy of Raymie Jones

A young Magic wins Best of Breed under breeder judge, Cathy Jones. Courtesy of Ruth McGuire

Black and white Muhammad x Magic puppy. Courtesy of Ruth McGuire

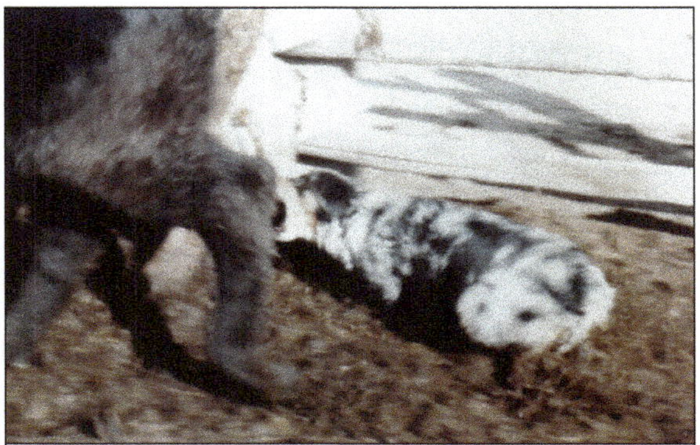

Ch Hemis Mountain Magic of J & J CDX STDcsd trying out her cattle working skills as a youngster. Courtesy of Ruth McGuire

Ch Ruff N Reddy Pee Dee Que working cattle. Courtesy of Raymie Jones

Cathy with Shadey Mountain Sumac and Ruth McGuire with Ch Ruff N Reddy Pee Dee Que taking Winners Bitch and Reserve Winners Bitch in Weatherford, Texas. Courtesy of Raymie Jones

Pee Dee Que fully mature and in good coat. Although she resembled Ted to a high degree she surpassed him in some qualities. An achievement which pleased Cathy greatly. Courtesy of Melissa Couch

angulation all around than Ted. Her temperament, like most of what Cathy produced, was slightly toned down from Ted's without losing the edge needed for stock work. To sum it up, Pee Dee Que was the dog Cathy had been trying to produce for a long time.

Although Pee Dee Que had a milder temperament than Ted, Casey Newman still remembers she had a penchant for dog fighting. When she was leased for a litter she did not fit well into a household of strange females and Casey was happy to return her once her litter was weaned. She does remember her as favoring Ted strongly in appearance with his head type and general conformation.

Pee Dee Que was bred sparingly at Ruff N Reddy, going back at least once to her sire to try to set that type and create more of the kind of dog Cathy was really trying to breed. Cathy kept Jones Stockyard Hustler from this cross and he bred on both at home and eventually in Alaska when Ruth McGuire moved north. He produced two top females for Leslie Sorensen—Colorado's Kodiak Kate, dam of Colorado's In Mahogany who produced Colorado's Mity Bludigger and he sired Colorado's Red Mountain Merle. In Alaska his influence was strongly felt at Carla Schroeder's North Star kennel where Tuba produced several good sires—North Star Charming Hustler, North Star Hemis Seme Savvy, and North Star Midnite Hustler. Pee Dee Que was born late in Ted's life so her next breeding was to the Ted son, Ruff N Reddy Mity Mean, producing the good sires Ruff N Reddy Ranger and Ruff N Reddy Mity Magician. When Casey Newman was looking for well bred females to breed to her Wood/Hemi bred male, Sprog she contacted Cathy and leased Pee Dee Que for a litter. Three females from this litter went on to produce for other kennels—Reniers Wood Rose STDds, HS&W Colorado Rose and HS&W Casey Lee

Ruff N Reddy Mity Magician, red merle male (Ruff N Reddy Mity Mean x Ch Ruff N Reddy Pee Dee Que] Courtesy of Melissa Couch

Ch Ruff N Reddy Pee Dee Que—a solid red female version of her sire Ted. Courtesy of Raymie Jones

Ruff N Reddy Mity Magician and a solid red littermate. Courtesy of Melissa Couch

Pee Dee Que as a nice young, correct dog. Her deep liver red color mirrored that of her sire, Ted. Courtesy of Raymie Jones

Pee Dee. Pee Dee Que was eventually sold to Annie Buffington where she was bred to two Hemi bred males.

A beautifully marked dark blue merle female from another of the Ted x Magic crosses was Ruth's pick, named Hemi's Ima Teddy Ten. Covered with large black splotches and carrying a beautiful correct coat, Ruth entered her in a Conformation Futurity being sponsored by CASA and won it all! She eventually sold this wonderful female but continued handling her in the show ring for a time.

Ruff N Reddy Mity Magician in the foreground and two of his littermates—a red merle and a blue merle in the background. Courtesy of Melissa Couch

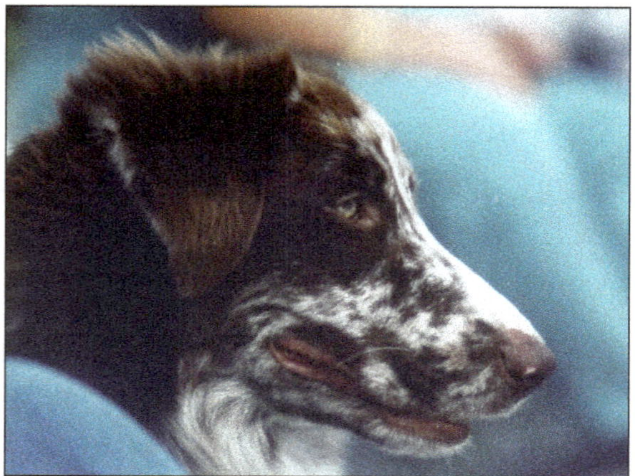

Ruff N Reddy Mity Magician, known as "Bubba" around home. Courtesy of Melissa Couch

Hemi's Ima Teddy Ten (Ted x Magic) at age 4-1/2 months. Courtesy of Ruth McGuire

Hemi's Ima Teddy Ten (Ted x Ch Hemis Mountain Magic of J & J CDX STDcsd) Courtesy of Ruth McGuire

Hemi's Ima Teddy Ten, blue merle female, (Ted x Ch Hemi's Mountain Magic of J & J CDX STDcsd) Courtesy of Raymie Jones

Although Ruth was not the biggest fan of Wood breeding in general, her mountain-type dog, Hemi, like Gunsmoke, Commanche and others of the era, turned out to be an excellent cross with Fletcher Wood's classic line. After experimenting with Magic, Ruth made a deal with Cathy to trade a breeding of Ruff N Reddy Mity Happy Hocker CD, STDd back to Hemi himself. Like so many of these trials, this one, too, worked beyond their wildest dreams. Cathy found the cross to Hemi added more obedience and tracking ability—a natural result one would expect since these were two of Hemi's greatest strengths. She also found the puppies moved down the spectrum in temperament, becoming better with strangers without losing the working ability she so highly valued. Her Wood bred dogs were one man dogs—fine at home but when strangers came around they would "let you know." Cathy appreciated the difference between the Wood dogs and Hemi and didn't consider one better than the other. She recognized they had been bred for different purposes and appreciated the things she was able to add from Hemi. She didn't make a habit of crossing many different types of dogs into her line however. She was used to the genetic predictability she had been getting all those years with the heavily bred Wood dogs and didn't like it if she started to lose it. Her buyers expected a guarantee Cathy's puppies would work cattle and do it well and she did not want to disappoint them. The top matron, Ruff N Reddy SeMe Sashay was one result of this cross and she proved to be a producer beyond compare.

Ruff N Reddy Ranger in an unguarded moment. Owner Mark Lane was living in Texas at the time and Ranger's red coat was suffering from the intense summer sun. His sound conformation is evident and his good type shines through his temporary rough exterior. Courtesy of Mark Lane

Ruff N Reddy Ranger, red and white male (Ruff N Reddy Mity Mean x Ch Ruff N Reddy Pee Dee Que) Pictured at about 18 months of age and owned by Mark Lane. Courtesy of Melissa Couch

In 1981, Cathy's beloved Boo Boo (Blue Tinker Toy) was bred to the old Wood dog, Olson's Blue Joe to produce a tiny working dynamo named Ruff N Reddy Too Small Jones. This little blue female became the best buddy of Sue Noll, who was no stranger to the Wood, especially Ted, bred dogs. In addition to her Search dog Tuffit she had a Ted son out of Ch Colorado's Ravishing Ruby named Ch Powder River Tuffernhel.

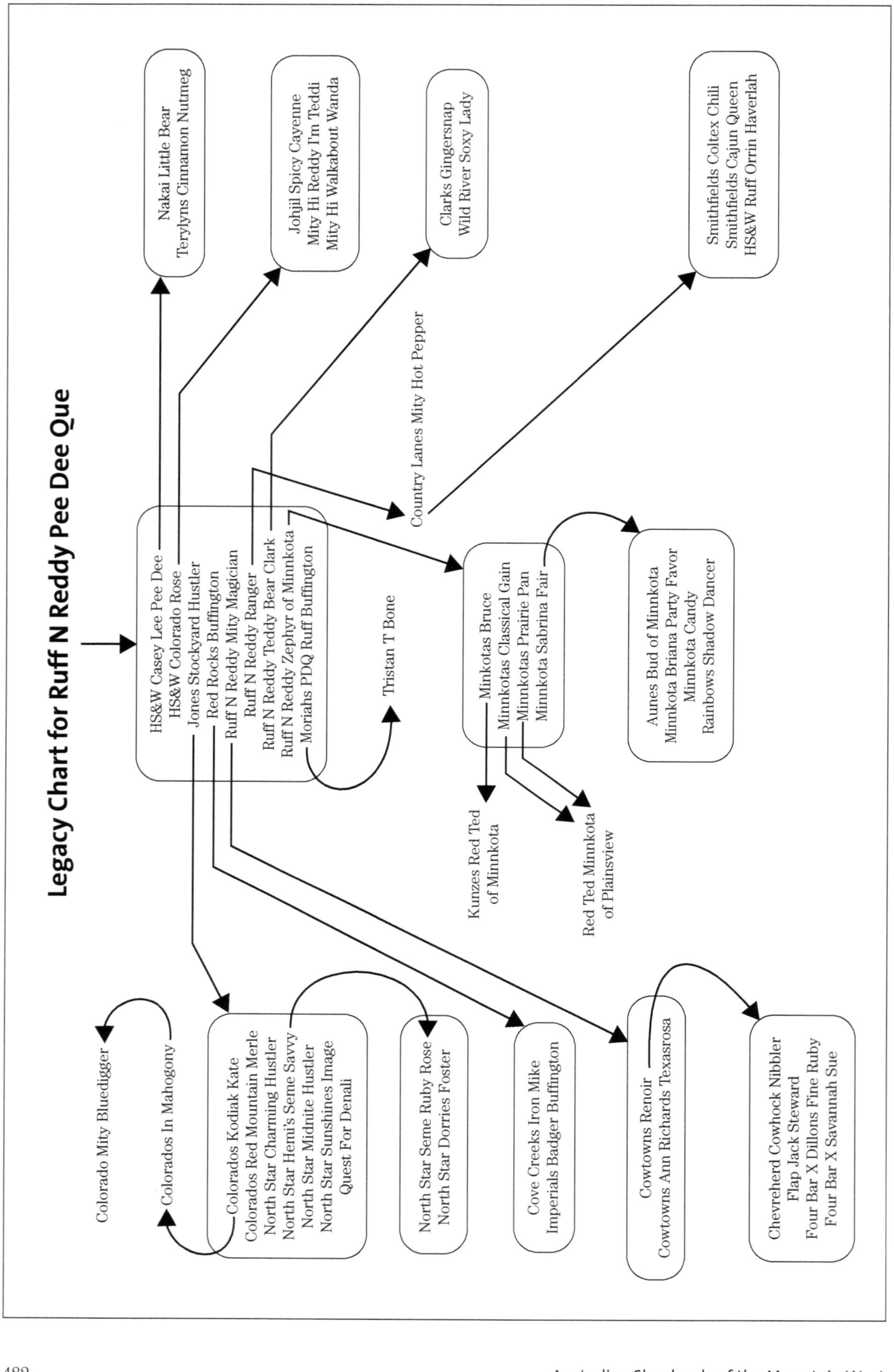
Legacy Chart for Ruff N Reddy Pee Dee Que
HS&W Casey Lee Pee Dee
HS&W Colorado Rose
Jones Stockyard Hustler
Red Rocks Buffington
Ruff N Reddy Mity Magician
Ruff N Reddy Ranger
Ruff N Reddy Teddy Bear Clark
Ruff N Reddy Zephyr of Minnkota
Moriahs PDQ Ruff Buffington
Nakai Little Bear
Terylyns Cinnamon Nutmeg
Johjil Spicy Cayenne
Mity Hi Reddy I'm Teddi
Mity Hi Walkabout Wanda
Clarks Gingersnap
Wild River Soxy Lady
Country Lanes Mity Hot Pepper
Smithfields Coltex Chili
Smithfields Cajun Queen
HS&W Ruff Orrin Haverlah
Tristan T Bone
Minkotas Bruce
Minnkotas Classical Gain
Minnkotas Prairie Pan
Minnkota Sabrina Fair
Aunes Bud of Minnkota
Minnkota Briana Party Favor
Minnkota Candy
Rainbows Shadow Dancer
Kunzes Red Ted
of Minnkota
Red Ted Minnkota
of Plainsview
Colorado Mity Bluedigger
Colorados In Mahogony
Colorados Kodiak Kate
Colorados Red Mountain Merle
North Star Charming Hustler
North Star Hemi's Seme Savvy
North Star Midnite Hustler
North Star Sunshines Image
Quest For Denali
North Star Seme Ruby Rose
North Star Dorries Foster
Cove Creeks Iron Mike
Imperials Badger Buffington
Cowtowns Renoir
Cowtowns Ann Richards Texasrosa
Chevreherd Cowhock Nibbler
Flap Jack Steward
Four Bar X Dillons Fine Ruby
Four Bar X Savannah Sue

Late in Blue Joe's life Cathy intended to repeat this breeding but Charlene lost Joe to old age before the cross could be made.

Mity Bite was bred back to Ted in 1981, producing Ruff N Reddy Katy Did. A smallish solid black female typical of the Ted x Mity Bite cross. She was neither pushy nor aggressive and had a nice stable temperament. She was owned by Annie Buffington, who bred her to Newmans Blue Sprog CD STDsc. Several females from this cross went east to contribute to several kennels breeding programs which needed an infusion of toughness for cattle work.

Ruff N Reddy SeMe Sashay "Bunny," blue merle, copper and white female. A very high quality puppy exhibiting her nearly flawless conformation and type. She had the genetics behind her to be a top producer, too. Courtesy of Mark Lane

For the next several years, Cathy and Ruth rode the success of the past, repeating the breedings of Magic to Ted and Ted to Mity Bite, experimenting with inbreeding by taking Pee Dee Que back to her sire, Ted, and crossing Ted with the Mity Bite daughter Shadey Mountain Mity Miss. The sheer number of high quality dogs being produced by Cathy and Ruth is something rarely seen before or since. Although Cathy bred a number of litters, she was continuing to cull for quality and the dogs which left the Ruff N Reddy kennel were few in number but consistently top of the line. A few of the better known dogs of this era were Ruff N Reddy Shadey Joe (Ted x Shadey Mountain Mity Miss), Hemi's Ima Magic Dolly (Ted x Magic), Jones Stockyard Hustler (owned by Ruth from the Ted x Pee Dee Que cross), and Mark Lane's beautiful dark blue merle female Ruff N Reddy Hunny Bear (Ted x Mity Bite).

Ruff N Reddy SeMe Sashay Blue merle and copper female. (Ch McGuire's SeMeGo Hemi UDT STDcsd x Ruff N Reddy Mity Happy Hocker CD STDd) Courtesy of Raymie Jones

In late 1982 Ted tried to clear a fence to get to a bitch, misjudged and ended up breaking one of his back legs when it hung up in the fence. Cathy rushed him to the vet, who put the leg in a cast. Cathy took him home and by morning the cast lay in ribbons on the floor. A return trip to the vet, another cast and another pile of cast crumbles by the next day. Cathy gave up trying to fight Ted's wishes but the leg did not heal well. After this incident, Ted was used sparingly on Cathy's best females only as breeding was difficult for him. In 1984, when

Ruff N Reddy SeMeSashay having a tryout at working cattle. She seems to be passing the test! Courtesy of Raymie Jones

Ruff N Reddy SeMe Lick'em CD STDdsc, blue merle, copper and white female born July 19, 1988—photo at 5 years. (Colorado's Talk of the Town x Ruff N Reddy SeMe Sashay) Bred by Cathy Jones. Photo by Carla Schroeder

Raymie Jones with Ch Hemi's Mountain Magic of J & J CDX STDcsdon the left and Ch Ruff N Reddy Blue Tinker Toy on the right. Courtesy of Raymie Jones

Ted was 11 years old, Cathy agreed to an outside breeding to Anita Laycock's nice female, Laycocks Cheyenne Maurine, a red merle J Bar D/Colorado/Las Rocosa cross to produce the dark red Ted-look-alike male Laycock's Mr Red Ted II.

The next generation was now beginning to take over at Ruff N Reddy as Ted neared retirement. Cathy was spending time training her young dogs and letting Ted enjoy the accolades of a long and successful career. He still worked and competed but only on occasion and winning wasn't as important as the enjoyment he and Cathy took in being together and playing their old familiar games. Cathy was beginning to plan the next generation in her breeding using the dogs she had bred from her impeccable foundations. In early 1986, the Ted grandson Jones' Little Redwood (Ruff N Reddy Shadey Joe x Ch Ruff N Reddy Blue Tinker Toy) was bred to his granddam Shadey Mountain Mity Miss. A well known successful cross to generations of breeders, this cross did not fail to produce as expected, producing several important females who would leave their marks—Ruff N Reddy Shadey Wood Ash and Ruff N Reddy Boots were two of the better known of these.

In April of 1986, at the age of 13 years, Ted was laid to rest. Ruth McGuire, adversary, mentor, competitor and friend took a day off work to be with Cathy as she said goodbye to the best friend she would ever have. This was a return of the same favor Cathy performed when Ruth said goodbye to her once-in-a-lifetime dog, Hemi. Ted joined Hemi and the Jones first Aussie, Smokey, in the dog cemetery on the Jones Ranch.

Ted's loss was not the end of Ruff N Reddy kennel, however. In fact, he had left so many superior offspring the future was bright for Cathy's breeding program. His son Mity Mean was brought into service, siring a litter with his half sister Pee Dee Que which again resulted in high quality progeny in the form of two top males, Ruff N Reddy Ranger and Ruff N Reddy Mity Magician and a nice red merle bitch named Hewitts Koko.

Cathy also had the superior male Ruff N Reddy Shadey Joe "Clank" who was to sire a number of top working litters for her and others. Clank was out of the two of the very best dogs of the day—Ted and Shadey Mountain Mity Miss. He was a dark red merle with a pretty head and a sweet temperament. When he retired to live with Patty Ennen he became a best buddy of her children and produced the same type of loving temperament in his puppies. Bred by Charlene Jeffreys, he was Cathy's pick of the litter for the stud fee. He was a good working dog, sharing Ted's low heeling ability on cattle but he never had the opportunity to compete for titles like his sire.

In spite of his lack of competitive experience and exposure, he was known by working breeders. The legendary WTCH Slash V Slide Me Sweet CD produced her second litter by this grand dog and three males from this cross spread his legacy from Texas to Alaska. He was bred to Cathy's own best females, Blue Tinker Toy and SeMe Sashay to produce another generation which carried on the high quality Ruff N Reddy represented.

After moving to Patty Ennen's he was bred back to his granddaugher Ruff N Reddy Shades of Ted and a Sprog x Pee Dee Que daughter, HS&W Colorado Rose, to produce several quality dogs which went to the east coast. Clank lived out a good life with Patty as a very beloved house pet just as Cathy preferred for all her dogs.

The other two males in Cathy's kennel after the passing of Ted were the Clank x Tinker Toy son Jones Little Redwood "Woody," and the

The grand old man, Olson's Blue Joe, blue male by Piz Joseph x Piz Lady II (by Piz Joseph) pictured in September 1978. Courtesy of Sorensen

Ch Ruff N Reddy Blue Tinker Toy, blue merle female in her prime. One of the most beautiful females Cathy ever owned. (Ted x Keefers Miss Nitzi) Courtesy of Raymie Jones

Ch Ruff N Reddy Blue Tinker Toy at approximately 12 years of age. Courtesy of Melissa Couch

Ch Ruff N Reddy Blue Tinker Toy bright eyed and still showing her stuff at 12 years of age. Courtesy of Melissa Couch

Ruff N Reddy Shadey Joe as a mature dog. Courtesy Melissa Couch

Litter of gorgeous silver blue merle puppies by Olson's Blue Joe and out of Ch Ruff N Reddy Blue Tinker Toy. Courtesy of Raymie Jones

Shadey Mountain Mity Miss at about 12 years of age, blue merle and white female (Jeffreys' Shade x Jones Mity Bite)

Two impressive puppies from the cross of Olson's Blue Joe and Ch Ruff N Reddy Blue Tinker Toy. Courtesy of Melissa Couch

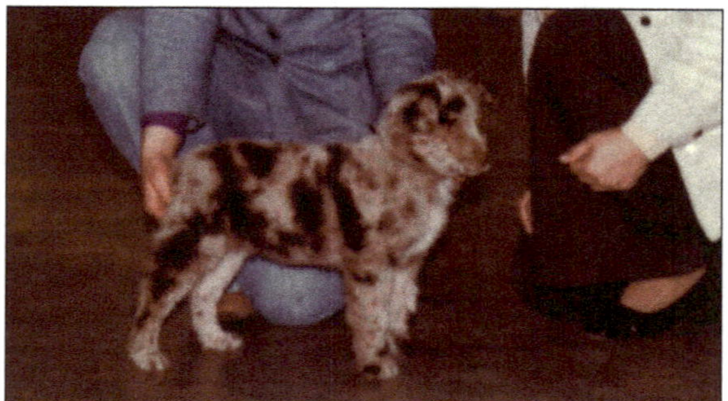

Ruff N Reddy Shadey Joe, "Clank," red merle male (Ch Jones Reddy Teddy CDX OTDdsc SchHA X Shady Mountain Mity Miss) as a young puppy. He was Cathy's stud fee puppy from this tremendous cross. Courtesy of Sorensen

double bred Ted and Hemi male, Jones Stockyard Hustler "Tuba." Woody was a medium sized red dog who strongly resembled his sire. Although he was a good quality dog he wasn't a dog who called attention to himself. Like his sweet mother he was laid back and good with people. He eventually made his retirement home with Mark Lane and Mark's Ted x Mity Bite daughter Ruff N Reddy Hunny Bear.

Ruff N Reddy Shadey Joe as a young dog. Very correct, very high quality, but how could he be otherwise with his breeding? Courtesy of Raymie Jones

Woody was a good producer and when bred back to his granddam Mity Miss produced a litter of truly superior females. Ruff N Reddy Boots was used sparingly with outcross males but produced the wonderful show and working dog, Ch Ruff N Reddy Ross STDdsc. Ruff N Reddy Shades of Ted became a good producer for Patty Ennen and the great Ruff N Reddy Shadey Wood Ash accompanied Ruff N Reddy Joe Willy Wang to Florida to produce Bearfoots Cinder N Ash and Bearfoots Red Man Chew Rossy—two males who made a big mark on the Pincie Creek kennel of top cattle working dogs. A steel blue and white female named Ruff N Reddy Mity Miss Shadey went to Texas to the cattle ranch of Melissa Couch. Melissa describes her as a pretty headed, overall nice dog with a lot of intelligence but a somewhat soft temperament. She was a good heading dog on cattle but never got much opportunity to work the Couch show cattle. She was a great guard dog when it came to coyotes. She had an inbred hatred for them and if they came near the barns or house she did not miss the opportunity sound the warning their nemesis was nearby. Only 18 to 18-1/2 inches tall, like most Wood bred dogs, Shadey turned nearly black in appearance as she got older. She was bred once to Melissa's Sisler bred male, Rowdy, for a top litter of cattle working dogs.

Ted winning Best of Breed under breeder judge Sandy Cornwell in Texas. Courtesy of Raymie Jones

Cathy's other stud dog, Tuba, eventually went to Alaska with Ruth McGuire to make his mark on the North Star kennel of Carla Schroeder. Tuba was a red dog who combined Ted and Hemi in a good working dog

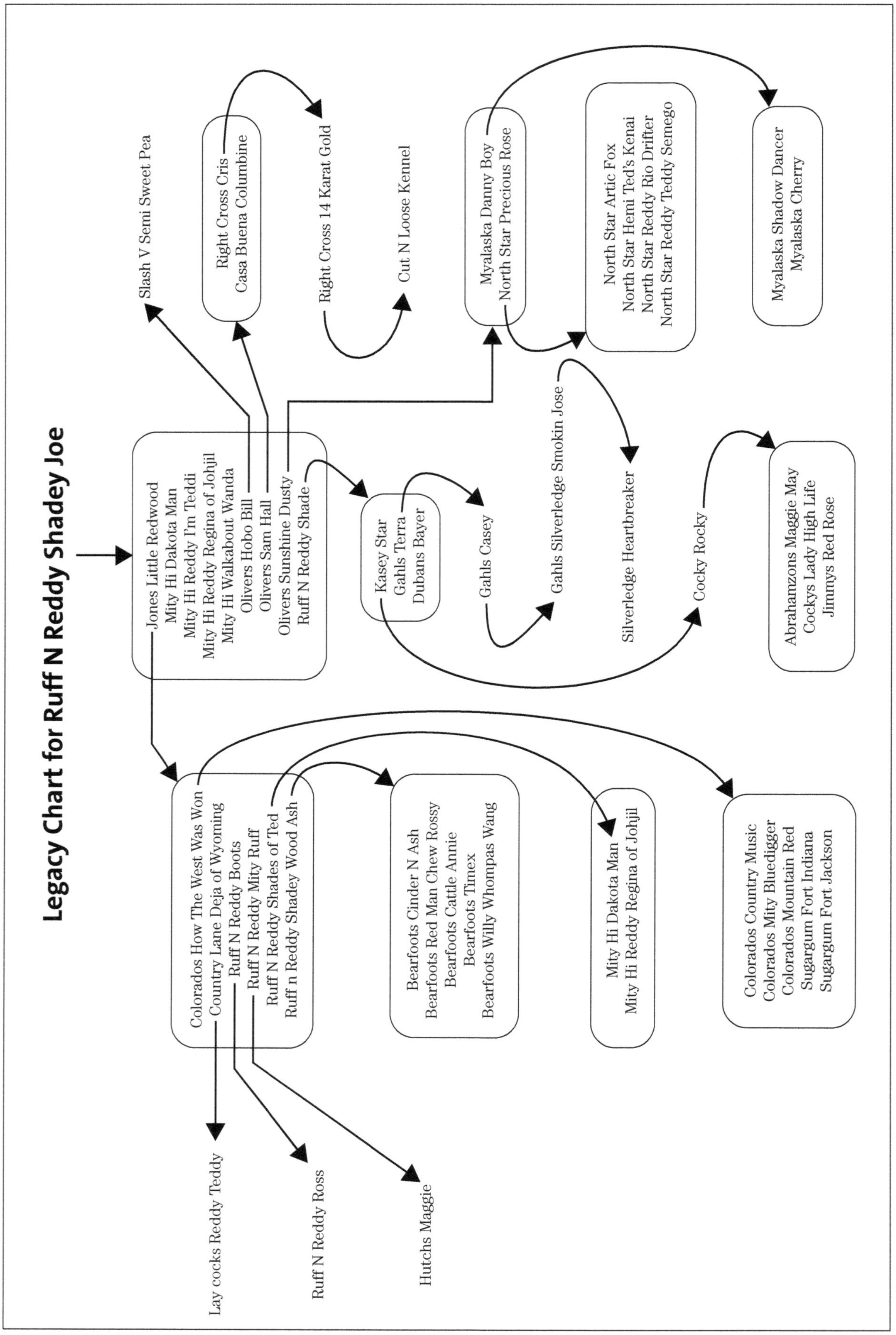
Legacy Chart for Ruff N Reddy Shadey Joe
Jones Little Redwood
Mity Hi Dakota Man
Mity Hi Reddy I'm Teddi
Mity Hi Reddy Regina of Johjil
Mity Hi Walkabout Wanda
Olivers Hobo Bill
Olivers Sam Hall
Olivers Sunshine Dusty
Ruff N Reddy Shade
Slash V Semi Sweet Pea
Right Cross Cris
Casa Buena Columbine
Right Cross 14 Karat Gold
Cut N Loose Kennel
Myalaska Danny Boy
North Star Precious Rose
North Star Artic Fox
North Star Hemi Ted's Kenai
North Star Reddy Rio Drifter
North Star Reddy Teddy Semego
Myalaska Shadow Dancer
Myalaska Cherry
Kasey Star
Gahls Terra
Dubans Bayer
Gahls Casey
Gahls Silverledge Smokin Jose
Silverledge Heartbreaker
Cocky Rocky
Abrahamzons Maggie May
Cockys Lady High Life
Jimmys Red Rose
Colorados How The West Was Won
Country Lane Deja of Wyoming
Ruff N Reddy Boots
Ruff N Reddy Mity Ruff
Ruff N Reddy Shades of Ted
Ruff n Reddy Shadey Wood Ash
Bearfoots Cinder N Ash
Bearfoots Red Man Chew Rossy
Bearfoots Cattle Annie
Bearfoots Timex
Bearfoots Willy Whompas Wang
Mity Hi Dakota Man
Mity Hi Reddy Regina of Johjil
Colorados Country Music
Colorados Mity Bluedigger
Colorados Mountain Red
Sugargum Fort Indiana
Sugargum Fort Jackson
Lay cocks Reddy Teddy
Ruff N Reddy Ross
Hutchs Maggie

with a nice temperament. The dogs he produced in Alaska were second to none.

In 1987 or 88 Cathy added an outside male from Leslie Sorensen to her stable of old Wood males. This dog was a show line outcross named Colorado's Toast of the Town. She had bred SeMe Sashay to this dog's sire and then used him on some of her own bitches. These dogs were all sold and several became influential in other breeding programs.

The solid red male Jones Little Redwood (left) and his dam Ch Ruff N Reddy Blue Tinker Toy (right). Courtesy of Melissa Couch

In 1989, Cathy Jean Jones was tragically killed, but her life's love—the breeding of some of the world's best cattle working Australian Shepherds—lived on in the hands of breeders all over the United States. Just as importantly, Cathy left a standard of training, judging, mentoring, teaching, and behavior as a breeder which were second to none. In the words of Leslie Sorensen she approached every project in exactly the same way—she never settled for less than a "thorough, complete and successful" outcome.

Ruff N Reddy Shadey Wood Ash, red merle and white female (Jones' Little Redwood x Shadey Mountain Mity Miss). Turning dark as she ages. Courtesy of Pincie Creek

She built a reputation for the Ruff-N-Reddy kennel name of dogs which were well known for consistency. Cathy had a firm idea of what she wanted to produce and did not compromise when it came to her selection of breeding stock. To pass muster the dog must have a good low heeling style and no fear when it came to heading cattle. If it had gathering ability and even some eye she valued that as well but the cattle working ability had to come first. She was beginning to see the trend among breeders for selecting dogs for trials instead of real ranch work. She believed crossing dogs with different styles resulted in puppies which were not very good at anything. They might work one way for awhile then switch to another style. This resulted in an unpredictability when breeding that Cathy would not tolerate. Cathy wanted to be able to guarantee her dog's ability to her buyers. She also saw a loss of toughness in dogs bred for trials. With the watering down of their style came a weaker dog not able to stand

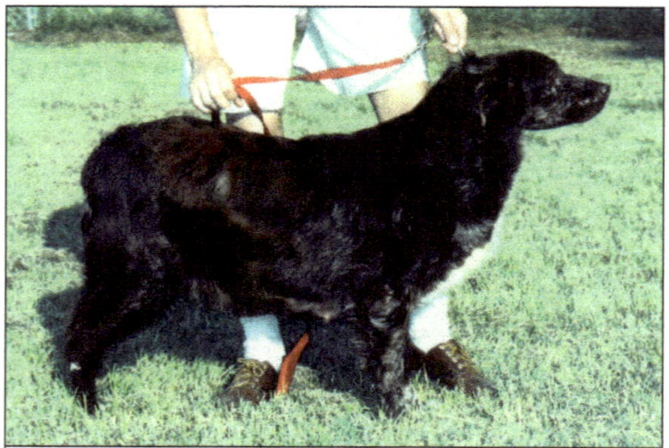

Ruff N Reddy Shadey Wood Ash as an aging matron. Courtesy of Pincie Creek

Ruff N Reddy Shadey Wood Ash. Courtesy of Pincie Creek

The beautiful, sweet expression of Ruff N Reddy Shadey Wood Ash. Courtesy of Pincie Creek

up to the tough conditions ranchers sometimes encountered.

As a result of her experience with outcrossing, losing predictability, and seeing the same results from other breeders, Cathy came full circle to the realization that dogs of Wood's breeding would provide the consistency she required. They had been bred as cattle dogs, their type and ability set by Fletcher Wood and his masterful close breeding program. It took her only a short time and a few breedings to discover the truth of her beliefs firsthand. This type of selection did not preclude getting dogs who could work sheep or do obedience, as she raised a number who were very good at both. But unless she kept as a priority the maintenance of a strong temperament with cattle ability, she believed she would quickly lose it and never regain it. Few other breeders had this strong focus. When they lost the cattle ability in their own dogs by outcrossing they knew they could get it back if they came to Ted or one of his offspring.

And so Cathy's recommendation was always to buy a dog from a proven working line to get a working dog—both when she was buying and when she was selling to others. This was her highest priority because it was the only way she could guarantee her dogs. She questioned potential buyers to find out what they wanted the dog for, what they could pay and whether they really wanted a cattle dog. She guaranteed her dogs for breeding soundness (lack of hip dysplasia and eye defects) and temperament if they intended to breed and she always wanted to hear from someone if a dog didn't work out for them. She was always available to help with problems and would replace a dog if she found it had a legitimate flaw. She recognized her dogs needed adequate socialization and did her best to educate buyers about their responsibility to put time and effort into this task. Anyone who complained about an unsocialized dog got little sympathy from Cathy.

Cathy was very generous with her time and knowledge and was a skilled teacher of humans. When asked a question her answers were often more questions which led her "pupil" to find the answer for themselves. Her dogs were not suitable for everyone and her diplomatic methods of learning who would be getting a dog from her and who wouldn't resulted in friendship regardless of whether she made a sale. Mark Lane recalls his first meeting with Cathy. He had seen a sign for her dogs at the local feed store and called her to inquire about a puppy. When he got to her place he might have expected a hard sell job but instead got exactly the opposite. He relates Cathy never bragged on Ted or tried to convince him one of her puppies was perfect for him. He got the same careful evaluation (probably without his knowledge!) by Cathy that she gave her dogs. Apparently he passed the test since he not only took home a superior female but in later years obtained two good males from Cathy, too.

Ruff N Reddy Shadey Wood Ash, although she is darkening with age, her clear merling is still visible through her body and on her head. Courtesy of Pincie Creek

Ch Ted as an older dog still winning Best of Breed. Photo by Sue Noll. Courtesy of Raymie Jones

Especially with novice owners, she would question them at length about what purpose they had in mind for the dog. She would know immediately if her dogs would work out for them. Anyone with an idea of showing and winning in conformation all over the country would not get a puppy under any circumstances. Although Cathy succeeded in this endeavor, it was always a tough proposition that few conformation exhibitors could tolerate. But anyone whose first priority was stock work would make the first cut and if she had the right puppy for that person she would allow them to have it. Being picky about who got her dogs and qualifying buyers in advance saved her a lot of heartache as she says, "…because if I sell a dog to someone who says, 'Yes, this dog is going to be a champion. He can do obedience; he can do all these things,' and I know good and well that the chances of that happening are next to none, I am going to have a problem about a year from now. I don't like that; it's not good for the breed. Qualify the owners and put them with the dog you feel is bred to do what they want to do.

A litter of young red Ruff N Reddy puppies probably having their first exposure to livestock. Courtesy of Mark Lane

Ch Ruff N Reddy Blue Tinker Toy. Courtesy of Melissa Couch

Clear blue merle of excellent conformation typical of the dogs Cathy produced. Courtesy of Raymie Jones

Sometimes I turn down sales because of that. I just know that the dogs or puppies I have at that time are not suited for that purpose."

Cathy had firm ideas about the right way to train dogs—mostly because these methods had brought her such success! Because of her emphasis on performance she was careful to keep her dogs fit. Her methods ranged from biking and hiking to working them from a 3 wheeler to using a horse hot walker when time was at a premium. At her peak, Cathy would keep 10-12 dogs in her kennel. Not all of them would be in heavy training at once but she still did not neglect their basic health and fitness. A pet peeve of hers was a fat dog and she was seeing more and more soft, flabby dogs as the show ring increased in popularity.

When Cathy had a goal she was training for, such as the old Gaines Obedience Invitationals, she spent a lot of time away working him in all situations. She learned you had to live with a dog to get the best performance out of them. A kennel dog would seldom have the bond with the trainer that a companion developed. She could bring out things in kennel raised dogs due to her experience and skill but she never felt the dog reached the potential it could have with a more personal upbringing. Cathy knew when to push hard with her training and when to give her dogs a day off. This type of intuition was one of the many benefits of living closely with the dog.

As time passed and Cathy's two children grew up she stopped traveling and competing as much and concentrated on just breeding a few litters each year. She gave up the cattle and ducks she had kept for many years for she and Ruth to train their dogs for trial work. But Cathy did not stop learning about stock training. A neighbor had some sheep which became her primary training livestock. Not only were they convenient but Cathy discovered when starting puppies that most of hers were more "brave than brains." Sheep were safer to start these young dogs on because they were less likely to get hurt if they made mistakes—not an uncommon occurrence! Sheep would also bolster the confidence of a young or novice dog and made it easier and safer to switch them to cattle later.

Cathy never started a young dog with an older dog, a technique used by some trainers of the day. She preferred to see what a dog would do all on it's own and what it's natural instincts were. She thought this was the only real way to know what you were getting in your breeding and what the genetics of the dog for working really were. She learned to know her line to know how much time to give a dog to start working. This experience prevented her from discarding some very good dogs which turned out to mature mentally a little more slowly—late bloomers. Her puppies would start working anywhere from 4 months to a year and half of age and she eventually learned to detect which dogs were never going to start working. She had begun to see a disturbing trend with the advent of stockdog futurities of young dogs being pushed too hard and harmed. She worked on the 1990 National Specialty and Futurity to be held in Loveland, Colorado to make this futurity for older dogs. She was able to add an extra year to the requirements from what the past futurities had required in order

Ruff N Reddy Mity Magician, red merle and white male (Ruff N Reddy Mity Mean x Ch Ruff N Reddy Pee Dee Que] Well loved companion of Melissa Couch. Courtesy of Melissa Couch

Cathy carefully starting an intense young blue merle puppy on cattle. When she didn't have sheep available or when first taking a young dog to cattle, the method used was to keep the dog on a long line for their own safety. Courtesy of Mark Lane

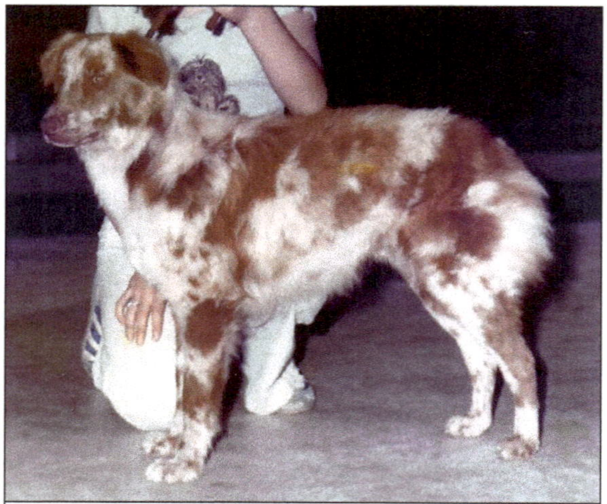

Raymie Jones with her red merle dog "Strawberry" Courtesy of Raymie Jones

The look of a working dog. Ch Jones' Reddy Teddy CDX OTDdsc SchHA taking first place at a trial. Courtesy of Raymie Jones

to let the dogs mature and not be pushed so hard. For her own purposes she knew how she liked a dog to work and wasn't impressed by futurity wins.

With Cathy's wide range of experience and success it was inevitable she would become a conformation show judge and she continued to educate and mentor others in the breed from this position. She had strong opinions and no qualms about telling it like it was when commenting on the direction of the breed. She heavily faulted fronts in the breed because she wanted good shoulder layback and had a hard time finding it when judging. She observed many breeders didn't care about watching this trend and fronts always seemed to suffer the most when breeders began to compromise. Along with this priority she also saw way too much elbowing out, pigeon toes and dogs that just overall were poorly balanced. Most often the poor fronts accompanied an overangulated rear, causing side winding (or crabbing) and the resulting inefficiency of movement so detrimental to a good working dog. No dog can go all day if they interfere with themselves just like the horses she knew so well. These judgments were based on the trends she was seeing in the show ring and her observations have turned out to be prophetic.

Cathy also believed temperaments were becoming a problem as more people who had no business with an Australian Shepherd were attracted to the breed. Lack of knowledge of bloodlines and pedigrees was another problem that came with the influx of novice owners and resulted in inappropriate choices of mates and production of too many mediocre and inferior puppies.

As a conformation judge, Cathy soon developed a philosophy to fit the state of the breed at the time. In the beginning she judged for a certain type of dog, but she soon found out "there is no such thing as an ideal Australian Shepherd. The breed has not had a standard long enough, and breeders haven't been breeding to the standard long enough to really have a set type. Every bloodline looks different." She was finding many good dogs that might have a different look in the eye, earset or color that she felt she couldn't criticize merely for these minor deviations from her own ideal. She refined her judging technique to looking first for an overall balanced dog, just as she did in horses. A functional animal had to be the first priority.

Two solid liver red puppies of Cathy's breeding showing the consistency and high quality which is possible when breeders keep their own standards high and their breeding choices based on fact and not emotion. Courtesy of Mark Lane

She had a very organized method of judging which allowed a quick and accurate evaluation of each dog. She first looked at the legs (her horse background coming to the fore once again) and how the dog moved. Judging from the ground up was the most natural method for her, just like it was with her horses. If they couldn't pass muster to do the job, they were no good "no matter how pretty they are." In her mind the dogs were intended to work all day long and that's where it started and ended. Nothing else mattered if that was missing and if it was—she moved on to the next dog.

Ted—the functional, sound type of dog Cathy always looked for, bred for, worked for. Courtesy of Raymie Jones

After the initial elimination based on leg structure and movement she could take a longer look at those who made the "cut." Good movers were always guaranteed in Cathy's first 3-4 places. However they also had to be within standard for other traits, too. She never minded dogs with blue eyes (she called them glass eyes) unless it made their expression hard to read. Of greater concern was the health of those eyes as genetic eye disease was beginning to be found in the breed. She did have concerns about light eyed dogs being bothered by the sun and she insisted on having good pigment around the eye rims to reduce the reflective effect of the sun.

An extremely perceptive observation by Cathy was that moderation had nothing to do with size. She knew that larger dogs occurred in the Wood line, yet they were not heavy or slow movers or lacking in agility or cattle working ability. Her personal preference for cattle work was a more medium sized dog but she was not prejudiced against larger or smaller dogs as long as they had the balance and moderation within themselves the standard infers. There is simply no substitute for the real world experience which allows a judge to recognize the difference between size and

Cathy had plenty of real world experience working livestock with Ch Jones' Reddy Teddy CDX, OTDdsc SchHA and other dogs to be an extremely qualified conformation judge. Courtesy of Melissa Couch

Dark liver red merle dog of Ruff N Reddy lines. Substantial without cloddiness and exhibiting a classically beautiful Wood type head. Courtesy of Mark Lane

moderateness. Cathy believed in the ideal that dogs in the show ring were the future breeding animals and she took their evaluation very seriously. Her experience at breeding gave her the perspective she needed to know which faults were truly serious in breeding and which were not.

So Cathy's Winners classes were always filled with excellent movers with good type. Once she had the cream of the day in her Winners classes she could focus on other things—type, color, coat texture, temperament. Her Best of Breed class theoretically held the top champions and the best movers from her Winners classes. Although this system is always dependent on what dogs exhibitors bring, Cathy was a popular and fair judge and she usually had plenty of good—and maybe not so good—dogs to choose from all day long.

As much as Cathy loved the breed, people were always more important than dogs to her. In the late 1980s more obedience and pet people were becoming involved in the breed and Cathy saw a disturbing trend beginning which was a change from what she had found when she started in the breed. The breed's exploding popularity was resulting in more and more shows, more and more competition for titles and awards and more and more cutthroat attitudes

Cathy didn't wait long to try out her puppies on stock—and most of the time they did not disappoint—like this nice red puppy showing a lot of maturity and self control on ducks. Courtesy of Mark Lane

among exhibitors. She was seeing the fun being lost and both dogs and people suffering from burn out from all the travel and animosity. Dogs were performance animals to Cathy, just like horses, and she believed they should be doing something more than just standing in a show ring. She hated to see dogs (and horses) go sour from this sort of treatment.

Most people in this era were owner handlers, which was something Cathy thought was good for the breed. But the down side to this was she also observed some bloodlines were becoming "specialty lines"—bred only for one thing. Along with the hyper competitiveness she was seeing the dogs becoming nothing but an ego boost to some people. As she called them, "…a piece of meat on the end of the leash." She acknowledged that everyone has some tendency to do this but it was the degree to which some people went that led to abuse of dogs.

The trial arena was beginning to trend the same direction. The dogs were becoming specialized as trial dogs versus actual ranch dogs and judges were entering the ranks with no experience working dogs except in a trial arena. Cathy sums it up this way, "We have evolved from getting stock people off ranches to be our judges to actually having trial specialists in handling and judging." These complaints have been heard ever since in the breed.

Cathy's legacy is one of sportsmanship, high standards and integrity. She tried to put her best on display at all times—from her dogs

Ch Jones Reddy Teddy CDX OTDdsc SchHA. Courtesy of Melissa Couch

Cathy and Skidder. She was accomplished because she truly loved her dogs. And they loved her, too. Courtesy of Melissa Couch

Ted showing the eye he had when working livestock other than cattle. Courtesy of Raymie Jones

to her own demeanor. She was an inspiration and a shining example of what every breeder should strive to be.

Ted was a prolific producer and a great many of his progeny went to ranchers who did not breed or may not have registered the dogs they owned or bred. Below is a partial list of dogs produced by Ted which were either bred, shown or recorded in Aussie historical documents. Following this list is an abbreviated Legacy Chart for Ted which gives a limited view of the influence of his more active progeny. More detailed Legacy Charts for his more prolific offspring can be found throughout this chapter.

- Almostheaven's Hillcountry Mesquite (Ted x Red Lady of Cattle Call—Malone's LaDego Diablo x Malone's Autumn Leaves)
- Almostheaven's Little Red Devil (Ted x Red Lady of Cattle Call)
- Almostheaven's Sweet Honesty (Ted x Red Lady of Cattle Call)
- Blue Isle Coffee W/Cream (Ted x Slash V Wild Firewood)
- Blue Isle Red Brio (Ted x Slash V Wild Firewood—Martins Tim Tim x Hoyt's Wildwood Flower)
- Blue Isle Red Guspacho (Ted x Slash V Wild Firewood)
- Blue Isle Red Hot Pepper (Ted x Slash V Wild Firewood)
- Blue Isle Slash (Ted x Slash V Wild Firewood)
- Blue Isle Teddie Christine (Ted x Slash V Wild Firewood)
- Blue Isle Teddy Bear (Ted x Slash V Wild Firewood)
- Briarbrooks Diamond Jim (Ted x Bar Lazy Diamond Lil of Las Rocosa)
- Briarbrooks Melbourne (Ted x Bar Lazy K's Diamond Lil of Las Rocosa)
- Colorados Brandy Bow (Ted x J Bar D Brandy)
- Colorados Reddy Raider (Ted x J Bar D Brandy)
- El Rojo Polvo De Almostheaven (Ted x Red Lady of Cattle Call)
- Greyoaks Cinder Flash (Ted x Parmenters Dixie of Bar Lazy K)
- Hemi Ima Magic Dolly (Ted x Hemis Mountain Magic of J & J)
- Hemis Ima Reddy Rascal (Ted x Hemis Mountain Magic of J & J)
- Hemi's' Ima Teddy Ten (Ted x Ch Hemis Mountain Magic of J & J)
- Hemi Rineharts Teddy (Ted x Hemis Mountain Magic of J & J)
- Hemis Ruffneck (Ted x Hemis Mountain Magic of J & J)
- Hemis Scarlet Samurai (Ted x Hemis Mountain Magic of J & J)
- Hemis Spice of Life (Ted x Hemis Mountain Magic of J & J)

- Hemis Texanna Teddy (Ted x Hemis Mountain Magic of J & J)
- Hotline Ruff N Tuff (Ted x Colorados Buckeye Barbie)
- Hotline Colorado Ruff N Tuff (Ted x Colorados Buckeye Barbie)
- Jones Stockyard Hustler (Ted x Ruff N Reddy Pee Dee Que)
- Kintyre Jeb Stuart (Ted x Davies' Blue Kola)
- Kintyre Reddy Roper (Ted x Davies' Blue Kola)
- Lambeth's Krispy Kritter (Ted x Parmenters Dixie of Bar Lazy K)
- Laycocks Mr Red Ted II (Ted x Laycocks Cheyenne Maurine)
- Lyman Chuckwagon Chili (Ted x Davies' Blue Kola)
- Mini Acre Becky Badger (Ted x Mini Acre Freckles)
- Mini Acre Sarah (Ted x Mini Acre Sour Jane)
- Nelsons Ben Ready (Ted x Nelsons Gin Fizz)
- Parmenters Latigo (Ted x Parmenters Dixie of Bar Lazy K)
- Parmenters Rags (Ted x Parmenters Dixie of Bar Lazy K)
- Powder River Brandywine of Hedgerow (Ted x Colorados Ravishing Ruby)
- Powder River Dusty Rose of Lazy (Ted x Colorados Ravishing Ruby)
- Powder River Hedgerows Charm (Ted x Colorados Ravishing Ruby)
- Powder Rivers Outlaw Annie (Ted x Colorados Ravishing Ruby)
- Powder River Powder Puff (Ted x Slash V Fudge Cookie)
- Powder River Mr T (Ted x Colorados Ravishing Ruby)
- Powder River Teddys Token (Ted x Colorados Ravishing Ruby)
- Powder River Terrace Ripper (Ted x Colorados Ravishing Ruby)
- Powder River Thunderbird (Ted x Colorados Ravishing Ruby)
- Powder River Tuffernhel (Ted x Colorados Ravishing Ruby)
- Reddy Teddy II of Almost heaven (Ted x Red Lady of Cattle Call)
- Rimrocks Brown Sugar of JRB (Ted x Moor's Rustlin Ruby)
- Ruff N Reddy BJ Bear (Ted x Davies' Blue Kola)
- Ruff N Reddy Blue Tinker Toy (Ted x Keefers Miss Nitzi)
- Ruff N Reddy Erik (Ted x Colorados Ravishing Ruby)
- Ruff N Reddy Flash Cattlac (Ted x Ruff N Reddy Mity Happy Hocker)
- Ruff N Reddy Hunny Bear (Ted x Jones Mity Bite)
- Ruff N Reddy Joe Willy Wang (Ted x Ruff N Reddy Mity Happy Hocker)

- Ruff N Reddy Katie Blue of Chianti (Ted x Ruff N Reddy Mity Happy Hocker)
- Ruff N Reddy Katy Did (Ted x Jones Mity Bite)
- Ruff N Reddy Loper (Ted x Mini Acre Mini Masterpiece—Kintyre Jeb Stuart x Colorado Mini Orphan Annie)
- Ruff N Reddy Magic Sparkle (Ted x Colorados Buckeye Barbie—Georges Buckeye Bobby x Sorensen's Cherry Brandy)
- Ruff N Reddy Megan (Ted x Ruff N Reddy SeMe Sashay)
- Ruff N Reddy Mity Mean (Ted x Jones Mity Bite)
- Ruff N Reddy Mity Muhammad (Ted x Pizs Sheba)
- Ruff N Reddy Pee Dee Que (Ted x Hemi's Mountain Magic of J & J)
- Ruff N Reddy Powder Puff (Ted x Slash V Frosted Snoberry—Slash V Frosted Fudge x Slash V Red Tick Fondue)
- Ruff N Reddy Ruff N Tuff (Ted x Colorados Buckeye Barbie)
- Ruff N Reddy Ruffian of Cowtown (Ted x Pizs Sheba)
- Ruff N Reddy Ruffneck (Ted x Hemi's Mountain Magic of J & J)
- Ruff N Reddy Samurai M (Ted x Davies' Blue Kola)
- Ruff N Reddy Seme Smokum (Ted x Ruff N Reddy SeMe Sashay)
- Ruff N Reddy Shadey Joe (Ted x Shadey Mountain Mity Miss)
- Ruff N Reddy Shake N Bake (Ted x Ruff N Reddy Mity Happy Hocker)
- Ruff N Reddy Sugar Magnolia of Chianti (Ted x Pizs Sheba)
- Ruff N Reddy Tar Baby (Ted x Davies' Blue Kola)
- Ruff N Reddy Teddys Blue Jay (Ted x Pizs Sheba)
- Ruff N Reddy Teddys Scarlet Harlot (Ted x Martins Red Fudge)
- Ruff N Reddy Teddys Tuffit (Ted x Colorados Buckeye Barbie)
- Ruff N Reddy Thiex Pepsi Cola (Ted x Davies' Blue Kola)
- Ruff N Reddy Tinker Toy of Mini Acre (Ted x Davies' Blue Kola)
- Ruff N Reddy Trapper M (Ted x Shadey Mountain Mity Miss)
- Ruff N Reddy Voodoo Queen (Ted x Pizs Sheba)
- Ruff N Reddy Woodchip Blue (Ted x Davies' Blue Kola)
- Ruff N Reddy Zephyr of Minnkota (Ted x Ruff N Reddy Pee Dee Que)
- Schultz Duke of Earl (Ted x Sugar)
- Shadey Mt Mity Blugoblin (Ted x Jones Mity Bite)

- Shorts Ole Blue Ricochet (Ted x Ruff N Reddy Mity Happy Hocker)
- Slash V Carbon Black (Ted x Nielsens Holly II)
- Slash V Easy Jet (Ted x Nielsens Holly II)
- Slash V Fudge Candee (Ted x Martins Red Fudge)
- Slash V Fudge Cookie (Ted x Martins Red Fudge)
- Slash V Fudgette (Ted x Martins Red Fudge)
- Slash V Hit Man (Ted x Nielsens Holly II)
- Slash V Ready Blue (Ted x Nielsens Holly II)
- Slash V Reddi Ina Minit (Ted x Slash V Frosted Snoberry)
- Slash V Reddy Frost (Ted x Slash V Frosted Snoberry)
- Slash V Reddy Maid (Ted x Slash V Frosted Snoberry)
- Slash V Scarlet Harlot (Ted x Martins Red Fudge)
- Slash V Semi Sweet (Ted x Martins Red Fudge)
- Slash V Sugar Bear (Ted x Slash V Frosted Snoberry)
- Slash V Tanna Teddy (Ted x Slash V Frosted Snoberry)
- Spindletop Kiss N Tell (Ted x Moor's Rustlin Ruby)
- Sun Shadows Silhouette (Ted x Shadey Mountains Sumac)
- Teddy's Rustlin Rudy of Staple G (Ted x Moor's Rustlin Ruby)
- Willadees Theodore Bear (Ted x Moor's Rustlin Ruby)

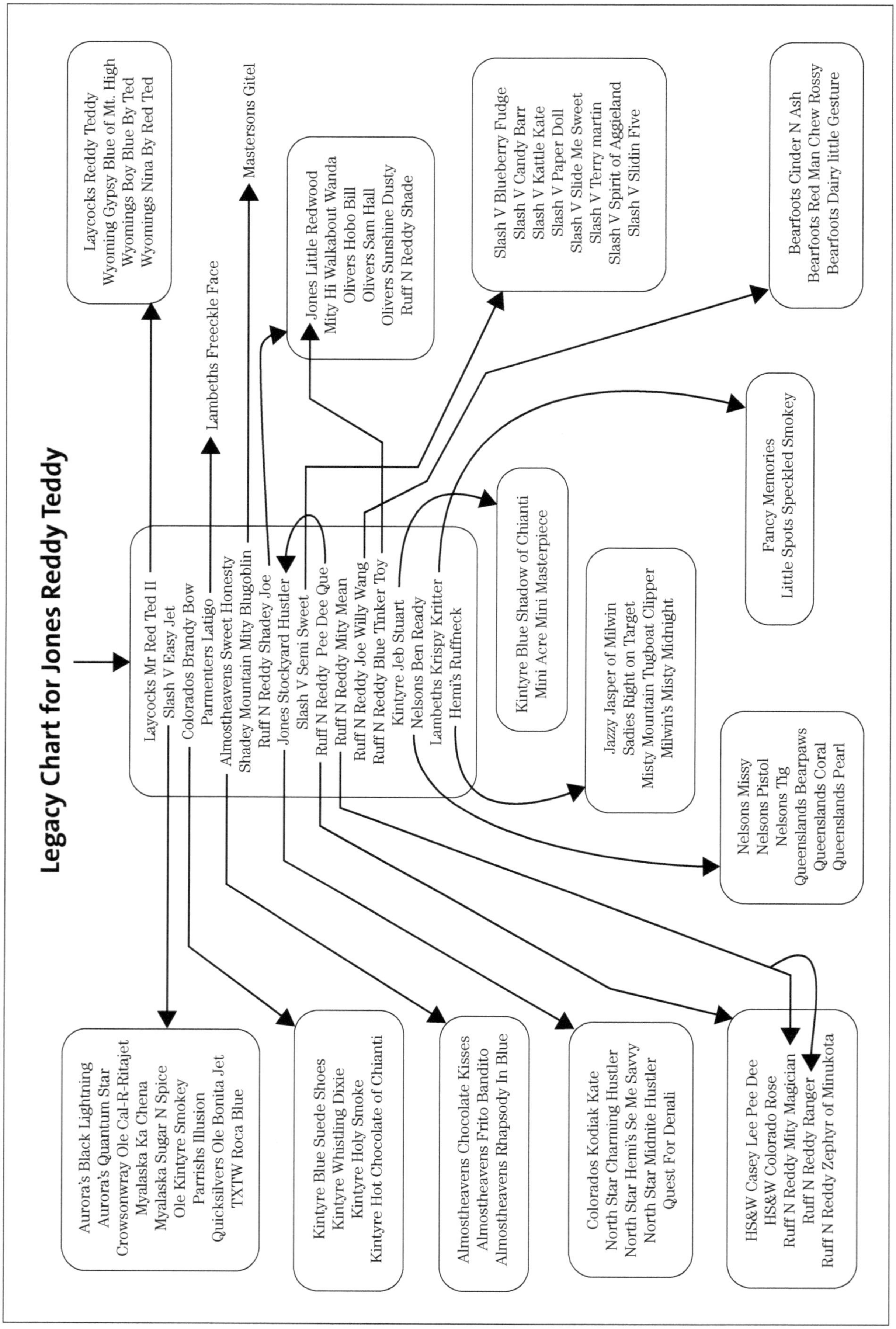
Legacy Chart for Jones Reddy Teddy
Laycocks Mr Red Ted II
Slash V Easy Jet
Colorados Brandy Bow
Parmenters Latigo
Almostheavens Sweet Honesty
Shadey Mountain Mity Blugoblin
Ruff N Reddy Shadey Joe
Jones Stockyard Hustler
Slash V Semi Sweet
Ruff N Reddy Pee Dee Que
Ruff N Reddy Mity Mean
Ruff N Reddy Joe Willy Wang
Ruff N Reddy Blue Tinker Toy
Kintyre Jeb Stuart
Nelsons Ben Ready
Lambeths Krispy Kritter
Hemi's Ruffneck
Laycocks Reddy Teddy
Wyoming Gypsy Blue of Mt. High
Wyomings Boy Blue By Ted
Wyomings Nina By Red Ted
Lambeths Freeckle Face
Mastersons Gitel
Jones Little Redwood
Mity Hi Walkabout Wanda
Olivers Hobo Bill
Olivers Sam Hall
Olivers Sunshine Dusty
Ruff N Reddy Shade
Slash V Blueberry Fudge
Slash V Candy Barr
Slash V Kattle Kate
Slash V Paper Doll
Slash V Slide Me Sweet
Slash V Terry martin
Slash V Spirit of Aggieland
Slash V Slidin Five
Bearfoots Cinder N Ash
Bearfoots Red Man Chew Rossy
Bearfoots Dairy little Gesture
Fancy Memories
Little Spots Speckled Smokey
Kintyre Blue Shadow of Chianti
Mini Acre Mini Masterpiece
Jazzy Jasper of Milwin
Sadies Right on Target
Misty Mountain Tugboat Clipper
Milwin's Misty Midnight
Nelsons Missy
Nelsons Pistol
Nelsons Tig
Queenslands Bearpaws
Queenslands Coral
Queenslands Pearl
Aurora's Black Lightning
Aurora's Quantum Star
Crowsonwray Ole Cal-R-Ritajet
Myalaska Ka Chena
Myalaska Sugar N Spice
Ole Kintyre Smokey
Parrishs Illusion
Quicksilvers Ole Bonita Jet
TXTW Roca Blue
Kintyre Blue Suede Shoes
Kintyre Whistling Dixie
Kintyre Holy Smoke
Kintyre Hot Chocolate of Chianti
Almostheavens Chocolate Kisses
Almostheavens Frito Bandito
Almostheavens Rhapsody In Blue
Colorados Kodiak Kate
North Star Charming Hustler
North Star Hemi's Se Me Savvy
North Star Midnite Hustler
Quest For Denali
HS&W Casey Lee Pee Dee
HS&W Colorado Rose
Ruff N Reddy Mity Magician
Ruff N Reddy Ranger
Ruff N Reddy Zephyr of Minukota

THE HEMI CONNECTION

Ch McGuire's SeMeGo Hemi UDT OTDc STDsd owned by Ruth McGuire was a true phenomenon. A mountain-type Aussie of typical old breeding of the area—Thornton, Gantenbein, a dash of Woods—he was a beautiful combination of brains, beauty and ability who had the good fortune to find a talented but untested trainer who made the most of both his and her own abilities. Although Ruth never developed a bloodline of her own, she used Hemi judiciously with other lines to produce dogs which had a wide influence in kennels throughout the breed.

The handsome, mature Hemi at 2 years of age. Courtesy of Ruth McGuire

Hemi is most closely associated with the breeding program of Cathy Jones/Ruff N Reddy due to the eventual close friendship between Cathy and Ruth. However it didn't begin that way! Ruth had been cleaning up in the obedience rings and showing most of the old timers "how it was done" for about 5 years before Cathy acquired Ted. Hemi was in his prime and Ruth had high ambitions for his future and no desire to share the limelight with two young, untested interlopers. In fact, she and Hemi likely served as inspiration and spur to Cathy to achieve the things she did with Ted. Cathy always shot for the stars and never settled for second best so if she was to get to the top she would have to compete against, and best, the Hemi and Ruth outstanding team. Challenging competition developed when Cathy and Ted first appeared on the horizon and began showing they were a force to be reckoned with. The competitive fires didn't allow for an instant friendship and the two teams fought neck and neck for awards for the first few years of Ted's career. One hotly contested prize was known as the "Gambler Trophy"—a large chrome tray engraved with the names of all the winners. The tray was donated by breeder Kitty Scherf in memory of her old dog "Gambler" and was

Hemi in November 1971. Courtesy of Ruth McGuire

Desserich's Ring, black and white male (Greens Kim x Greens Dingo) Sire of Gantenbeins Mike. Courtesy of Ruth McGuire

PEDIGREE OF Ch McGuires SeMeGo Hemi, UDT, OTDc, STDsd

Born: ____________ **Sex:** ________
Owned by: ____________________

Parents	Grandparents	Great-grandparents	Great-great-grandparents
Sire: Gantenbeins Mike	Sire: Desserichs Ring	Sire: Greens Kim	Sire: Rogers Red Dog
			Dam: Rogers Blue Merle
		Dam: Greens Dingo	Sire: Woods Jay
			Dam: Elys (Sislers) Blue
	Dam: Gantenbeins Blue Muffett	Sire: Thorntons Blue Smokey	Sire: Harpers Old Smokey
			Dam: Harpers Nita
		Dam: Gantenbeins Sugar Blue	Sire: Harpers Old Smokey
			Dam: Thorntons Blue Blue
Dam: Gantenbeins Lady Blue	Sire: Harpers Half Pants	Sire: Harpers Joker	Sire: Breazales Bob
			Dam: Dons Judy
		Dam: Harpers Blue Austina	Sire: Elys Sage
			Dam: Shannons Dingo
	Dam: Harpers Pepino	Sire: Harpers Old Smokey	Sire: Tucsons Joe
			Dam: Tucsons Sis
		Dam: Harpers Jinx	Sire: Sislers Shorty
			Dam: Beckmans Queenie

Bred by: ____________ **Address:** ____________ **Phone:** ____________

The details given above are correct to the best of my knowledge. Signed: ____________(Breeder)

Hemi's final triumph as a competition dog—winning the Gambler Trophy, which he won three times with progeny from 5 different females to retire. It was won once by Ted. Not only was Hemi the ultimate performer, he was also the ultimate producer. Courtesy of Ruth McGuire

awarded in a special class in which two progeny of the same sire were shown together for that sire. The sire had to win the class three times to retire the trophy and retain permanent possession of it.

After a long life of achievement and awards, winning the Gambler trophy became Ruth's last goal for Hemi—the most fitting cap to a stellar career. In 1982, a year before his death, Hemi won it for the third and final time. The progeny shown for him were six different dogs from five different females—a credit to his ability to pass on his high quality to the next generation. His challenger, Ted, and the venerable Wood's Stubbie were two of many well known dogs to compete for this coveted prize for the 10 years it was offered. Kitty Scherf says this in a letter to Ruth after Hemi retired her trophy, "Better than 10 years ago, you and Hemi got me started in something that has lingered since and will remain for years to come—the enjoyment from being involved in showing and breeding

Hemi showing his typical smooth, balanced style on ducks. Courtesy of Ruth McGuire

Hemi with his litter, September 26, 1968. Courtesy of Ruth McGuire

Hemi shortly after coming home to the McGuire's. Age 2-1/2 months. Courtesy of Ruth McGuire

Aussies…I'm proud that you chose the Gambler Award as a goal, it makes the donating all that much more meaningful. I guess you might say that HEMI inspired it from the beginning, so its only fitting that he should own it. Congratulations."

Eventually Cathy's strong desire to learn from the best overcame any hesitancy she might have had to go to her toughest competition to learn. Somehow her sweet, outgoing personality broke down any reluctance Ruth might have had to help her competition and the two discovered a mutual love for training and the breed. Their shared high standards and talent for training proved to be a stronger bond than a barrier and a lifelong friendship was born.

Aussie activity in Colorado was on the rise when Ruth decided to get her first Aussie. She had owned dogs prior to this but Hemi was to teach her a lot more than all those others combined. When Hemi was 10, Ruth summarized an astounding career and gave tribute to a dog who had no more worlds to conquer in this way, "The Chrysler Hemi engine was, in 1968, considered THE top engine. It was noted for its excellence in perfection performance. Our race car was a Plymouth, and we called it SEE ME GO. So, it only seemed natural that I name my new puppy McGuire's SeMeGo Hemi. I had picked him out of a litter of 11 when they were 3 days old and fortunately it was a very good litter of excellent breeding. (Knowing nothing about choosing a puppy, Ruth's novice criteria was, 'This one's a male—I'll take him'.) At the time I didn't know very much at all about the Australian Shepherd, but I was soon to learn." In hindsight Ruth was to marvel at her luck in getting Hemi, saying, "Now to look back and wonder at how that particular pup and I came together—not knowing then where we'd go and what he would accomplish, what he would mean to my life, and what he would do for his breed. Luck? Perhaps. He was born on September 26, 1968, and I took him home when he was seven weeks old. I was fascinated by his intelligence. He learned to 'come,' 'sit,' 'down,' and 'stay' before he was

9 weeks old and with my non-experience as a trainer too!! I thought he was the prettiest dog on the face of the earth.

"When he was almost six months old the Aussie club in Colorado held sub-novice training classes and I very eagerly attended. I loved to work with him and so upon graduation he earned his first blue ribbon. Shortly thereafter I entered him in his first real show, a Specialty, and he earned another blue ribbon. I was hooked! From there we went to training classes in Denver under the supervision of noted trainer Sally Terroux. I soon learned that Hemi was NOT the ordinary run-of-the-mill dog. He learned so quickly, his eagerness to please was surpassed only by his great intelligence. He responded to praise and I never once used the choke collar on him. Again, he earned a first place upon graduation." In addition to Hemi's talent at obedience training, his natural Aussie guardian instincts were beginning to emerge at this age, too. Mail coming through Ruth's mail slot was an invasion of his territory he would not tolerate. Ruth once had a great deal of trouble explaining to her bank the toothhole-riddled tax refund check she was trying to cash!

Hemi as a 4 month old puppy. Courtesy of Ruth McGuire

Hemi hitting his "teenage" stage at 6 months. Courtesy of Ruth McGuire

Hemi wasn't the low heeler Ted was but he was a tough cattle working dog nonetheless. Courtesy of Ruth McGuire

"By this time we had given up drag racing and so all of my spare time was spent with Hemi. I learned about dog shows (for the Australian Shepherd the 'matches' were the real thing) and I learned that the International Australian Shepherd Association was awarding obedience titles according to the AKC rules. His first 'leg' on his CD was earned easily with a score of 198-1/2 and first place. I was told that I looked like a mummy in the ring, pale white and stiff as a board, altho I felt like my legs would give out on me at any second—I was so nervous. But, Hemi took it all in stride, his heeling was perfect and he responded immediately to my every command. He never took his eyes off me and pranced happily at my side.

Hemi's dark soulful eyes spoke volumes. Courtesy of Ruth McGuire

Hemi was so well trained he would work for anyone. Here he is at the 1972 Arizona Specialty winning first place and High In Trial with Lori Acierto. Courtesy of Ruth McGuire

"The day before the show to try for the second 'leg', my husband had left to tow a car over the mountain passes back to Denver. March in the Rockies means a lot of snow. I expected him back about midnight and when he failed to return, naturally I became worried. The hours ticked by, I called the State Patrol, hospitals, but I could get no information. All thru the night I worried. He finally got home at ten in the morning, unhurt but very tired having driven thru a bad snow storm. So, needless to say I was in bad shape, but off I went to the dog show! I remember asking Hemi to 'just qualify'. When it was all over and the scores were announced, Hemi had indeed qualified—with a perfect 200! I laughed and cried and loved my beautiful blue dog a little bit more.

"Hemi's final leg on the CD was earned with a score of 199. The judge said, 'I never give perfect scores, but that dog was as perfect as I've ever seen!" Thus, a reputation and legend was formed.

"We immediately went into Open training, again with Sally Terroux, and it was at those classes that I first heard about Tracking. I was invited to join a small group to see if Hemi 'had a nose'. Hemi was a little over a year old. So, one cold but bright February Tuesday morning I drove Hemi out east of Denver to a very large open field. Jackie McClure instructed me to 'lay the track' and she handled Hemi. I hid behind some bushes and watched as Jackie brought Hemi out and pointed to the ground, his nose went down and he began to track. Not wasting any time and dead on the track he found me. Jackie was so excited 'He actually tracked, the very first time out!!' That was the beginning and from there he progressed rapidly, soon doing full length tracks and then on to 'aged' tracks. By June we were a part of the Colorado Canine Search Group and on our first search!

"People don't get lost where the terrain is nice and they don't think as one would think they would. That first search was quite a lesson, I learned to trust my dog as he knew far more than I. I won't go into the details here, but the lost victim was found dead, suicide. When the phone would ring at one in the morning we knew it was a search. Hemi loved it—get out the tracking equipment and he knew we were leaving shortly. His 'nose' was respected by the search group and by the groups we worked with. He was a capable 'air scenter' and an excellent 'tracker'.

A mature Hemi showing his nearly faultless conformation. Photo by Sue Noll, Courtesy of Ruth McGuire

"It was during those three and a half years that Hemi became my most trusted, loved and closest friend. The rapport we gained cannot be equaled. He learned to eat fresh orange slices and raisins which I carried in my back pack. He learned to drink water from my cupped hand and he taught me when we were just out for a stroll and when there was scent—however weak it might be. He was actively involved in saving lives of human beings—and what greater gift can any dog give?" The only fright the team experienced was their first airplane ride to a search. The pair were equally terrified with Hemi pressed against Ruth on her lap, shaking, and Ruth, pale as a ghost, clinging for dear life to him for comfort!

Hemi in an iconic pose seen many times in ads during his career. Courtesy of Ruth McGuire

"We were involved in a search at the 13,000 foot level of Mt Shavano, Hemi had a badly broken toe on his right front foot. We were flown in by Army helicopter, which made a crash landing, scaring the wits out of me, but Hemi went on working a seven square mile area, above timberline, looking for the body of a baby lost after the crash of a small aircraft.

"Then there was the search for 11 year old Kevin Dye in the mountains west of Casper. Again the going was rough, but the search was a success. I can't remember how many searches we were on, but I remember them fondly because my little blue Hemi was truly living up to his name.

"During our search years, we continued in obedience. Hemi earned his CDX easily in three shows with scores of 197-1/2, 198-1/2, and 197. His last leg was earned at the National Specialty Show of the International Australian Shepherd Association in Napa, California, October 1970. Hemi was two years old. On the way back to Denver we were involved in a bad car accident. Hemi was thrown thru the rear window and was lost for two days. The Civil Air Patrol was contacted and they were about to start an air search for him, when I found him wandering along an irrigation ditch. He was badly injured so we rushed him to the nearest Veterinarian in Grand Junction, Colorado. For over an hour the Vet worked on him, while I stood by and watched him still breathing under the anesthetic. When we got back home and the bandages came off it was clear that there was severe damage both to his right rear leg and his right front leg. All of the tendons on the front leg were severed clear through. And the large tendon on the rear leg was cut but not severed. After consulting with several Vets and friends, I took Hemi to Fort Collins to Colorado State University Vet School. It was there, under the direction of Dr Smith, that the tendons were repaired. He was also x-rayed for hip dysplasia and checked for eye abnormalities. The x-ray was good, the eye check was good and thank God—the surgery was a success! However, after I brought him home from the five day stay at CSU, he began coughing. I immediately took him to my local Vet, who advised me that my Hemi had distemper. For five days he received massive doses of vitamins and protein directly into the bloodstream. He was one sick dog—but again his desire to live took over and he came thru with flying colors. It wasn't until almost a year later that I learned that there was kidney and liver damage, thanks to the treatment of the distemper. A pill a day for the rest of his life, and gladly—he was alive."

Another win for the great Hemi. 1977 Arizona Specialty First Place in Open Obedience and High in Trial. Courtesy of Ruth McGuire

Hemi at the 1977 Arizona Specialty. Courtesy of Ruth McGuire

During Hemi's recovery from the tendon repair in 1970, a Denver television station called Ruth's obedience instructor, Sally Terroux, about doing a story on dog training. Hemi was still in a cast but Sally wanted him in the story and, in Ruth's words, "That night Denver saw an Australian Shepherd, cast and all, doing a complete obedience routine. They watched him gently roll a raw egg around with his cast until it was just right for him to pick it up in his mouth and bring it to me—uncracked." Following this performance, Hemi made a stop at the vet to have his cast removed and his biggest fan, instructor Terroux, called Ruth. She had lit a candle for Hemi and wanted to know, "was his leg okay?" Fortunately, yes, the leg had healed well.

"Three months after his surgery he earned the first leg on his UD with a score of 184-1/2—a TEN POINT handler error didn't help!! His second leg earned him a score of 198, the same day getting a score of 197-1/2 in Open, for Highest Scoring Dog in Trial and Highest Combined Score PLUS I launched his conformation career. Just entered him in breed on the spur of the moment and it did pay off—he went first in his class, Winners Dog and Best of Breed for a five point major."

Featured in the July 23, 1977 Rocky Mountain News for his accomplishments, Hemi was a goodwill ambassador for the breed to dog people and the general public alike. Courtesy of Ruth McGuire

So, what does a superstar and his faithful sidekick do after they have achieved nearly all there is to achieve? In the case of Hemi and Ruth, they continued to enjoy their partnership in obedience while still looking for more challenges to conquer. An obedience demonstration at a shopping mall was the scene of an embarrassing episode in the team's career. Ruth had sent Hemi out for the directed jumping and told him to sit at the opposite end of the ring. Hemi apparently misinterpreted this command as "S---t!" and promptly complied! It wasn't all work and no play for Hemi and Ruth.

Hemi in his prime, the first time shown in a conformation class. Courtesy of Ruth McGuire

"A week later Hemi earned that 'Golden T'. On a hot August morning in a field so very dry, Hemi did his track in beautiful style, becoming the first Australian Shepherd to earn the TD. And just a few short weeks after that, on his third birthday he finished his UD with a score of 193, making him the first Aussie UDT. Six months, and five shows later he became the first Champion UDT." What makes these achievements even more monumental is the realization that shows were few and scattered over a huge area of the west from Colorado to Oregon, Washington, Idaho, Utah, Wyoming, California, Arizona, Kansas, and Louisiana. Travel was more difficult and finding opportunities was

Another superlative day for the Ruth and Hemi team. On September 26, 1971, Aspen, Colorado, Hemi 's third birthday, he won First Place in Utility to finish his UDT and won Best of Breed in conformation. Courtesy of Ruth McGuire

much more of a challenge. Trainers had to be ready to go at any moment, keeping their dogs in peak form year round!

"But, still my Aussie was not doing what he was bred to do—work stock! Oh, how I wished that I had some stock to work him on. We moved to Oregon when Hemi was 5-1/2 years old and it was there that he got his chance to work. Seven of the wildest sheep in the west, soon learned that that dog was not one to fool with." The trial was in Bend, Oregon in 1973 and the stock dog program was in it's infancy. Clubs either had very little choice in the selection of livestock or did not know how to find appropriate stock for trials. The dogs did have a group of seven head to work but it was a good thing they were tough and the handlers took the stock they were given with good humor and no complaints. Ruth and Hemi entered the Bend arena to find themselves face to face with a ram named "Rocky Mountain High." He was not only a ram but also 7/8ths bred Big Horn Sheep! Mr. High and his six wild cohorts learned a number of lessons from the dauntless Hemi, who emerged triumphant from his "run"!

"When we returned to Colorado almost 2 years later, The Australian Shepherd Club of America had become the parent breed club and Hemi was awarded new certificates from them.. They had also begun a Working Stock Dog Program with stock dog trials being held in several states. Not long after I had begun teaching obedience classes, I met some very fine people who offered their cattle to me to train Hemi on. Not knowing the first thing about training a dog to work stock, I just let him 'do his thing'. His intelligence along with his natural 'low heeling' instinct proved all that was necessary." With less than 10 hours total training time under his belt, Ruth set out with Hemi to earn those coveted stock titles. "He did not qualify at his first trial, but at the tender age of 8 years old, we traveled to Utah and he earned his first qualifying score on cattle. Two weeks later in Kansas he qualified again and placed in the ribbons, to earn his Open working certificate making him the first Aussie to hold the 'Triple Crown'."

"The next year, again we traveled to Utah and I tried him on sheep—again he qualified! His last qualifying score was earned in his home state at the Colorado Australian Shepherd Association Specialty held over Memorial Day weekend in 1978...at nine years of age, he became double certified on stock and placed second in the class." Hemi still wasn't quite finished. At the age of 12, he and Ruth set out to finish what they had started in the stock arena and earn that last title on ducks. He earned his Started Duck title, giving him titles in all three types of livestock offered in ASCA and was a "Triple Crown" winner once again.

"My beautiful Hemi is now retired, he enjoys his days playing with his daughter and grandson and his nights sleeping on my bed. He has proven that the Australian Shepherd is a truly versatile breed and that HE is a cut above the rest. Over the years he earned twenty seven Highest Scoring Dog in Trial honors, several High Combined Score and Best of Breed wins AND has won or placed in several stock dog trials. His sons, daughters, grandsons and granddaughters have also proven themselves to be very worthy progeny of their great sire." A celebration of Hemi's life was held on his 13th birthday with a party in his honor. Cathy Jones supplied the huge steak for the guest of honor with a chocolate cupcake for dessert. After singing Happy Birthday to Hemi he was allowed to work a group of Cathy's cattle one last time. Ruth remembers, "He thought that he was young again, low heeling, full of

Cattle work was a high priority for breeders of Australian Shepherds in the early years. Hemi shows his style. Courtesy of Ruth McGuire

1978 ASCA National Specialty, Fort Collins, Colorado. Winners of the Working Dog Class—Hemi was the male winner and Ch Colorado's Ravishing Ruby was the female winner. In addition, Hemi won Best Working Dog. Photo by Sue Noll. Courtesy of Ruth McGuire

Hemi taking control of his sheep at a trial. Stock titles crowned his many achievements making him the first "Triple Crown" Australian Shepherd in the breed—winning titles in all three venues offered at the time—conformation, obedience and stock work. Courtesy of Ruth McGuire

Hemi finishing his Started Sheep title with a second place at the 1978 ASCA National Specialty in Fort Collins, Colorado, under judge Dick Sorensen. He did this feat nursing a rather severe injury to one foot. Courtesy of Ruth McGuire

The Great Ch McGuire's SeMeGo Hemi UDT STDcsd. Courtesy of Ruth McGuire

Hemi surrounded by just a few of his many trophies. Courtesy of Ruth McGuire

Hemi at his 13th birthday party. Courtesy of Ruth McGuire

want-to, his eyes sparkling. My heart filled with pride and love."

"Hemi will be long remembered by the Aussie fanciers and by those of us mere mortals who have had the pleasure of knowing him and seeing him in action. He broke the frontier in many areas to give the Aussie recognition. CHAMPION McGUIRE'S SEMEGO HEMI, UDT, OTD cattle, STD sheep—how appropriate that 10 years ago I named a little fluffy blue puppy after the Hemi engine. Both are proven THE ULTIMATE OF PERFECTION PERFORMANCE. I love you Hemi, there will NEVER be another like you!!"

After so many years of public success, Ruth never lost sight of the fact Hemi was first and foremost her best friend and constant companion. His fame in the show world and as as search dog earned him a large following of fans, many of them complete strangers to Ruth. Over and over she would be asked "How is Hemi?" when she was out in public. He had been seen not only at shows and trials but also on television and in newspapers, becoming a true legend in his own time. But the partnership between Hemi and Ruth was the true legend. He was Ruth's constant guardian, saving her life at least once. Nobody would dare touch her truck if Hemi was on duty inside. He once broke through her back door, cleared the dining room table and stopped a man from pulling the trigger of a gun at Ruth's throat! In Ruth's words, "(He was) a dog to whom fame and winning meant nothing. He did it all because I asked him and because the ability was inside him—in his head and in his heart. How many times I asked him 'One more time Hemi—do it for me.' He always did."

If anything, Ruth's summary of Hemi's life is still somewhat understated. His personal achievements are a matter of record but another record speaks to his genetic worth and legacy. Breeders have always had a tendency to give more weight to a dog's titles than might be prudent. If the dog is poor quality or from a poor background but in the hands of a gifted trainer this can cause great harm to a breed. The dog can be bred many times, passing along inferior genes which drag down the quality of the breed in a very short time. Fortunately, in Hemi's case his genetic background was strong, proven and worthy of preservation.

Like Sorensen's Gunsmoke and his son Commanche, Hemi proved to be a good cross with the predominantly Wood bred dogs which were fairly common in the area at the time. Like Gunsmoke and Commanche he brought

Hemi in 1981 at his 13th birthday party. Still ready for action. Courtesy of Ruth McGuire

Blue, copper and white puppy by Hemi x Cache La Poudre Cameo. Courtesy of Ruth McGuire

Blue merle, copper and white puppy by Hemi x Cache La Poudre Cameo. Courtesy of Ruth McGuire

"Jesse James," blue merle, copper white son of Hemi x Cache La Poudre Cameo. Courtesy of Ruth McGuire

Blue, copper and white puppy by Hemi x Cache La Poudre Cameo. Courtesy of Ruth McGuire

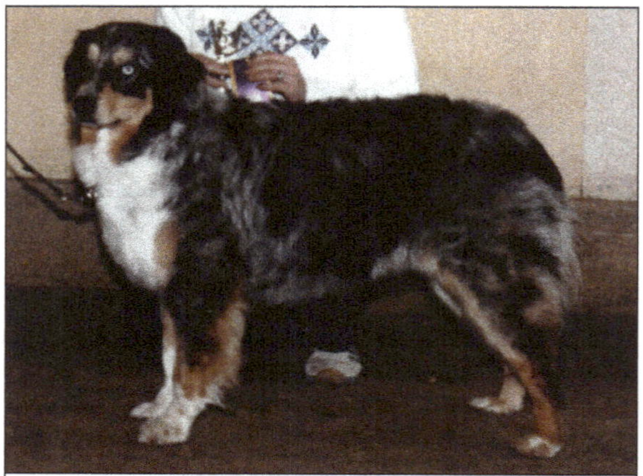

Magic showing off her correct conformation and breed type. Courtesy of Ruth McGuire

Hemi's Travelin Trashman, blue merle, copper and white puppy by Hemi x Piz's Sheba. Courtesy of Ruth McGuire

The Hemi son, Hemi's Atlas, black and white male, winning First Place in the 9-12 month puppy class at the ASCA National Specialty. Photo by Sue Noll. Courtesy of Ruth McGuire

Magic brings up the stragglers on the way to the repen. Courtesy of Ruth McGuire

biddability and a new set of working instincts to dogs with already strong working ability and intelligence. Cathy Jones and Ted had become the predominant Wood representatives by that time so it was only natural the Hemi/Ted combination would eventually be made.

Hemi's natural working instincts were to head and heel on cattle but he had little to no eye. Ruth recalls he worked differently from the Wood dogs but his style was a compliment to the Wood style and produced dogs which would excel on all types of livestock without losing their cattle working ability—the one trait still most highly valued among stockmen of the region.

Another trait valued by both the Wood breeders and by Ruth was the ability of the dog to think for themselves. Stock dog training for trials was still in its infancy and prior to the advent of trials, real working ranches didn't have the time or inclination to spend time carefully training a dog to work. The dogs figured out the job as they went along and even those who were competing in trials still largely depended on the natural ability of the dog to learn the job itself. Handlers learned to just trust the dog and allowed them to make their own decisions a lot of the time. Sometimes this was a mistake! But working a dog that was able to think for itself, encouraged to think for itself and bred

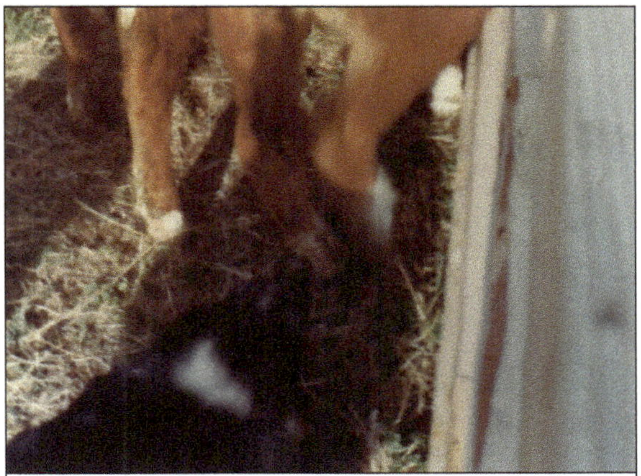

Hemi, low heeling those calves. Courtesy of Ruth McGuire

Hot Smokin' Hemi, blue, copper and white puppy by Hemi out of Harmony of Good Fortune CD as a young puppy. Courtesy of Ruth McGuire

Hemi gathering sheep. Courtesy of Ruth McGuire

Hot Smokin' Hemi, blue, copper and white puppy winning Best Puppy at the ASC of Utah show. Photo by Sue Noll, Courtesy of Ruth McGuire

Hemi's Command Performance, red merle and white female by Hemi x Mastersons Dotclijee Shunka—a Wood/Gantenbein cross bitch and foundation for HS&W kennel. The only red merle female ever produced by Hemi. Courtesy of Ruth McGuire

Hot Smokin' Hemi, blue, copper and white male as an adult. (Hemi x Harmony of Good Fortune CD) Courtesy of Ruth McGuire

A top quality blue and white dog by Hemi out of Gilsons Kandee Koquett. Courtesy of Ruth McGuire

Ch Hemi's Mountain Magic of J & J CDX STDcsd, blue merle, copper and white female (Hemi x Trixie Bear of J & J) Courtesy of Ruth McGuire

Ch Hemi's Regal Request STD as a young dog. Courtesy of Ruth McGuire

to think for itself did not result in as many bad outcomes as one might predict. Ruth likened this to tracking where trusting the dog becomes vital to success. Those who were working the Australian Shepherd on livestock were learning this valuable lesson as they learned the best way to work their talented dogs.

So, due to the good fortune of a background of good genetics, when it came to breeding, Hemi triumphed with nearly every cross which was tried. His popularity with every type of breeder ensured a wide variety of females were crossed with Hemi and he produced well with every one. Probably his most famous son was Ch Hemi's Regal Request STD (Hemi x Ch Snows Regal Cactus Berry CD, a Flintridge and old Arizona line female). "Hymie" as he was known was a nearly a clone of his beautiful sire and founded a dynasty of great show dogs in the southwest. Leslie Sorensen considered him the most perfect Aussie she had seen.

A dog who followed in the performance footsteps of his famous sire was Ch Merledales Zack CDX, a very dark blue dog, who was not only a well proportioned and good moving breed champion but also a top obedience performer. Owned by Corky Shaha, Zack and Hemi became friendly rivals in the performance rings. Corky and Zack had traveled to a Kansas stock trial with Ruth and Hemi when they were chasing the "Triple Crown"—a Breed Championship, a CDX and a CWD (Certified Working Dog—what the original stock dog titles were called). Hemi lacked only one leg on cattle to become the very first Triple Crown in the breed. However, his other rival, Ted, only needed his last points to finish his breed Championship to become the first Triple Crown. The two were at different shows in different states and the fate of timing would deem Hemi the winner of the honor "First Triple Crown Australian Shepherd"! Ted's final Championship points were won 2 hours after Hemi had qualified for his final cattle leg!

Some sires are known as producers of great sires and some are famous for their ability to produce top brood bitches. Hemi had the ability to produce both. With the wide variety of females bred to him, nearly all of them outcrosses, Hemi could consistently add very high quality to his progeny that bred on in later generations. This is not only a credit to this great dog but also to his breeder, Eleanor Gantenbein, and her knowledge and experience as a breeder of the mountain-type Aussie.

Hymie lived in Arizona and his popularity in that area generated a lot of interest from breeders in Hemi. A number of good females of old Arizona and California breeding were bred to Hemi producing foundations for established and new kennels in the area. The female Snows Regal Cactus Berry produced not only Hymie but also the Champion female Chiltons Tapestry of Hazelwood. A number of lesser known dogs of this cross also became influential in area kennels.

A Hymie daughter bred back to Hemi, Molly of Windmill Hill, provided the foundations for the California Windmill Hill kennel. Windmill Hill Sweet Nell, Hemis Encore of Windmill

Ch Hemi's Regal Request STD, probably Hemi's best son. Out of the Champion female Snow's Regal Cactus Berry CD. Courtesy of Ruth McGuire

Like many Hemi offspring, Ch Merledales Zack CDX excelled as both a conformation and obedience dog. Courtesy of Ruth McGuire

Blue and white littermate to Ch Hemi's Mountain Magic of J & J CDX STDdsc winning BOS Puppy at the ASC of Utah show. Courtesy of Ruth McGuire

Hemi and Hymie in 1972. Courtesy of Ruth McGuire

"Mindy" black and white puppy by Hemi x Molly of Windmill Hill CD. Courtesy of Ruth McGuire

"Shine," blue, copper and white puppy by Hemi x Molly of Windmill Hill CD. Courtesy of Ruth McGuire

"Cory" blue, copper and white puppy by Hemi x Molly of Windmill Hill CD. Courtesy of Ruth McGuire

Hill and Hemi's Delight of Windmill Hill are just a few of the dogs who made their mark in early California breeding programs.

Ch Deines Bobbi Bear was already a proven cross with Hemi, having produced the Ch Merledales Zack CDX, when Marty Acierto first saw her and fell in love with her. The Acierto's eventually purchased this lovely female and she became the foundation of the Bluecrest kennel. Sticking with success, Bobbi was bred back to Hemi to produce a number of top females for them. In addition, Bluecrest's old Arizona and Flintridge cross female Foster's Duchess was bred to Hemi to bring in even more of the successful genes from Hemi.

For a time Ruth owned a daughter of Rowe's Hondo Warrior named Ch McGuire's Blu-Duster CD (out of Williams Sissy, a double bred Desserich's Ring female). She finished the Championship and CD on this beautiful female and bred her at least once to Hemi producing the striking black and white female Hemi's Double Trouble of Windsor. She also produced Hemi's Challenger of Blu-Ridge, Hemi's Dinky Diamond of Bluecrest, Hemi's Spittin Image and Hemi's Tonk Saga Kahotek. She was eventually sold to Lloyd and Linda Sloan (Wilson).

When Hemi was nearing retirement and Ruth was looking for a replacement, she chose a blue merle, copper and white Hemi daughter from a breeding to Trixie Bear of J & J, a Gantenbeins Mike granddaughter. Ch Hemi's Mountain Magic of J & J CDX STDdsc became nearly as accomplished as her famous sire under Ruth's guidance, earning her Championship, CDX and STDcsd. A beautiful blue merle, copper and white, Magic easily earned her Championship with her correct, moderate conformation and flowing, efficient movement.

Her inheritance from her sire was a strong working instinct and high intelligence. Excelling in obedience, she achieved her CDX

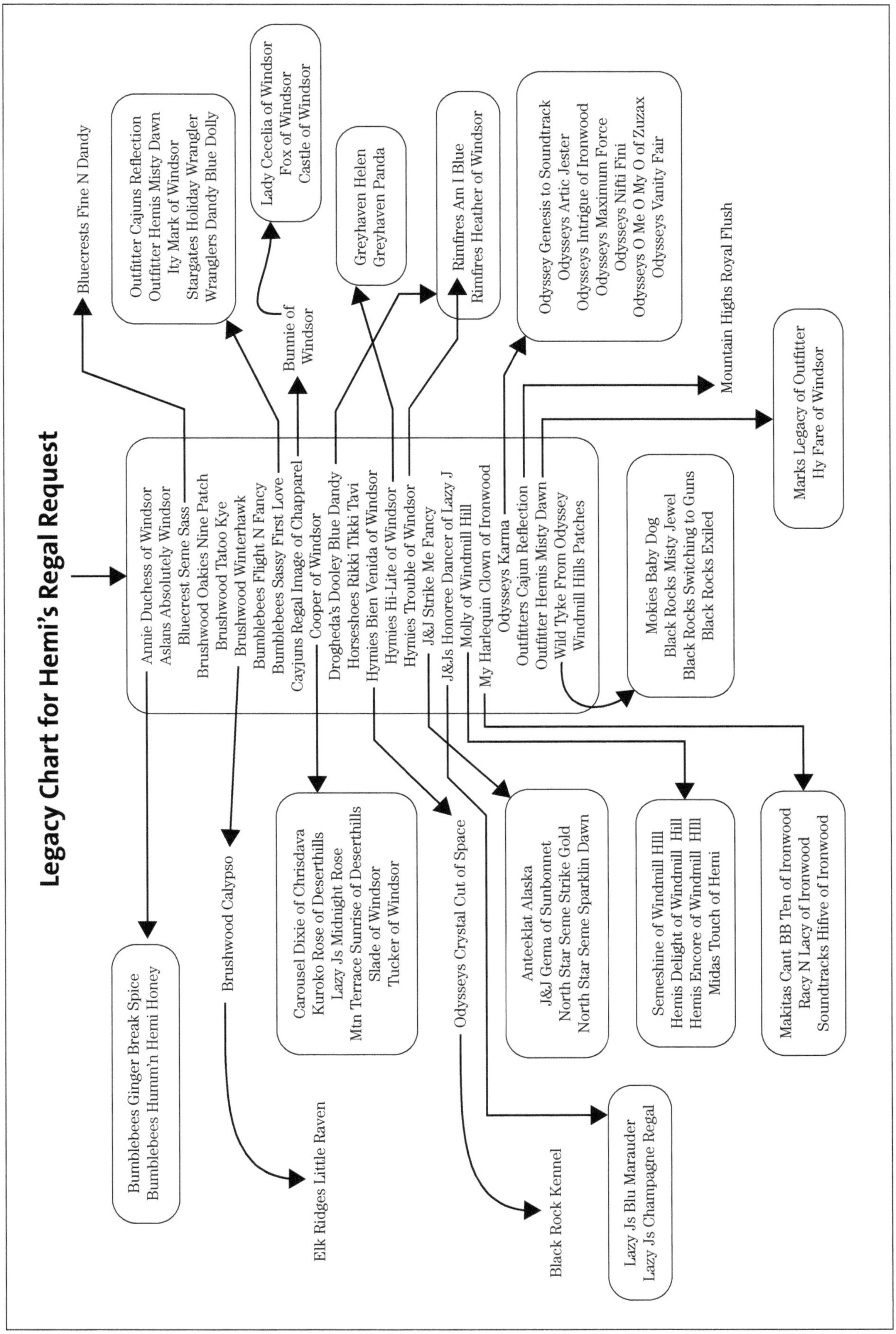
Legacy Chart for Hemi's Regal Request
Annie Duchess of Windsor
Aslans Absolutely Windsor
Bluecrest Seme Sass
Brushwood Oakies Nine Patch
Brushwood Tatoo Kye
Brushwood Winterhawk
Bumblebees Flight N Fancy
Bumblebees Sassy First Love
Cayjuns Regal Image of Chapparel
Cooper of Windsor
Drogheda's Dooley Blue Dandy
Horseshoes Rikki Tikki Tavi
Hymies Bien Venida of Windsor
Hymies Hi-Lite of Windsor
Hymies Trouble of Windsor
J&J Strike Me Fancy
J&Js Honoree Dancer of Lazy J
Molly of Windmill Hill
My Harlequin Clown of Ironwood
Odysseys Karma
Outfitters Cajun Reflection
Outfitter Hemis Misty Dawn
Wild Tyke From Odyssey
Windmill Hills Patches
Bluecrests Fine N Dandy
Outfitter Cajuns Reflection
Outfitter Hemis Misty Dawn
Ity Mark of Windsor
Stargates Holiday Wrangler
Wranglers Dandy Blue Dolly
Lady Cecelia of Windsor
Fox of Windsor
Castle of Windsor
Bunnie of Windsor
Greyhaven Helen
Greyhaven Panda
Rimfires Am I Blue
Rimfires Heather of Windsor
Odyssey Genesis to Soundtrack
Odysseys Artic Jester
Odysseys Intrigue of Ironwood
Odysseys Maximum Force
Odysseys Nifti Fini
Odysseys O Me O My O of Zuzax
Odysseys Vanity Fair
Mountain Highs Royal Flush
Marks Legacy of Outfitter
Hy Fare of Windsor
Mokies Baby Dog
Black Rocks Misty Jewel
Black Rocks Switching to Guns
Black Rocks Exiled
Bumblebees Ginger Break Spice
Bumblebees Humm'n Hemi Honey
Brushwood Calypso
Elk Ridges Little Raven
Carousel Dixie of Chrisdava
Kuroko Rose of Deserthills
Lazy Js Midnight Rose
Mtn Terrace Sunrise of Deserthills
Slade of Windsor
Tucker of Windsor
Odysseys Crystal Cut of Space
Black Rock Kennel
Anteeklat Alaska
J&J Gema of Sunbonnet
North Star Seme Strike Gold
North Star Seme Sparklin Dawn
Lazy Js Blu Marauder
Lazy Js Champagne Regal
Semeshine of Windmill Hlll
Hemis Delight of Windmill Hill
Hemis Encore of Windmill Hlll
Midas Touch of Hemi
Makitas Cant BB Ten of Ironwood
Racy N Lacy of Ironwood
Soundtracks Hifive of Ironwood

PEDIGREE OF Ch Hemi's Mountain Magic of J&J, CDX, STDcsd

Born: ______ Sex: ______
Owned by: ______

Parents	Grandparents	Great-grandparents	Great-great-grandparents
Sire: McGuires SeMeGo Hemi	Sire: Gantenbeins Mike	Sire: Desserichs Ring	Sire: Greens Kim
			Dam: Greens Dingo
		Dam: Gautenbeins Blue Muffett	Sire: Thorntons Blue Smokey
			Dam: Gantenbeins Sugar Blue
	Dam: Gantenbeins Lady Blue	Sire: Harpers Half Pants	Sire: Harpers Joker
			Dam: Harpers Blue Austina
		Dam: Harpers Pepino	Sire: Harpers Old Smokey
			Dam: Harpers Jinx
Dam: Trixie Bear of J&J	Sire: Schmeigs Blue Bear	Sire: Deines Spark	Sire: Hartnagles Hud
			Dam: Townsends Miss Dixie
		Dam: Deines Nuisance	Sire: Halls Smokey Blue
			Dam: Deines Lucky Lady
	Dam: Schmeigs Sugar Bear	Sire: Gantenbeins Mike	Sire: Desserichs Ring
			Dam: Ganteubeins Blue Muffett
		Dam: Etans Joker	Sire: Oxbrow Freckles
			Dam: Davis Blue Bambi

Bred by: ______ Address: ______ Phone: ______

The details given above are correct to the best of my knowledge. Signed: ______(Breeder)

easily, then moved on quickly to Ruth's new love—working livestock. Magic was a versatile working dog, showing good power on cattle and sheep with quiet control of ducks. Her talents led her to win both the 1981 and 1982 CASA All Around Dog Awards!

By the time Magic was old enough to breed, the Cathy Jones friendship was well cemented and the two began to experiment with crossing their lines. Since Ruth had shown and finished Ch Ruff N Reddy Mity Muhammad CD STDds for Cathy and Patty, falling completely in love with him in the process, she chose him to sire Magic's first litter. It was inevitable Magic would be bred to Ted and, in fact, the cross worked so well, Ruth did it 3 times! Ch Ruff N Reddy Pee Dee Que was the most well known puppy from this cross.

Ch Hemi's Mountain Magic of J & J CDX STDdsc, blue, copper and white female by Hemi x Trixie Bear of J & J. Courtesy of Ruth McGuire

Pee Dee Que was nearly a clone of her famous sire, not only in looks but in personality. She was bred to a number of males, one of which was Ted and one of Ruth's last dogs was a puppy from this cross named Jones Stockyard Hustler. A solid red male called "Tuba" by his friends, Ruth had moved to Alaska and Tuba became an important influence in the new kennel being formed by Carla Schroeder named North Star. Even in the Alaska wilderness Tuba earned duck and cattle titles. Ruth describes him as having a good style and being tough on cattle. Apparently livestock training of dogs was in it's early stage in Alaska and Ruth states many of the sheep in the area had been spoiled by being worked with weak dogs. Tuba quickly reformed the stock everyone was using for training! He gripped and had eye like Ted and, like Ted, especially used eye when working ducks.

Magic working with the quiet, well controlled style typical of Hemi. Making the second panel at an early trial. Courtesy of Ruth McGuire

Magic was not only a nice working dog but a good moving one as well. Courtesy of Ruth McGuire

Another Ted x Magic puppy was named Hemi's Ima Reddy Rascal and was sold to a breeder in Texas. When that breeder abandoned her dogs, she came to Melissa Couch, who had to catch the frightened dog before she could take her home. When she did get her home she found

Hemi's Ima Reddy Rascal, black and white female. (Ted x Ch Hemi's Mountain Magic of J & J CDX STDdsc), rescued from a bad situation with her good temperament and working instincts intact. Courtesy of Melissa Couch

Solid black Ted x Magic pup. Courtesy of Ruth McGuire

Hemi's Ima Reddy Rascal. Her kind face reflects intelligence and good temperament that survived intact in spite of a poor early start in life. Courtesy of Melissa Couch

she had a dog with a good, sweet temperament although she states Rascal was never a good watch dog! At the time Melissa and her husband were raising cattle and it was here Rascal discovered her many talents for working cows. When not supervised she could spend all day (and probably all night!) gathering cows and holding them in a corner. When she tired of that position she would easily move them up a fence to another corner and hold them there for awhile. She was less than a year of age and when Melissa learned of her new profession, she began confining her unless she could keep an eye on her.

Rascal was also an excellent chute dog. She would low heel or hit heads. The chutes at the Couch ranch were solid metal with a walkway alley used to vaccinate so nobody had to get into the chute. Rascal figured out entirely on her own how to go down the walkway, jump up and lean in on the cattle to move them. Her early life had evidently taught her to use her own mind very effectively in learning new tasks on her own and she had a strong desire to please!

After Rascal had lived with Melissa for a few years, Melissa became divorced and Rascal went to live with Patty Ennen. Patty had a son of her Mity Muhammad and bred Rascal to him but she was not able to carry the litter to whelping. Her second litter was by Clank, who also lived with Patty at this time and, again, Rascal was unable to carry the litter to whelping. It was a real loss to the breed this good little working dog was unable to reproduce and preserve her independent thinking and all around working ability. She lived out her life as a loved member of Patty's family and died around age 11 or 12.

A big winner from the Ted x Magic cross was a magnificent blue merle female named

Hemi's Ima Teddy Ten. Covered almost entirely with black and showing only small stripes of merling through her coat, she was blessed with beautiful conformation to go with her striking appearance. Ruth showed her to a big win in a conformation futurity and continued showing her to complete her breed Championship quickly and easily.

Another successful Jones/McGuire cross was that of Hemi to Ruff N Reddy Mity Happy Hocker CD STDd. From this cross came a blue female named Ruff N Reddy SeMe Sashay, known as "Bunny." Her daughter by Colorado's Toast of the Town, Ruff N Reddy SeMe Lick'em also went to Alaska with Ruth and also contributed greatly to the North Star breeding program.

Quest for Denali was Ruth's last Australian Shepherd. A solid black male with the tiniest amount of white, he was from the cross of Jones Stockyard Hustler x Ruff N Reddy Seme Lick'em, he had two crosses to Ted and two crosses to Hemi, keeping the legacy of Ruth's foundations alive. She remained active with Denali while in Alaska and he earned two 5 point majors at shows under breeder judges Sharon Rowe and Jeanne Hartnagle although Ruth never got around to finishing his Championship. Ruth has good memories of Denali, calling him a "very good dog" but he was to be just a companion as Ruth's competitive days came to a close.

Like Ted, Hemi was bred to a wide range of females, spreading his influence from coast to coast. Following is a partial list of many of his progeny and a Legacy Chart showing his continuing influence of many of his more important offspring. Legacy Charts of two of his most famous offspring, Magic and Hymie, can be found elsewhere in this chapter.

Hemi's Ima Teddy Ten, CASA Best in Futurity winner. Courtesy of Ruth McGuire

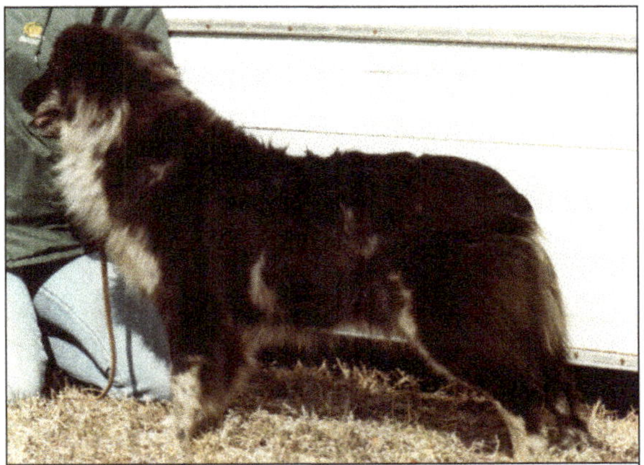

The striking blue merle female, Hemi's Ima Teddy Ten. Courtesy of Ruth McGuire

Left: North Star Reddy Teddy SeMeGo, solid red male (North Star Hemi's SeMe Savvy CD, STDds x North Star Precious Rose). Right: "Striker" blue and white male (J & J Strike Me Fancy x Ruff N Reddy SeMe Lick'em) Photo by Carla Schroeder

- Coberly's Gypsy Smoke (Hemi x Agnew Red Squaw)
- Bumblebee Hummdinger Hemi (Hemi x Annie Duchess of Windsor—Hemi's Regal Request x Depindet Kembla)
- Bumblebee Handsome Hemi Devil (Hemi x Annie Duchess of Windsor)
- Bumblebee Beautiful Beebe (Hemi x Annie Duchess of Windsor)
- Merledale's Copper Chauncey (Hemi x Deines Bobbi Bear)
- Adair's Black Bear (Hemi x Dotclijee Shunka)
- Hemis Mountain fortune (Hemi x Harmony of Good Fortune)
- Dizneyland Le Me Go Wicha (Hemi x Las Rocosa Flecka)
- Hemi's Spittin Image (Hemi x McGuire's Blu-Duster)
- Hemi's Tonk Saga Kahotek (Hemi x McGuire's Blu-Duster)
- Blu Ridge Fury (Hemi x McGuiredsBlu-Duster)
- Farmer Dusty Angel (Hemi x McGuire's Blu-Duster)
- Patton's Hemi Trash Man (Hemi x Pizs Sheba)
- Hemi's Encore of Windmill Hill (Hemi x Molly of Windmill Hill—Hemi's Regal Request x Windmill Hill Sweet Nell—Sorensen, J Bar D Rhynard)
- Mida's Touch of Hemi (Hemi x Molly of Windmill Hill)
- Semeshine of Windmill Hill (Hemi x Molly of Windmill Hill)
- Ruff N Reddy Remember Me (Hemi x Ruff N Reddy Mity Happy Hocker)
- Ruff N Reddy Legend of Mike (Hemi x Ruff N Reddy Mity Happy Hocker)
- Hemis Silver Drifter (Hemi x Ruff N Reddy Mity Happy Hocker)
- Regal Reflection of hazelwood (Hemi x Snows Regal Cactus Berry—Evans Regal Spice x Jarvis Matilda)
- Blue Minx of Hazelwood (Hemi x Snows Regal Cactus Berry)
- Silver Medley (Hemi x Snows Regal Cactus Berry)
- Merledales Zack (Hemi x Deines Bobbi Bear)
- Chiltons Tapestry of Hazelwood (Hemi x Snows Regal Cactus Berry)
- Hemi's Regal Request (Hemi x Snows Regal Cactus berry)
- Hemi's Double Trouble of Windsor (Hemi x McGuire's Blu-Duster)
- Deine's CJ (Hemi x McGuire's Blu-Duster)
- Hemi's Challenger of Blu-Ridge (Hemi x McGuire's Blu-Duster)

- Hemi's Dinky Demon of Bluecrest (Hemi x McGuire's Blu-Duster)
- Crystal Blues Sir Barrimore (Hemi x Gilson Kandee Koquett—McConkeys Tiger Britches x Gasaways Kotton Kandee)
- He Mi Go Jody (Hemi x Gilsons Kandee Koquett)
- Dyers Ironside of Hemi (Hemi x McGuire's Blu-Duster)
- Hemi's Regal Duke (Hemi x Fosters Dutchess—Evans Regal Spice x Jarvis Matilda)
- Hemi's Lagrima Negra De Chaparrel (Hemi x Fosters Dutchess)
- Bluecrest Elite Dolca De Hemi (Hemi x Hemis Lagrima Negra De Chaparrel)
- Bluecrest Hiwan Hemi (Hemi x Fosters Dutchess)
- Bluecrest Poco Bueno (Hemi x Fosters Dutchess)
- Bluecrest Fancy Bear of J & J (Hemi x Fosters Dutchess)
- Hemis Mountain Magic of J & J (Hemi x Trixie Bear of J & J)
- Bluecrest Seme Fox Ya (Hemi x Deines Bobbi Bear—Gantenbeins Mike granddaughter)
- Hemi's Atlas (Hemi x Pizs Sheba)
- Hemi's Mountain Harmony (Hemi x Harmony of Good Fortune)
- Hot Smokin Hemi (Hemi x Gajewski & Binders Harmony)
- Harmonys Black Gold (Hemi x Harmony of Good Fortune)
- Bumblebees Humm'n Hemi Honey (Hemi x Annie Duchess of Windsor)
- Dizneylands Spinner (Hemi x Las Rocosa Flecka)
- Hemi's Delight of Windmill Hill (Hemi x Molly of Windmill Hill)
- Ruff N Reddy Seme Sashay (Hemi x Ruff N Reddy Mity Happy Hocker)

Legacy Chart for Hemis Mountain Magic of J&J

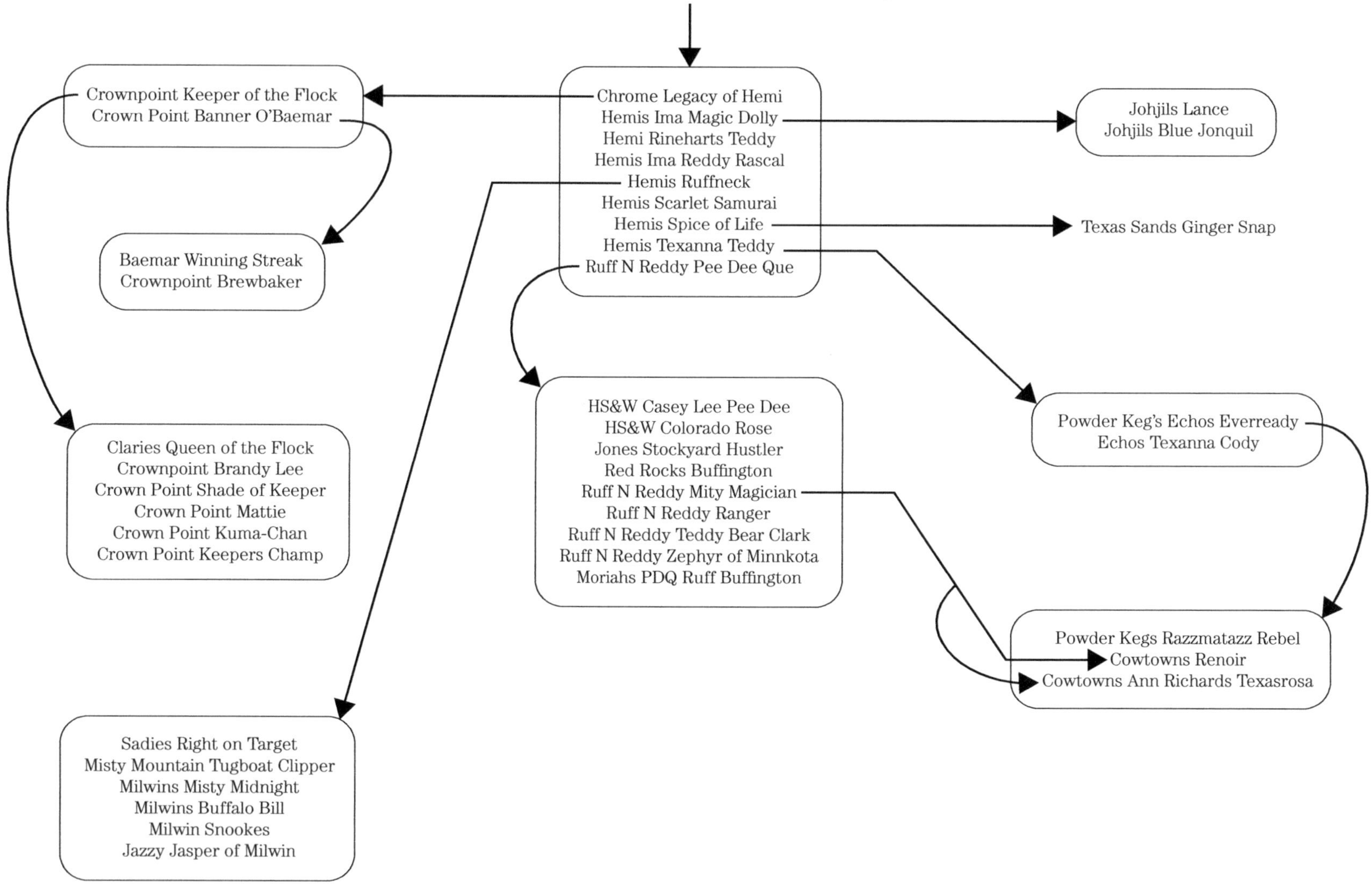

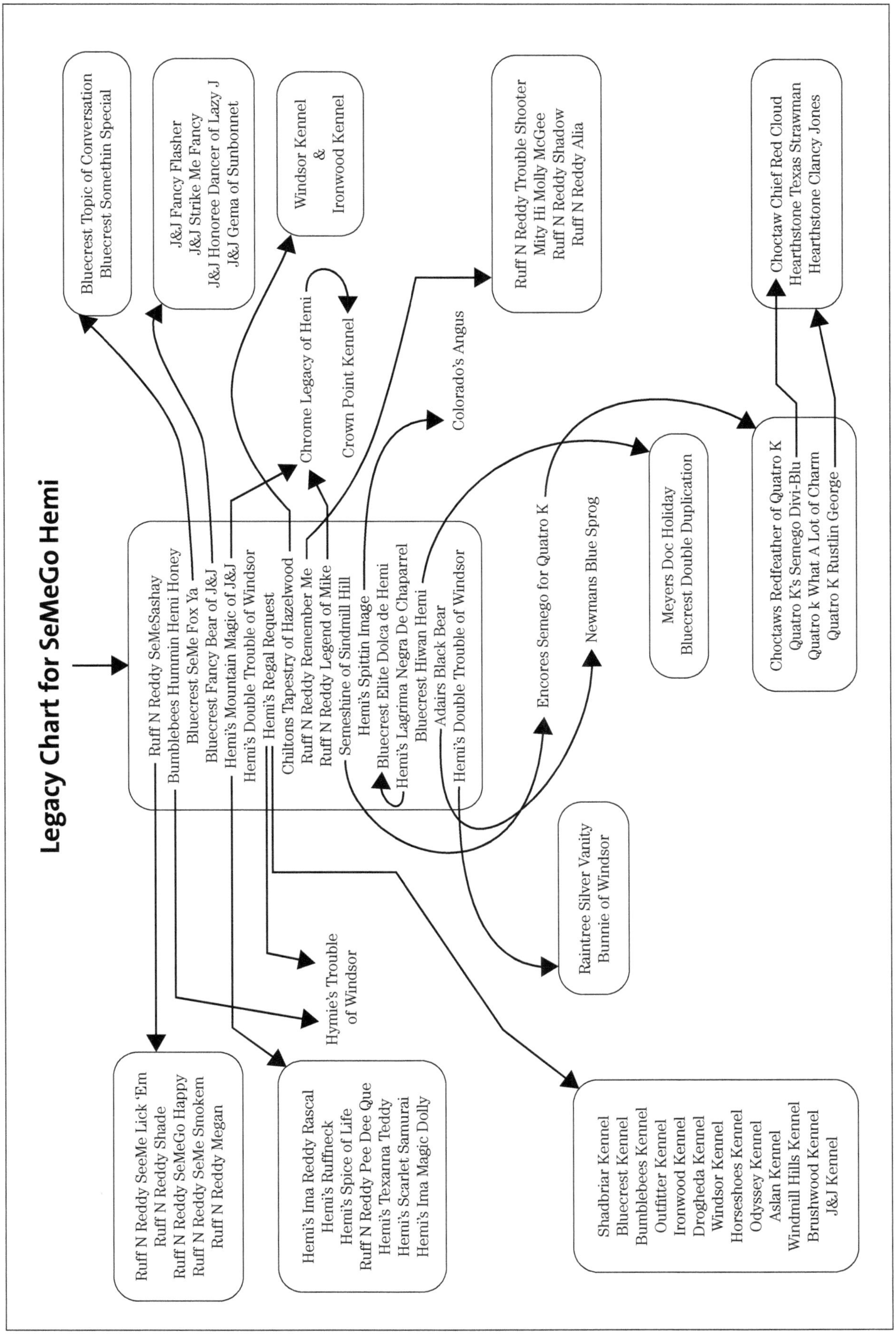
Legacy Chart for SeMeGo Hemi
Ruff N Reddy SeMeSashay
Bumblebees Hummin Hemi Honey
Bluecrest SeMe Fox Ya
Bluecrest Fancy Bear of J&J
Hemi's Mountain Magic of J&J
Hemi's Double Trouble of Windsor
Hemi's Regal Request
Chiltons Tapestry of Hazelwood
Ruff N Reddy Remember Me
Ruff N Reddy Legend of Mike
Semeshine of Sindmill Hill
Hemi's Spittin Image
Bluecrest Elite Dolca de Hemi
Hemi's Lagrima Negra De Chaparrel
Bluecrest Hiwan Hemi
Adairs Black Bear
Hemi's Double Trouble of Windsor
Bluecrest Topic of Conversation
Bluecrest Somethin Special
J&J Fancy Flasher
J&J Strike Me Fancy
J&J Honoree Dancer of Lazy J
J&J Gema of Sunbonnet
Windsor Kennel
&
Ironwood Kennel
Chrome Legacy of Hemi
Crown Point Kennel
Ruff N Reddy Trouble Shooter
Mity Hi Molly McGee
Ruff N Reddy Shadow
Ruff N Reddy Alia
Colorado's Angus
Choctaw Chief Red Cloud
Hearthstone Texas Strawman
Hearthstone Clancy Jones
Encores Semego for Quatro K
Newmans Blue Sprog
Meyers Doc Holiday
Bluecrest Double Duplication
Choctaws Redfeather of Quatro K
Quatro K's Semego Divi-Blu
Quatro k What A Lot of Charm
Quatro K Rustlin George
Raintree Silver Vanity
Bunnie of Windsor
Hymie's Trouble
of Windsor
Hymie's Trouble
of Windsor
Ruff N Reddy SeeMe Lick 'Em
Ruff N Reddy Shade
Ruff N Reddy SeMeGo Happy
Ruff N Reddy SeMe Smokem
Ruff N Reddy Megan
Hemi's Ima Reddy Rascal
Hemi's Ruffneck
Hemi's Spice of Life
Ruff N Reddy Pee Dee Que
Hemi's Texanna Teddy
Hemi's Scarlet Samurai
Hemi's Ima Magic Dolly
Shadbriar Kennel
Bluecrest Kennel
Bumblebees Kennel
Outfitter Kennel
Ironwood Kennel
Drogheda Kennel
Windsor Kennel
Horseshoes Kennel
Odyssey Kennel
Aslan Kennel
Windmill Hills Kennel
Brushwood Kennel
J&J Kennel

Ch Ruff N Reddy Mity Muhammad CD STDds x Ch Hemi's Mountain Magic of J & J CDX STDcsd puppy. Solid black with a small patch of white on the chest. Courtesy of Ruth McGuire

Kintyre Silver Shadow, blue merle Jeb Stuart puppy. September 1979. Courtesy of Sorensen

Red merle male by Ruff N Reddy Mity Magician x Powder Keg's Echo's Eveready STDdsc. Courtesy of Melissa Couch

Red merle male by Ruff N Reddy Mity Magician x Powder Keg's Echo's Eveready STDdsc. Courtesy of Melissa Couch

Red female by Ruff N Reddy Mity Magician x Powder Keg's Echo's Eveready STDdsc. Courtesy of Melissa Couch

Ruff N Reddy Mity Mean. Blue merle male who has turned nearly black in old age. Courtesy of Melissa Couch

Ruff N Reddy Mity Mean. Courtesy of Melissa Couch

Ruff N Reddy Mity Magician. Courtesy of Melissa Couch

Ruff N Reddy Mity Magician, dark red merle male (Ruff N Reddy Mity Mean x Ch Ruff N Reddy Pee Dee Que) Courtesy of Melissa Couch

Ruff N Reddy Mity Magician, red merle and white male (Ruff N Reddy Mity Mean x Ch Ruff N Reddy Pee Dee Que) Courtesy of Melissa Couch

Ruff N Reddy Mity Magician. Courtesy of Melissa Couch

Powder Keg's Echo's Eveready STDdsc, blue, copper, and white with blue eyes. (Tindol's Echo x Hemi's Texanna Teddy—Ted x Magic) Courtesy of Melissa Couch

6 month old red and white male by Ruff N Reddy Mity Magician x Powder Kegs Echo's Eveready STDdsc Courtesy of Melissa Couch

6 month old red and white male by Ruff N Reddy Mity Magician x Powder Kegs Echo's Eveready STDdsc Courtesy of Melissa Couch

6 month old red and white male by Ruff N Reddy Mity Magician x Powder Kegs Echo's Eveready STDdsc Courtesy of Melissa Couch

6 month old red and white male by Ruff N Reddy Mity Magician x Powder Kegs Echo's Eveready STDdsc Courtesy of Melissa Couch

Red and white female by Ruff N Reddy Mity Magician x Powder Keg's Echo's Eveready STDdsc. Courtesy of Melissa Couch

Red and white female by Ruff N Reddy Mity Magician x Powder Keg's Echo's Eveready STDdsc. Courtesy of Melissa Couch

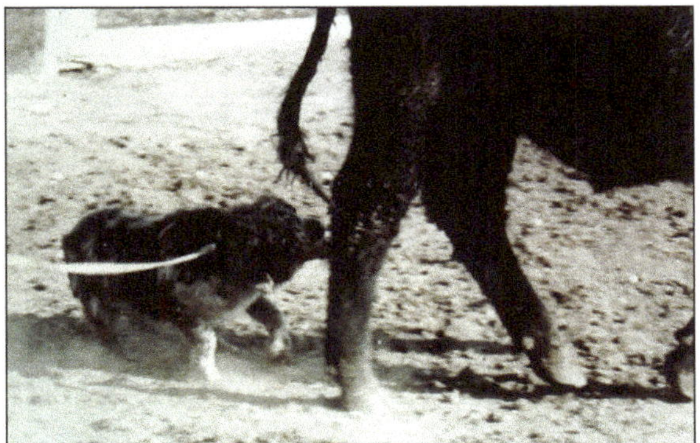

Ruff N Reddy Shadey Joe as a young dog learning the ropes on cattle. Courtesy of Mark Lane

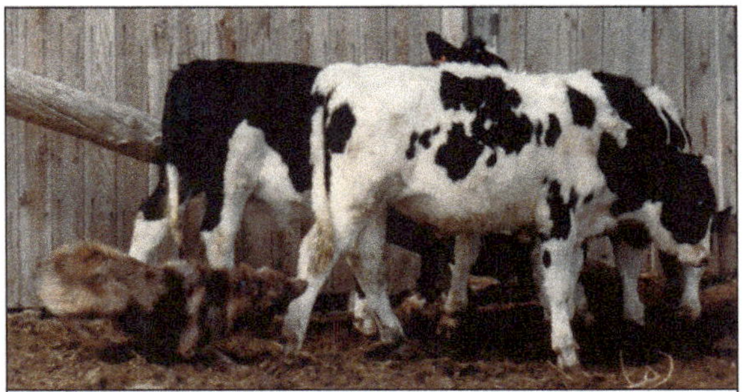

Ruff N Reddy Shadey Joe "Clank" Red merle male (Ted x Shadey Mountain Mity Miss) A good working dog showing the style Ted was famous for producing. Courtesy of Raymie Jones

Ruff N Reddy Shadey Joe as a young dog—under one year of age. Courtesy of Melissa Couch

Ruff N Reddy Shadey Joe "Clank" as a handsome youngster. Courtesy of Melissa Couch

WTCH Oliver's Sam Hall CD (left) and Oliver's Hobo Bill (right), double bred Ted males, both red merles. (Ruff N Reddy Shadey Joe x WTCH Slash V Slide Me Sweet CD) Courtesy of Raymie Jones

Magic turns a stray with confidence and authority. Courtesy of Ruth McGuire

A young Ch Hemi's Regal Request STD, blue and white male by Hemi x Ch Snows Regal Cactus Berry CD. Courtesy of Ruth McGuire

Ruff N Reddy Shadey Joe. Courtesy of Melissa Couch

Ruff N Reddy Shadey Joe—an intense, intelligent expression was a hallmark of the Ruff N Reddy dogs. Courtesy of Melissa Couch

An early lesson on cattle for a young Magic. Courtesy of Ruth McGuire

Ch Ruff N Reddy Blue Tinker Toy at age 12. Courtesy of Melissa Couch

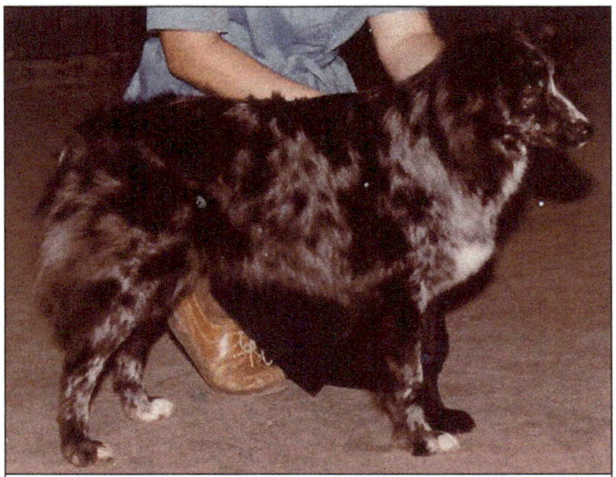

Ch Ruff N Reddy Blue Tinker Toy. Photo by Sue Noll. Courtesy of Raymie Jones

Gahl's Terra, blue and white female who was invaluable to Joe Gahl on his hog farm (Ruff N Reddy Shade x Tina Horejsi. Shade was a son of Ruff N Reddy Shadey Joe x Ruff N Reddy SeMe Sashay. Tina Horejsi was old Sorensen Colorado bloodlines) Bred to Newman's Blue Sprog CD STDsc she produced a single female (via fresh shipped semen) named Gahl's Casey. Photo courtesy of Casey Masterson Newman

Clank as a young dog. Beautifully marked, conformationally correct in every way with excellent head type and working ability. Courtesy of Mark Lane

Ruff N Reddy Shadey Joe "Clank" as an older dog. Red merles darken with age just like blues and he has lost a lot of the contrast in his coat he had as a young dog. Courtesy of Melissa Couch

Ruff N Reddy Shadey Joe demonstrating the working style that was his birthright. Courtesy of Melissa Couch

Jones' Little Redwood, red and white male (Ruff N Reddy Shadey Joe x Ch Ruff N Reddy Blue Tinker Toy) as a young dog. He is out of coat in this photo but his type and good moderate conformation are clearly evident. Courtesy of Melissa Couch

Jones' Little Redwood, a young dog with the good conformation to mature out nicely. Courtesy of Melissa Couch

Jones' Little Redwood in 1990. Matured and in good coat—a lesson in caution about judging young dogs too harshly. Photo by Patty Ennen. Courtesy of Melissa Couch

Jones' Little Redwood. Courtesy of Melissa Couch

Jones' Little Redwood (left) with his dam Ch Ruff N Reddy Blue Tinker Toy. Courtesy of Melissa Couch

Ruff N Reddy Shadey Wood Ash, dark liver red merle female (Jones' Little Redwood x Shadey Mountain Mity Miss) at about 1 year of age. Courtesy of Raymie Jones

Ruff N Reddy Shadey Wood Ash as an older dog. Her Wood breeding shouts from her beautiful type head. Courtesy of Pincie Creek

Ruff N Reddy Shades of Ted, red merle and white female (Jones' Little Redwood x Shadey Mountain Mity Miss). There is no mistaking her heritage with her dark liver color and Ted type head. A high quality female of good type. Courtesy of Melissa Couch

Ch Ruff N Reddy Ross STDcsd, blue merle male. (Colorado's Toast of the Town x Ruff N Reddy Boots). An outcross male of show lines and old Ruff N Reddy Wood-based lines. Boots was linebred Reddy Teddy and Shady Mountain Mity Miss and the old type came through strongly in this beautiful male. Pictured here as a young dog just beginning his show career. Courtesy of Sorensen

Ruff N Reddy Mity Miss Shadey , blue and white female. (Jones' Little Redwood x Shadey Mountain Mity Miss) Courtesy of Melissa Couch

Ruff N Reddy Mity Miss Shadey. Courtesy of Melissa Couch

Blue merle and white female from Cowtown Mity Rowdy x Ruff N Reddy Mity Happy Hocker CD STDd. Courtesy of Melissa Couch

Ruff N Reddy Mity Miss Shadey. Courtesy of Melissa Couch

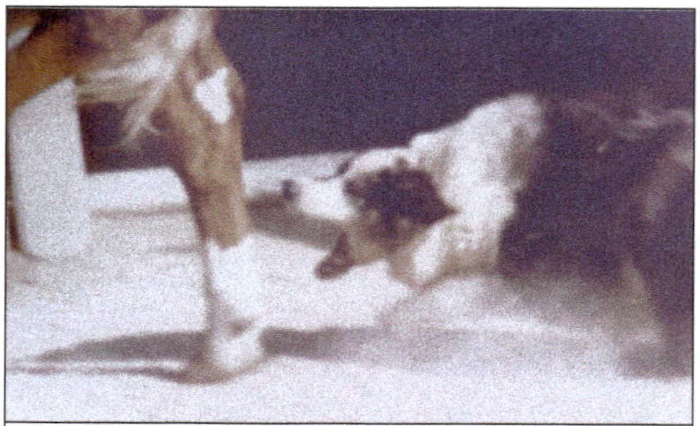

Moreland's Old Charter, blue and white male (Penns Jim x Penns Tuffy). The Penn dogs were all directly acquired from Jay Sisler so this dog was 100% Sisler bred. Photo by Cookie Moreland. Courtesy of Melissa Couch

Moreland's Old Charter, Sisler bred sire of Cowtown Mity Rowdy. Photo by Cookie Moreland. Courtesy of Melissa Couch

Jones' Stockyard Hustler "Tuba", red and white male (Ted x Ch Ruff N Reddy Pee Dee Que). A great influence in Alaska, both as a breeding dog and a working dog. Ruth McGuire owned this dog and won two majors with him before she stopped showing. He also was a strong working dog who retrained the spoiled sheep being used by trialers in Alaska! Courtesy of Melissa Couch

Cowtown Mity Rowdy, black and white male (Moreland's Old Charter x Ruff N Reddy Mity Happy Hocker CD STDd), an incredible dog with close up Sisler bloodlines through his sire and one of the best Ruff N Reddy females in his dam. Rowdy took the "one man dog" trait to extremes and nearly died of starvation once when he had to be boarded away from Melissa for a period of time. Courtesy of Melissa Couch

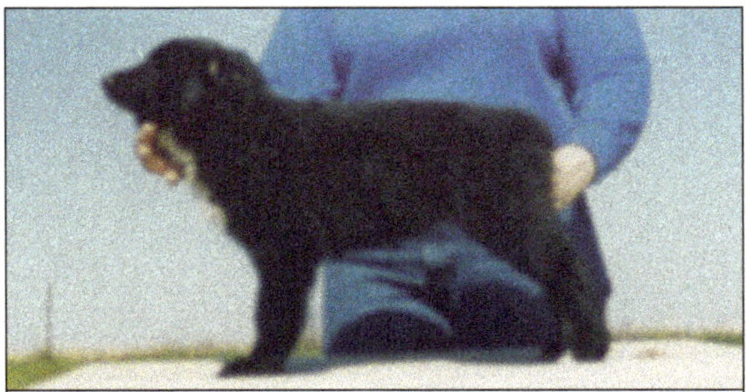

Black and white puppy from the cross of Cowtown Mity Rowdy x Ruff N Reddy Mity Happy Hocker CD STDd. Courtesy of Melissa Couch

Ch Ruff N Reddy Ross STDcsd was that rare individual that was a winner in both the show and working arenas. Shown here going Best of Winners on July 19, 1992 he also won High in Trial Herding at the USASA National Specialty. Courtesy of Sorensen

Blue merle and white female from Cowtown Mity Rowdy x Ruff N Reddy Mity Happy Hocker CD STDd. Courtesy of Melissa Couch

Born: ____________ Sex: ________
Owned by: ____________________

PEDIGREE OF Colorados Toast of the Town

Parents	Grandparents	Great-grandparents	Great-great-grandparents
Sire: Colorados Talk of the Town	Sire: Colorados Clint	Sire: Dorns Poudre Lakohta	Sire: Iacorettas Buck
			Dam: Dressel's Duchess
		Dam: Colorados Perky Miss	Sire: Sorensens Gunsmoke
			Dam: Nichols Little Bonnie
	Dam: Posey Patch of Blue Mist	Sire: Fieldmaster of Flintridge	Sire: Herdsman of Flintridge
			Dam: Savor of Flintridge
		Dam: Overbeys Heather of Bonnie-Blu	Sire: Wildhagens Dutchman of Flintridge
			Dam: Thistle of Flintridge
Dam: Colorados Christmas Posey	Sire: Slash V Wild Timber	Sire: Martins Tim Tim	Sire: Taylor's Whiskey
			Dam: Taylors Buena
		Dam: Hoyts Wildwood Flower	Sire: Georges Red Rustler
			Dam: Quaglino's Miss Pooh
	Dam: Posey Patch of Blue Mist	Sire: Fieldmaster of Flintridge	Sire: Herdsman of Flintridge
			Dam: Savor of Flintridge
		Dam: Overbeys Heather of Bonniw-Blue	Sire: Wildhagens Dutchman of Flintridge
			Dam: Thistle of Flintridge

Bred by: ____________ Address: ____________ Phone: ____________

The details given above are correct to the best of my knowledge. Signed: ____________(Breeder)

Ruff N Reddy Shadow, black tri female (Colorado's Toast of the Town x Ruff N Reddy Remember Me—Hemi x Ruff N Reddy Mity Happy Hocker CD STDd) Courtesy of Casey Newman

Ted and Cathy. Courtesy of Melissa Couch

Ruff N Reddy Shadow. Courtesy of Casey Newman

The Ted granddaughter, Couch's Lacy Mae CD (Briarbrooks Melbourne—Ted x Bar Lazy K's Diamond Lil of Las Rocosa—out of Windhavens Brisbane) Black tri (1978) at approximately 8 years of age. Courtesy of Melissa Couch

One of the finest groups of Australian Shepherds ever assembled. L to R: Jeffreys' Shade, Jones' Mity Bite, Ch Ruff N Reddy Mity Muhammad CD STDds, Piz's Sheba, Ch Jones' Reddy Teddy CDX, OTDdsc SchHA, Davies' Blue Kola, Clare Olson's Blue Joe, Ch Ruff N Reddy Blue Tinker Toy. Courtesy of Sorensen

Couch's Lacy Mae CD showing her good Aussie head type. Courtesy of Melissa Couch

Couch's Belle Starr CD—full sister to Couch's Lacy Mae CD. Dark self blue merle female. Cowtown Mity Moses, Black male (Ch Ruff N Reddy Mity Mohammad CD STDds x Couch's Belle Star CD) Courtesy of Melissa Couch

Couch's Lacy Mae CD, a beautiful, typical, excellent type of Aussie. Courtesy of Melissa Couch

Bar Lazy K's Diamond Lil of Las Rocosa, red merle female bred to Ted and granddam of Couch's Lacy Mae CD and Couch's Belle Starr CD. Courtesy of Melissa Couch

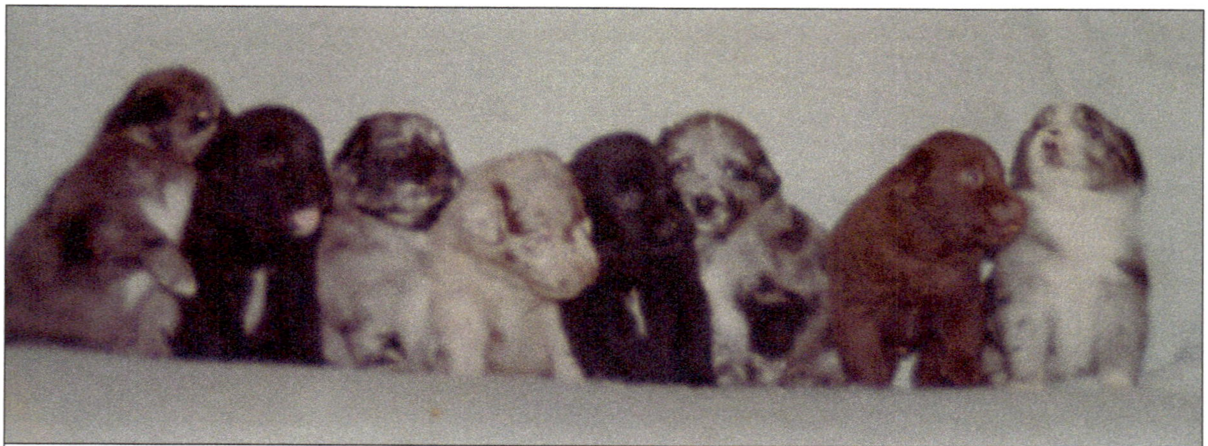

Litter by Ch Ruff N Reddy Mity Muhammad CD STDds x Couch's Belle Starr CD. All four colors were produced from this cross. Courtesy of Melissa Couch

Bar Lazy K's Diamond Lil of Las Rocosa, red merle and white female. Courtesy of Melissa Couch

Unknown black and white dog of Cathy's breeding showing the good type and quality the Ruff N Reddy dogs were known for. Courtesy of Raymie Jones

Unknown red merle dog of Wood breeding working cattle in typical style of the line. Courtesy of Mark Lane

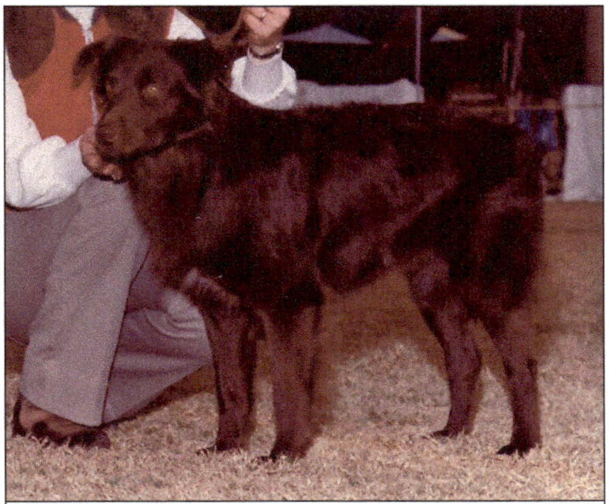

Ted winning as a handsome young dog. Courtesy of Raymie Jones

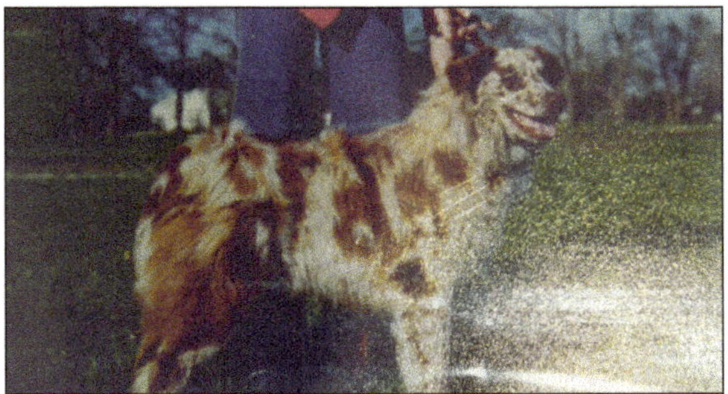

Briarbrooks Diamond Jim, red merle and white male, littermate to Briarbrooks Melbourne (Ted x Bar Lazy K's Diamond Lil of Las Rocosa] Courtesy of Melissa Couch

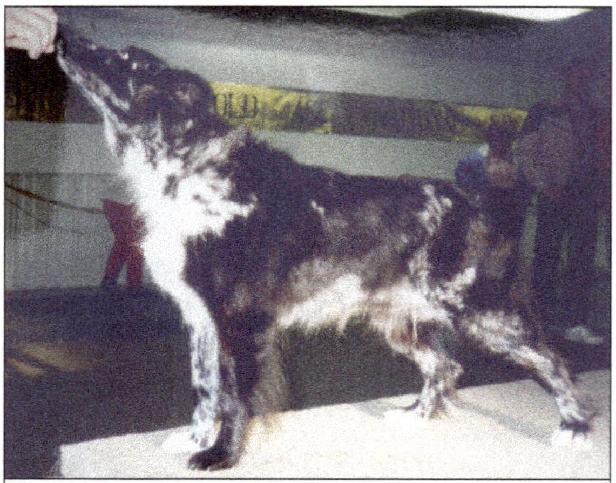

Ch Ruff N Reddy Blue Tinker Toy being shown at the 1990 National Specialty in the Parade of Titleholders. She was shown as a special tribute to Cathy. Courtesy of Melissa Couch

Ted was a beautifully correct dog. Every now and then Cathy allowed him to be photographed to show this! Courtesy of Ruth McGuire

It takes a mature, fit, and experienced dog to work cattle. This exciting photo illustrates the physical ability needed for this tough work. Ted showing his style in another popular photo used often by Cathy in her advertising. Courtesy of Raymie Jones

Unknown dog of Ruff N Reddy type—possibly a Ted son. Very minimally marked red merle. Athletic, correct and showing the head type typical of Ted. Cathy Jones photo courtesy of Mark Lane

Ch Jones Reddy Teddy CDX OTD SchHA and Ch Ruff N Reddy Blue Tinker Toy. Cathy knew what winners should look like. Courtesy of Raymie Jones

Gantenbein's Blue Muffett, blue, copper and white female (Thornton's Blue Smokey x Gantenbeins Sugar Blue). Dam of Gantenbeins Mike. Courtesy of Ruth McGuire

Hemi's sire, Gantenbeins Mike, black and white male (Desserichs Ring x Gantenbeins Blue Muffett). Courtesy of Ruth McGuire

Gantenbein's Blue Muffett, dam of Gantenbeins Mike. Courtesy of Ruth McGuire

Gantenbeins Mike at age 9 months. Courtesy of Ruth McGuire

Gantenbein's Blue Muffett, dam of Gantenbeins Mike. Courtesy of Ruth McGuire

Gantenbeins Mike at age 10 years. Courtesy of Ruth McGuire

Hemi's dam, Gantenbein's Lady Blue, blue and white female. (Harper's Half Pants x Harper's Pepino) Lady was a donated as a puppy by Janeane Harper to the ASC of Colorado promotional booth to be raffled off as a draw to the booth. The winners of the puppy lived in town and didn't want her, so in a fateful move, Eleanor Gantenbein purchased her from them, and, as they say, the rest is history. Courtesy of Ruth McGuire

The strikingly marked blue and white male Ch Merledales Zack CDX (Hemi x Ch Deines Bobbi Bear CD) owned by Corky Shaha. Courtesy of Ruth McGuire

Gantenbeins Lady Blue at age 2 years. Courtesy of Ruth McGuire

Gantenbins Lady Blue at age 6 years. Courtesy of Ruth McGuire

Ruth's pick puppy and next versatile Aussie, Ch Hemis Mountain Magic of J & J CDX STDdsc. Blue merle, copper and white female (Hemi x Trixie Bear of J & J) Courtesy of Ruth McGuire

Hemi winning First Place in Open Obedience at a Napa, California, show in 1970. Courtesy of Ruth McGuire

Hemi competing in Started Sheep at the 1972 Arizona Specialty, handled by Chic Acierto. Courtesy of Ruth McGuire

A consistent winner like her sire Hemi, Magic is shown here winning Best of Breed. Courtesy of Ruth McGuire

Magic hurries to get in position to control her sheep. Courtesy of Ruth McGuire

Hemi, Hymie and an unknown Hemi son winning the Get of Sire class at the Arizona Specialty 1972. Courtesy of Ruth McGuire

Hemi triumphing at a 1970 IASA show in Napa, California. First place in the Open Obedience class competing with some of the top dogs of the day. Courtesy of Ruth McGuire

Hemi's intelligence and good breed type shows clearly in this lovely head shot. Courtesy of Ruth McGuire

One of the most well known photos of Hemi. A beautiful dog from any angle. Courtesy of Ruth McGuire

Hemi continues his winning ways. Photo by Sue Noll. Courtesy of Ruth McGuire

Hemi carefully brings in the rogue duck! Courtesy of Ruth McGuire

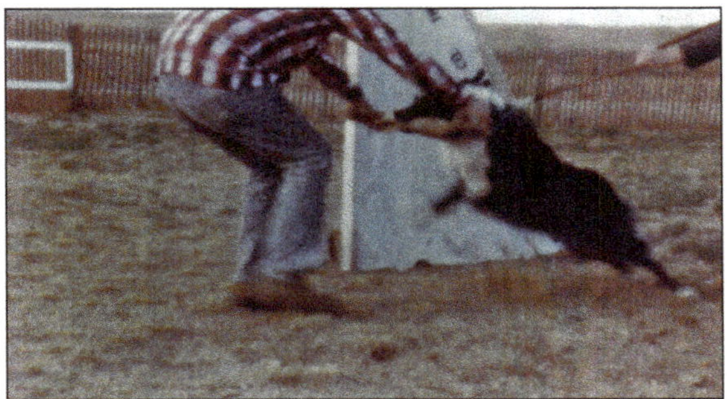

When Hemi was 8 or 9 years of age he tried Schutzhund for the first and last time. Ruth's comment, "He liked it way too much." Courtesy of Ruth McGuire

Happy 13th Birthday Hemi! Cathy Jones donated steak for the birthday boy, everyone sang Happy Birthday, then a quick trip to Cathy's place so Hemi could work cows one last time. Courtesy of Ruth McGuire

Hemi enjoying his 13th birthday party! Courtesy of Ruth McGuire

Hemi, still sound and beautiful at 13 years of age. Courtesy of Ruth McGuire

"Jemima" and "Snapper," black tri puppies by Hemi x Molly of Windmill Hill CD. Courtesy of Ruth McGuire

Magic demonstrating a big reason she was a big winner in the show ring. Courtesy of Ruth McGuire

A very promising youngster who fully lived up to her heritage. Courtesy of Ruth McGuire

A young Magic showing the quality which would make her a big winner throughout her career. Courtesy of Ruth McGuire

Ch Hemis Regal Request STD "Hymie" at the ASC of Utah trial earning his first leg on Started Sheep. Courtesy of Ruth McGuire

Hemi winning Best of Breed and Ch Deines Bobbi Bear winning Best Opposite Sex in 1974. Courtesy of Ruth McGuire

"Mack," blue, copper and white puppy by Hemi x Molly of Windmill Hill CD. Courtesy of Ruth McGuire

"Shine," blue, copper, white puppy by Hemi x Molly of Windmill Hill CD. Courtesy of Ruth McGuire

The unusually marked blue merle and white dog, Bluecrest Hiwan Hemi (Hemi x Fosters Duchess) Courtesy of Ruth McGuire

The Hemi son "Lancer" out of Fosters Duchess. Courtesty of Ruth McGuire

Hemi and Ch Deines Bobbi Bear the day Bobbi finished her Championship. Courtesy of Ruth McGuire

The Hemi/Woods cross male, Hemis Blue Dandy, blue, copper and white (Hemi x Ch McGuire's Blu-Duster CD) Courtesy of Ruth McGuire

Two blue, copper and white puppies from the litter by Hemi out of Fosters Duchess. Courtesy of Ruth McGuire

Ruth and Magic prepare to start another stock run. Courtesy of Ruth McGuire

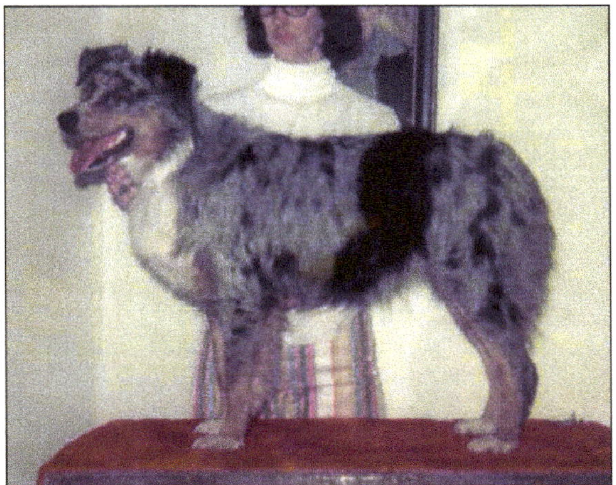

Mohr's Hemis Spittin Image, blue merle, copper and white female (Ch McGuire's SeMeGo Hemi UDT, STDds OTDc x Ch McGuire's Blu-Duster CD). Courtesy of Sorensen

A young Magic learns how to heel cattle. Courtesy of Ruth McGuire

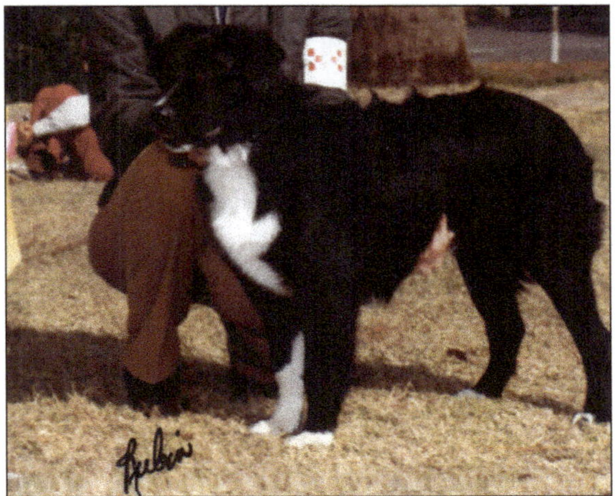

Another impressive Hemi/Wood cross female, Hemis Double Trouble, black and white (Hemi x Ch McGuire's Blu-Duster CD). Courtesy of Ruth McGuire

Looking a lot like his grandsire Gantenbeins Mike, Hemi's Atlas, black and white male is a wonderful example of the Hemi/Wood cross that worked so well. Courtesy of Ruth McGuire

Another win for Hemi—Arizona Specialty Get of Sire Class. From left to right: Hemi, Ch Hemi's Regal Request STD, Ch Chilton's Tapestry of Hazelwood, unknown dog, Hemis Double Trouble and unknown dog. Courtesy of Ruth McGuire

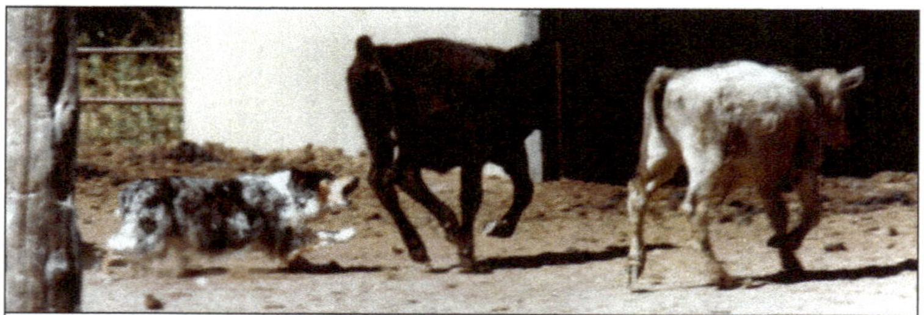

Magic could work those cows, too! Courtesy of Ruth McGuire

The 1982 All-Around Dog, Ch Hemi's Mountain Magic of J & J CDX, STDcsd. Courtesy of Ruth McGuire

Like her sire, Hemi, Magic excelled at obedience competition. Showing here in Open with Ruth. Courtesy of Ruth McGuire

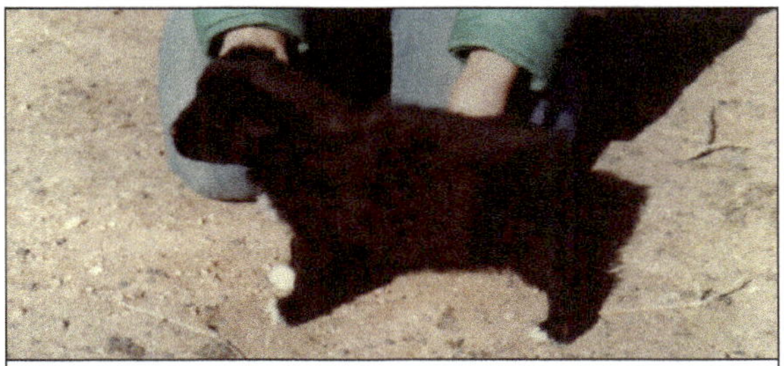

Ch Ruff N Reddy Pee Dee Que, red and white female (Ted x Magic) as a tiny puppy. Courtesy of Ruth McGuire

North Star Charming Hustler, black and white female born August 12, 1992 (Jones Stockyard Hustler x Odysseys Vanity Fair—a double Hemi bred female). Bred by Carla Schroeder. Photo by Carla Schroeder

North Star Midnite Hustler, solid black male born March 26, 1992 (Jones Stockyard Hustler x Myalaska KaChena—by Slash V Easy Jet) Bred by Carla Schroeder. Photo by Carla Schroeder

Hemis Ima Reddy Rascal, a substantial, well made female. Courtesy of Melissa Couch

Ch Hemi's Mountain Magic of J & J CDX STDdsc, a youngster's early attempts at working ducks. Courtesy of Ruth McGuire

Magic working ducks—a little quickly but not roughly. Courtesy of Ruth McGuire

Balance, control and power, Magic combined all the best traits of a versatile stock dog. Courtesy of Ruth McGuire

Magic brings home the ducks at a Colorado trial. Courtesy of Ruth McGuire

Approaching a tough situation in training with a good attitude. Magic pulling some big Suffolks out of a tight corner. Courtesy of Ruth McGuire

Magic earning a leg in cattle with Cathy Jones handling. Courtesy of Ruth McGuire

Ruth's last Aussie, the double bred Ted and double bred Hemi black and white male, Quest for Denali (Jones Stockyard Hustler x Ruff N Reddy SeMe Lickem). Courtesy of Ruth McGuire

North Star SeMe Ruby Rose, red merle and white female, born September 7, 1991—photo as a 2 year old (North Star Hemi's SeMe Savvy x Myalaska Sunshine Freckles—Ruff N Reddy SeMe Smokum x Myalaska KaChena. KaChena was by the Ted son Slash V Easy Jet) Bred by Carla Schroeder. Photo by Carla Schroeder

North Star Hemi's SeMe Savvy CD, STDds, blue merle and white male born July 19, 1990—photo as a 3 year old. (Jones Stockyard Hustler x Ruff N Reddy SeMe Lick'em) Bred by Carla Schroeder. Photo by Carla Schroeder

North Star Precious Rose, red merle and white female, born July 29, 1988 (Oliver's Sunshine Dusty x North Star Dark Rose. Dusty was by Ruff N Reddy Shadey Joe x Slash V Slide Me Sweet. Dark Rose was Ted and Hemi linebred) Bred by Carla Schroeder. Photo by Carla Schroeder

North Star Reddy Teddy SeMeGo, solid red male, born May 3, 1994 (North Star Hemi's SeMe Savvy x North Star Precious Rose) Bred by Carla Schroeder. Photo by Carla Schroeder

The double bred Ted/Hemi female "Ellie" (Jones Stockyard Hustler x Ruff N Reddy SeMe Lickem) Courtesy of Ruth McGuire

"Ellie" showing the quality of the Ruff N Reddy line holding true through the generations. Courtesy of Ruth McGuire

Quest for Denali at home in Alaska. Courtesy of Ruth McGuire

chapter 18

Casey Masterson Newman

HS & W

A horse crazy teenager grew into the founder of a kennel of old mountain-type Australian Shepherds through a love of the type and a life spent in the heart of their development. After a childhood of travel as a Navy family, Casey Masterson Newman's parents decided to settle in Boulder, Colorado upon her father's retirement. Their timing was such that she grew up in the hotbed of Colorado Aussie activity in the mid 1960s. In 1965 she was finally able to buy a horse and was boarding her Appaloosa gelding at a stable near Boulder. In that day everywhere you saw horses you saw Australian Shepherds—barns, rodeos, horse shows, literally anywhere horses were found. The boarding stable owner, Dick Ayres, had a dark blue female with no white named Mitch who came to work with him every day. Casey describes her as a moderately sized dog, typical of the Wood type of dogs. Mitch shadowed Dick all day long, politely refrained from bothering anyone who didn't bother her first and ran behind the horses. Casey remembers these dogs as being very "one man" type dogs used to guard stalls, tack and trucks. You didn't try to pet them without asking and even then they would tolerate you only so long as their owners insisted on it. Casey was smitten by the intelligence and good sense of Mitch who strongly reminded her of a bob tailed version of Jimmy Brown—her Dad's favorite dog while growing up on her grandfather's Montana ranch. In later years, Casey discovered Mitch actually came from Fletcher Wood. Her fate was set from the beginning to be forever tied to the Wood dogs.

Even after Casey was grown, married and had a family, horses and dogs remained an important and FUN part of her life. Left to right: Casey, her kid's horse Cookie, kids Kate and Jess, Sprog hitching a ride on the rumble seat and Blaze bringing up the rear. Courtesy of Casey Masterson Newman

She began noticing more of these dogs around town but they were not the same as Mitch and did not appeal to her like this sound, quiet, reserved, polite self merle. She had never been able to have a dog before because military families have to be ready to make big moves at a moments notice and dogs could not be part of the equation. But once settled in Boulder she went to school with kids who had Aussies—Joan Beien, Jeanne Hartnagle and others—and soon became acquainted with them and the dogs they had. Although she was quickly getting a priceless education on the early Colorado Aussie she never saw another dog that appealed to her like Mitch. She says, "Some were crazy, some didn't have bottom but the ones I liked were just plain dogs, steady, sound and tough. They wouldn't bother you if you didn't bother them or what was theirs. Most of them were named Blue, King, Stubby, Freckles. Just a blue dog. Everyone had one laying in front of their horse trailer. 'Don't pet him. He smiles when he bites.' That was pretty standard advice. If you got bit, well it was your fault and the dog was just doing his job. That was before lawyers and folks suing each other. Dogs didn't have to be on leash, you just watched out for them. While I saw a lot of Aussies, I didn't see any that spoke to me like old Mitch had done. Jeanne had their Las Rocosa lines, Joan had gotten her red Wood's dogs from Joe Piz."

Joan Beien's red and red merle dogs often had blue eyes and Joan would tie them to the fences while she was working her paint horses at the stable. She had an impressive self red merle named Beien's Tuffy tied to the lunging ring near the parking area one day when the Mastersons came to ride. Casey's mother was driving and was so startled by the two blue eyes on this outstanding dog she forgot what she was doing and drove through the lunging ring fence! Fortunately she missed Tuffy but she had to pay to rebuild that fence.

Casey's life was about to change forever the day Joan arrived at the stables with a new blue merle female she had gotten from Charlene Jeffreys named Poudre Condesa, who she called "Tess." Casey confirms, "It was love at first sight. That was the dog I wanted a pup from." Bred by Eleanor Gantenbein and sired by Dessericks Spook out of Gantenbein's Naughty Blue Britches, Tess was true mountain breeding at it's best. Dessericks Spook was by Dessericks Ring who was from the legendary cross of Green's Kim and Green's Dingo, the two top working dogs of Ken and

Leila Green. Ring also sired Gantenbein's Mike, who would later sire the celebrated Ch McGuire's SeMeGo Hemi UDT STDsd OTDc. Spook's dam was Wood's Slate, dam of Fletcher Wood's immortal Wood's Stubbie, giving Spook 3 crosses to Wood's Jay similar to his half brother. Tess' dam was a granddaughter of Ring so she was strongly Wood's bred with dashes of other mountain-type dogs scattered through her pedigree.

Tess was Joan's third dog, joining Tuffy and another female named Crystal who was a Piz bred dog. Tess was the typical dark Wood blue with little white, a moderate size and very intelligent. She was a good working dog and Joan would take Tess with her when she visited breeder Charlene Jeffreys to work cattle in the mountains. Tess and Crystal were with her one day when the two tangled with a tough old cow. While making their escape, Tess snagged her shoulder on a barb wire fence and peeled off a large slab of skin that was left hanging. Joan was too far from help to load her up for veterinary care, so she and Charlene used an old ranchers trick, packed the wound with pepper, sewed the skin flap back in place and Tess healed up perfectly! Tess never suffered any ill effects from this rude treatment and continued working just as before.

In 1973 Casey was old enough to make her own decisions about her future and she knew she wanted one of Tess' puppies to accompany her to college. She also knew Joan was getting ready to breed Tess so she got on the waiting list for a puppy. Joan's choice of a sire for the litter was Jeffreys' Pooh Bear "a big tough son of a gun" in Casey's words, and Charlene Jeffreys was to have first choice of the puppies as a stud fee. Charlene was living on a ranch west of Boulder at the time and all her dogs had come from Fletcher Wood or were dogs of his bloodline. Casey describes them like this, "...they were big, dark blue and tougher than heck. Hard biting, bull headed dogs not afraid to take on man nor beast." As intimidating as this sounds, she continues, "Although these dogs had a fearsome reputation, they were family dogs, friendly and faithful to their families and those they knew. I guess you had to see that loving gentle side to believe it, but to me, they had a big soft side."

Jeffreys' Pooh Bear, the legendary sire of Masterson's Dotclijee Shunka. Courtesy of Casey Masterson Newman

So in 1973 the breeding that was to start Casey on her journey with the Australian Shepherd was made. Joan was quite the eccentric and the deal with Casey was on her terms. Tess whelped a good sized litter but in those days breeders culled heavily mostly for color. Solids were undesirable because few people recognized them as Australian Shepherds. Red dogs,

Masterson's Dotclijee Shunka, the classic Wood type dog. Courtesy of Casey Masterson Newman

Shunka was always hanging out close by ready for action even if it meant riding on the wood pile. Courtesy of Casey Masterson Newman

especially solid reds, had little chance at all of living to weaning age as it was believed they were not even an accepted color. Although Joan was obviously aware of the existence and value of red dogs from her association with Joe Piz, general knowledge among puppy buyers of their acceptability was non-existent. Jay Sisler's red dog troupe of rodeo performers is proof they were part of the foundation of the breed but awareness of this fact was lacking in the breed. So, after culling Tess' litter, four puppies remained. Of the two surviving female puppies born to Tess, Charlene chose one she later named Pepper (or Peppy) while Casey took the dark blue second pick bitch when she was only four weeks old due to Joan's family problems—she cared for her disabled mother alone and was raising a 4 year old nephew. Her baby blue dog cost her $150, an enormous sum at that time considering Casey's horse had only cost her $100 and she was making the impressive sum of $2.50 an hour working at the Dixie Cream Donut shop! Enrolled at the University of Colorado at the time and living with a Lakota Indian roommate she came up with the name Mastersons Dotclijee Shunka which translates to "Little Turquoise Dog." In the end, Masterson's Dotclijee Shunka grew up to become the better of the two females in the opinions of both Casey and Charlene so the stage was set for success for Casey's future breeding.

Shunka was the image of her dam, with a little more size from her sire Pooh Bear. As a puppy she never missed a college class with Casey, sleeping quietly under Casey's seat during lectures, then meeting and greeting everyone she met around campus. If she wasn't allowed in a class or lab she was left tied to a bush outside—a risky option today! She was with Casey constantly, running hundreds of miles behind her horse when they would go riding, even outrunning a forest fire on one memorable ride. In Casey's words "Shunka was the perfect dog."

Shunka was an individual crash course in dog behavior and the unique qualities of the Wood/mountain-type Aussie. Casey studied Animal Behavior in college, working with coyotes, foxes and wolves. One project she was personally responsible for was the raising of two McKenzie River Valley/Tundra wolf cross puppies taken from their parents before their eyes were open. Raising these pups on a bottle was not an easy task. Casey claims Shunk was not all that maternal but her intolerance for incompetence overcame her distaste for babies and she took over caring for them when Casey's efforts weren't up to her standards. From cleaning to keeping them warm to discipline, Shunk became their blue merle aunt. The behavior project ended after about a year and the two pups were slated for euthanasia. Casey was getting married and moving to a homestead in the mountains so after the wedding—where Shunk served as Best Dog, carrying the flowers—the young couple built a thirteen foot high fenced enclosure and obtained a zoo permit to keep the pair. When Casey bred Shunk, payback for raising them fell to the two grown "wolf aunts" and they loved all of Shunk's puppies and helped to raise and spoil them—although their style was a little bigger and rougher than any "dog aunt" would have given them!

Masterson's Dotclijee Shunka, blue and white female (Jeffreys' Pooh Bear x Poudre Condesa) in the spring of 1974. She is about one year of age in this photo. Courtesy of Casey Masterson Newman

The Casey and Shunka adventures only began with the raising of wolf cubs. Casey always took for granted Shunka could and would do anything she asked and Shunka always could and did. Her in-laws had horses and her sister-in-law enjoyed roping calves. Shunka would help with the chores of gathering and moving but was heeling the calves way too high. One finally kicked and found its mark, rolling Shunka across the arena and leaving her seeing stars. She staggered to her feet and shook herself off while casting a thoughtful eye at the retreating calves. Since she seemed fine, Casey continued using her. On the next go round, as Shunka approached a calf Casey told her to lie down, which she did AS she heeled. This time her hit was in the right spot and she came back spitting dirt instead of reeling from a kick!

Shunka as a mature dog. Photo courtesy of Casey Masterson Newman

Like most training at that time, stock dog training became a learn-as-you-go affair for Casey and Shunka. Heeling and driving cattle with dogs in various situations was their most common use at the time. The dogs were allowed to figure out the job as long as they weren't endangering anyone's life. Shunka had learned her heeling lesson on the roping calves and accompanied Casey when she rode her horse in the pasture across the road from the boarding stable. This field housed a herd of fairly wild cattle and it was never safe to dismount around the bulls. This was not a problem when Shunka was along. She sized up the situation and kept the bulls at bay for the riders.

Shunka. Courtesy of Casey Masterson Newman

John Masterson, Casey's father, in 1930 with his favorite dog of all time—an Australian shepherd named Jimmie Brown. The Masterson ranch outside Miles City, Montana was a self sufficient operation raising everything from chickens to horses, including a huge sheep flock, cattle and hogs. Jimmie was a typical ranch dog, a dark blue merle who helped out at any job where he was needed. Jimmie Brown was named after one of the sheepherders who worked on the ranch for a time. Courtesy of Casey Masterson Newman

While in college, Casey lived in a small house with a roommate. Her roommate's brother was in the Navy and dropped by one weekend to visit his sister. The two girls decided to go shopping one morning, leaving the Navy man home alone with Shunka. The two were getting along well when the Navy man decided to get a bowl of cherries and settle in on the couch for a little TV watching. Unbeknownst to him, Shunka adored cherries and followed him to the couch to beg for her share. However she had an unusual method of begging for them. She would stare intently and growl until the fruit was delivered. The Navy man obliged, splitting the cherries with Shunk, but after they were gone she continued to hold her position, staring and growling for more. Unsure of his safety if he dared move, Navy man held his position on the couch for the entire morning until the two girls returned!

After her marriage to Rick Newman, Casey decided to follow her grandfather's example and bought a small flock of sheep to show and raise for meat. Her grandfather, Jim Masterson, was a Montana rancher, artist and author. In the 1890s, Masterson came west from Chicago to begin his life in Big Sky Country. He was mostly raising sheep to feed the masses settling the west but quickly added cattle and horse herds to his land. Self sufficiency was the order of the day and he had to take care of himself and his family on his own.

Casey's father grew up on the ranch eating a steady diet of lamb and working the ranch. Helping with chores were a number of Australian Shepherds, one a dark blue merle with a long tail named Jimmy Brown. His stories of life in Montana stayed with his red haired daughter and she longed to raise her own sheep just as her grandfather had once done. Starting with a few Suffolk lambs, she and Shunka quickly learned this breed had a lot of "attitude" and Shunka learned to treat them as roughly as she did calves and bulls in order to keep them in line. Fearing for the welfare of both her sheep and her dog, Casey decided to cross these belligerent ovines with the more amenable Columbia and Corriedales. Convincing Shunka these crosses were more co-operative was another issue.

John Masterson, father of Casey Masterson Newman, with his favorite dog, Jimmie Brown, and his parents. Although the dogs weren't known yet as Australian Shepherds (late 1920s) it is easy to see the similarity to these ranch dogs and our Aussies. This type of ranch dog was common in the area and had no known recorded background although they did have a fairly consistent type. Courtesy of Casey Masterson Newman

After several less than successful experiences, Casey became more and more interested in learning to work livestock correctly. She had shown sheep for many years and with her own flock to handle now she wanted some help. Like many early attempts of Aussie owners to work stock, some of Casey's early experiments are somewhat comical in hindsight. Without a clear understanding of the goal of herding, but a good understanding of what was going wrong with their current methods they were willing to try anything to train their dogs. One such experiment involved pairing the sheep by tying them together at the neck, then hobbling them so they could not run away from the dogs! Of course, this method slowed things down a good deal but the handlers still didn't know where to go from there using instinct they didn't yet understand.

Shunka at HS&W ranch. Courtesy of Casey Masterson Newman

One good use Casey found for her budding stock dog was discovered while caring for some of her neighbor's cattle. Working cattle was what she was really bred for and her natural talent was second to none. One morning she found a little auction yard heifer which had calved unexpectedly. She broke her pelvis calving, couldn't get up and her calf could not nurse. Not wanting to have to bottle raise the calf but not wanting to lose it, either, it was up to Casey to figure out a way to set the situation right. Milking the heifer while she was lying down was a failure—-milk just doesn't flow well horizontally. Trying to get the heifer on her feet was another bad idea, even with the help of a tractor. Finally it became apparent the heifer needed help keeping her hind feet together in order to stand. So she hobbled her hind feet and sicced Shunka on her with a "Git her up!." In Casey's words, "Shunk hit that cow so hard that she levitated up, swaying on her feet, the ropes holding her from splaying down and splitting the pelvis again." Once she was up, Casey gave Shunk a quick "Back off!" and Shunk returned to her post under a shady cottonwood nearby.

So the heifer was now on her feet and on the prod but she was showing no interest in the calf. Even though they had dried off the calf it had been quite some time between the calving and when the heifer was found and the two beasts were not really aware of their relationship. Casey tells the rest of the story this way, "The problem was to get her to take the calf without killing any of us. This is where the difference between a dog and a good dog come in being. Any dog can chase a cow. Getting them to quit chasing the cow can be the hard part. A good dog instinctively knows how to play that cow like a fiddle. We shoved that calf over as close to the cow as we could get but she was just looking at Shunk. The trick was to get her to notice and protect the calf without trampling it or the dog. In and out, back and forth, Shunk worked that heifer until she was up on that bawling, hungry calf. Then far enough back so the calf could nurse while the cow glared at Shunk. Then way back so she could lick the calf and relax. Almost done except for now we had to move them to a safe place so the other cows couldn't hurt or separate the calf and the heifer could eat the hay pile we'd given her in peace. We slowly moved them, the calf riding on the tailgate of the pickup. Shunk carefully driving the hobbled cow behind. We had to come twice a day to get that heifer up on her feet (kept the rope tying her legs together, this kept her from splitting again). Shunk had to do the job calmly so the cow didn't get riled and somebody got hurt or the calf trampled. The ending to the story is a good one. The cow's hips healed enough that we took off the hobbles and turned her out with the rest of the herd. She weaned the calf successfully but went to the packers herself as she would have had trouble calving the next time. But the bottom line is she made weight, raised a calf and didn't end up dead. That's what counts for a rancher. And it was thanks to a good dog."

Shunka. Courtesy of Casey Masterson Newman

During Casey's early years of marriage she and Rick lived a rugged life in the mountains on their new homestead. Casey was raising sheep for wool, using it for spinning and trying to establish a reputation for good sheep and high quality fleece. She needed a name for this budding small business and decided on Homestead Sheep and Wool. Although the sheep and wool business never became profitable, the name was abbreviated to HS & W and today stands for top quality Wood bloodline Australian Shepherds. The only housing for the young couple during this time was a one room log cabin with a small wood stove. The two slept on the floor in sleeping bags with Casey huddled between Shunka and Rick. When they finally graduated to a real bed Shunka still considered herself a part of the marriage and would pretend not to hear when Rick demanded she get off his side!

The two took some time to work out an arrangement for sharing Casey on a day to day basis. Shunka managed to express her opinion of Rick's interference in her life in a number of ways. While working on renovating and rebuilding the burned out main house which stood on their property, Rick was carrying bags of cement. Shunka saw her chance to make a statement and delighted in knocking him down as often as she could get away with it.

Over time Shunka soon realized Rick was there to stay and she accepted him as part of her family although she still did not feel the same sort of responsibility for his welfare as she did Casey. The Newmans once took a vacation to Yellowstone Park in an elderly vehicle. The old clunker made it to the park then promptly died. A park ranger discovered their predicament but only offered a tow to the nearest camping area as help. Yellowstone was a haven then, as now, for parasitic bears who learned humans were easy pickens when it came to providing tasty snacks. They were aggressive about demanding food and more than a little ticked off when nothing was forthcoming. Unfortunately the ranger had left Rick and Casey's car near the park dumpsters where bears were especially persistent in their search for food. Since tent camping seemed a little risky in this environment, the Newmans chose to spend the night in the car with Shunka on guard. Although the three were really in this pickle together, Shunk believed Rick was capable of defending himself so she concerned herself only with protecting Casey and warding off bears which threatened her!

She eventually accepted Rick and considered his property worthy of her protection. Rick once made a trip to the airport in his Power Wagon pickup to pick up Casey's father and Shunka went along for the ride. The pickup was full of Rick's contracting tools but he didn't worry about their safety with Shunka along. When he returned to the truck he found the doors to the cab wide open and immediately panicked thinking Shunka had escaped! His worry quickly became relief when he found her reclining on the floor of the cab with an innocent look on her face. Apparently the would-be thief left in such haste he didn't even have time to slam the door!

She proved herself a worthy varmint dog on many occasions at their mountain home, chasing off coyotes on a regular basis. The wily coyote packs in the mountains were not something to be taken lightly however as their cunning extended to luring unsuspecting dogs into the woods to be surrounded and dispatched by the hidden members of their pack. Shunka was never taken in by these antics and she learned her limitations when it came to evicting these pests. She never had any trouble confronting two coyotes but if there were three she would always wait for human backup before taking them on…and she COULD count to three. Any more than that and she would employ a tactical retreat in order to wait for more inviting odds. Mountain lions and bears were not common in the area because they were not protected at that time and had been hunted extensively. But stray dogs could be as troublesome as coyotes and Shunka took care of ridding their property of these pests as well.

Shunka and Rick eventually formed a bond when she decided she enjoyed following him around when he used his construction equipment! Photo taken in October 1981. Courtesy of Casey Masterson Newman

Even though she was a tough and sometimes very hard-headed dog, Shunka also displayed a tremendous sense of justice. More then once Casey was forced to use some form of punishment in order to get through to Shunka that she was not to behave in unacceptable ways. It was on one of these occasions Casey learned a valuable lesson in fairness. She lost her temper at Shunka over some legitimate behavioral transgression and had her on the ground attempting to get through to her. Shunka was well aware of what she had done wrong and was taking the correction she knew she deserved. But at some point Casey continued punishing her well beyond the point where Shunka understood and had given into to Casey's will. This display of temper from Casey was not going to be tolerated by Shunka. She reached up, grabbed Casey's arm and gripped her tightly until Casey realized her mistake and stopped what she was doing. Shunka had no intention of harming Casey and held no grudge but she was not going to be a rug to be beaten up merely to exhaust someone's ire! From this point on Casey knew how far she could justifiably go and she knew exactly where to stop in order to make her point without crossing that line into an unjustified tantrum of her own. Shunka was willing to take what she knew she had coming, but no more.

Her sense of justice extended to training, too. She would do anything for Casey once. Being a hard headed dog, Casey would tell her what to do instead of asking her. Shunk would grudgingly do it one time to prove to Casey that she could if she wanted to. After that, she would do it her own way. In some cases Shunk's way was better, in some cases it wasn't. The partnership of two hard headed beings functioned well on this give and take basis once the ground rules were established.

As the adventures and experiments of the Casey/Shunka team continued, Casey became interested in breeding. When Shunka was 5 years old, her Mom decided she wanted a puppy so Casey started looking at potential sires at several area dog shows. Caring little for conformation shows, she was drawn to the obedience rings where dogs were being judged on ability and brains. Although she loved her pig headed blue teammate, she wanted a little more trainability in her puppies. At the time (the late 1970's), the dominant presence in obedience competition for the entire breed was Ch McGuire's SeMeGo Hemi UDT, OTDc, STDsd. Hemi, too was a mountain-type dog with a strong Wood and Sisler heritage coupled with infusions of blood from other early dogs which weren't as common, such as the Thornton dogs. His sire, Gantenbein's Mike, as mentioned earlier, was a son of Desserick's Ring, who was from the legendary cross of Green's Kim and Greens Dingo. Mike's dam was Gantenbein's Blue Muffett, an inbred Harper's Old Smokey daughter and granddaughter. Muffett was also a granddaughter of Thornton's Blue Blue, a daughter of the immensely influential Wood's Jay daughter Lighter's Asta.

Masterson's Dotclijee Shunka in 1978 pregnant with her litter by Ch McGuire's SeMeGo Hemi UDT, OTDc STDsd. Courtesy of Casey Masterson Newman

Hemi's dam, Gantenbein's Lady Blue, was bred by Janene Harper and from her Kansas breeding program of working farm dogs. With a number of Ely crosses, Lady Blue also could claim one line to Sisler's Shorty through his daughter, Lady Blue's granddam, Harper's Jinx. So Hemi had the foundation credentials to cross well with the other dogs of the day,

Shunka with the 3 blue females from her first litter by Hemi. From left to right: Shunka, Dusty, Shamber, Gantenbein's Annie O. November 1979 when the puppies were about 10 months old. Courtesy of Casey Masterson Newman

Masterson's Dotclijee Shunka in 1978 at 5 years of age. Courtesy of Casey Masterson Newman

the individual quality which made him a superb breeding dog and the slight addition of outcross blood which made him valuable to breeders looking to bring in additional genetic diversity to their bloodlines. For Casey, crossing Shunka to him was an excellent linebreeding of which she was blissfully unaware.

Hemi was the epitome of the versatile Aussie in any era. His show career is evidence of this versatility—he was the first "Triple Crown" Australian Shepherd in history, eventually becoming Ch. McGuire's SeMeGo Hemi, UDT, OTDc, STDsd—but he was also able to pass along his best traits of loyalty, intelligence, protective instincts and working ability. This, as much as his show record, was what eventually drew Casey Masterson Newman to breed her founding female, Shunka, to Hemi for her first litter.

Like nearly every event in Casey's dog owning life, Shunka's first breeding to Hemi was not without some humor and colorful occurences. Ruth was interested in doing the breeding after seeing Shunka's pedigree and told Casey Hemi was an experienced sire and would know when Shunka was ready. Although Ruth was not a fan of the Wood dogs she did like the crosses to the Gantenbein dogs in Shunka's pedigree and arranged with Casey to take a puppy back from the litter in exchange for the breeding—a red puppy if there was one. Ruth insisted on having Shunka's hips and eyes checked so after she received the "all clear" in these respects the breeding was on. Casey delivered Shunka when she came into heat but Hemi's famous ability to know when a girl was ready failed him this time. Shunka let him know his instincts were off and Hemi wore some tooth holes in his face for awhile as a result. Casey left her girl in a crate with her own coat and Ruth's promise to get the breeding done. But Shunk's attitude toward Hemi was similar to her attitude toward Ruth. She would not let Ruth touch her and ended up spending two solid days in the crate before Casey was called to come get her. A week later the timing was correct and Casey returned with Shunka to get her breeding to Hemi.

Triumph nearly turned to tragedy after Casey and Rick picked Shunka up from Ruth's during a blizzard. The roads were slick with snow and ice and an illegal alien hit their car, smashing them into a heavy pipe fence holding a field of buffalo and leaving the couple for dead. A snowplow driver witnessed the crime and called an ambulance and both Casey and Rick were rushed to the hospital. Rick was hospitalized for a month but Casey regained consciousness in the ambulance on the way to the hospital and her first thoughts—after checking Rick—were of Shunka and her fate. The ambulance driver called her parents on Casey's insistence and they went to the scene to search for Shunk. The snowplow driver had noticed the dark blue dog who was thrown out of the car into the field of bison and was able to let them know where to look for her. Fortunately she was waiting by the wreck for a friendly face to rescue her and Casey's folks were able to find her and bring her home until Casey got out of the hospital the next day. The two were reunited then seemingly no worse for the experience.

When it was time for Shunk to whelp her due date came and went. Casey had other obligations and couldn't stay with her so she left Shunk at her Mom's house in the morning. It just happened to be Christmas Day and when Casey finally arrived back at her Mom's Shunk began whelping almost immediately! She had deliberately put off having those puppies until Casey's return. Four gorgeous girls, a red merle one for Ruth and 3 blue merles plus one black male were raised on Casey's mountain ranch with the help of Shunk's two wolf stepdaughters and eventually placed in homes with friends and family. The male traveled to Kiowa, Colorado to work on a ranch that raised beefalo—a buffalo/cattle cross.

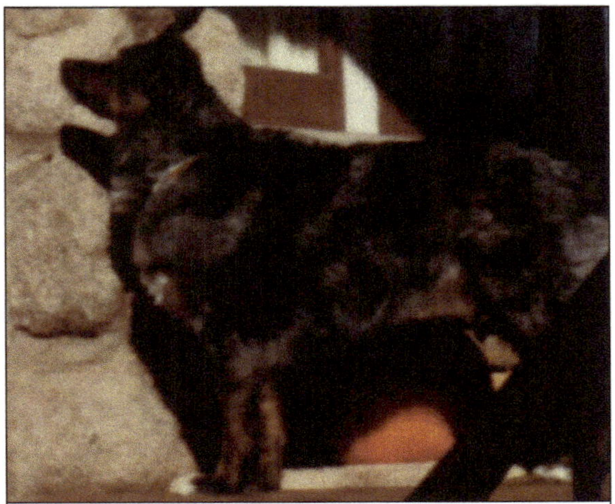

Dusty, blue merle female from Hemi x Shunka. Beloved companion of Casey's mother. Courtesy of Casey Masterson Newman

Dusty (Hemi x Shunka) seems to be helping out on a construction project. As long as they are with their people, these dogs are happy. Courtesy of Casey Masterson Newman

Hemi's Command Performance, the only red merle Hemi daughter. From Casey's first litter out of Masterson's Dotclijee Shunka. This puppy was Ruth's pick from the litter. Courtesy of Ruth McGuire

Shunka with her three 3 blue daughters by Hemi. From left to right—Shunka, Shamber, Dusty and Gantenbeins Annie O. Photo taken in December 1982. Courtesy of Casey Masterson Newman

Hemi's Command Performance "Mandy," Ruth's red merle, copper and white female stud fee puppy from (Hemi x Shunka). Photo taken in January 1980. Courtesy of Casey Masterson Newman

Shunka and her 3 blue girls. Shunka (lying down on the left), then left to right Shamber, Dusty and Gantenbeins Annie O. Photo taken December 1980. Courtesy of Casey Masterson Newman

When the puppies were only a few days old, Ruth offered to come over to remove tails and dewclaws but this just reinforced Shunka's dislike for her and she had to be locked in the basement until Ruth had left. Ruth eventually received the red merle female from this litter, with a blue female going to the Gantenbeins with the intent of breeding her back to their Mike—Hemi's sire. Sadly, Mike died before this cross could be made and this nice blue girl was never bred.

Another of the blue females was sold but Casey later discovered the dog was unsocialized and being mistreated and took the dog back. She was eventually named Dusty and given to her mother and became a wonderful loyal companion. Despite her early bad treatment she was an honest dog who would do anything to work for her owner. She learned many useful skills and could be trusted to deliver a candy bar to someone in another room without ever thinking of taking it for herself!

The lone solid black male was named Adair's Black Bear. The Adair family who acquired him were ranchers who raised beefalo. He not only was a valuable livestock working dog but also was devoted to the children of the family and provided guard services on the ranch. He was also to sire one litter later in life which would preserve his name for all time.

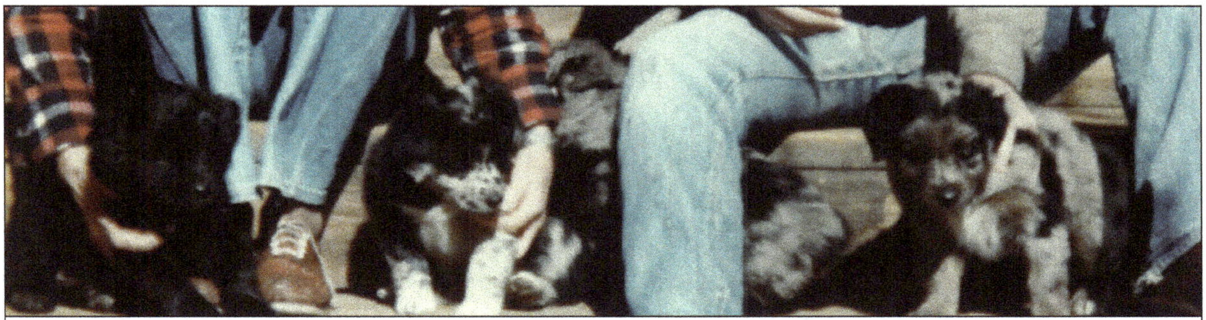

Shunka's first litter by Hemi. Left to right: Adair's Black Bear, the only male and future sire of Sprog, Annie, blue female owned by Eleanor Gantenbein, Shamber, blue female owned by a friend of Casey's, and Dusty, blue female eventually owned by Casey's Mom. Photo taken January 1980. Courtesy of Casey Masterson Newman

Casey did not keep a puppy from her first adventure into dog breeding but it had piqued her interest in doing more to preserve the bloodlines of her superior bitch. Jeffreys' Pooh Bear, Shunk's sire, was still alive but about 12 years old at the time and Casey decided to try an inbreeding to him hoping to produce another Shunka. Although the breeding was successful, only four puppies were produced, including one double merle white. The three remaining puppies failed to thrive and were all lost despite Casey's best efforts to save them.

After this sad occurrence, merle to merle breedings were off the table and Casey began looking for a black dog for her blue female. Shunk was now 10 years old and Casey was determined to only use a black male. She contacted Cathy Jones about using her black Reddy Teddy son, Ruff N Reddy Joe Willy Wang. Cathy had advised her to breed to Ted but Casey's prejudice against red dogs reared its head and she foolishly rejected this suggestion. So Cathy approved the cross to Joe Willy Wang and when Shunka came into heat again she made a visit to Cathy's in the dead of night. Casey remembers very little about this attempt because it was a poorly lit night and Joe Willy and Shunka were nearly invisible in the dark! As before with Hemi, Shunk was not very willing and although the breeding was completed Shunka failed to conceive from this breeding and Joe Willy was sold to Florida before they could try the cross again.

Shunk had been right and within a month came into a true heat in the middle of a blizzard. Cathy again suggested Ted but Casey was dead set on getting a blue puppy and considering Shunka's age she was determined to use a black dog with the right bloodlines. She says, "I should have known that a good dog can't be the wrong color but what the heck, I was young and dumb. Remember that pup I'd sent out to Kiowa?" With few other options available, solid black dogs were not popular and often put down because nobody wanted them, Casey turned to the black male

The Adair family with Adair's Black Bear as a puppy. Courtesy of Casey Masterson Newman

Teddy was a superior individual who had a slightly different temperament—different enough to change the history of HS&W's breeding program. If Teddy had had the mental characteristics of Sprog he would have been Casey's first choice to keep from the litter. Courtesy of Casey Masterson Newman

from Shunk's first litter, Adair's Black Bear. He had grown into a very handsome and superior type of dog and, even though this would be an extremely close inbreeding, Casey was confident she would get what she was looking for without any problems. She describes Bear like this, "Adair's Black Bear. Bear was a solid black dog. A tough dog, a fierce guard dog, Bear kept all callers in their trucks until they passed approval. A rustlin' dog, he liked nothing better than to slip to the neighboring ranch, round up a few head of cows and bring them home to add to his own herd. Because of that he spent quite a bit of time on a chain." So, she jumped in the car and drove through another blizzard out to Adair's and brought Bear home with her to complete the task. The result, born July 15, 1983, was four beautiful puppies—3 males and 1 female.

In the beginning Casey was enamored of the black male named Licorice but he turned out to be too shy for her and he found a great home with a 4H child. Her next choice was a blue male she called Teddy but when he developed the habit of hiding from her under the porch and refusing to come when he was called she decided that wasn't the temperament she wanted, either. Teddy was sold to a mechanic who was discovered to be beating the dog for stealing food and Casey finally bought him back. The dark slate blue merle with blue flecked eyes soon became her mother's next dog and was her loyal if somewhat naughty companion for the rest of his life.

The female puppy, eventually named Casey, was sold to a Trakehner breeder friend because she didn't want to keep another female with her mother, Shunka. The puppy was extremely smart but described as "mouthy" by her breeder. She loved working sheep but was a bad wool puller and if corrected would talk back to whoever was trying to work her. She also had the dangerous habit of swinging from the tails of horses but seemed to be blessed with

PEDIGREE OF Newman's Blue Sprog CD STDsc

Born: ______ Sex: ______
Owned by: ______

Parents	Grandparents	Great-grandparents	Great-great-grandparents
Sire: Adairs Black Bear	Sire: McGuires SeMeGo Hemi	Sire: Gantenbeins Mike	Sire: Desserichs Ring
			Dam: Gantenbeins Blue Muffett
		Dam: Gantenbeins Lady Blue	Sire: Harpers Half Pants
			Dam: Harpers Pepino
	Dam: Mastersons Dotelijee Shunka	Sire: Jeffreys Little Pooh Bear	Sire: Jeffreys Shade
			Dam: Sorensens Blue Mist
		Dam: Poudre Condesa	Sire: Desserichs Spook
			Dam: Gantenbeins Naughty Blue Britches
Dam: Mastersons Dotclijee Shunka	Sire: Jeffreys Little Pooh Bear	Sire: Jeffreys Shade	Sire: Woods Stubbie
			Dam: Jeffreys Missy
		Dam: Sorensens Blue Mist	Sire: Sorensens Gunsmoke
			Dam: Rowe's Dingo
	Dam: Poudre Condesa	Sire: Desserichs Spook	Sire: Desserichs Ring
			Dam: Woods Slate
		Dam: Gantenbeins Naughty Blue Britches	Sire: Phillips Blue Smokey Mist
			Dam: Hogans Robin Blue

Bred by: ______ Address: ______ Phone: ______

The details given above are correct to the best of my knowledge. Signed: ______(Breeder)

Teddy was good at retrieving, a common trait among this line of dogs. Courtesy of Casey Masterson Newman

Newman's Blue Sprog CD STDsc, blue and white male. Intelligence and drive were obvious in his eyes and his attitude. He never failed to give any task his full effort. Courtesy of Casey Masterson Newman

an abundance of luck and was never seriously injured from it. She was bred only once to a show bloodline dog.

So if Casey wanted to keep a male puppy from this cross it was going to boil down to the little blue merle male that was left, a dog she named Newman's Blue Sprog. She admits,""Sprog had that 'can do' attitude. Sometimes, despite all your own ideas, the right dog picks you." This "leftover" puppy was to become the foundation of Casey's kennel and the ideal she continues to breed for today.

Sprog was the smallest of the litter but nicely built with a wonderful temperament and extreme intelligence. Casey began working with him on obedience training when he was only a tiny puppy and he learned all his commands in Chinese—a language she had studied in college. He was born a rich dark blue merle but darkened as he aged as many of his ancestors did. His nearly black brown eyes did not miss a thing and he and Casey became a team.

As he grew she began taking him to the barn and teaching him how to work around her sheep. She was raising colored show sheep at the time—a Columbia/Suffolk cross—and they were not the best type for working dogs. In the lambing pens he learned to be quick when urging the ewes out before they had a chance to trap him against the fence. This tendency to be a little defensive in the pen had to be retrained later when Casey was training him for stock dog trials. Along with the sheep the Newman ranch housed a few horses, goats, hogs and other small livestock so Sprog's working dog education included plenty of variety. As he matured, his mother began taking more of a back seat and "retired herself" at age fourteen, leaving Sprog to take over the duties of "right hand man" with duties around the ranch. Like his father, he had been raised with the wolves and doted on them and they on him.

When Sprog later had the opportunity to work cattle he showed his true talent and breeding. His background said he should be a superior cattle dog and he lived up to it fully. He never pushed cattle very hard

Casey and Sprog getting a little professional instruction working sheep. Sprog's excellent early training in obedience and his experience with Casey's own sheep are showing in his steady approach and control. Courtesy of Casey Masterson Newman

and would only heel as needed, then drop back to push with his presence. Although he still tended to be rough on sheep he adored working ducks even though Casey despised them. She once had some Pekin ducks and a friend dropped off a goose to join the flock when she was away from home. The goose complicated an already difficult situation by refusing to stay with the ducks when she tried to work them together and she eventually found a new home for this oddball.

Sprog had a talent for working ducks—and tolerance for a little unsolicited help. Courtesy of Casey Masterson Newman

By now, stock dog trials were becoming more popular and methods for training dogs were being refined and perfected by breeders and trainers. Casey had attended a few trials and knew her dog had the talent to do well so she sought out a trainer to help her understand her dog's instinct and develop it. During this time the only trials for Aussies were ASCA and many people had the idea a dog should bite cattle excessively to show their style and skill. But there were those who disagreed with these methods and Casey was one of those. She taught Sprog to hit only when necessary, then quietly control his cattle without abusing them.

Casey's philosophy of only using bite on cattle when needed resulted in a dog who could control cattle and keep them together without turning the job into a rodeo. Sprog moving a big herd at a comfortable pace completely in control. Courtesy of Casey Masterson Newman

At one of the team's first trials, the HPASC trial on July 5, 1986, marking only his third time working cattle, Sprog tied for first place out of eleven dogs on Started cattle! He soon earned his Started Cattle title in ASCA demonstrating this wise method of handling stock. Their casual training schedule and Casey's interest in many other facets of working dogs precluded earning any further titles on cattle but it is likely Sprog would have easily earned Advanced titles with only a little more experience and direction.

At the same time Sprog was learning proper stock management he was also teaming with Casey in obedience class and entertaining 4-H groups and others with his wide repertoire of tricks. After graduating at the top of his obedience class it was time to earn another title—-that of Companion Dog. Sprog excelled in the obedience ring just as he did at everything, quickly earning his CD with good scores. At the same time he was learning formal obedience Casey dabbled in a variety of other occupations with Sprog, including carting, Schutzhund, tracking and what we would now call agility! He had an aptitude for everything she threw at him.

Sprog with his oversize Schutzhund dumbbell. Courtesy of Casey Masterson Newman

Sprog never did things the easy way. Taking a big solid jump carrying his Schutzhund dumbbell! He would carry Casey's brother-in-law's 10 pound steel workout dumbbell when the family went for a visit. Courtesy of Casey Masterson Newman

Sprog's short time spent in the Schutzhund world was due to some amusing occurences. At the time Cathy Jones and Jim Foster were dabbling in the sport and their dogs, Ted and City Slicker, were showing good aptitude for it. Casey's competitive instincts (or maybe it was just plain old pride) led her to join with a group of friends who had German Shepherds to give this a try. But Schutzhund titles were not in the cards for Sprog. His high intelligence was almost insulted when asked to run a search pattern when he already knew where the decoy was hiding on the field. Bite work—where the dog was expected to work a sleeve and nothing else—was boring. Why tug on this big toy when one could get a more satisfying reaction from one's adversary by sneaking around to his backside and administering a good bite on the butt? The helpers soon began declining to work Sprog when he progressed to work involving them running away from him. It was a real chore for them to acquire any running speed looking over their shoulders and waving the sleeve across their hindquarters to protect themselves from this clever student!

As Casey met more Aussie owners and learned more about the breed she became more interested in trying to preserve the type of Aussie she had. Sprog was everything Shunka had been and much more. He was every bit as tough and smart but somewhat more biddable than his dam. Casey had the opportunity once to meet Fletcher Wood and learn firsthand about Sprog's ancestors and the breed's history. Mr. Wood told her Sprog was just a smaller version of Wood's Jay and shared many stories of his dogs which sounded all too familiar to Casey. She also met more breeders who had the type and bloodlines she was looking for, including Cathy Jones.

She had assumed all Aussies were like her dogs, thinking an Aussie was an Aussie. But she soon learned they weren't. And her type of dog was getting scarcer all the time. She says, "The good old common sense ranch dogs had gradually been replaced by something else. If I was going to show people what a real dog was, I had better put my money where my mouth was. I started working hard and put a couple of titles on Sprog. It took us years as we only went to one trial and show a year. But we did get our CD and STDcs. Then I started having a family. Kids take a lot of your time. While Sprog wasn't one of those dogs who just love little kids, he had a highly developed sense of responsibility toward family. He added babysitting duties to his ranch chores. He brought diapers, retrieved bottles and made sure little kids didn't stray too far. I decided to maybe try to breed a few dogs to keep the old fashioned kind of dog I liked around."

Of course Sprog was endowed with the line's renowned guardian abilities, too. Many visitors to Casey's home learned quickly that coming into her fenced yard if Casey wasn't around was not a wise move. However, a problem existed for the visitor in getting Casey's

Sprog at about 3 years of age. Mature physically and showing the quickly darkening coat typical of the Wood type dogs. Courtesy of Casey Masterson Newman

Newman's Blue Sprog CD STDsc, solid Wood type with the mental traits both Casey and Fletcher Wood valued. Courtesy of Casey Masterson Newman

There really wasn't any job Sprog wouldn't tackle. Wearing a sled dog harness he hauls the family Christmas tree to the house. Courtesy of Casey Masterson Newman

attention to notify her of their arrival since Sprog would not bark at anyone coming to the gate! His style was more to silently watch and wait for the opportunity to handle the situation himself. After a time, a method was devised whereby the visitor would lift their own small child over the gate and send them to find Casey to deliver the message she had company. Sprog recognized the harmlessness of a child and would allow a small intruder any time without harming them!

Casey knew Cathy from her attempt to breed Shunka to Cathy's black male, Ruff N Reddy Joe Willy Wang. She now went back to Cathy in order to find some top quality females to breed to Sprog. She never intended to become a prolific breeder, just have enough litters to ensure the type of dog she wanted would be around for years to come. Cathy had several very well bred and proven females she no longer needed and Casey was able to lease several to use with Sprog. Unfortunately many of them were quite old and either no longer able to conceive or they had very small litters.

The first breeding Sprog ever did was to an outside female owned by Claudia Nelson

Not only were Casey's own children perfectly safe with Sprog, so were the kids of others! Courtesy of Casey Masterson Newman

As long as he was being useful, he was happy. Sprog giving Casey's two oldest children a ride down the road. Courtesy of Casey Masterson Newman

Few photos exist of Mt Blue Christian's Praise. This group photo includes her and a collection of other good Sprog kids and grandkids. Left to right: Casey with Sprog and Gitel under her arms and Blaze lying in front of her, Claudia Nelson with Shoni lying in front of her, Cheyenne (Shoni x Molly) under her arm and Praise sitting behind her, Lisa Renier with Stitch (Sprog x PDQ) lying next to her and Sam (Sprog x Pepper) standing on her left. Courtesy of Casey Masterson Newman named Misty Mountain Blue CD, STDdc, OTDs. Obviously a blue merle, she was by Las Rocosa Sydney out of Stonehenge Phantasy—a Rowe's Commanche Warrior daughter out of Shannahan's Phantom. On July 1, 1987 she whelped a litter of 9 puppies of all four colors—4 females and 5 males.

Claudia kept a red merle male out of this litter she named Mt Blue Shoshone Chief STDc who was later bred back to Casey's female HS&W Blazing Star to produce HS&W Cyborg Fang. Shoni's strong working ability echoed that of his sire, Sprog. He was a strong cattle working dog who didn't run cattle or bite without need. And like his sire he earned his Started cattle title with placements of 1st and 3rd on his first try at trialing and the first time he had seen cattle. Shoni was also the first Sprog offspring to trial and earn titles. A blue merle female from this litter named Mt Blue Christians Praise was later bred to Queenslands Wibi Windwalker, producing J Bar D Mystic Warrior HS STDcd OTDs, known as Mikey, owned by Jerry and Sharon Rowe.

Before she was able to obtain several females from Cathy she returned to Charlene Jeffreys looking for potential mates. Two of the first females Casey obtained were Shady Mt Mity Blu Goblin and Shadey Mountain Flirt. Flirt was a half sister to Sorensen's Gunsmoke, by Williams' Laddie and out of Shady Mountain's Lady II. Lady II was a Jeffreys' Shade daughter out of Shunka's dam, Poudre Condesa. Flirt was approximately 10 years of age when Casey obtained her from Charlene Jeffreys. She had spent much of her life on lease to a man in Kansas as well as having been used by Charlene for breeding. Unfortunately this wonderfully bred and superior individual was too old to conceive and Casey was not able to preserve her bloodlines. Records of any produce from her time in Kansas or with Charlene are not available.

Mt Blue Shoshone Chief STDc, red merle and white male (Sprog x Misty Mountain Blue CD STDdc OTDs). A very high energy, high intelligence dog of beautiful type and conformation. Courtesy of Casey Masterson Newman

Shoni, a good sire only lightly used. Photo by Author

HS&W Cyborg Fang, blue and white male (Mt Blue Shoshone Chief STDc x HS&W Blazing Star), a double bred Sprog male used a few times for breeding but his puppies were never used so his line has been lost. Courtesy of Casey Masterson Newman

Shady Mountain Flirt at about 10 years of age. Her excellent conformation and type are apparent in this photo. Courtesy of Casey Masterson Newman

Shadey Mountain Flirt looking a lot like her half brother Sorensen's Gunsmoke. She had a similar head type with the dark blue Wood look from her dam. Courtesy of Casey Masterson Newman

Goblin, too, came from Charlene Jeffreys and was something of an afterthought when Casey picked up Flirt. She was very well bred herself but a somewhat unattractive individual who had never been bred. With a dent in her head from being kicked, a permanently dilated eye and a caved in hip—again due to a kick—Goblin needed a savior. Due to difficult personal circumstances of her owner, Goblin had been ignored and her situation was not likely to improve. Casey decided she had room for this lonely and sweet tempered dog and brought home two bitches that day. As fate would have it, Goblin became the only one of the two who would produce when bred to Sprog. Goblin's lineage was impeccable, by Ch Jones' Reddy Teddy CDX OTDdsc SchHA out of Jones' Mity Bite. This combination produced many of Ruff N Reddy's very best dogs, including Ruff N Reddy Hunny Bear, Ruff N Reddy Katy Did (a female later bred to Sprog, also), and Ruff N Reddy Mity Mean. When she entered Casey's life she was approximately 7-8 years of age but her only puppy by Sprog, Masterson's Gitel, a dark blue female born in early 1988, proved to be a superb foundation matron. Both Flirt and Goblin lived out the remainder of their lives with Casey and are buried at Casey's mountain home.

Masterson's Gitel was a dark blue female with an outstanding temperament and second to none in conformation. She was working livestock at a very young age and was every bit as talented and intelligent as her sire, Sprog. Bred only twice in her life, two of her daughters became the most important producers for both HS & W and Kristone. Her first breeding was made to a blue merle male named Hangin' Tree Huckleberry—a very controversial dog even to this day. Casey had seen Huck at a National Specialty and was looking for a dog to breed to Gitel. She wanted to bring a little more biddability to her line to temper a tendency to independence and sometimes just plain hard headedness in her dogs. From a letter sent to Betty Nelson by Jean Taylor comes this observation about Casey's dogs,"You'll

PEDIGREE OF Mastersons Gitel

Born: ____________ Sex: ____________
Owned by: ____________

Parents	Grandparents	Great-grandparents	Great-great-grandparents
Sire: Newmans Blue Sprog	Sire: Adairs Black Bear	Sire: McGuires SeMeGo Hemi	Sire: Gantenbeins Mike
			Dam: Gantenbeins Lady Blue
		Dam: Mastersons Dotclijee Shunka	Sire: Jeffreys Little Pooh Bear
			Dam: Poudre Condesa
	Dam: Mastersons Dotclijee Shunka	Sire: Jeffreys Little Pooh Bear	Sire: Jeffreys Shade
			Dam: Sorensens Blue Mist
		Dam: Poudre Condesa	Sire: Desserichs Spook
			Dam: Gantenbeins Naughty Blue Britches
Dam: Shadey Mountain Mity Blugoblin	Sire: Jones Reddy Teddy	Sire: Piz Little Joe	Sire: Piz Joseph
			Dam: J Bar D Choctaw Chilli
		Dam: Davies Ginger Red	Sire: Montoyas Rowdy
			Dam: Montoyas Snipper
	Dam: Jones Mity Bite	Sire: Jeffreys Cinnamon Man	Sire: Beiens Red Tuffy
			Dam: Jeffreys Cinnamon Bear
		Dam: Pizs Sheba	Sire: Piz Joseph
			Dam: Pizs Lady II

Bred by: ____________ Address: ____________ Phone: ____________

The details given above are correct to the best of my knowledge. Signed: ____________(Breeder)

Gitel benefitted from Casey's experience training Sprog on livestock. She handled tough stock with ease and confidence and without unnecessary roughness. Pictured here at only 8 months old in October of 1988, her natural ability is evident. Courtesy of Casey Masterson Newman

Masterson's Gitel, blue and white female, (Newman's Blue Sprog CD STDsc x Shady Mt Mity Blu Goblin) at about a year of age, a perfect blend of the Hemi and Ted cross which had proven to be so successful over the years.

be glad to know there are quite a few people out there interested in preserving the true Aussie. Casey Newman (Sprog's owner) is one of them, although she's of the old school—some of her dogs are pretty hard-headed, or so the grapevine says. My dogs have always been very tuned-in and biddable, so I can handle some of that. Personally I prefer one you don't have to stomp on."

Huck was a Montana ranch dog working well over a thousand head of sheep every day for his owner, Jenny Johnson. Casey knew the risks of out crossing to this degree but decided to take the chance and sent Gitel to Montana. It was winter and Jenny had only Huck and one other dog to help her with the sheep. So Gitel was pressed into service and helped handle the stock on the ranch while waiting to be bred. The winter weather made shipping Gitel very difficult and Jenny was not able to make the trip back to the airport with Gitel so she left her with some of her friends. The friends eventually were able to get Gitel on a plane but somewhere along the way the airlines lost all records of her. Casey arrived at the airport in Denver to pick up her dog, only to be told she was not to be found. After a few frantic hours and phone calls Gitel was located and found her way safely home.

From this litter came 4 puppies—-two who went to pet homes and two to breeders. One of those breeders was Casey who kept what she thought for years was a black female she named HS&W Blazing Star. She later discovered Blaze was a phantom merle, her only merle markings being a small line across her forehead and several hairs on her legs. Casey also made an interesting discovery from this litter which shocked her when she first encountered it. Her line of dogs would have the usual number of natural bob tails and long tails in each litter. This was by the book as far as everything Casey knew about the breed. But she had never had a litter of puppies which had dewclaws. Every puppy from her original lines was born clean legged. It was only when she began out crossing that she began to see these added appendages and the first time was quite a shock! Not knowing what she was seeing she assumed these 5 toed puppies had some sort of deformity! It was only after talking to other breeders she discovered these were, in fact, dewclaws.

Blaze was a different type of dog temperamentally from what Casey was accustomed to. She was softer but much easier to train and work, combining the intelligence and talents of the dogs she had always had with the increased trainability and co-operation she had hoped to get from Huck. Blaze became the perfect family dog, doing duty as a sheep herder with Casey then going with any of Casey's three children to training class, tolerating mistakes and confusion with good humor and sweetness. Casey also relates that Blaze was about 20 inches in height—which she considered small—and had a softer coat than she got with her Wood dogs

Although milder tempered than her grandsire, Sprog, and great-granddam, Shunka, Blaze was no pushover. She had an innate aversion to racoons and would seek them out for harrassment at every opportunity. Big game was not out of her league, either. She and Casey were once taking an evening walk up a mountain road near their home at dusk. Both Casey and Blaze were pursuing their own personal thoughts with little awareness of their surroundings. Suddenly they heard the "ruff! ruff! ruff!" sound of a bear issuing threats from the brush. They both froze in their tracks attempting to locate the sound. Out of the corner of her right eye, Casey spotted two small cubs playing near the road. Out of the corner of her left eye she spotted the mama bear rising to her hind legs. Many years of life in the mountains gave Casey the knowledge of how to handle this confrontation but she wasn't too sure about her young dog, who happened to be standing between her and Mama Bear. She knew if she could just remove herself from between the bear and her cubs the situation would likely resolve peacefully.

Masterson's Gitel. Photo by and courtesy of Melissa Couch

Gitel. There is no mistaking her heritage. Photo by and courtesy of Melissa Couch

HS&W Blazing Star, a cute and correct blue and white female. She fulfilled every goal Casey had for her except one—she never had the opportunity to breed her back to her grandsire Sprog. Courtesy of Casey Masterson Newman

HS&W Blazing Star. The narrow strip up the middle of her face and the white markings on her chest are normal white trim. The horizontal marking over her right eye is actually a merle spot which gave away her "phantom merle" status. Photo by Author

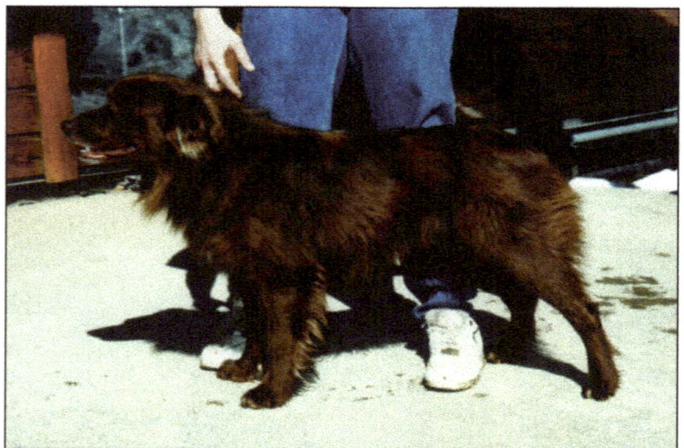

"Tigger," solid red male by Shoni x Blaze. Photo taken in 1996. Courtesy of Casey Masterson Newman

Casey faced Mama, spread her coat trying to look bigger, began whacking the ground with her leash yelling "NO! NO!" and began slowly backing away. She silently hoped Blaze would keep her cool and not begin barking, which would have sent Mama into full attack mode. If that happened, Casey had decided she probably could do nothing to save Blaze but hoped if she had to give her life, she would have done it saving Casey! But Blaze's good judgment kept her silent and both she and Casey backed slowly away from Mama. Casey states, "Blaze had shown the common sense I prized, she stayed with me, growling but not running or attacking the bear, either one of those would have gotten us in big trouble." Soon the bear spotted her cubs, collected them quickly and disappeared back into the brush. Casey and Blaze hurried home grateful to be alive!

Blaze had more than one experience with bears. At one time Casey had an orphan lamb she was keeping near the house to make constant feedings a little more convenient. The lamb lived in the house at night but was left in the yard surrounding her house most of the day until bedtime. It was evening and darkness was just falling when Blaze and Sprog began pacing from door to door to window becoming very agitated. Soon Casey heard the sound of the lamb bleating and the sound was circling the house. She looked out the sliding glass door to her bedroom and saw the lamb streak by with a large black bear in hot pursuit. The dogs saw him, too, and immediately threw themselves at the door, obviously convinced they could take him on and save the lamb! Casey called for her husband to get the gun—which was locked in the gun safe far from their quick and convenient grasp! Meanwhile the dogs were just waiting for their opportunity to dash out the door and take on the bear that was after "their" lamb while the lamb continued doing laps around the yard one jump in front of the bear. The only thing Casey could do was call the lamb, hold on to the dogs and hope the lamb would come to her. On the next lap the lamb "saw the light" of the barely open sliding door and dashed through while Casey slammed the door nearly on the nose of the bear! Thankfully the bear didn't try to break down the door and when Rick finally arrived breathlessly carrying a gun the adventure was over. Eventually the bear left but the lamb was guarded more closely from that day forward!

Blaze was not short on working ability either and turned out to be tougher than Casey expected. She had acquired some Jacob sheep for her ranch—these are the big, spotted four horned sheep. The ram she had was always trying to use those lethal horns on something and Blaze had to learn how to handle him without becoming a shish kabob. Sprog managed to teach her the tricks of that trade and she became skilled at moving the group without being skewered.

HS&W Cyborg Fang (Shoni x Blaze) Courtesy of Casey Masterson Newman

Soon the time came to breed Blaze and her first mate was the red merle male from Sprog's first litter out of Claudia Nelson's foundation bitch, Misty Mountain Blue CD STDcd OTDs, named Mt Blue Shoshone Chief STDc, called Shoni. Shoni was a multi-talented red merle who was a very strong working dog as well as a superior obedience dog even though he was never shown for any titles. Blaze and Shoni produced a large litter of about 10 puppies who were given to friends and acquaintances of Casey. Casey's pick from this litter was a blue merle male named HS&W Cyborg Fang. Borg was a kid dog from the beginning, becoming a companion to all of Casey's children and he eventually joined Gitel as the beloved companion of Casey's mother. He even sired one accidental litter with his dam, with all the puppies being placed in pet homes.

Half brothers out of Blaze. (Left) Remuda Bar Taylormade at Topeka RE (Silverledge Slide Me to Da Bar CD STDs x HS&W Blazing Star) blue, copper and white male. (Right) HS&W Cyborg Fang (Mt Blue Shoshone Chief STDc x HS&W Blazing Star) blue and white male. Photo by Author

Casey's plan was to eventually breed Blaze back to her grandsire Sprog but in the meantime she was still attempting to find other top quality females to breed to him. Ruff N Reddy Katy Did, a solid black full

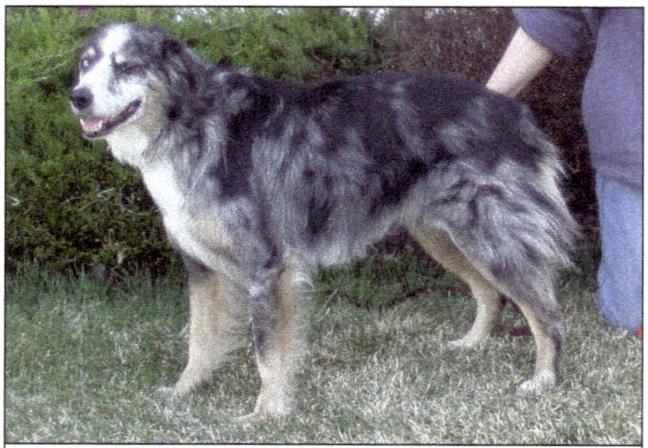

Taylor poses for posterity. Photo by Author

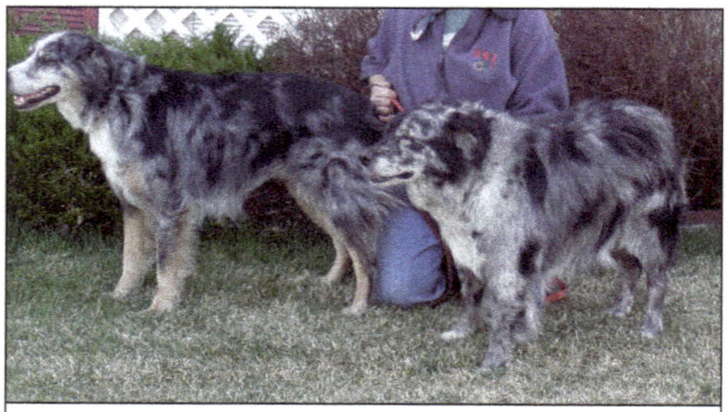

Taylor and Borg—alike in some ways, and showing the characteristics of their different sires in others. Photo by Author

Shady Mountain Mity Miss, blue merle and white female (Jeffreys' Shade x Jones Mity Bite). This royally bred bitch was obtained by Casey from Cathy Jones in her old age and bred to Sprog for her last litter, producing the superb stud dog, HS&W Ace in the Hole BC, CD. Shown here at about 13 years of age. Photo by Author

sister of Goblin (Ted x Mity Bite), was bred to him once while on lease to Annie Buffington in western Colorado and these puppies went mostly to ranches.

One of the best producers ever to come from the Jeffreys' Shady Mountain kennel was a silver blue female named Shady Mountain Mity Miss. Missy had been bred to a number of top males in her life and had produced some of the best Wood type dogs in the history of the breed. She was a half sister of Goblin through their dam Jones' Mity Bite and her sire was the great Jeffreys' Shade. She was a beautifully made dog with a stunning head and a sweet and loving temperament. Her produce are a Who's Who of the Australian Shepherd breed—Ruff N Reddy Shady Joe, Ruff N Reddy Shady Wood Ash, Ruff N Reddy Boots among others. While these names may not be well known for their own achievements, a look at what they produced tells the tale of their value.

Missy is another female who came to Casey from Cathy Jones as an old dog with Casey hoping she could get one last litter from her by Sprog. Fortunately she was able to conceive and produced two puppies, both solid blacks, a male and a female. Both puppies were superior, as their pedigree would decree and Casey intended to keep the female and place the male into a carefully selected home. Sadly, at about 4-5 months of age, her prize black female was stricken with parvo and died, leaving only the black male to carry on the line. This male was registered as HS & W Ace in the Hole BC and grew into a magnificent example of his line. Knowing she would have problems with Ace and his father, Sprog, living together in her household she took some time finding the right home for Ace where he would be appreciated and available to her for breeding in the future.

While growing up, Casey made sure he got plenty of socialization and tried him out in several training venues. He worked her own sheep following his sire's example and proved to have a nice gather and biddability. He went to obedience classes and was "trained" by a variety of adults and children, learning the basics while being passed around and socialized among the group. Casey even explored Schutzhund training at a local facility. Although he had great enthusiasm for the sleeve he became bored with this game and eventually sought other targets on the helper's body. Fortunately the man was wearing a full body suit but after Ace began choosing some tender targets, he refused to work him again!

He traveled to Missouri at about one year of age with a Colorado exhibitor heading to a dog show on a cold January day. He was picked up and taken to his new home during a miserable ice storm and began to learn about his new life. Later that day he needed to go out and his owner attached his leash and headed out the door. As soon as she stepped onto the icy sidewalk she slipped and fell, spooking Ace who jerked the leash from her hand and ran off into the darkness. Not knowing his new owners and being young and frightened, he would not come back when he was called and his owner desperately called Casey to tell her the bad news.

Casey's second child had been born only a few months prior and she was nursing the baby. However she knew her dogs and realized she

HS&W Ace in the Hole BC CD, black and white male (Sprog x Shadey Mountain Mity Miss) at about 6-7 weeks of age. Even at this young age it is easy to see the trademark broad, beautiful head Mity Miss always gave her puppies. Courtesy of Casey Masterson Newman

Ace as a mature dog. He inherited the same beautiful head Missy stamped on all her puppies, the common sense and intelligence Sprog was known for and excellent conformation from both his parents. Photo by Author

Ace showing the typical early working ability of this line. Pictured here at about 4-5 months of age and getting a little help from dad (far right). Courtesy of Casey Masterson Newman

HS&W Ace in the Hole BC at about 9 months of age. He inherited the same good broad square head Mity Miss gave all her puppies regardless of the sire of the litter. Photo by and courtesy of Melissa Couch

HS&W Ace in the Hole BC heeling a lagging sheep during one of his first real herding lessons. Photo by and courtesy of Melissa Couch

Sprog plays with his son Ace. Ace is about 10-12 weeks old in this photo. Courtesy of Casey Masterson Newman

had only 24 hours to find Ace before he would decide to head for home on his own. Calling her mother and packing up her newborn the trio headed for the airport to catch the first flight to Missouri. Arriving the following morning, Ace's new owner picked up Casey, her daughter and her mother at the airport, along with her own sister-in-law and everyone else she could muster in order to drive the countryside searching for the lost dog. Casey left one of her own shirts at the owner's home in case Ace decided to come back and the group began driving around the countryside calling and searching. After several hours, someone in the van spotted a black animal huddled behind a large round bale of hay in a field. Casey's heart leapt to her throat—she knew it was Ace! Pulling to the shoulder, Casey got out of the van and began slowly walking toward the frightened youngster, talking to him and calling him to her. It took Ace only a minute to recognize Casey and come bounding to her with relief!

After a stern lecture to the new owners, Ace lived with them for the next several years without any further incident. The family moved to the east coast and contact with them was lost until Ace was about 3 years old. By sheer chance it was discovered they were having problems with Ace and were in the process of looking for a new home for him. Casey was notified and reached them, reminding them they were not to dispose of him without her permission! Casey was not able to take him back but Kris Toft was able to take him and keep him for several years until Casey was able to get him back in later years.

Ace was a beautiful, glossy solid black with an arrogant but often silly and always friendly temperament. Of incredible intelligence, his active mind was his downfall when it came to those who were not able to deliver the discipline and structure he needed. Although not a particularly aggressive dog he enjoyed getting into and causing trouble just for his own entertainment. A superior obedience dog he eventually earned a CD title and was used for awhile as a tracking dog due to his superior scenting ability. Most of all he was an enjoyable companion. Although three of his owners were women he showed an amusing preference for men and adored any man whose attention he could attract—being willing to leave his female owners if he could find a man to take him!

Ace in September 1990 at about 8 months of age shortly before coming to his new home in Missouri. Photo by Author

Ace spent several years in Kris Toft's home until the day his sire passed on and Casey could take him back. Although a sometimes difficult dog, he was always a source of amusement and amazement. He once passed through the door on his way inside when he made a quick leap in the air and neatly plucked a half eaten oatmeal cookie from Toft's fingers and swallowed it before she even realized what happened! Despite his preference for the company of men he was an intuitive guardian of home and property. Kris Toft recounted an episode: "An unwelcome salesman once arrived at my gate after dark. The front and inside of my house was dark and I hoped the salesman would leave. Ace and I sat on the front steps in the dark watching his actions. When Ace began to growl I released him and he ran to the gate barking and menacing the intruder. I regretted letting him go, wondering if this man was up to no good and possibly armed. But I needn't have worried. The headlights of his car formed a half circle of light in my yard. Ace ran to the edge of the lights and began trotting back and forth at the very edge, coming no closer to give the trespasser an easy shot if that was his intent and staying in motion to make his black form difficult to see in the dark. Ace would eventually tire of this game and return to me on the steps where I would quietly praise him lavishly! Then he would run back to the arc of light and continue warning this stranger to keep his distance while keeping himself safe in the process. After this incident I never again worried about his willingness to protect me."

Ace as a mature dog. He was a nearly solid black dog with only a small white spot on his lower chest. Photo by Author

Ace, a truly masculine dog in the tradition of Sisler's Shorty, Wood's Jay and Jeffreys' Shade. Photo by Author

Ace, in actuality an amusing, goofy, fun-loving dog, he still had all the old qualities of guardiness, common sense and seriousness when the situation called for it. This sound, stable temperament and ability to judge a situation and react appropriately is becoming more and more rare in the Australian Shepherd. Photo taken in 1995—Ace as a mature 5 year old. Photo by Author

Kris further stated: "I once lost my temper with Ace after he attacked an old dog of mine. The attack was unprovoked and the sort of trouble he enjoyed causing at times. After checking the welfare of the old dog I was still angry at Ace and determined to expend my anger at his expense. I found a long pole that I intended to hit him with and went outside to find him. Ace innocently ran up to me and I hit him with the pole. Of course he ran away which caused my temper to flare even more. I followed him around the yard until he suddenly stopped and sat looking at me. As I got ready to hit him again he jumped in the air, grabbed the pole and jerked it from my hand then sat down again where he was before. I quickly realized the unfairness of my actions and what he was telling me! When relating this incident to Casey she remembered the story about Ace's granddam, Shunka, when she, too, had lost her temper and was unfairly punishing Shunk. Casey's change in outlook was similar to mine—we both realized we were being unfair and that our dogs, in fact, were extremely tolerant and explicit in relaying that fact to us!"

When Ace returned to Colorado at about the age of 7 or 8 he immediately attached himself to Casey's husband, Rick, and the two "boys" were fast pals to the end of Ace's life. Ace rode in the bucket of a front loader when Rick was working at a construction site. He never missed a chance to accompany Rick to work, riding shotgun in the front of his car. He enjoyed priority sleeping arrangements in the house and loved being around kids again for the rest of his life. He died peacefully at home around the age of 15.

During his life, Ace was bred to only 4 bitches, but he produced a number of superior puppies which are breeding on today. He was bred to the Gitel daughter, Blaze, producing 8 puppies, only one which was a merle, but only one was needed to confirm that Blaze was, indeed, genetically a merle. Most of these puppies were sold to ranch homes where they proved to be top quality working dogs which were, unfortunately, never bred. Ace was also bred to a bitch of J Bar D lines and produced a number of nice puppies from this cross. Again, none of these were ever bred and the lines have been lost.

Black male by Ace x Queenslands Spunky Haverlah (1997) Courtesy of Casey Masterson Newman

While staying with Kris Toft he was bred to one of her females of mostly Fairoaks breeding, producing 5 puppies. Although none of these were ever bred during their lifetimes, two males were collected and will hopefully contribute to these lines via frozen semen in the future. Ace's last litter was when he was an old dog and was bred to a great-granddaughter of Gitel. He produced two stunning females, HS & W Ebony and HS & W Kelbe Heidali Tobias, who are contributing to the future of HS & W kennel.

Kristone Anna Lightfoot, red merle, copper and white female (HS&W Ace in the Hole BC CD x Echo Ridge No Regrets CD). Shown here at 15 months of age. Photo by Author

Kelbe and Ebony's dam, Kristone HS & W Pipssisiwa, was a blue merle descended from the second breeding of the Sprog x Goblin daughter, Gitel. After crossing to Huck, Casey had what she wanted to ensure the future of her bloodline but she felt Gitel should be bred again due to her superior quality. Kris Toft leased her to breed her to an aging Fairoaks male Toft had, Ch Fairoaks Medicine Wolf, CD. So Gitel once again took to the air and landed in Missouri to meet her next mate. She soon whelped but there was only one puppy in her litter, a solid black female of very high quality. Although Kris would have liked to repeat this cross, Gitel's owner, Casey's brother in law, wanted her to come home and to spay her and so this ended her breeding career.

Fortunately the little singleton puppy, Kristone Mity Black Medicine CD, proved to be the epitome of her bloodlines. She was highly intelligent and in need of a home which would appreciate her and give her the attention she deserved. The decision was made to place her with friends who could handle a tough, quirky dog. Shadow, as she was called, proved to be a most unique dog.

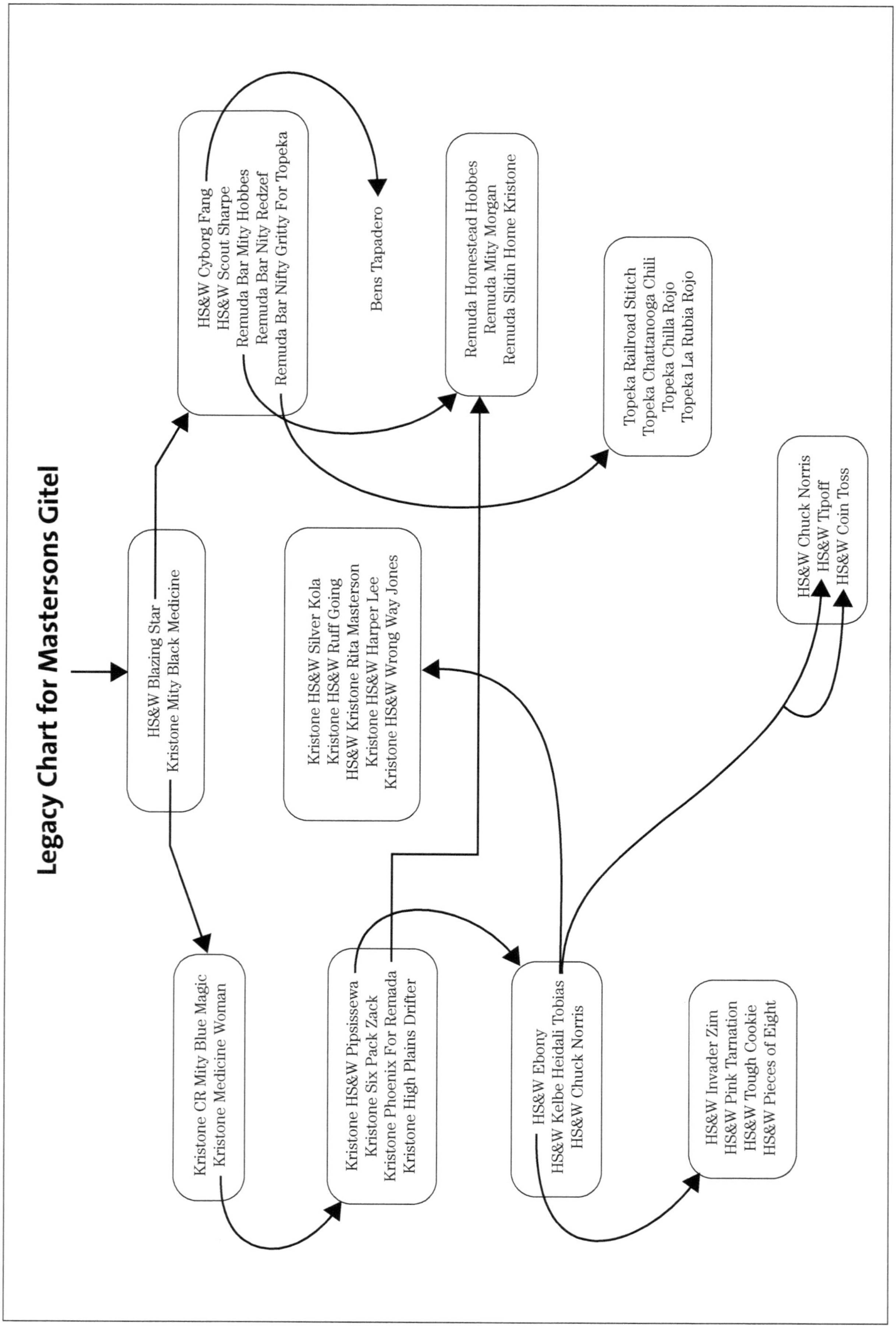
Legacy Chart for Mastersons Gitel
HS&W Blazing Star
Kristone Mity Black Medicine
HS&W Cyborg Fang
HS&W Scout Sharpe
Remuda Bar Mity Hobbes
Remuda Bar Nity Redzef
Remuda Bar Nifty Gritty For Topeka
Bens Tapadero
Remuda Homestead Hobbes
Remuda Mity Morgan
Remuda Slidin Home Kristone
Topeka Railroad Stitch
Topeka Chattanooga Chili
Topeka Chilla Rojo
Topeka La Rubia Rojo
Kristone HS&W Silver Kola
Kristone HS&W Ruff Going
HS&W Kristone Rita Masterson
Kristone HS&W Harper Lee
Kristone HS&W Wrong Way Jones
HS&W Chuck Norris
HS&W Tipoff
HS&W Coin Toss
Kristone CR Mity Blue Magic
Kristone Medicine Woman
Kristone HS&W Pipsissewa
Kristone Six Pack Zack
Kristone Phoenix For Remada
Kristone High Plains Drifter
HS&W Ebony
HS&W Kelbe Heidali Tobias
HS&W Chuck Norris
HS&W Invader Zim
HS&W Pink Tarnation
HS&W Tough Cookie
HS&W Pieces of Eight

Like many dogs of these lines, she enjoyed provoking others (dogs and humans) for her own entertainment. Kris remembered a memorable incident that occurred: "When I was attempting to obtain a CD title on Shadow, she had managed to earn two legs (while keeping me on my toes on both occasions with her antics) and I was hoping to finish this title before she gave me a heart attack with her unpredictable shenanigans. During the figure eight exercise she reached over and bumped the leg of one of the 'posts' with her nose as we passed by. I thought at first she might have nipped the woman but since neither she nor the judge reacted I continued walking. However, I strongly suspected on our second pass Shadow would escalate her behavior and actually nip the leg of this 'post', earning us a disqualification from showing! I was not about to allow this and made a hugely exaggerated circle on the second pass which I was sure would result in a non-qualifying score but at least would prevent a disqualification. The entire way, Shadow darted quick glances at the enticing leg but was unable to reach it without pulling me much closer. We managed to complete the exercises and earned a qualifying score that day but she was returned home after this and any dreams of higher titles were ended as the tension of showing her was just too great!"

Shadow at about one year of age. Photo by Author

Shadow with two half siblings. These three dogs were all sired by Ch Fairoaks Medicine Wolf CD but out of completely different females. The Wood type of Shadow's dam, Gitel, comes through strongly in her solid black daughter. Photo by Author

Shadow often accompanied her owners on walks in the woods near their home which was in a secluded area near the Missouri River. They once headed down the path to the river when Shadow took off ahead of them and ran out of sight. One of the owners sat down on a log across the path to tie her shoe when she began to hear the sound of hoofbeats. She looked up just in time to see a deer running full tilt toward her with Shadow gleefully bringing up the rear. Her reaction was to leap to her feet to fend off the deer before she was run down and as Shadow passed by the look on her face

Tigger x Shadow litter in 1997. Photo by Author

Kristone CR Mity Blu Magic, blue merle and white male (Silverledge Slide Me To Da Bar CD STDs x Kristone Mity Black Medicine CD). Fletcher Wood would have easily recognized "Charlie" as a dog of his old breeding. He is about one year of age in this photo. Photo by Author

was pure disgust at her ungratefulness for the "livestock" she had found and brought to her!

Shadow was bred only once to the great sire, Silverledge Slide Me To Da Bar CD STDs, "Tigger." She produced 6 puppies, 2 males and 4 females, which figure largely in the HS & W and Kristone kennels today. One daughter, the black female Kristone Medicine Woman, called "Quinn," became the dam of a litter by J Bar D Mystic Warrior HS STDcd OTDs (a Sprog grandson). Two of the females from this litter, Kristone HS & W Pipsissewa and Kristone Phoenix for Remuda have produced another generation of top dogs.

The one female from this litter who was bred, Kristone Medicine Woman "Quinn," was responsible for producing the two females which are carrying on the HS&W and Kristone tradition today. After being sold as a puppy she was reclaimed when it was learned her home was not working out. Her grit and working ability were intact although they had been mishandled. Her forgiving nature allowed some correction of the mistakes made with her but she never had the opportunity to earn any titles. Quinn was a very correct female who inherited an extra dose of beauty from the Silverledge side of the family. Her intuitive nature was a gift from both sides of the family tree. A search was begun to find the perfect mate for her and good fortune shone on the search when Michelle Murvai discovered a Sprog grandson living with Jerry and Sharon Rowe, J Bar D Mystic Warrior STDcd OTDs. "Mikey" was out of a litter sister to the Sprog son, Shoni, named Mt Blue Christian's Praise. His sire was a wonderful mix of old J Bar D and Reddy Teddy lines from the Queensland kennel. This was the perfect mix of linebreeding and outcross with a line well known to "nick" with the

Woods/Hemi lines. And it did not disappoint. Two females, Pip and Phoenix, have produced a number of dogs who are breeding on today and semen has been collected on two of the best males for future use.

After these early attempts to produce more dogs which she liked, Casey leased two younger bitches in order to increase her chances of live litters. She returned to Cathy Jones to find Sprog's next mate, bringing home the solid red Ruff N Reddy PeeDeeQue. PDQ was one of Cathy's favorites, probably because of her strong similarity in both looks, breeding and temperament to her great Reddy Teddy. Casey was looking for the connection back to Hemi in PDQ's pedigree to cross back to her own Hemi grandson. In May of 1989 PDQ produced a litter of 7 puppies of all four colors—6 females and one male—from Sprog and was later returned to Cathy. Three females were eventually used for breeding to outcross males.

Kristone Medicine Woman "Quinn" was a feminine version of her sire, Silverledge Slide Me to Da Bar CD STDs in head type. Photo by Author

Kristone Six Pack Zack, blue merle, copper and white male (J Bar D Mystic Warrior HS STDcd OTDs x Kristone Medicine Woman) a very strong type Wood dog shown here at about 10 weeks of age. Photo by and courtesy of Michelle Murvai

Casey then contacted Mark Lane and was able to lease a daughter of his Ruff N Reddy Hunny Bear (full sister to Goblin), the red merle Country Lane's Mity Hot Chili Pepper (Ruff N Reddy Ranger x Ruff N Reddy Hunny Bear). Pepper was similar in type and temperament to Shady Mountain Mity Miss—a fairly large female with a wonderfully sweet temperament. Pepper successfully whelped a litter of five outstanding puppies in 1993, all but one of which were placed on ranches and became topnotch working dogs.

One of these dogs was a red merle male named HS & W Ruff Orrin Haverlah who went to a south Texas cattle ranch. On more than one occasion Orrin rescued his owner from attack by an angry Santa Gertrudis cow. He was used sparingly with the owner's females but the puppies were sold to ranches and have not been used for breeding.

In addition to these good Wood bred females which Casey was able to obtain and breed to Sprog, he also produced some excellent offspring from several outside bitches. As discussed in the chapter on Silverledge kennels, the Taylor's last great daughter of Charlie Russell, Silverledge

Gahl's Terra, the good working female owned by hog producer Joe Gahl of Minnesota. Photo courtesy of Casey Masterson Newman

Just A Jock, blue, copper, and white male. (1995) Photo by Author

Black Magic, was bred to Sprog to produce the beautiful blue merle Silverledge Freckles. Joe Gahl, a Minnesota hog farmer, bred his priceless blue female from Ruff N Reddy lines—Gahl's Terra—to Sprog via fresh shipped semen, producing one excellent female—Gahl's Casey—who later produced several nice puppies when bred to a Pincie Creek male of predominantly Wood bloodlines. One of these Sprog grandsons, Gahl's Silverledge Smokin' Jose, came to Jean Taylor and became a good obedience dog and sire for her.

The earlier mentioned Misty Mountain Blue CD STDcd OTDs was a granddaughter of Rowe's Comanche Warrior and bred to Sprog early in his life. She produced a nice litter, including Mt. Blue Shoshone Chief STDc, who sired HS & W Blazing Star's first litter. Also in this litter from Sprog and Misty was a female named Mt. Blue Christian's Praise, a blue merle later bred to Queensland's Wibi Windwalker. Windwalker was predominantly old J Bar D and Jones Reddy Teddy breeding and produced one of the last J Bar D dogs tracing to the Rowe's original foundation dogs, J Bar D Mystic Warrior HS STDcd OTDs, called "Mikey." Mikey was bred late in life to the Silverledge and HS & W bred bitch Kristone Medicine Woman (Quinn) to produce Kristone HS & W Pipsissewa and Kristone Phoenix for Remuda. Another Queensland bred female, the blue merle Queensland's Spunky Haverlah was bred to the Sprog son, Ace, producing a number of very nice dogs which, like too many dogs of this line, were sold or lost and never reproduced.

Sprog was also bred to a Colorado's Toast of the Town x Ruff N Reddy Remember Me (Hemi daughter) bitch named Ruff N Reddy Shadow. Shadow was owned by a 4H child and was a good working dog who had earned firsts in obedience and showmanship at the Adams County Fair and blue ribbons in both classes at the Colorado State Fair. She produced some excellent puppies which, also, were placed in pet homes where they were never used for breeding. One blue merle who looked remarkably like his sire was named Just A Jock.

After being placed in a poor home by the breeder, Casey reclaimed Jock and placed him with a woman who had had another dog from her

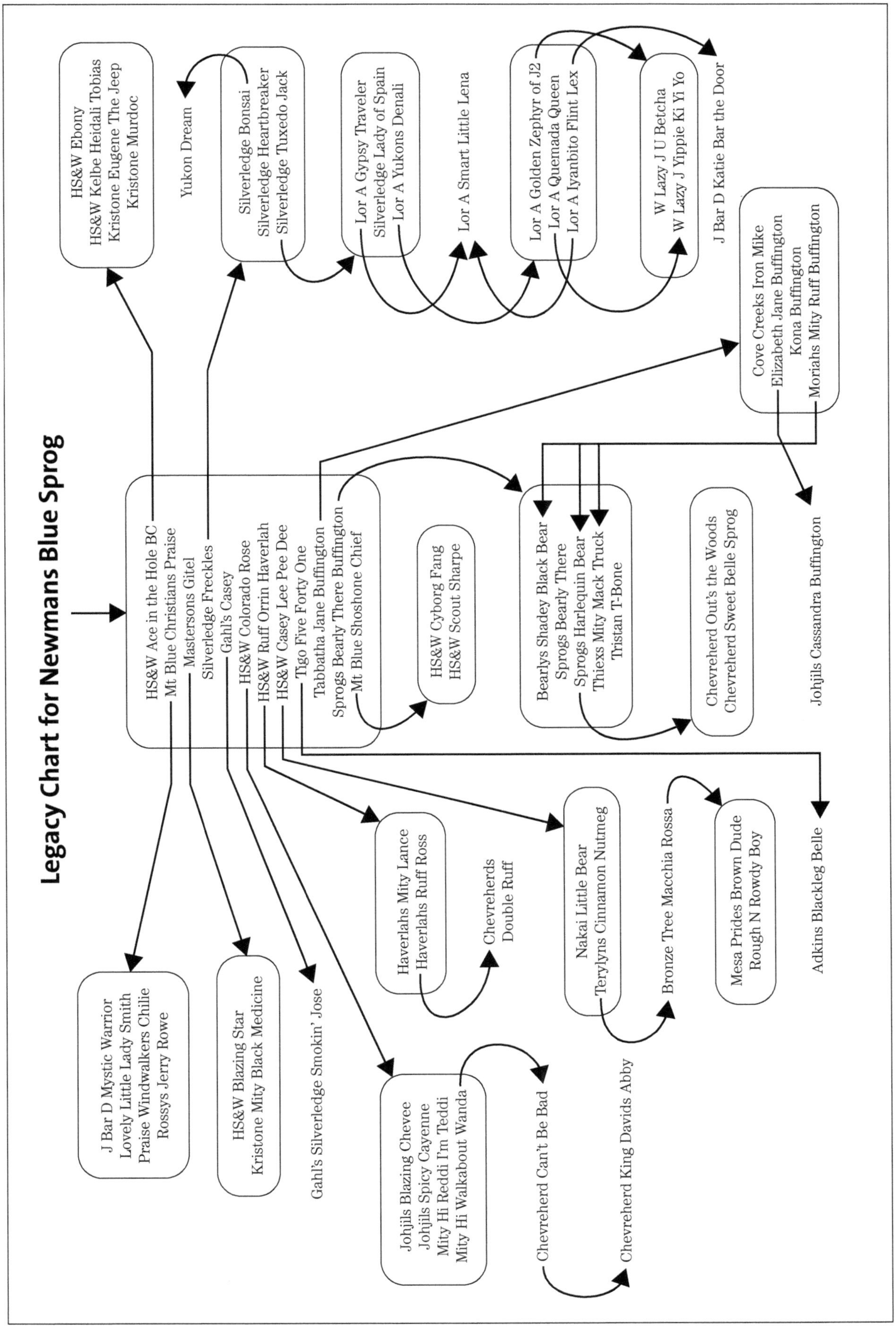
Legacy Chart for Newmans Blue Sprog
HS&W Ace in the Hole BC
Mt Blue Christians Praise
Mastersons Gitel
Silverledge Freckles
Gahl's Casey
HS&W Colorado Rose
HS&W Ruff Orrin Haverlah
HS&W Casey Lee Pee Dee
Tigo Five Forty One
Tabbatha Jane Buffington
Sprogs Bearly There Buffington
Mt Blue Shoshone Chief
HS&W Ebony
HS&W Kelbe Heidali Tobias
Kristone Eugene The Jeep
Kristone Murdoc
Yukon Dream
Silverledge Bonsai
Silverledge Heartbreaker
Silverledge Tuxedo Jack
Lor A Gypsy Traveler
Silverledge Lady of Spain
Lor A Yukons Denali
Lor A Smart Little Lena
Lor A Golden Zephyr of J2
Lor A Quemada Queen
Lor A Iyanbito Flint Lex
W Lazy J U Betcha
W Lazy J Yippie Ki Yi Yo
J Bar D Katie Bar the Door
Cove Creeks Iron Mike
Elizabeth Jane Buffington
Kona Buffington
Moriahs Mity Ruff Buffington
HS&W Cyborg Fang
HS&W Scout Sharpe
Bearlys Shadey Black Bear
Sprogs Bearly There
Sprogs Harlequin Bear
Thiexs Mity Mack Truck
Tristan T-Bone
Chevreherd Out's the Woods
Chevreherd Sweet Belle Sprog
Johjils Cassandra Buffington
J Bar D Mystic Warrior
Lovely Little Lady Smith
Praise Windwalkers Chilie
Rossys Jerry Rowe
HS&W Blazing Star
Kristone Mity Black Medicine
Gahl's Silverledge Smokin' Jose
Haverlahs Mity Lance
Haverlahs Ruff Ross
Chevreherds
Double Ruff
Johjils Blazing Chevee
Johjils Spicy Cayenne
Mity Hi Reddi I'm Teddi
Mity Hi Walkabout Wanda
Chevreherd Can't Be Bad
Chevreherd King Davids Abby
Nakai Little Bear
Terylyns Cinnamon Nutmeg
Bronze Tree Macchia Rossa
Mesa Prides Brown Dude
Rough N Rowdy Boy
Adkins Blackleg Belle

Just A Jock, blue, copper and white male (Colorado's Toast of the Town x Ruff N Reddy Remember Me) at his home in Missouri. Except for his markings he strongly resembled Sprog in size, type and temperament. Photo taken in August 1991 when Jock was about 1-1/2 years old. Photo by Author

Masterson's Gitel. Sprog produced a number of top quality dogs but Gitel was one of the best. Shown here as a young dog and out of coat. Photo by and courtesy of Melissa Couch

earlier. The woman's husband traveled a lot for his work and with several small children at home Jock became her companion and guardian when she was home alone. More than once he stood between the family and undesirable prowlers. When the family was finally able to purchase a small farm the chores mostly fell to the wife. She had been to the barn one night and was heading to the house in her truck. Jock was in the cab as usual and when she got out to close the gate somehow the truck slipped out of gear and started rolling toward her. Her back was to the vehicle and if it hadn't been for Jock's barked warning she would have been crushed between the gate and the truck!

All told, Sprog was bred to nine bitches in his lifetime, producing only 1 or 2 puppies in many of those litters (due to advanced age of several of these females). The majority were of very high quality and would have made significant contributions if they had been used more extensively. A few of his offspring have contributed to the future of several working dog breeding programs but many have been lost.

One female who was slated to be another Sprog mate was never bred to him due to his sudden death after being severely injured by a ram. The Gitel daughter Blaze was bred by Casey to be crossed back to Sprog to intensify his genes and preserve much of what she loved. She was waiting for Blaze to mature and for the most convenient time to make this cross. But while she was waiting, Sprog met a challenge he could not overcome. He was about 12 years of age and wandered into the barn where Casey's ill-tempered ram was penned. Nobody saw the accident but Casey speculates he was near a wall when the ram pinned him between his horns and the wall. After suffering internal injuries a valiant attempt was made to save him but he eventually succumbed. He lies on the mountain side behind Casey's barn with his ancestors and descendants.

Since his loss, he has remained the ideal Casey has tried to recreate through selectively breeding as closely as possible to the lines of the

PEDIGREE OF Remuda Homestead Hobbes

Born: ____________ Sex: ________
Owned by: ____________

Parents	Grandparents	Great-grandparents	Great-great-grandparents
Sire: Remuda Bar Mity Hobbes	Sire: Silverledge Slide Me To Da Bar	Sire: Silverledge Charlie Russell	Sire: Taylors Luke the Drifter
			Dam: Colorados Am I Blue
		Dam: Silverledge Black Magic	Sire: Silverledge Charlie Russell
			Dam: Silverledge Sasparrilla
	Dam: HS&W Blazing Star	Sire: Hangin Tree Huckleberry	Sire: Hangin Tree Red Zephyr
			Dam: Hangin Tree Blue Lady
		Dam: Mastersons Gitel	Sire: Newmans Blue Sprog
			Dam: Shadey Mountain Mity Blugoblin
Dam: Kristone Phoenix for Remuda	Sire: J Bar D Mystic Warrior	Sire: Queenslands Wibi Windwalker	Sire: Queenslands Bear Paws
			Dam: Queenslands Cahuilla Cayenne
		Dam: Mt Blue Christians Praise	Sire: Newmans Blue Sprog
			Dam: Misty Mountain Blue
	Dam: Kristone Medicine Woman	Sire: Silverledge Slide Me To Da Bar	Sire: Silverledge Charlie Russell
			Dam: Silverledge Black Magic
		Dam: Kristone Mity Black Medicine	Sire: Fairoaks Medicine Wolf
			Dam: Mastersons Gitel

Bred by: ____________ Address: ____________ Phone: ____________

The details given above are correct to the best of my knowledge. Signed: ____________(Breeder)

Remuda Homestead Hobbes, black and white male (Remuda Bar Mity Hobbes x Kristone Phoenix for Remuda). Photo by Author

Remuda Slidin Home Kristone "Skid" red and white female (Remuda Bar Mity Hobbes x Kristone Phoenix for Remuda) Hobbes litter sister and a different type of dog from her brother in both looks and personality. Photo by Author

breed's founders. In pursuit of that goal, Blaze was crossed to the Sprog sons Mt Blue Shoshone Chief STDc and HS & W Ace in the Hole BC CD. She was also bred once to Silverledge Slide Me To Da Bar CD STDs for a litter born in quarantine in England. Later her grandson Remuda Homestead Hobbes out of the Gitel granddaughter Kristone Phoenix for Remuda returned home to continue the legacy of Wood dogs in this country.

Hobbes is a solid black male with a sweet, kind temperament who has proven to be the producer his pedigree dictates. Bred to his aunt, Kristone HS&W Pipsissewa, he produced a number of excellent service dogs, including a fun loving female named Maya who works with deaf children and a serious blue merle male named Captain Hook who assists a wheelchair bound man. Hobbes was also bred to the Ace x Pip daughter, HS&W Kelbe Heidali Tobias to produce two outstanding females who will be instrumental in preserving the HS&W/Kristone lines into the future.

The blue merle J Bar D Mystic Warrior x Kristone Medicine Woman daughter, Kristone HS & W Pipsissewa, became the matron of the next generation of excellent dogs produced at HS & W. A kind sweet dog, she became the companion of Casey's elderly mother for many years, joining Borg as a spoiled house dog. Her apparent lack of enough grit to suit Casey was not entirely true. Although laid back the majority of the time she tolerated no foolishness or challenges from other dogs. Her natural guardian abilities were also just beneath the surface and she served Casey's mother well for many years as a best friend and alert caretaker.

Bred twice to her nephew Remuda Homestead Hobbes, she has produced a number of superior dogs including one who serves as an Assistance Dog to a handicapped man. Hobbes and Pip represent a slightly different variation of the bloodline in that their temperaments are somewhat milder and not the hard headed dogs difficult to place in homes today. Without losing the basic genetics of the line they are producing puppies which the more average dog owner can handle. Yet these dogs still show glimpses of the old tendencies when absolutely necessary. They are the perfect combination of old and new without the sacrifice of outcrossing and losing or diluting the genetic heritage of the Wood dogs. Through selection the tougher dogs can be resurrected in a generation if needed or desired through the "throwbacks," such as Ebony who demonstrate the old independent, tougher type of temperament. There is a dog for every situation and as a breeder it is important to acknowledge these differences, accurately evaluate potential owners, and place the right dog in the right situation.

HS&W Chuck Norris, blue and white male (Remuda Homestead Hobbes x Kristone HS&W Pipsissewa). Companion and service dog to Casey's mother and upcoming sire of the next generation at HS&W. Courtesy of Casey Masterson Newman

Bred to HS & W Ace in the Hole BC CD, Pip has produced two females which are breeding on for HS & W—HS & W Ebony and HS & W Kelbe Heidiali Tobias. This cross intensifies the genes of Sprog, Silverledge and the foundation dogs behind them producing dogs which still exhibit all the great qualities that Fletcher Wood valued over 50 years ago.

HS&W Ebony, black and white female (HS&W Ace in the Hole BC CD x Kristone HS&W Pipsissewa) Her working style is similar to Sprog's—quiet, confident, controlled and forceful only when needed. Photo by Author

Ebony is Rick and Casey's special dog who is the closest thing to recreating Sprog and Shunka Casey has yet achieved. A solid black with deep brown eyes she has flawless conformation and an intelligent mind. Extremely protective, she has a special talent at fending off trespassing bears. A troublesome bruin once climbed the fence into Casey's yard and Eb was the first to sound the alarm. When Casey opened the door, the 40 pound Eb treed the bear in short order. Casey called her back into the house so the intruder could leave. He must not have been convinced of Eb's ownership of this territory and returned a short time later to take a lamb from Casey's small flock. Eb, again, was the first to discover the marauder, was released and tore into the bear a second time. When he turned to defend himself, the lamb escaped and by now the Newmans had loaded a gun and managed to scare him off permanently by firing in his direction.

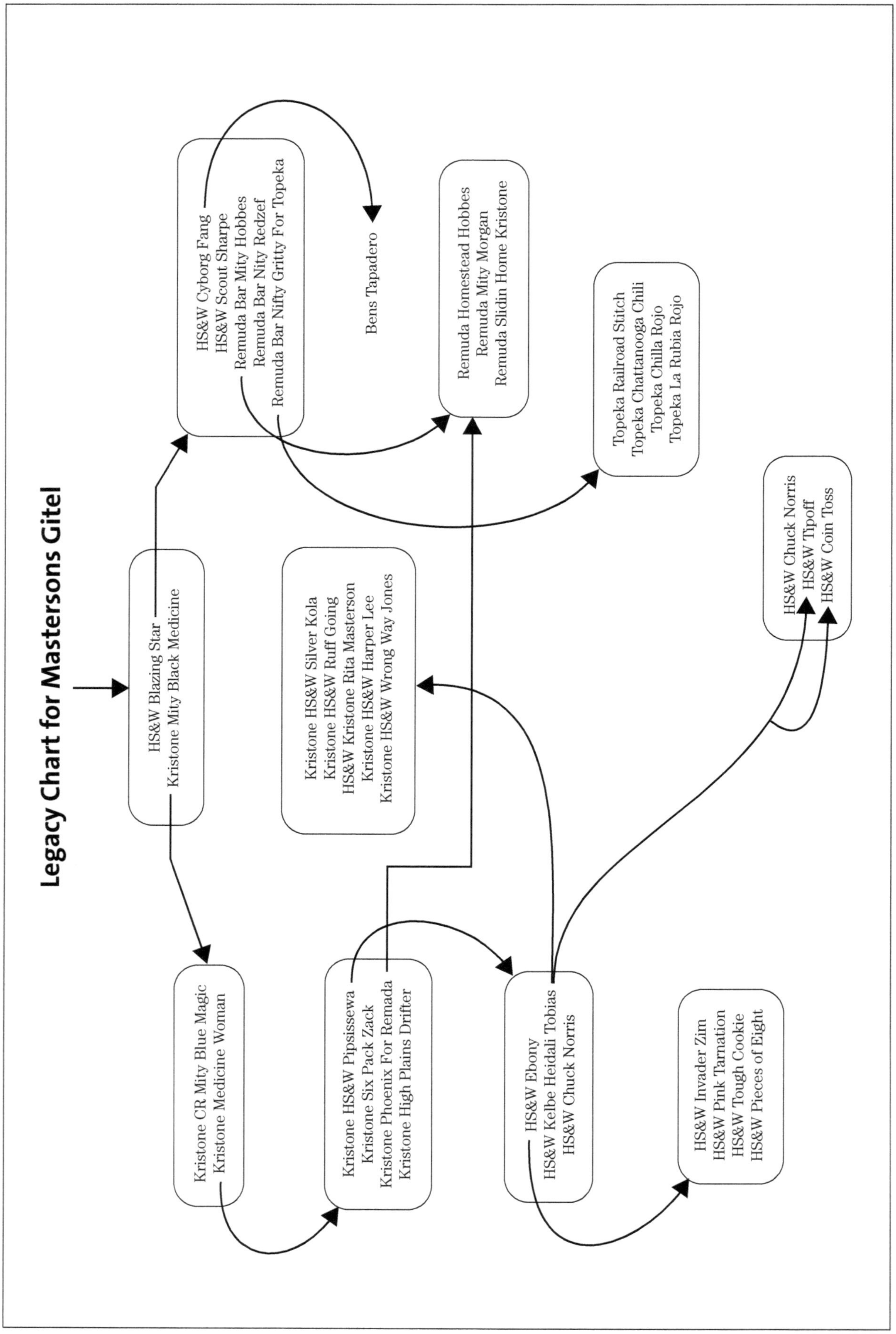
Legacy Chart for Mastersons Gitel
HS&W Blazing Star
Kristone Mity Black Medicine
HS&W Cyborg Fang
HS&W Scout Sharpe
Remuda Bar Mity Hobbes
Remuda Bar Nity Redzef
Remuda Bar Nifty Gritty For Topeka
Bens Tapadero
Remuda Homestead Hobbes
Remuda Mity Morgan
Remuda Slidin Home Kristone
Topeka Railroad Stitch
Topeka Chattanooga Chili
Topeka Chilla Rojo
Topeka La Rubia Rojo
Kristone HS&W Silver Kola
Kristone HS&W Ruff Going
HS&W Kristone Rita Masterson
Kristone HS&W Harper Lee
Kristone HS&W Wrong Way Jones
HS&W Chuck Norris
HS&W Tipoff
HS&W Coin Toss
Kristone CR Mity Blue Magic
Kristone Medicine Woman
Kristone HS&W Pipsissewa
Kristone Six Pack Zack
Kristone Phoenix For Remada
Kristone High Plains Drifter
HS&W Ebony
HS&W Kelbe Heidali Tobias
HS&W Chuck Norris
HS&W Invader Zim
HS&W Pink Tarnation
HS&W Tough Cookie
HS&W Pieces of Eight

Ebony doesn't confine her guardian abilities to four footed felons. Rick and Casey were inspecting a property they owned one afternoon when they surprised a burglar. Fortunately the crook took off rather than fought but Eb was having none of that. He disappeared for awhile but Eb tracked him a mile down the train tracks and held him while the police were called to haul him off.

"Trixie" black female by Chuck x Ebony. She is her mother's daughter, with the strong temperament and good sense of her line. Photo by Author

Ebony has produced two litters so far. The first by her half brother HS&W Chuck Norris (Remuda Homestead Hobbes x Kristone HS&W Pipssisewa) resulting in two puppies which are in good pet homes.

Her second litter was by Chevreherd Smokin' Flintlock (J Bar D Smoke Signal x Chevreherd Cowhock Nibbler—a Cowtown Renoir daughter). Renoir was a Ruff N Reddy Mity Magician x Powder Keg's Echo's Eveready son—essentially the old proven cross of Ted and Hemi). Four puppies resulted from this litter and all are waiting in the wings for their turn to preserve the line into the future.

HS&W Valentino, red merle and white male (HS&W Chuck Norris x HS&W Ebony). Brother to Trixie with a softer more easy going temperament. Even the easy going ones retain innate characteristics such as guardian instincts which surface when needed. Photo by Author

One of Hobbes and Pip's daughters, HS & W Heidali Sheba, was once out running on the ranch of her owner with her housemates, two sled racing huskies. Coyote packs were becoming more and more dangerous in the mountains and have become very cunning at killing dogs. One member of the pack will appear in an open area seeming to be alone. A dog or several dogs will be lured into thinking they can attack the lone coyote and begin chasing it. The loner will head for the trees where the rest of its pack are lying in ambush. Once the dog or even several dogs reach the trees the pack surrounds and kills the unsuspecting canines. This trick fooled one of Sheba's husky buddies on the day they were roaming the ranch. By the time she and the two huskies were found, one husky was nearly disemboweled but the other two dogs were unharmed. The injured dog was still alive and rushed to a veterinarian where she was put back together and survived. The only conclusion that

HS&W Heidali Sheba, blue and white female (Remuda Homestead Hobbes x Kristone HS&W Pipsissewa) a beautiful dog Jean Taylor said reminded her strongly of some of her original dogs. Photo by Author

HS&W Kelbe Heidali Tobias, red merle and white female (HS&W Ace in the Hole BC CD x Kristone HS&W Pipsissiwa) a beautiful dog with a great sense of humor and the ability to do anything asked of her. Photo by Author

can be drawn is that Sheba and the other husky heard the cries of their pal, ran to the rescue and somehow managed to defeat the pack intent on making a meal of their friend.

Sheba and Kelbe were often used as substitute huskies in the team of a local teenager who enjoys sled racing. While out training one afternoon in the mountains a quick winter storm blew in. The teen decided she had better get off the mountain quickly before she was stranded alone out of range of help. While unhitching the dogs and loading them in her car, Kelbe bolted into the thick, blowing snow and was gone. Her young driver called for her but could not launch a search of her own without endangering her own life. She left without Kelbe hoping to be able to come back the next day to find her.

She enlisted the help of several people the next day, including Casey. They all left articles of clothing at the last place she was sighted in case she returned to find them and then spread out to cover the surrounding territory. As Casey headed up one canyon a group of snowmobilers passed by and stopped to find out what she was doing. She explained the situation and asked them to watch for the missing dog. They promised to help out and sped off. 30 minutes later Casey heard the sound of the returning machines and riding happily behind one driver was Kelbe! They had found her far up the path where she had likely sought shelter from the storm.

Kelbe has produced two litters so far, one by her uncle Remuda Homestead Hobbes. This litter resulted in two superior females who will contribute to the future of HS&W and Kristone. Her second litter was a landmark cross with the old Ruff N Reddy bred dog, Bearfoots Red Man Chew Rossy "Chewy" (Ruff N Reddy Joe Willy Wang x Ruff N Reddy Shadey Wood Ash). The wonders of modern technology and the generosity of Pincie Creek kennel allowed this cross to be made even though "Chewy" had been dead for over 10 years. 6

stunning puppies are now growing up and waiting for their turn to make their mark as some of the few remaining high percentage Wood dogs in the world!

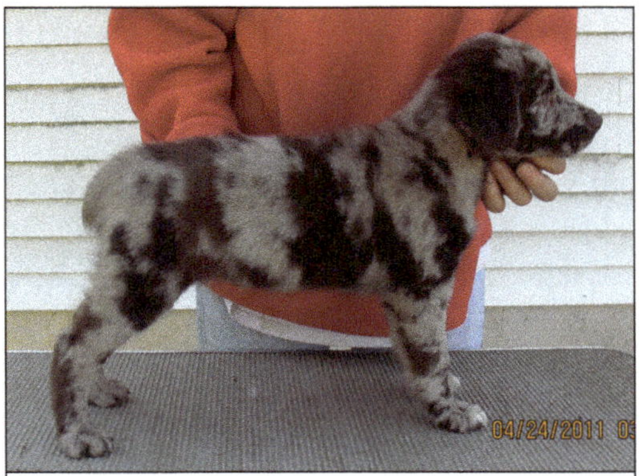

Kristone HS&W Ruff Going, red merle and white male (Bearfoots Red Man Chew Rossy x HS&W Kelbe Heidali Tobias) Part of the future of HS&W, Kristone and the preservation of the Wood bloodline. Photo by Author

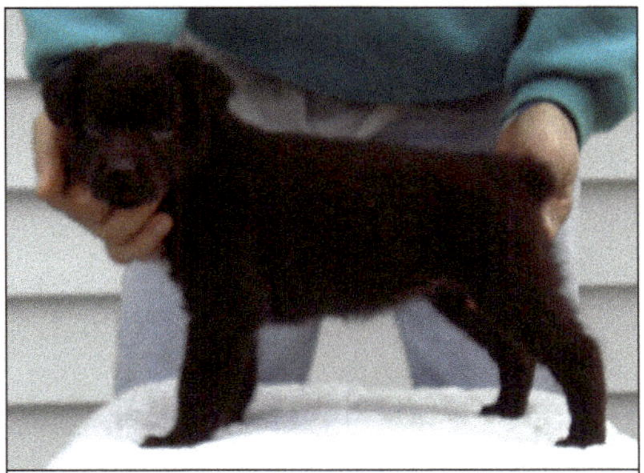

Kristone HS&W Harper Lee, red and white female (Chewy x Kelbe) Photo by Author

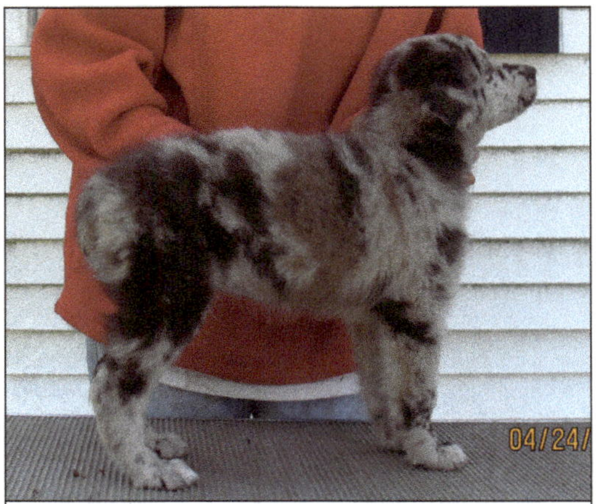

Kristone HS&W Wrong Way Jones, red merle and white male (Chewy x Kelbe) Photo by Author

Kristone HS&W Silver Kola, red merle and white female (Chewy x Kelbe) Photo by Author

Pincie Creek "Obie," red merle and white female (Chewy x Kelbe) Photo by Author

HS&W Kristone Rita Masterson, red merle and white female (Chewy x Kelbe) Photo by Author

The two blue merle brothers, Teddy (left) with Sprog (right) in January 1984 as youngsters. Courtesy of Casey Masterson Newman

Sprog worked hogs with the same calm, strong style he showed on cattle. Courtesy of Casey Masterson Newman

Teddy as a young dog after his return to HS&W. Courtesy of Casey Masterson Newman

Mother and son, Shunka (left) and Sprog (right) in January 1986. Courtesy of Casey Masterson Newman

Sprog's litter brother Teddy living the good life with Casey's mother. Courtesy of Casey Masterson Newman

Shunka (top) and Sprog (bottom) in Casey's barn waiting to be needed. January 1986. Courtesy of Casey Masterson Newman

Teddy showing his classic Wood head type.
Courtesy of Casey Masterson Newman

Sprog's brother "Teddy" an excellent dog who could have been the patriarch of HS&W kennel but for a chance behavior as a puppy!
Courtesy of Casey Masterson Newman

Naturally correct, Sprog poses for a quick picture between stock working sessions.
Photo by and courtesy of Melissa Couch

Sprog had a quirky affinity for rabbits and considered them his special wards. One has strayed from the group and he is trying to figure out how to gently bring it back!
Courtesy of Casey Masterson Newman

Sprog getting some real world experience working cattle, learning the cover a big herd and keep them together and moving.
Courtesy of Casey Masterson Newman

Sprog was a tough cattle dog but had a soft spot for working smaller livestock. He never harmed them or pushed them too hard.
Courtesy of Casey Masterson Newman

Sprog was comfortable with cattle at the head or the heel. Proper instruction on handling cattle added to his natural confidence to make him a useful and skilled working dog. Courtesy of Casey Masterson Newman

Kids CAN be fun! Sprog on the left with Blaze and two of Casey's three children having fun in a treehouse. Courtesy of Casey Masterson Newman

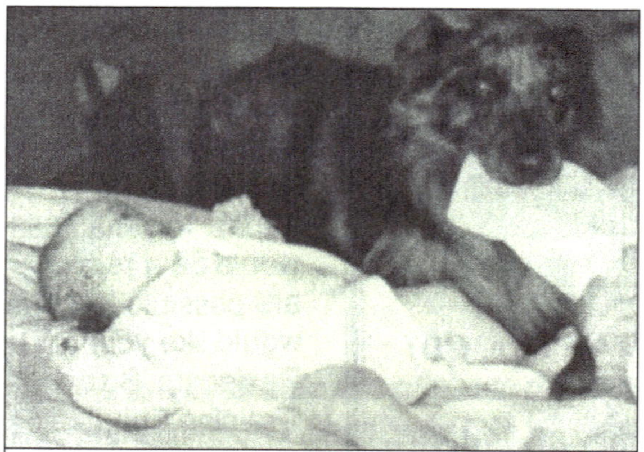

Of course kids involve a little work, too. Sprog fetches a clean diaper for a new baby. Courtesy of Casey Masterson Newman

Sprog showing his agility ability. Notice that he is standing on a regular upside down 5 gallon bucket! Photo taken spring 1991. Courtesy of Casey Masterson Newman

Sprog during a break while working stock. He was always up for a little performance of his wide repertoire of tricks! Photo by and courtesy of Melissa Couch

Blue and white female from the Sprog x Misty litter. Photo by Author

Blue and white female from Sprog x Misty. Photo by Author

Blue and white female from Sprog x Misty. Photo by Author

White turkeys or brown ones, it was all the same to Sprog. Courtesy of Casey Masterson Newman

Sprog didn't care what kind of stock he worked. He always did the same competent job. Casey tried her hand at raising turkeys for a short time, Sprog took it in stride. Courtesy of Casey Masterson Newman

Mt Blue Shoshone Chief STDc, red merle, copper and white male (Sprog x Misty Mountain Blue CD STDdc OTDs) a nice blend of the Wood/Hemi/old J Bar D type dogs. Photo by Author

"Borg" was a dog of great character and many talents! Courtesy of Casey Masterson Newman

Borg was a pleasing dog of good type. Photo by Author

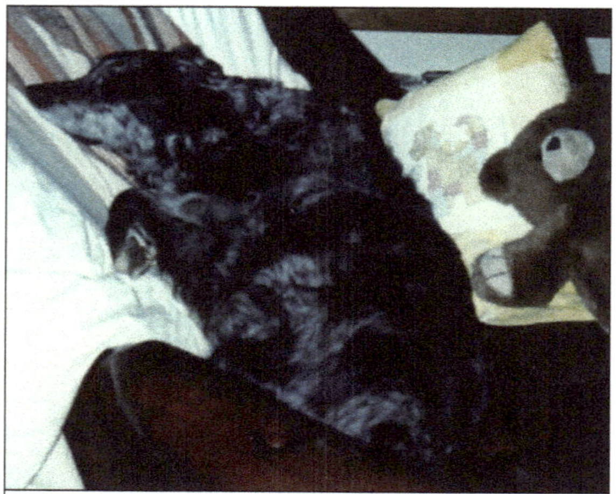

Shadey Mountain Flirt, blue merle and white female (Wiliam's Laddie x Shadey Mountain Lady II) a beautiful superior female with the genetic background to have contributed greatly to the preservation of the mountain-type dog. Sadly she was unable to conceive after Casey obtained her. Fortunately she enjoyed a comfortable retirement! Courtesy of Casey Masterson Newman

Shadey Mountain Flirt looking much younger than her years. Courtesy of Casey Masterson Newman

Shadey Mountain Flirt—a top female in any age. Courtesy of Casey Masterson Newman

Shadey Mountain Flirt. Courtesy of Casey Masterson Newman

Shady Mt Mity Blu Goblin, very dark blue merle and white female (Ch Jones' Reddy Teddy CDX OTDdsc SchHA x Jones' Mity Bite). A lucky find for Casey who proved that genetics are everything when it comes to breeding. Courtesy of Casey Masterson Newman

A happy ending! Several of the elderly dogs Casey collected early in her breeding career to try to preserve the old type of Wood dogs. The star crossed Goblin was eventually placed with this older couple to live out her life in luxury as a well loved only dog. Pictured here from left to right are: Masterson's Gitel, Pawsies (black puppy on man's lap from Sprog x Ch Ruff N Reddy Pee Dee Que), Shady Mt Mity Blu Goblin—renamed "Dinkum" by her new owners, Shadey Mountain Mity Miss, and lying in front, Masterson's Dotclijee Shunka. Courtesy of Casey Masterson Newman

Gitel, good looking, good moving, good intelligence, good working ability, she was the total package. Photo by and courtesy of Casey Masterson Newman

Gitel showing the early working ability treasured and bred for by Cathy Jones. She inherited all the best qualities of the Ruff N Reddy line from Goblin. Courtesy of Casey Masterson Newman

Gitel as a baby. Compare this photo to that of Ruff N Reddy Hunny Bear as a puppy. The two could have been clones. Courtesy of Casey Masterson Newman

Masterson's Gitel, blue and white female, the only puppy ever produced by Shady Mt Mity Blu Goblin. Shown here as a young puppy. A female of the highest quality who produced the two best females ever for HS&W and Kristone kennels. Courtesy of Casey Masterson Newman

Masterson's Gitel as a young, mature female. Her resemblance to her aunt Ruff N Reddy Hunny Bear continues to hold true. Courtesy of Casey Masterson Newman

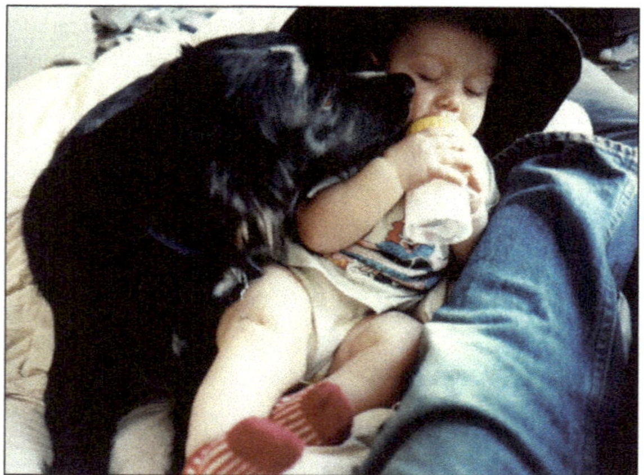

Blaze was a consummate "kid dog." Her blue merle marking is easily seen in this photo going sideways across her forehead. Courtesy of Casey Masterson Newman

Masterson's Gitel seems to remember some of the lessons she learned in Montana helping on the Johnson sheep ranch while being bred. She is rating her sheep well and is under control. Photo by and courtesy of Melissa Couch

HS&W Blazing Star, phantom blue merle and white female (Hangin Tree Huckleberry x Masterson's Gitel). An outcross to bring in more biddability on stock, Blaze still displayed many strong Wood traits along with a bit softer temperament and more co-operative attitude when working livestock. Courtesy of Casey Masterson Newman

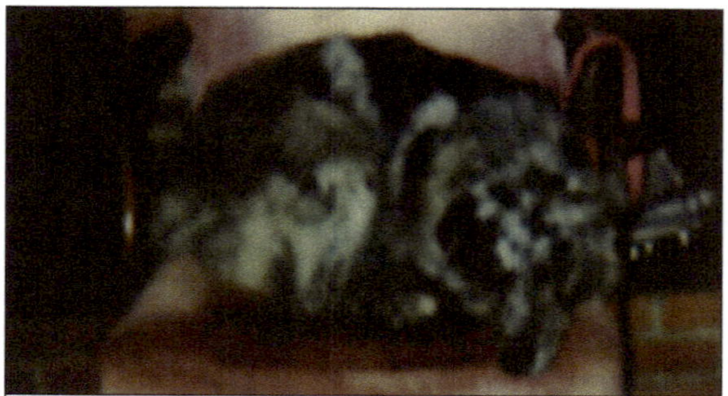

"Taz," sister to Blaze sold to Texas. Blue merle female. Courtesy of Casey Masterson Newman

Blaze was a solid, sound, well made female. Photo by Author

"Tigger" (Shoni x Blaze) Courtesy of Casey Masterson Newman

HS&W Cyborg Fang, blue and white male, at about 4-5 months of age (October 1994) Courtesy of Casey Masterson Newman

Borg was a solid, athletic, well made dog. Photo by Author

Mity Miss had two striking light blue eyes in her dark face. She never failed to turn heads due to her overall quality and superb temperament. Photo by and courtesy of Melissa Couch

Shadey Mountain Mity Miss as an active teenager. She still got to go along with the crowd to training days even though she was just a spectator. Photo by and courtesy of Melissa Couch

Mity Miss enjoying a day out with the crowd! Even an old grandma likes to be included in the fun. Photo by and courtesy of Melissa Couch

Ace as a young pup going to the head on sheep. Courtesy of Casey Masterson Newman

Ace pictured against a "backdrop" of his dam Mity Miss. He is about 9 months old in this photo. Photo by and courtesy of Melissa Couch

Ace at about 9 months of age getting one of his first real working lessons on sheep. Photo by and courtesy of Melissa Couch

Everybody learns how to do their job around the ranch. Ace learning to pull Sprog's dogcart and give the kids a ride. In this photo Casey says, "Ace had worn a sled harness one time before and pulled a piece of wood. We harnessed him in the cart harness and attached the cart. Jess jumped on it and I said 'he's too young to pull you!' Jess pleaded, 'just to the mail box' and Ace was great, just like he knew what he was doing." Courtesy of Casey Masterson Newman

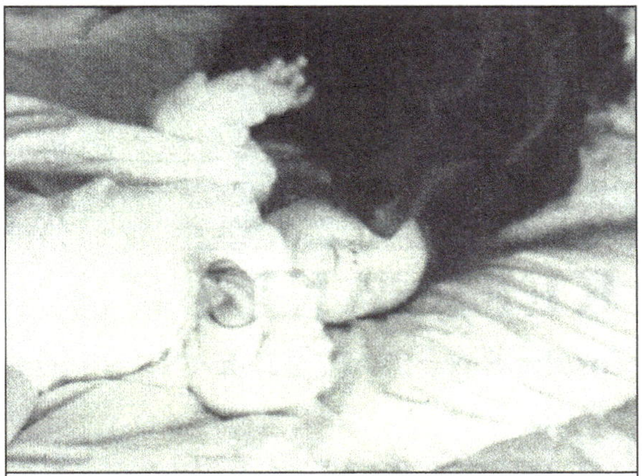

Ace helping take care of the baby—somebody has to clean up! Courtesy of Casey Masterson Newman

HS&W Ace in the Hole BC CD. Photo by Author

Blue female by Ace x Queensland's Spunky Haverlah (1997) Courtesy of Casey Masterson Newman

HS&W Ace in the Hole BC CD, a superb individual who produced as well as his pedigree dictated. Photo by Author

Blue female by Ace x Queensland's Spunky Haverlah (1997) Courtesy of Casey Masterson Newman

Black male by Ace x Queensland's Spunky Haverlah (1997) Courtesy of Casey Masterson Newman

Black male by Ace x Queensland's Spunky Haverlah, 10 weeks old. Courtesy of Casey Masterson Newman

Kristone Anna Lightfoot, a pleasing blend of Wood and Fairoaks breeding. Photo by Author

Kristone Anna Lightfoot at about a year of age. Photo by Author

Kristone Eugene the Jeep, red merle and white male (HS&W Ace in the Hole BC CD x Echo Ridge No Regrets CD) This male resembled his grandsire Sprog in many ways. Photo by Author

Jeep at home. Like his grandsire Sprog he was not a trained "show dog" but was naturally correct. Photo by Author

Gitel during her stay in Missouri to be bred for a second time. She is a beautiful mature matron at this time. Photo by Author

Kristone Mity Black Medicine, solid black female (Ch Fairoaks Medicine Wolf CD x Mastersons Gitel) Called "Shadow," like her mother she was a singleton puppy who fortunately was a very high quality dog. She is pictured here at about 10 weeks of age. Courtesy of Casey Masterson Newman

Jeep had the right combination of sweet friendliness with a guardiness just under the surface ready when needed. Photo by Author

Kristone Canta Dulce "Dulce" black and white female (HS&W Ace in the Hole BC CD x Echo Ridge No Regrets CD) a beautiful blend of her two bloodlines. Photo by Author

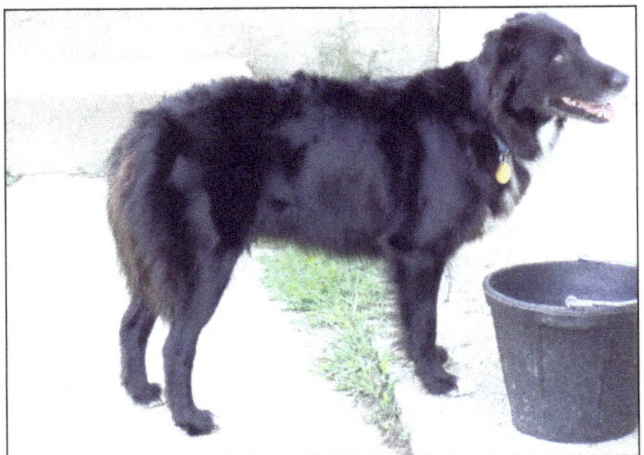

Dulce showing off her lovely conformation. Photo by Author

Jeep strongly favored the Wood side of his pedigree. Photo by Author

Jeep was a correct and balanced dog showing the athleticism typical of his line. Photo by Author

Mom Gitel with her single black puppy Shadow. Courtesy of Casey Masterson Newman

Shadow at approximately 9 weeks of age. Photo by Author

Shadow at about one year of age. Photo by Author

Two priceless genetic treasures. (Left) HS&W Ace in the Hole BC CD (right) Kristone Mity Black Medicine CD. Pictured in February 1997. Photo by Author

Shadow as a young dog. Photo by Author

Tigger x Shadow puppies about 5 weeks of age. Photo by Author

Charlie as a gangly teenager but still showing all the classic type of his Wood breeding. Photo by Author

Charlie as a mature dog. Always a breathtaking dog at any age. Photo by Author

Charlie at maturity. It is easy to imagine him working Fletcher Wood's cattle ranch. Photo by Author

Charlie in old age, approximately 12 years. His Wood type became even more pronounced as he matured. Photo by Author

Charlie as a 12 year old. He has darkened to near black from his youthful blue merle color. Photo by Author

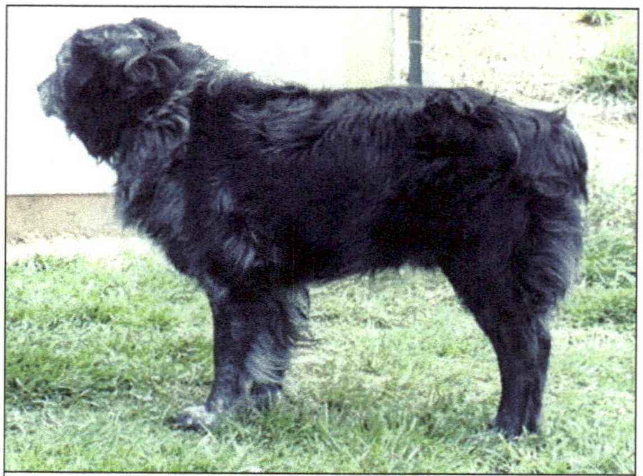

Soundness and good structure stand up to time. Charlie at 12 years of age. Photo by Author

The broad, square head type that is the trademark of the Wood dogs. Charlie would never be mistaken for any other bloodline. Photo by Author

Kristone Colt Tyler, "Colt" solid black male (Silverledge Slide Me to Da Bar CD STDs x Kristone Mity Black Medicine CD). A slightly different head type from his brother Charlie but another typical Wood type dog. Photo by Author

Colt showing a more Ruff N Reddy head type similar to his dam. Photo by Author

Brick showing the head type of the female side of her family. Photo by Author

Colt, a big solid dog of the old "thoroughbred" type favored by Fletcher Wood. Photo by Author

Kristone Mity Screamin' Mimi "Mimi" black and white female (Silverledge Slide Me to Da Bar CD STDs x Kristone Mity Black Medicine CD). A remarkable female who inherited her sire's intuitive nature about humans. Photo by Author

Kristone Mity Bad Penny, "Brick" red merle and white female (Silverledge Slide Me to Da Bar CD STDs x Kristone Mity Black Medicine CD). Brick had a strong resemblance to her granddam, Gitel and her great-granddam Goblin. Photo by Author

Mimi, a high quality female who was sadly never bred. Photo by Author

Kristone Medicine Woman, solid black female (Silverledge Slide Me to Da Bar CD STDs x Kristone Mity Black Medicine CD) a superior type and intense working dog. Shown here at about one year of age. Photo by and courtesy of Michelle Murvai

Kristone Medicine Woman "Quinn" showing her good working style on sheep. Photo by and courtesy of Michelle Murvai

"Quinn" as a mature female. She is pregnant with her first and only litter in this photo. Photo by Author

Quinn, like most dogs of this line, was a very correct dog who needed no training to in order to "pose"! Photo by Author

Quinn epitomized the breed standard in many ways, including her good head type. She had the small ears of her sire which tended to stand upright as she aged. Photo by Author

Zack at about 10 weeks. A clear blue with distinct black patches he would turn nearly black as he matured. Courtesy of Michelle Murvai

Kristone Six Pack Zack at about 14 months. An excellent blend of Wood, J Bar D and Sprog breeding. Photo by and courtesy of Michelle Murvai

Zack, darkening with age to look nearly black. His dark merling is still visible on his head. Courtesy of Michelle Murvai

Zack as a mature dog looking very much like the old Queensland type dogs. Courtesy of Michelle Murvai

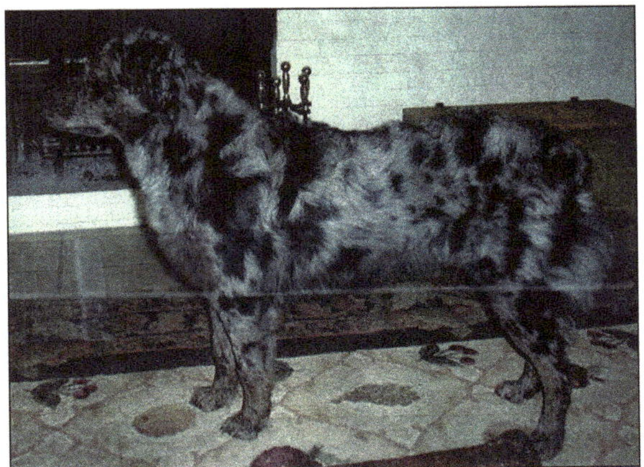

"Rocky," blue and white male (J Bar D Mystic Warrior HS STDcd OTDs x Kristone Medicine Woman) This nice male is very reminiscent of the old J Bar D and Queensland dogs. Photo by and courtesy of Casey Masterson Newman

Zack, a large stout dog of the type Fletcher Wood selected for and produced. Courtesy of Michelle Murvai

Kristone High Plains Drifter CD JHD HTAD1 "Cowboy," blue merle and white male, (J Bar D Mystic Warrior HS STDcd OTDs x Kristone Medicine Woman) a strikingly marked but typical Wood type dog. His working ability is second to none. Photo by and courtesy of Hope Stewart

Cowboy showing off his gorgeous head and expression. Courtesy of Hope Stewart

Cowboy had the good movement needed for performance in any venue. Courtesy of Hope Stewart

Cowboy was a smooth and controlled working dog. He earned several AHBA herding titles and two legs on his AKC started sheep title before a back injury ended his performance career. Courtesy of Hope Stewart

"Kit" as a young dog. Photo by Author

Cowboy—curious, intelligent, friendly, an overall good dog who represents his bloodline very well. Courtesy of Hope Stewart

"Kit" strongly resembled her sire, Mikey, in head type. Photo by Author

Kristone Griti Iti Biti Kiti, "Kit" blue merle and white female (J Bar D Mystic Warrior STDcd OTDs x Kristone Medicine Woman) as a young dog. Photo by Author

Kristone Mity Intrepid. Photo by Author

J Bar D Mystic Warrior "Mikey" HS STDcd OTDs

Kristone Mity Intrepid "Trip," black and white male (J Bar D Mystic Warrior STDcd OTDs x Kristone Medicine Woman) a dog of great character, intelligence and working ability. Photo by Author

Trip was a beautiful, masculine dog with a fearless outlook on life. Photo by Author

Renier's Wood Rose, red merle and white female known as "Stitch" (Sprog x Ch Ruff N Reddy Pee Dee Que) doing a nice job of working sheep in one of her few opportunities to work. Photo taken in October 1991. Courtesy of Casey Masterson Newman

Renier's Wood Rose as a young dog. A nice blend of the Ted and Hemi type. Photo by and courtesy of Melissa Couch

Renier's Wood Rose showing the typical good conformation of this bloodline. Photo by Author

Renier's Wood Rose "Stitch" as a mature female. Photo by Author

"Stitch" was a good blend of her Sprog and Ruff N Reddy heritage. Photo by Author

"Stitch" was a similar type to her dam Pee Dee Que in many ways, including head type. Photo by Author

"Pawsies," black and white female (Sprog x Ch Ruff N Reddy Pee Dee Que). Working at a young age continued to come through from the Ruff N Reddy side of the family as well as the Wood side. Photo taken February 1989. Courtesy of Casey Masterson Newman

Country Lane's Mity Hot Chili Pepper, red merle and white female shown shortly after weaning her Sprog litter. Linebred on the wildly successful cross of Ted x Mity Bite and comparable in quality to the great Shady Mountain Mity Miss. Photo by Author

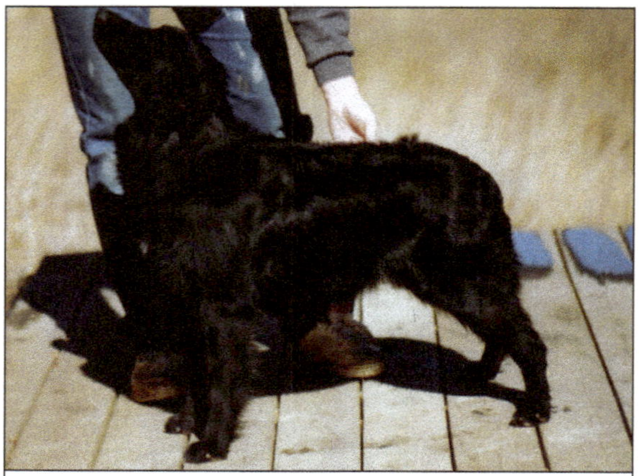

"Cinder" solid black female and the only female from the litter Sprog sired out of Country Lane's Mity Hot Chili Pepper. A beautiful female who was never bred. Photo by Author

Consistency of type is obvious in this photo! (Top) Sprog with his daughter Gitel (right) and one of his mates, (left) Country Lane's Mity Hot Pepper (Ruff N Reddy Ranger x Ruff N Reddy Hunny Bear). Gitel and Pepper were out of full sisters and their close breeding is evident! Courtesy of Casey Masterson Newman

"Cinder" (Sprog x Pepper) another typical Wood type dog. Photo by Author

"Cinder"—photo taken when she was about a year of age. Photo by Author

Two red merle males from Sprog x Pepper. Foreground is "Sam," behind him is HS&W Ruff Orrin Haverlah. Photo by Author

"Sam" showing his good head type, typical of his heritage. Photo by Author

"Sam" (Sprog x Pepper) red merle male as a mature dog. Photo by Author

Sprog with his litter of puppies from Country Lane Mity Hot Chili Pepper. The red male is closest to him. This litter was very consistent in quality and type. Photo by Author

"Sam" although placed in a pet home, was a high quality dog and a credit to his breeding. Photo by Author

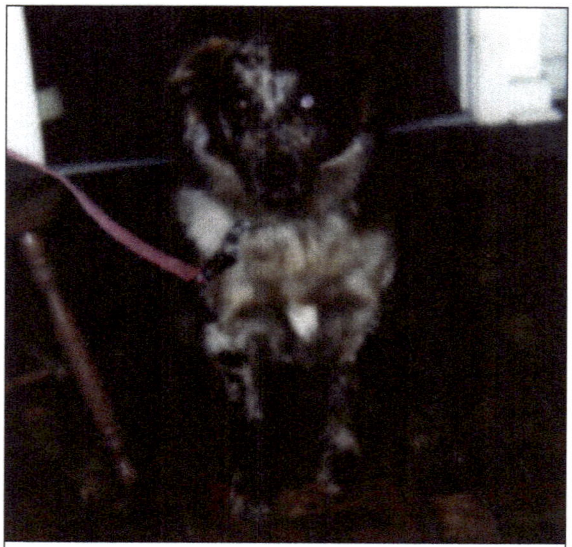

HS&W Ruff Orrin Haverlah, red merle and white male (Sprog x Country Lane's Mity Hot Pepper) A top male who sired a number of litters in Texas. Photo taken February 1993. Courtesy of Casey Masterson Newman

HS&W Ruff Orrin Haverlah. The resemblance to his dam is striking. Courtesy of Casey Masterson Newman

HS&W Ruff Orrin Haverlah as a young dog in February 1993. Courtesy of Casey Masterson Newman

Orrin—working dogs seldom strike "show poses"! (1993) Courtesy of Casey Masterson Newman

HS&W Ruff Orrin Haverlah. Courtesy of Casey Masterson Newman

Black male by Ace x Queensland's Spunky Haverlah. Courtesy of Casey Masterson Newman

Ruff N Reddy Shadow, black tri female (Colorado's Toast of the Town x Ruff N Reddy Remember Me) Courtesy of Casey Masterson Newman

Just A Jock, his resemblance to his sire is obvious.(1995) Photo by Author

Just A Jock. (1995) Photo by Author

Just A Jock (1995) Photo by Author

Sprog entertaining neighborhood children on the mountain roads. If Casey asked, his answer was always "yes." Courtesy of Casey Masterson Newman

Remuda Homestead Hobbes, black and white male in 2010. Photo by Author

Pip at home showing off her good type and structure. Photo by Author

Just A Jock—a high quality male sadly never used for breeding. Photo by Author

Just A Jock working sheep with Claudia Nelson before his move to Missouri. His nice balance and control is all natural. Courtesy of Casey Masterson Newman

Just A Jock covering his sheep well and quietly. Photo taken in October 1991. Courtesy of Casey Masterson Newman

Kristone HS&W Pipsissewa, "Pip" blue merle and white female (J Bar D Mystic Warrior STDcd OTDs x Kristone Medicine Woman). A high quality female from a high quality litter, Pip has been the foundation of a new generation of HS&W and Kristone dogs. Photo by Author

Pip—"just" a typical Wood type dog. Photo by Author

Hobbes at home in Colorado. Photo by Author

Skid, although different from her brother Hobbes is nonetheless a quality dog of extreme athletic ability and phenomenal trainability. Photo by Author

Skid, an athlete deluxe and a dog who harkens back to the old Ruff N Reddy dogs in type. Photo by Author

Chuck, another nice example of the good conformation and excellent type still found in the old Wood type dogs. Photo by Author

Shadey Mountain Mity Miss, with her half sister Goblin, were the two treasures Casey wisely bred at the end of their lives to preserve the best of the old Wood genetics into the future. Photo by and courtesy of Melissa Couch

"Flint" at about 4-5 months of age working sheep. Courtesy of Terri Daniel

Ebony working sheep. With only minimal training she instinctively finds balance and control. Photo by Author

Ebony ready to take on the challenge. Photo by Author

HS&W Ebony, intense, intelligent, correct, athletic, everything an Aussie should be. Photo by Author

Good movement often goes with good working ability. Eb takes her sheep back to Casey. Photo by Author

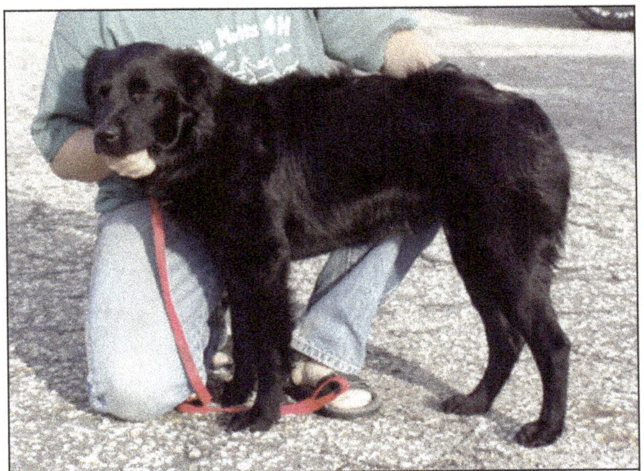

Ebony as a young dog at a little over a year of age. Photo by Author

Ebony as a beautiful mature dog. She has the good head always thrown by Mity Miss and the strong correct structure typical of the entire line. Photo by Author

Cheverherd Smokin' Flintlock, "Flint" blue merle and white male (J Bar D Smoke Signal x Chevreherd Cowhock Nibbler) Photo from about 2005 Courtesy of Terri Daniel

"Flint" at about 1 year of age. Courtesy of Terri Daniel

"Flint" at about 4-5 months of age showing excellent instinct on sheep. Courtesy of Terri Daniel

HS&W Pieces of Eight "Calypso," blue and white female (Chevreherd Smokin Flintlock x HS&W Ebony) Photo at 4-5 months of age. Courtesy of Casey Masterson Newman

Calypso at about one year of age. Courtesy of Melissa Couch

Calypso at about one year of age. Her stunning head type is just the beginning of her many good qualities. Courtesy of Melissa Couch

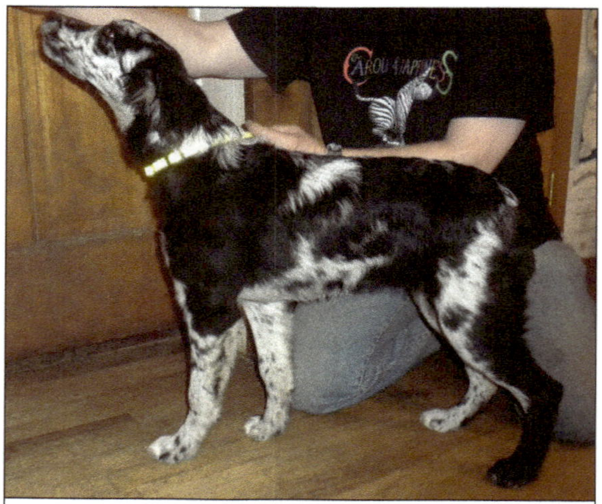

HS&W Tough Cookie "Brain," blue and white female (Chevreherd Smokin Flintlock x HS&W Ebony). Photo at 4-5 months of age. Courtesy of Casey Masterson Newman

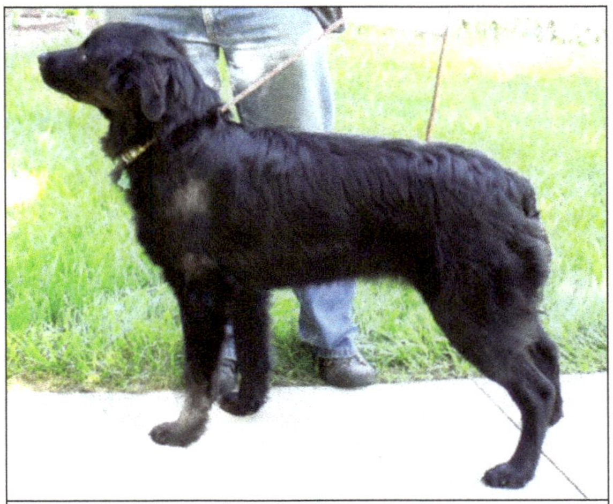

HS&W Invader Zim, black and white male (Chevreherd Smokin Flintlock x HS&W Ebony) at about 9-10 months of age. He inherited good type from his dam and a sweet temperament from his sire. Photo by Author

Zim developing into a handsome young male. About two years of age. Courtesy of Casey Masterson Newman

HS&W Invader Zim. His kind expression reflects his good temperament and his dam's excellent type. Photo by Author

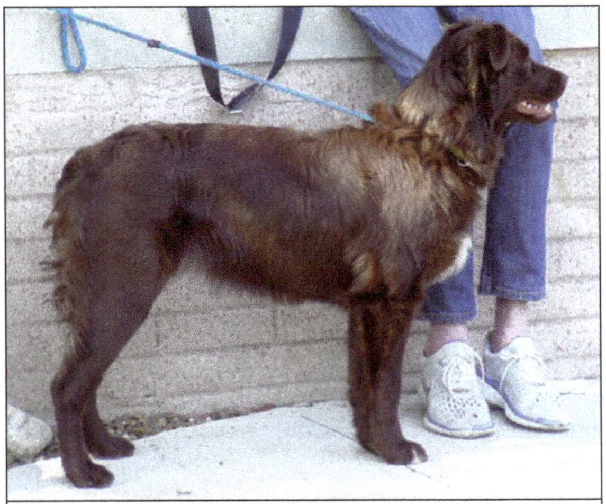

HS&W Pink Tarnation, phantom red merle and white female, (Chevreherd Smokin Flintlock x HS&W Ebony) athletic, intelligent with the good common sense Casey so values in her dogs. Photo by Author

HS&W Pink Tarnation. Her small red merle marking can be seen on the left side of her nose. Photo by Author

HS&W Pink Tarnation "Pinky." Her head type is nearly identical to her dam's. Photo by Author

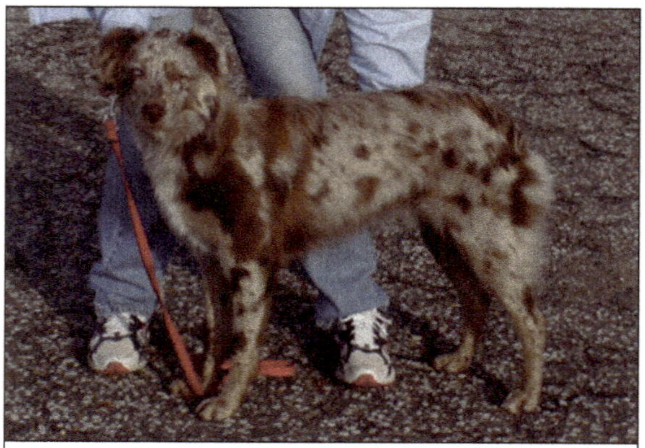

Kelbe as a youngster, a little over one year of age. Photo by Author

Even in her summer coat, Kelbe can hardly be faulted. Photo by Author

The two females from Kelbe's first litter by Remuda Homestead Hobbes at about 6 weeks of age. The black and white female is HS&W Coin Toss "Meg" and the blue merle is HS&W Tipoff "Bella." Photo by Author

Sheba as a young, out of coat teenager. Her solid quality is very evident. Photo by Author

HS&W Tipoff, "Bella" blue merle and white female (Remuda Homestead Hobbes x HS&W Kelbe Heidali Tobias) the new generation of HS&W dogs. "Bella" is linebred on all the important dogs of the past—Sprog, Tigger, Shadow, Gitel and others. Photo by Author

Kelbe, a female of the highest quality, good type, intelligence and trainability. Photo by Author

Bella, sweet but reserved and alert. Photo by Author

"Bella" maintains the type and character of dogs from the early days of the Australian Shepherd. A goal of HS&W and Kristone is preservation of the original type Aussie into the future. Photo by Author

The Sprog Chronicles

A note by author, Kris Toft.

I think it is safe to say this book would not exist if not for a special blue dog named Sprog. Although for years I had admired the Wood's dogs, I owned dogs of other bloodlines which I liked and never pursued getting into a new line. The wonders of hindsight make me wish I had not waited so long but thankfully Casey Newman decided to tell the tales of her wonderful Sprog on the pages of the *Aussie Times*. In a stroke of luck I happened to be visiting a friend in the Denver area and while sitting around bored one day I suggested we call this woman with the Wonder Dog and go see her.

Casey graciously invited us to her home in the Rocky Mountains and we set out on an adventure. She lives so high we drove through snow on our way up even though it was midsummer "down below"! Shunka was still alive at this time, though in her teens and she welcomed us, along with Casey and Sprog. We spent a thoroughly enjoyable afternoon lunching with Casey and her oldest son (who was only about 2 years old at the time), petting these two superb dogs and learning about the line and the breed.

But it was Sprog who convinced me to abandon the road I was on in Australian Shepherds and convert exclusively to loving and raising Wood line dogs.

Late in the afternoon we said our goodbyes to Casey, Sprog, and Shunka with promises to keep in touch. We walked down the mountain and got in my truck to leave only to discover I had left the headlights on when we passed through the snow and my battery was now dead. No problem. Several of Casey's vehicles were parked beside my truck, all we had to do was get a jump from one of them.

Sprog, who started this all—for me anyway. Courtesy of Casey Masterson Newman

I trudged back up to her house. Casey's house has a glass enclosed porch extending around the entire front half of it. Of course there was no doorbell and no way to knock on the actual front door without entering this porch. She had left Sprog on the porch after we left so I walked up to the sliding door as he watched carefully. He was standing quietly just eyeing me and, knowing my own dogs, I assumed he would remember our pleasant afternoon which had ended only minutes before. I thought if I just talked to him and made him aware I was okay he would let me come in to knock on the inside door. Good plan if he had been just about any other dog.

Sprog at home in the mountains of Colorado. A nearly faultless dog who could do nearly anything he was asked. Photo by Author

But this was Sprog. He listened politely as I spoke to him and when he showed no signs of aggression I assumed he would let me come in. I was standing on the small step in front of the door and placed my hand on the door handle in order to slide the door open. It's possible I managed to open the door about 3 inches, but also possible it was only 3 millimeters. Regardless of the size of the gap, it was enough for Sprog. He erupted into a savage, vicious, protective beast, flinging himself at the miniscule opening and showing me in NO uncertain terms he was willing to eat me alive if I dared move another inch!

Shocked by this impressive display I fell backwards off the step, slamming the door as I went. Instantly, Sprog resumed his quiet stance, waiting for my next move. WOW! The only thoughts going through my head were how thrilling it was to see a dog with such judgment and discipline!! It

It was no wonder everybody fell in love with Sprog—kids and adults alike—even though it might have been for different reasons! Courtesy of Casey Masterson Newman

His kind and friendly expression could change in an instant when necessary! Photo by Author

took me several minutes to decipher Sprog's thinking about me but eventually I realized he was not going to fall for some sweet talking sap trying to convince him my intentions were purely innocent. He had been left alone on the porch, in charge of security, and until the boss told him otherwise, I was surely up to no good. That one show of higher intelligence, good judgment, common sense, and self determination sold me on the spot. My own dogs would have let someone back in. Here was a dog who was not going to slavishly fall for a con without confirmation from Casey that my sweet words were actually true!

As I stood there in the cold admiring the dog who was changing my entire view of the Australian Shepherd breed I began to remember exactly what had brought me back to Casey's door. I still had a dead truck at the bottom of the hill. So, now, how was I going to attract Casey's attention? She had obviously not heard Sprog's outburst and I was not about to risk losing a limb to my new hero in order to stir him up again. Mountain homes are built solid and tight to repel the bitter winter winds. It took a lot of yelling on my part to attract Casey's attention.

But persistence paid off and she eventually emerged from the front door and started toward the sliding glass door currently keeping me safe from Sprog. My relief in seeing her suddenly turned to raw fear as she began to open that sliding door. I was just opening my mouth to warn her to hold onto Sprog when the two walked out to greet me as friendly as old pals. I don't remember what Casey said to me because I was still in shocked amazement this dog did not eaten me alive when he finally got the chance! My expression must have puzzled Casey because all I could babble to her was how Sprog had kept me off the porch and how cool I thought it was that he didn't fall for sweet talk but kept me at bay until she came to say it was okay. Now it was her turn to be puzzled. "So? Don't all Aussies do that?"

"No!!" I cried! She took this rare behavior for granted. I thought I knew this breed. But this was something entirely new to me.

And so began a friendship and a cause I am still pursuing today—the preservation of a bloodline which defined this breed over half a century ago. I have never regretted a minute of it. Casey, who knew so many of the early dogs and is finally realizing how rare they are today, has written a number of amusing anecdotes of her adventures with Sprog, some were published in the *Aussie Times*, some never made it into print. Along with the series of cover ads she bought in that magazine to promote Sprog, which were a story in themselves, these stories are preserved in this book in her own words.

The following was written in the *Aussie Times* to accompany a "show pose" photo she had managed to take to show people Sprog's conformation. She usually had him doing something in her photos—which is what these dogs are for—doing—not just standing around looking pretty.

Never trained to be a "show dog," Sprog was just naturally correct and alert. Photo by Author

SPROG

by Casey Masterson Newman

For all you folks who've always wondered what Sprog REALLY looks like here he is. It's not that he's usually a blur of speed, working his tail off. It's just that I never thought to take a picture of him standing still. In fact, as he doesn't know how to pose, it was tough.

First we tried putting him on sheep and asking him to stop. That worked, except he was crouched, ready for action. And he got frustrated. Were we herding sheep or what? If we were, would I please let him work and quit telling him to stop here, move there, or look at me?

One of Sprog's many tricks—notice he is balancing on only one foot! Photo by Kris Nelson, Courtesy of Casey Masterson Newman

Next we tried baiting him. He sat up. He stood up, then he laid down and rolled over. We must want him to do tricks, what else was I doing waving a munchy in front of his nose? At last he figured out we just wanted him to stand. Boring.

And so for all you folks who've always wondered if Sprog does all his ads himself or hires a stunt dog, yes, Sprog does all the work you see in the ads and more. He's smart, tough, and versatile. He looks like a plain blue dog until he rips open his dog suit and you see the big red S on his chest. That's when he's faster than a speeding Suffolk, stronger than a bull, and leaps corral fences in a single bound. After all, he's an Aussie.

Soon after Shunka died, Casey wrote the following about her, which was published in the *Aussie Times*:

This is a eulogy to Shunka. The cover is of me and Jess (her oldest son) with Sprog, Gitel and Shunka. Shunka died mid September of old age and I wanted to write something about her. Shunka was my first dog and therefore special. I

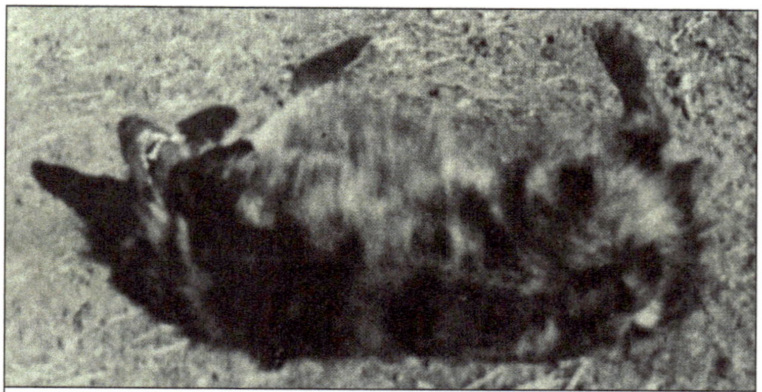

How do you get a smart, trained dog to just stand still and put up his ears? Photo by Kris Nelson, Courtesy of Casey Masterson Newman

chose her before she was born because I liked her mother. I never tried to train her beyond "good manners." I was in college and her job was to be my friend, companion and protector. She went to class with me by day (that's where she learned Chinese) and to work with me at night. She trotted a million miles behind my horse and kept cattle off that would take a person on foot. She raised my adopted wolf pups. She passed approval on my husband and carried the bouquet at our wedding. When we got Suffolk sheep, she kept the ram in line but would only bump the lambs with her nose. She would try to swim any river and retrieve any stick. Shunka lived a long and happy life, queen of the house for more than seventeen years. To me Shunka was the smartest, toughest dog in the world. She was my friend.

Casey also wrote several stories of the adventures of Sprog, Shunka and whatever other dogs happened to be hanging around her place at the time. They follow for your enjoyment!

Of course sitting up and begging was part of Sprog's repertoire. Photo by Kris Nelson, Courtesy of Casey Masterson Newman

Shunka's granddaughter, Gitel, a true gift to the breeding program at HS&W and Kristone. Photo by Author

Casey on the step of the one room cabin she shared with her husband, oldest son and three dogs for many years. From left to right: Sprog, Casey, Jess, Gitel and Shunka. Courtesy of Casey Masterson Newman

SHEEP TALES

by Casey Masterson Newman

Seems like if you have sheep, people are always asking you about them. They want to borrow a few sheep for the Christmas play or a lamb for Easter, or (heaven forbid), have you do a shearing demonstration. Local elementary schools also call on you. The third graders at the nearby school were studying ranching and a friend of ours was persuaded to give a talk.

Tom is a cowboy from Kansas who settled here after Vietnam. He works for the county driving the road grader, has a big old quarter horse, three cows and a blue heeler. He has a couple of kids too, which is how he got into this.

The idea was that Tom would bring his horse and a calf to school and do a show and tell. As there are sheep ranchers as well as cattle ranchers, he asked to borrow a lamb. No problem. On the appointed day, I picked out a month old Suffolk ram lamb, (the best one, I wanted everyone to see what good sheep I had). I put a dog collar and lead rope on him and loaded him in the stock trailer with old Baldy and the calf, and waved good bye.

About two hours later the phone rang—it was Tom. "What are you doing, Casey?" As I was getting ready to leave for Boulder I suspiciously said "Why?"

"Well we have a little problem." (Funny how WE had a problem). Seems one of the kids had unsnapped the lambs lead and he had taken off. And they couldn't find

Is it any wonder Sprog was a favorite at schools? Photo by and courtesy of Melissa Couch

One of many amusing photos taken by Casey to advertise her Wonder Dog in the *Aussie Times*. This orphan lamb was really nursing while Sprog held the bottle! Courtesy of Casey Masterson Newman

him. "Bring your dog cause I think we're going to need him" were Tom's parting words.

So I tossed 2 dogs and the baby in the car and left. All the way to town I was trying to figure out how much to charge for a lamb. When I got to the school, I was met by Tom in full regalia (you have to dress the part when you give school talks, so he was wearing boots, spurs, chaps, hat and a genuine silk neckerchief). He was sweating blood. Handing the baby to one of the teachers, we started up the hill to where the lamb had crossed the highway, Tom told me what had happened. "I didn't know lambs could run that fast. He just took off. (I hadn't told him I had the fastest sheep in the west). I tried to catch him but he got across the highway. So I came back and got the horse but I still couldn't find him. (I was going to make a snide remark about roping the poor lamb but Tom was pretty worried and I know you can see farther on horseback than on foot). He continued, "So I got Skipper, my dog. We saw the lamb and Skip chased him but by the time I caught Skipper the lamb was gone. I guess he doesn't like sheep much."

Sprog, intense, intelligent, sensible, loyal, he had every positive temperamental trait and Aussie should have. Courtesy of Casey Masterson Newman

By now we were heading up the mountainside into a subdivision. I was pretty sure that we would find the lamb if we could get above where he'd last been seen and work down from there. That was as long as none of the subdivision dogs hadn't found him first. Meanwhile we turned back a couple of teacher's aides who'd been beating the brush in vain. They had panicked and called the sheriff's dept to help look for the lamb. (The last thing I needed was them to show up with lights and sirens blazing, and maybe shoot the lamb, Sprog or me by mistake!)

I had turned Sprog loose, telling him to "Find the sheep." He was scouting around but it sure didn't look like he was tracking any sheep to me.

I turned to Tom and said, "This is going to sound dumb but I'm going to start baaing." Tom then confessed to having tried this already with no success. (And it was hard for a cowboy to admit to baaing for a lost sheep!) Most likely the lamb had been too scared and upset to answer him or maybe he just didn't sound right because after about 4 baas, we heard a faint baa answering from down the hill. Sprog took off and we split up, hoping to surround the critter. As you might suspect, one frightened 4 week old lamb will just run blindly, they don't herd. We would just have to find and corner him.

I kept calling as I headed down hill. Then I got another answer, I turned towards the sound. There was the lamb about 50 feet away. There was Sprog 20 feet behind him, walking slowly up. I calmly asked him to down (Well, if not calmly at least quietly as I was out of breath from hiking around a mountainside, baaing my head off!) He downed. (I'd say of course! But that's not always the way it is). I baa'd again and the lamb charged towards me, skidding to a halt about 6 feet in front of me. He looked pretty surprised to see me and not his mom! Sprog got up and was walking up on the lamb. I hissed "Down!" at him as the lamb turned to see what was back there. While Sprog laid there staring at the lamb, I snuck up and grabbed a hind leg. Caught him!!

Tom came whooping up the hill and with Sprog in front and Tom behind we got a rope on the lamb and carried him down the mountain (he was very stressed out). As the third graders reassembled to check out the lamb, Tom told them that the ONLY way to work stock was with a stock dog. He pointed to Sprog, who was surrounded by kids petting him and said, "There's the hero, he found the lamb." Actually I thought I deserved equal credit but somehow it doesn't seem as romantic to go down in history as sounding like an old ewe.

MISADVENTURES WITH SHEEP

by Casey Masterson Newman

It all happened because I was in a hurry. We were going out to dinner with my in-laws and I was supposed to meet Rick and his folks at a restaurant about 20 miles down the canyon. So I was all dressed for dinner and packing a month old baby when I went up to feed and lock up the stock for the night.

Everything was in except 5 ewes in the little pasture. I opened the gates and sent Sprog out to get them. In they came and spooked at me standing there in my good clothes (trying not to step in the manure). They blew past the gate. I have to explain that our old log barn sits in the middle of a meadow and if the sheep miss the gate they can take off for the neighboring National Forest and keep going 'til they hit Wyoming. So I hollered at Sprog to "git around" and bring them back. Well, he caught 4 out of 5 and brought them. So I sent him out for the last one while I locked up the four. I didn't think Old Green Tag'd gone too far by herself but she'd had a lead on the rest of the girls and had headed for the woods. A couple of minutes later, Sprog came back—with no sheep! I sent him out again and kept on feeding stock. He shows up again, no sheep! I was determined that he was going to find that sheep. After all, he'd lost her and I wasn't dressed to go tramping thru the forest looking for a stupid sheep. Sprog comes to the gate again, still no sheep! I figured he hadn't cast out far enough and by now she was in the next county. Wishing for a Border collie with a 2 mile outrun, I spoke thru gritted teeth "Go get that sheep"! (Actually I think I might have said a few other things too). I turned up a bucket and sat down with the kid. By now it was a matter of principal, I was NOT going to help look for that dumb sheep. Sprog had lost her, he could darn well find her if it took all night! A few minutes later there is a big clatter and here comes Old Green Tag. She runs past the gate and disappears around the back of the barn. Sprog shows up at the gate and looks past me for the sheep. By now I'm seething but I say in a calm but tense voice "Go get that sheep." He takes off. Shows up again, no sheep. Off again. This time he comes in thru the back barn window. I thought "Darn him, he's lost her and he's hiding from me!" So I whistled him up and sent him again. Now I hear him on the roof. He comes back, no sheep! I give up. Collecting the baby, we go off to find that darn sheep. Sprog is bouncing ahead, glad that I am finally going to find the sheep. As we round the back of the barn, he jumps to the top of the shed and up on to the barn roof.

Sprog was always ready and willing to take on any job—just tell him what you want him to do! Photo by and courtesy of Melissa Couch

One of the unusual circumstances found while living on a mountainside is a barn that has a normal height roof from the front but nearly touches the sloping ground in back. Dogs and livestock both discover this fairly quickly, resulting in sometimes humorous, sometimes troublesome situations! Sprog is trying to figure out how to get these turkeys to turn around and not fall off the front of the roof! Courtesy of Casey Masterson Newman

Can you find Old Green Tag in this group? Courtesy of Casey Masterson Newman

The sheep is on the roof! This is pretty amazing as I have never had one do this before. Old Green Tag is standing right at the edge of the roof with about a nine foot drop below her, and poor Sprog is sneaking carefully along the front edge, trying to bring her to me. She'd sidle over closer to the brink each step he'd take. I called him off and decided to try myself (with baby in tow). Sprog figured I couldn't handle it and got back on the roof too.

That was the last straw. Old Green Tag took one look at everybody closing in on her and jumped off the roof. There was a thud and not another sound.

We climbed off the barn and walked around to the corral. There was Old Green Tag standing calmly by the gate, looking over her shoulder, baaing to be let in with the other sheep. No broken legs or neck (yet). With murderous feelings in my heart, I opened the gate, locked her up and left for dinner. We were late.

Guess I should have listened to Sprog the first time he said he needed help. Next time he comes back without the sheep I'll follow and check the barn roof.

THE GREAT PIG CHASE

by Casey Masterson Newman

It was September 8th and we had six inches of snow! The dogs and I ran up to the barn to feed quickly (Leaving the baby asleep in the house). We fed the horses, sheep, turkeys, and ducks. Checked the rabbits and discovered we'd lost two new litters to freezing. It really upsets Sprog to see little cold rabbits laid out as he thinks they're his personal pets. While the dogs were inside the barn commiserating with the rabbits, I went out to feed the pigs. They were looking kind of cold, wet, and miserable, so I brought out new bedding.

If you've never seen a bale of shavings, it's a giant plastic bag packed with 60 lbs of wood chips. It's very awkward because I can't quite get my arms around it and it tears if you drag it. As I was lugging the shavings into the pig pen, one of the pigs panicked at the sight. He ran around the pen and bolted out the gate. Over the top of me and the shaving bale. I was underneath the bale. This was a mixed blessing. It saved me from having 175 pound pig hoof prints down the front of me, but then again I was on the bottom, pushed into the pig manure. I let out a yell. Where were those darn dogs? It's their job to be guarding the gate while it's open, not in the barn counting rabbits.

Sprog with his personal pets—baby rabbits. Courtesy of Casey Masterson Newman

The three dogs burst out of the barn. Two of them immediately took after the pig. Shunka (who's 15) came over to see what I was doing laying there in the manure. I got up and shut the gate on the second pig. Then I started yelling for the dogs. Sprog is fine, he can herd pigs. Gitel is a 7 month old pup with lots of enthusiasm but not much finesse. It took a little yelling and whistling before they came back. I shut Shunka and Gitel in the sheep corral and sent Sprog after the pig.

Sprog "counting" his rabbits and checking for any missing ones. Courtesy of Casey Masterson Newman

About ten minutes later the pig appeared, pushed slowly by Sprog. I was so proud! He worked that pig back and forth to keep him coming, nice and slow. It was beautiful. Just as he got him to the edge of the woods, a blue bullet streaked past me. Gitel had dug out under the corral gate and raced by to help Sprog. Not that she was much help. The pig headed north for Wyoming.

After a lot of hollering and whistling the two dogs appeared. I caught Gitel (who thought pig herding was great fun), and sent Sprog back out after the pig.

Sprog working a pig in the barn corral at HS&W ranch. Courtesy of Casey Masterson Newman

Now normally I would have locked the sheep in the barn, opened the corral and put out food for the pig. Then I could sneak up and shut the gate, catching him without all this excitement. (In fact I bet you're wondering why I didn't do this). But pigs don't like cold, wet and stress. It gives them pneumonia. I was afraid he would lie down out in the woods and die, and there'd go my profit on hogs for 1988. I had to get him in where it was warm and dry.

So I locked Shunka and Gitel in an escape-proof lambing pen and headed back out. (Shunka was too stout to squeeze under the corral gate, so she'd been barking loud encouragement to the others).

Darned if Sprog didn't come back with that pig! I was so proud! He worked that pig through the woods, only breaking to a slow trot to head the porker off when he would try to turn back. Not that the pig was moving fast. He was heaving for breath and froth was dripping from his open mouth. I called softly "Bring him to me, Sprog." And he tried. Sprog worked the pig patiently back and forth to the edge of the meadow where Gitel had surprised them earlier, but he couldn't get the pig out into the open. He worked and worked. He tried heeling the pig to force him out of the woods. Nothing worked. That pig wasn't going to risk being attacked by the pup again.

Okay, people are smarter than pigs. I told Sprog to hold the pig and I ran back to the barn. Time to try the old bucket trick. I got a bucket with some grain and ran back to the pig. Sprog had him cornered in some downed timber. I offered the pig the grain, planning to lure him slowly to the barn with Sprog bringing up the rear.

The pig didn't even look at the grain, he just lay there panting. I dumped some grain in front of him. He ignored it. If I left him here he was going to die. I HAD to get him back. I remembered someone telling the way to move a stubborn pig was to put a bucket over it's head and back it to where you want it to go. Sounded like a good idea. I pushed the bucket over the pig's face. He backed right out of the bucket and took off (north to Alaska?). Maybe you need a runway for this idea to work. Or another person to keep the pig pushed into the bucket. Of course if I had a runway the pig wouldn't be loose and I wouldn't have this problem. And there is NEVER anyone else around when you need help. So Sprog cornered the pig again while I thought. I had a book on livestock behavior and handling, what did it say about moving pigs? I vaguely recalled a drawing of a pig with a rope around it's snout.

Back to the barn! I grabbed a thin rope sheep halter and an old lead rope. I ran back to the pig. He was still laying there panting. I made a noose, slipped it over his jaw and tightened it. He started to scream. (Pigs can make a lot of noise. Naturally nobody came to see if I was being murdered up on the back forty. I wish they would because I could sure use the help!) Holding the rope in one hand, I grabbed the pig by the tail and said "Get up." Sprog hit him on the heels. The pig took off squealing and the noose slipped down to the end of his nose. He was gasping for breath as I shut off his air, but the strangled screaming never stopped. He lunged backwards. Sprog hit him. The noose slid further down his snout. It looked as if it was going to pinch his nose off. Every time I'd pull on the rope, the pig'd go backwards, Sprog'd hit him, he'd lunge forward, and we'd go in a circle. Between the pig squealing, and me yelling at the dog "get off, get up, get back, get out!" it was pretty bad. The rope finally slipped off the end of the pig's snout and he took off.

Try again. Sprog cornered the pig. This time he wouldn't get away from me. I tied the lead rope around my waist and got the slip noose over his jaw. He started screaming. I took advantage of his open mouth to feel inside it. Ah hah! His teeth had been clipped! No wonder the rope kept sliding off! (They didn't mention this in the book). I threw a half hitch over his nose to secure this noose. The volume of noise increased. I grabbed his tail and hollered "Get up"! Sprog hit him. The pig shot forward. So did I. Remember how I tied the rope around my waist?

Have you ever tried running through the forest on six inches of sloppy snow tied to a pig who outweighs you? The trees were dense, the branches low and sometimes we were on opposite sides of the same tree. The pig was screaming, I was yelling and Sprog was nipping the pig's head and shoulders, trying to head him off. I tripped. I was skidding through the slushy snow on my stomach, holding on to the rope with both hands. I was actually relieved when the rope slipped off and he got away.

I was soaked and was wondering who would die of pneumonia first, me or the pig? I seriously considered leaving him to take his chances. The other option was to go down to the house and get a gun and shoot the sucker. Either way the meat would be a total loss after all this running around. I hate to give up on something I've started. I went after the pig.

I'm a quick learner and don't often make the same mistake twice. (These were all different ones). This time I tied the rope around the pig's waist! (Well, pigs don't have waists, behind his armpits). Taking the other end of the rope, I made a couple of real tight half hitches over his snoot. Now I had sort of a luggage handle arrangement. I had a tight grip of the rope in the middle, grabbed his tail and said "Get up." Sprog hit him and the pig 'got up'. Fast. Fortunately the pig was getting tired (so was I). Between me steering and Sprog motivating, we made it back to the pig pen.

Getting him into the pig pen was complicated by the other pig wanting to come out and join the fun. I couldn't take any more. I backed through the gate hauling the pig by brute strength. I hollered "HIT HIM"!!! Sprog hit him. The pig shot forward over the top of me. I slammed the gate and crawled to my feet. Picking up the sack of shavings, I dumped it over the pig who had collapsed in the corner.

Muttering under my breath, I got the dogs and went down to the house to take a hot bath…The pig survived to become 250 lbs and now lives in the freezer as pork chops. He is delicious.

The ultimate revenge is mine.

SPROG'S ADVENTURE AS A SHOW DOG IN THE BIG TIME

After several years of front cover advertising in the *Aussie Times*, Sprog had gained a fair amount of "fame" as a unique character in the breed. My friendship with Casey had grown during this time and when the ASCA National Specialty was to be held in Loveland, Colorado I begged Casey to enter Sprog in conformation so the "world" could see what a nice quality dog he really was. She had no interest in conformation showing and it took some fast talking on my part to not only convince her to enter her special dog but to allow me to handle him for her in the ring.

When the show rolled around Casey was nearing her due date with her second child. She had arranged with the show committee to be part of the Opening Ceremonies Parade with Sprog pulling his cart loaded with her son and other children. Hoping she could make it through the parade without adding a baby to Sprog's load, we had decided to play

Another of Casey's creative photos taken for an *Aussie Times* ad. Her caption was "No dog is an island but it can sure feel that way sometimes" The Wood dogs are becoming an island in today's world but dogs like Sprog still survive. Letting the world know about them is one goal of this book! Don't you wonder how she got Sprog to go out in that lake, climb on the rock and stay while she took the photo? Courtesy of Casey Masterson Newman

it safe and she would leave Sprog with me afterward so she could go home to be close to the doctor. This arrangement allowed me to get better acquainted with Sprog and teach him the basics of being a show dog.

He had no trouble learning the "baiting" portion of the routine since Casey had always used treats to train him for all his tricks and other duties. The problem was getting him to just stand rather than run through his entire repertoire every time I pulled out the goodie! We eventually reached a truce whereby I would hand stack him and hold him in position while stuffing food into his mouth for staying. Gaiting was not a problem and could that little dog move! He floated smoothly and quickly over the ground with me just trying to keep up.

The day of his debut came and we lined up with the Open Blue class looking as out of place as a lumberjack at a tea party. But we proudly entered the ring and started turning heads as we took our first lap with the group. Since we weren't at the head of the line we had

Sprog with his cart. He could pull it loaded with as many small children as Casey could fit on it. He also helped with chores around home using the cart and performed at every opportunity in the area—parades, fairs, the ASCA National Specialty! Courtesy of Casey Masterson Newman

Sprog was a handsome and correct dog regardless of whether he "fit" in the show ring or not. Photo by and courtesy of Melissa Couch

plenty of time to practice his lessons of the day before while in the ring. The old pro was actually in his element from all the performances he had been in through his life. After insisting on a few free stacks with ears up (the last was never a problem) I let him show me a few of his many talents in return for a few extra treats.

Of course we never expected to win, never even expected to get much of a "look" from the judge, but we did want the Aussie world to get a first hand look at the dog they had been reading about for several years. We just wanted to have fun showing the conformation world they didn't have a corner on correct and well built dogs. A friend was video taping our performance and upon viewing the tape later, it was obvious the audience was sitting up and taking notice. I even heard a number of comments as we were waiting our turn in line while in the ring! People *knew* who he was without opening a catalog or having to ask their neighbor.

When our turn came for examination, Sprog behaved well like the gentleman he was. He was such a

Just before the 1990 ASCA National Specialty, Casey gathered Sprog and some of his kids for this photo. Left to right: Sprog, Gitel, Casey, Stitch and Ace. Courtesy of Casey Masterson Newman

nicely made dog he didn't require much "stacking" so I just needed to steady him as the judge went over him. His expression during the exam was amusing as this stranger got increasingly personal as he moved toward his rear! He kept his opinions to himself until the exam was complete, then gave one last disdainful look over his shoulder at the judge as we stepped off to show his movement. He flew around the ring, making a perfect show dog free stack when we returned to the judge, then had me sprinting like an Olympic 100 meter runner as we made our final lap around! As much as I would like to say we triumphed in this large class that day, it didn't happen. But I'm sure a number of biases against the working dogs came down that day, and that, after all, was really why we were there.

Sprog. A good dog is never a bad color, even in the show ring. Photo by and courtesy of Melissa Couch

Conclusion

The Colorado equation was one of the most successful in the history of the Australian Shepherd breed. The combination of Fletcher Wood's dogs with area dogs of that general type created dogs with the highest levels of intelligence, working ability, protectiveness and versatility. Their loss and relegation to endangered status in the breed is something of a mystery. Was it merely the societal changes of their era which has led to their near demise? Was it a lack of understanding by the general public of what an Australian Shepherd is supposed to be? Was it a desire for profit which led breeders away from the ideal? Or was it merely there was no longer a need for such dogs? Most likely a combination of these and other factors has brought these dogs to the rare status they face today. A number of older breeders are now realizing the value of many bloodlines which have already disappeared from the breed forever. Once lost, these lines and genetics can not be regained.

Perhaps this is a fate all breeds face with the advancement of urban living for the majority of people. When the jobs for dogs disappear, the dogs become extinct, too. It has been my desire to preserve for history some of the tales of these outstanding individuals and their well earned place in the founding of this breed. I am thankful to all those who contributed and allowed me access to their memories.

Historical truth is elusive. If those who live it do not record it (or record it inaccurately), recreating it is left to those who come later. Sometimes the veil of time reveals those things which were to be significant but not obvious, immediately. Sometimes personal bias skews the facts to achieve an end unrelated to the truth. In the case of a dog breed, history is seldom seen as important when it is occurring. It is easy for myths and legends to fill this vacuum.

—Kris Toft
[found in author's archives]

Bibliography

I have made every effort to thoroughly document every fact in this book. This documentation has been in the form of the actual words of those about whom I write or the words of people I interviewed who were actually around during these historic times to witness dogs, events and people. At times I needed to summarize certain events but the full text from which the summary was taken can be accessed by any reader who wishes to peruse the resources listed below. While I do not believe that the written word is necessarily false, as implied by some, I have attempted to bring the readers of this book as much first hand information and actual quotes of those who have passed on as possible. I have interviewed directly as many founding breeders as possible. This book is not complete but it is as complete as I was able to make it. Enjoy it and consider it an alternate and factual view of the Australian Shepherd.

Chapters One and Two

Sheepwagon by Nancy Weidel

A Picture or Two biography of Ray Eyerly by Beverly Blodgett

Personal interviews with Jerry and Sharon Rowe

Personal interviews with Les Burlile

Personal interviews with Dick and Leslie Sorensen

Personal interviews with Mary Wood

ASQ interview with Fletcher Wood

The Working Kelpie by Tony AD Parsons

Dogs of Australian by Kennel Control Council of Victoria

Pyrenean Dogs by Duconte and Sabouraud

The Border Collie in Australasia by Peidje Vidler

Australian Barkers and Biters by Kaleski

ASCA Yearbook 1957-1975

Chapters Three, Four, Five, and Six

The Western Horseman interview, July 1969, "Jay Sisler How He Went to the Dogs"

Western Horseman article on Run Appaloosa Run

ProRodeo Hall of Fame website:

Jay Sisler, The Man Behind Those Famous Blue Dogs, an interview with his wife Joy and daughter Maggie by Andrea Scott

Dinah Shore website: www.tv.com/the-dinah-shore-chevy-show/show/8238/episode_guide.html?season=5

Film—*Stub, the Best Cowdog in the West*

Film—*Run, Appaloosa, Run*

Film—*Greta, the Misfit Greyhound*

Website on Wetback Hound and Man Called Horse

http://basquemuseum.com/content/jausoro-jim-jimmy

Jay Sisler home movies

Aussie Times photo with Wilbur Plaugher and Doc—also references in home video to Plaugher plus most likely training dogs brother Gene took on the road.

The Australian Shepherd's Origin and History by Gwen Stevenson

Personal conversations with Sisler nephew, Les Burlile

The Wonderful World of Disney Television by Bill Cotter

ASCA Yearbook 1957-1977

Australian Shepherd Quarterly

Swiss National Library, Patricia Highsmith Collection, Don Coates Donation

National Cowboy and Western Heritage Museum

ProRodeo Hall of Fame

Riding High: Colorado Ranchers and 100 Years of the National Western Stock Show by Thomas J. Noel

1960 National Western Stock Show program

http://www.ruralroots.org/About/newsletter_winter_2005.pdf

Independent Press Telegram 3-16-69 *Larry Lansburgh: Film Maker on Horseback* by Lorraine Gauguin

Tri-City Herald 8-26-56

Walla-Walla Union Bulletin 12-4-56 *Sisler and Dogs in Short Subject*

The Salt Lake Tribune 5-24-64 *Lansburgh Loves Excitement*

Winnipeg Free Press 7-18-70 *Top Acts Coming to Big M*

Walla Walla Union Bulletin 8-2-68

Corpus Christi Caller Tines 2-3-63 *Animals Can Become Temperamental Stars*

Syracuse Herald Journal 9-23-56 *Cow Dog: Idea Came Off Range*

Florence Morning News 12-7-74 *Stars Reunited After 17 Years*

The Independent 4-8-56

Montana Standard 7-4-69 *Jay Sisler and his educated dogs to show at Butte rodeo*

Cedar Rapids Gazette 7-26-53

San Antonio Light 2-18-55 *Jay Sisler's Australian Dogs Will Perform as Special Event*

Modesto Bee and News-Herald 4-8-63 *Oakdale Rodeo Adds Dog Act For Weekend*

Chilicothe Constitution-Tribune 10-2-65

Albuquerque Journal 4-1-55 *Dogs to Perform At Fiesta Show*

Hobbs Daily News-Sun 5-23-60

El Paso Herald-Post 2-6-52 *Schedule Trained Dog School Show To Boost Rodeo*

El Paso Herald-Post 2-3-58 *Broken Ankle Turns Hobby Into Rodeo Job*

El Paso Herald-Post 1-23-61, "Educated Dogs to Perform At Championship Rodeo"

Tales From the Dog Listener, by Jan Fennell, Foreword by Monty Roberts

Dodge News magazine, November 1967

Always Standing by by J. Burnett Smith

Andy Little—Idaho Sheep King by Louise Shadduck

Rodeo Idaho by Louise Shadduck

Colorado Springs Gazette Telegraph, July 31, 1965, "Trainer Feels That Dogs Are the Closest Animal to Human Beings"

Colorado Springs Gazette Telegraph, July 31, 1960, "Sisler and Dogs Will Be Pikes Peak Rodeo Feature"

The Idaho Statesman, April 24, 1995, "Gone to the Dogs"

(Idaho) Messenger Index, May 2, 1990 "Jay Sisler recalls dog-act career started with horse"

The Idaho Statesman, July 10, 1995, "Famous dog trainer dies with pal at his side" by Colleen LaMay

Idaho Press-Tribune, July 10, 1995, "Rodeo entertainer, dog trainer dies" by Dave Wilkins

Personal communication from William Cotter

Personal communication from Tanna Autry, ProRodeo Hall of Fame

The Ranch That Dogs Bought by Lyle Orwig, published in National Livestock Producer

"Sisler Breaks a Leg, Becomes Top Trainer of Trick Dogs" by Marilyn Tammany: July 28, 1963, *Wyoming State Tribune-Cheyenne*

"The Senior's Story" by Rose Spicuzza, *Aussie Times*

The Herder, 1973, Interview with Jay Sisler

Ogden Standard-Examiner, June 16, 1953

Twin Falls Times News, September 1, 1961

Twin Falls Times News, September 2, 1951

Chapters 7-13

August 24, 2006, *Golden News* "Rodeo itself remains the same as Fletcher Wood Memorial revamps" by Emily Hois

Australian Shepherd Quarterly interview

http://www.co.jefferson.co.us/ctb/resolutions/91/cc91-860.htm—Dedication of Fairground Arena to Fletcher H. Wood

http://www.hillsideequestriancenter.com/legend.html (Fletcher Wood Memorial Rodeo/obituary)

Personal interviews with Mary Wood

Personal interviews with Jean Taylor

Personal interviews with Casey Newman

Personal interviews with Dick and Leslie Sorensen

ASCA Yearbook 1957-1977

Chapter 14

Personal interviews with Dick and Leslie Sorensen

Personal interviews with Jean Taylor

Winter 1987-88 *ASQ* interview with Leslie Sorensen

Chapter 15

Personal interviews with Sharon and Jerry Rowe

Personal notes from the files of Betty Nelson

Chapter 16

Personal interviews with Jean Taylor

1995 Australian Shepherd Annual

2009 Australian Shepherd Annual

2010 Australian Shepherd Annual

Personal letters from Polly Bjork to Betty Nelson

Personal letters from Jean Taylor to Betty Nelson

Chapter 17

Australian Shepherd Quarterly Interviews Summer 1989 and Fall 1989 issues

November-December 1992 *Aussie Times* "Ch Jones Reddy Teddy CDX OTDdsc SCH HA" by Terry Martin page 36

Personal letters from Cathy Jones to Betty Nelson from Nelson Archives

Personal interviews with Leslie Sorensen

Personal interviews with Melissa Couch

Personal interviews with Ruth McGuire

Chapter 18

Personal interviews with Casey Masterson Newman

Australian Shepherd Quarterly, Winter 1984-85, ASQ Centerfold story, pages 32-34

www.ingramcontent.com/pod-product-compliance
Lightning Source LLC
LaVergne TN
LVHW071344210926
848486LV00003B/227
* 9 7 8 1 9 3 9 0 5 4 1 8 0 *